Mycotoxins Problem and its Management

S.M. Reddy
S. Girisham
V. Krishna Reddy
M. Surekha

Department of Botany and Microbiology
Kakatiya University
Warangal-506 009

Published by:
Scientific Publishers (India)
5-A, New Pali Road, P.O. Box 91,
Jodhpur – 342 001 (India)

E-mail: info@scientificpub.com
www.scientificpub.com

Print : 2020

ISBN: 978-81-7233-671-4

Lasertype set : Rajesh Ojha

Binding at : Rajasthan Law Book Binding Works, Jodhpur
Printed in India

PREFACE

Mycotoxins, secondary metabolites of storage moulds, have been receiving increasing attention in view of their undisputed role in public health. Many books appeared, conferences in different parts of the globe are being conducted which clearly indicate direct relation between the incidence of mycotoxigenic fungi, degree of mycotoxin contamination and their prevalence revealed their relation to some of the human ailments. Out of several mycotoxins, aflatoxins, ochratoxins A and fumonisins are posing serious health hazards specially in Asian countries. Inspite of intensive and extensive studies related to mycotoxins the problems becoming intricate and intriguing and decisive conclusions could not be drawn and clear cut solutions are provided.

There is no way of establishing the incidence of human mycotoxicoses and their degree of incidence. Further, there are no epidemiological studies which enable us to draw decisive conclusions but we are forced to draw conclusions only on circumstantial evidences. It is now visualized that the natural incidence of mycotoxins contamination and health problems are not to the same degree to that of *in vitro* studies. This has been attributed to the complexity of environment in which we are living. Thus what we conclude based on laboratory studies may not hold true to that happening in nature. One has to be cautious in deriving conclusions based on *in vitro* studies.

In order to high light the possible eventualities and in the context of Indian conditions, it was felt the need of assessing and prepare a comprehensive account so that person interested may get clear understanding of the problem and probe further in this direction to get better insight.

Present attempt is made to assess developments in the subject and project the future problems in the light of changing conditions with special reference to India. This book is divided into two parts. First part deals with general aspects of mycotoxins covering areas of natural incidence, taxonomy of mycotoxigenic fungi, analytical techniques, and mechanism of mycotoxins, toxicity epidemiology and management of mycotoxins. The second part deals with individual mycotoxins. The detailed account of producing organisms, detection, biological activity and management practices specific to that of mycotoxins are discussed in individual chapters. An attempt has been made to discuss mycotoxins problem with reference to Indian conditions. In view of voluminous literature only comparatively recent and relevant literature is quoted and may be some important works might have been omitted.

The task of this magnitude cannot be accomplished unless and until help and encouragement received from different corners. Our heartfelt thanks are to our late revered teacher Prof. K.S. Bilgrami who introduced me to this area of research and motivated for further research. Our thanks are due to band of students who did stupendous work and struggled in the field which formed the basis for this book. We are also thankful to authors and publishers from which we have borrowed the material. Our profuse thanks are due to research scholars specially Dr. Supriya Sarkar. Mr. Ranjith Kumar R. and Shilpa, P. who toiled in the preparation of the manuscript. The grant received from Department of Science and Technology, New Delhi in the form of users programme is gratefully acknowledged.

Date:
Warangal **Authors**

CONTENTS

Preface *iii*

PART-A

Chapter 1: INTRODUCTION **1-38**

1.1 Genesis of the problem 1
1.2. Natural incidence of mycotoxins in different food commodities 21
1.2.1. Cereals 24
1.2.2. Pulses 27
1.2.3. Oil seeds 28
1.2.4. Dry fruits and spieces 29
1.2.5. Milk and milk products 30
1.2.6. Feeds 30
1.2.7. Fodders 30
1.2.8. Sea foods 31
1.3. High risk areas for mycotoxins in India 31
1.4. Weather conditions and mycotoxins 31
1.5. Management of mycotoxins 32
1.5.1. Prevention of mould growth 33
1.5.2. Detoxification 34
1.5.3. Dietary modifications 35
1.5.4. Resistant varieties 36
1.5.5. Biological control 36
1.6. Limits and Regulations for Mycotoxins 37
1.7. Current research Scenario 37

Chapter – 2: TAXONOMY OF MYCOTOXIGENIC FUNGI **39-111**

2.1 Introduction 39
2.2. Key to the identification of mycotoxin producing fungi 42
2.3. Morphology of mycotoxigenic fungi 50
2.3.1. *Aspergillus* 50
2.3.2. *Penicillium* 64
2.3.3. *Fusarium* 83
2.3.4. *Alternaria* 98

2.3.5. *Stachybotrys* 100
2.3.6. *Myrothecium* 103
2.3.7. *Trichoderma* 105
2.3.8. *Trichothecium* 107
2.3.9. *Pithomyces* 107
2.3.10. *Phoma* 107
2.3.11. *Phomopsis* 108
2.3.12. *Diplodia* 108
2.3.13. *Chaetomium* 109

Chapter – 3: ANALYTICAL TECHNIQUES **112-133**

3.1. Sampling strategies 114
3.2. Methods of Analysis 114
3.3. Units of Concentration 118
3.4. No Detectable Amounts (NDA) 118
3.5. Chemical Analysis 119
3.5.1 Extraction 119
3.5.2 Clean-up 119
3.5.3 Work-up 120
3.5.4 Detection and Quantification 121
3.5.5 Confirmation 122
3.5.6 Determination of purity and concentration of mycotoxin standards 123
3.6. Immunological techniques 125
3.6.1. Radio immunoassay 126
3.6.2. Enzyme immunoassay 127
3.6.3. Immunoscreening methods 129
3.6.4. Complementation of immunoassays with chemical methods 131
3.6.5. Other newly developed immunochemical methods 132

Chapter 4: TOXICOLOGY **134-151**

4.1. Acute and chronic toxicity 134
4.2. Cytotoxicity 134
4.3. Carcinogenicity 137
4.4. Mutagenicity 137
4.5. Teratogenicity 137
4.6. Tremoregenicity 137
4.7. Immunotoxicity 137
4.8. Haematotoxicity 137
4.9. Induction of apotoxicity 138
4.10. Antitumor properties 139
4.11. Antimicrobial activity 139

4.12. Insecticidal activity 139
4.13. Phytotoxicity 139
4.14. Mechanism of action of mycotoxins 140
4.14.1. Receptors 140
4.14.2. Metabolic activation and detoxification 141
4.15. Nucleic acids as receptors 141
4.16. Lesions on proteins 144
4.17. Target sites and biological effects 144
4.18. Inhibition of energy production 143
4.18.1. Cellular resperation 144
4.18.2. Uncoupling of oxidation phosphorylation 144
4.18.3. Alteration of membrane permeability 144
4.19. Inhibition of macromolecular synthesis 145
4.19.1. Inhibition of DNA synthesis 145
4.19.2. Alteration in RNA metabolism 146
4.19.3. Inhibition of protein synthesis 146
4.19.4. Alteration of immune responses 146
4.19.5. Inhibition of gluconeogenesis 147
4.19.6. Effects on lipid metabolism 147
4.20. Effect on central nervous system 147
4.21. Modifications of cytoskeletons 148
4.22. Hormonal effects 148
4.22.1. Activity of steroid hormones 148
4.22.2. Estrogen effect of zearalenone 149
4.23. Carcinogenic effects 149
4.23.1. Initiation 150
4.23.2. Promotion 150

Chapter 5: EPIDEMIOLOGY **152-164**

5.1. Ergot 154
5.2. Aflatoxins 155
5.3. Trichothecenes 157
5.4. Fumonisins 159
5.5. Patulin 161
5.6. Zearalenone 162
5.7. Ochratoxins 162

Chapter 6: MANAGEMENT **165-198**

6.1. Prevention of mould contamination and growth 167
6.1.1. Primary prevention 168
6.1.2. Secondary prevention 168

6.1.3. Tertiary prevention 168
6.1.4. Methods of prevention 169
6.2. Improved farm management 172
6.3. Separation of infested grains 172
6.4. Volatile compounds 172
6.5. Food preservatives 173
6.6. Antioxidants 173
6.7. Polyamine biosynthetic inhibitors 174
6.8. Antagonists 174
6.9. Genetic engineering approach 177
6.10. Rapid screening techniques 178
6.11. Control of environmental conditions 178
6.11.1. Moisture 178
6.11.2. Temperature 179
6.11.3. Modified Gaseous Atmosphere 180
6.12. Detoxification of toxic products 180
6.12.1. Physcial methods 182
6.12.2. Chemical methods 183
6.12.3. Biological methods 184
6.13. Quality control and regulatory measures 185
6.14. HACCP in mangament of mycotcoxins 193
6.14.1. Pre-harvest and Post harvest (Field) situation 193
6.14.2. Codes of practice 195
6.14.3. Post-harvest control 196
6.14.4. Post-harvest –CCP's 196
6.15. The integrated approach 197

PART-B

Chapter 7 CARCINOGENIC MYCOTOXINS **199-249**
7.1. Aflatoxins 199
7.1.1. Physico-chemical characteristics 199
7.1.2. Organisms 202
7.1.3. Biosynthesis and Metabolism 204
7.1.4. Isolation and Detection 204
7.1.5. Natural incidence 219
7.1.6. Biological activity 219
7.1.7. Factors affecting AFB_1 toxicity 226
7.1.8. Management 226
7.2. Fumonisins 227
7.2.1. Physico-chemical characteristics 228

7.2.2. Organisms 229
7.2.3. Biosynthesis 230
7.2.4. Isolation and Detection 230
7.2.5. Natural Incidence 236
7.2.6. Biological activity 240
7.2.7. Management 245

Chapter – 8. NEPHROTOXIC MYCOTOXINS 250-279

8.1. Ochratoxin A 250
8.1.1. Physico-chemical characteristics 250
8.1.2. Biosynthesis of Ochratoxin A 252
8.1.3. Organisms 252
8.1.4. Isolation and Detection 254
8.1.5. Factors influencing ochratoxin A production 261
8.1.6. Natural Incidence 262
8.1.7. Biological Activity 265
8.1.8. Pharmacokinetics and Metabolism 269
8.1.9. Management 270
8.2. Citrinin 271
8.2.1. Physico-chemical characteristics 271
8.2.2. Organisms 271
8.2.3. Biosynthesis 272
8.2.4. Isolation and Detection 272
8.2.5. Natural Incidence 273
8.2.6. Factors influencing citrinin production 275
8.2.7. Biological Activity 275
8.2.8. Management. 278

Chapter 9: CYCLOPIAZONIC ACID 280-288

9.1. Introduction 280
9.2. Physico-chemical characteristics 280
9.3. Organisms 281
9.4. Biosynthesis 282
9.5. Isolation and Detection 282
9.6. Natural Incidence 284
9.7. Factors Influencing CPA production 285
9.8. Biological Activity 286
9.9. Management 287

Chapter – 10: TREMORGENIC MYCOTOXINS 289-307

10.1. Introduction 289

10.2. Physico-chemical characteristics 289
10.3. Organisms 291
10.4. Isolation of and quantification 291
10.4.1. Penitrems 291
10.4.2. Janthitrems 294
10.4.3. Aflatrem 295
10.4.4. Paxilline 295
10.4.5. Paspaline, Paspalicine and Paspalinine 295
10.4.6. Paspalitrems A and B 296
10.4.7. Territrems 296
10.4.8. Fumiteremorgin 296
10.5. Natural Incidence 297
10.6. Structure and Activity relationship 297
10.7. Biological Activity 297
10.8. Penitrems 299
10.8.1. Physico-chemical characteristics 299
10.8.2. Organisms 299
10.8.3. Isolation and Detection 300
10.8.4. Natural incidence 300
10.8.5. Production 300
10.8.6. Biosynthesis 302
10.8.7. Biological activity 303
10.8.8. Management 306

Chapter – 11: PATULIN **308-321**

11.1. Physico-Chemical properties 308
11.2. Organisms 309
11.3. Biosynthesis 310
11.4. Isolation and Detection 312
11.4.1. Chemical methods 312
11.4.2. Paper chromatography 312
11.4.3. Thin-layer chromatography 312
11.4.4. Gas liquid chromatography 314
11.4.5. High performance liquid chromatography 314
11.4.6. Bioassay 316
11.5. Natural Incidence 316
11.6. Biological activity 318
11.6.1. Animal diseases 319
11.6.2. Phytotoxicity 320
11.6.3. Health hazards in man 320
11.7. Pharmacokinetis 320

Chapter – 12. CYTOTOXIC MYCOTOXINS **322-356**

12.1 Isolation and detection 327
12.1.1. Cytochalasins A and B 327
12.1.2. Cytochalasins C and D 328
12.1.3. Cytochalasin E 329
12.1.4. Cytochalasin F 331
12.1.5. Cytochalasin G 331
12.1.6. Cytochalasin H and J 333
12.1.7. Cytochalasins K, L and M are produced by *Chalara microspora* 333
12.1.8. Zygosporins D, E, F and G 333
12.1.9. Aspochalasins A, B, C and D 334
12.1.10. Deoxaphomin, proxiphomin and protophomin 335
12.1.11. Chaetoglobosins and Acetylchaetoglobins 336
12.2. Biological activity 340
12.2.1. Cytokinesis 342
12.2.2. Locomotion and Cytoplasmic streaming 342
12.2.3. Mutagenicity and Teratogenicity 342
12.2.4. Cytophysiological effects 343
12.2.5. Attachment and Adhesion 343
12.2.6. Morphogenesis 343
12.2.7. Endocytosis 343
12.2.8. Exocytosis-I 346
12.2.9. Immunological functions 346
12.2.10. Viral infectibility 347
12.2.11. Metabolism and Macromolecular Synthesis 347
12.2.12. Alterations in membrane structure and function 348
12.2.14. Structure and function 349

Chapter – 13: ESTROGENIC MYCOTOXINS **357-375**

13.1. Zearalenone 357
13.1.1. Physico-chemical characteristics 357
13.1.2. Producing organisms 359
13.1.3. Biosynthesis 359
13.1.4. Isolation and Detection 361
13.1.5. Natural Incidence 366
13.1.6. Biological Activity 368
13.1.7. Mechanism of action 371
13.1.8. Management 374

Chapter - 14: GASTROINTESTINAL MYCOTOXINS **376-425**

Trichothecenes 376
14.1. Physico-chemical characteristics 378
14.2. Isolation and Detection 383
14.3. Factors influencing 399
14.4. Biological activity 407
14.5. Metabolism and pharmacokinetics 422
14.6. Management 423

Chapter 15: MISCELANEOUS MYCOTOXINS **426-443**

15.1. Alternaria toxins 424
15.1.1. Organisms 426
15.1.2. Physico-chemical characteristics 426
15.1.3. Alternariol, alternanol and altrnariol monomethylether production 427
15.1.4. Isolation and detection 428
15.1.5. Tenuazonic acid 432
15.1.6. Production 432
15.1.7. Isolation and detection 432
15.1.8. Natural incidence 434
15.1.9. Biological activity 434
15.2. Sporidesmins 435
15.2.1. Chemistry of Sporidesmins 436
15.2.2. Natural Incidence 438
15.2.3. Biological activity 438
15.3. Sterigmatocystin 439
15.3.1. Physico-chemical characteristics 439
15.3.2. Organisms 439
15.3.3. Isolation and Detection 440
15.4. Slaframine 441
15.4.1. Extraction and detection 441
15.4.2. Biosynthesis 443
15.4.2. Biological activity 443

REFERENCES **444-515**

CHAPTER 1

INTRODUCTION

1.1. GENESIS OF PROBLEM

Mycotoxins are the low molecular weight, non-antigenic secondary metabolites of moulds which mostly live primarily as saprophytes and sporulate heavily on variety of substrates. These are mostly xerophills with a minimum (a_w) requirement may be as low as 0.6 and thrive well comparatively at elevated temperatures. These moulds mainly infest foods, feeds and fodders at different stages of harvest, storage and consumption and are called storage moulds. These fungi not only cause huge loss due to their deterioration activity but also may convert those substrates into toxic. The presence of mycotoxins in food is an inevitable and unavoidable consequence of the existence of fungi. It is now widely agreed that approximately 25% world food crops are affected each year by variable levels of mycotoxins which can have considerable economic consequences for the crop, live stock producers, grain handlers, processors, consumers and indeed national economics (Muller *et al.*, 1998). The food poisoning and death of human beings is an age old problem. There are several instances of outbreak of diseases of unknown etiology which probably may be due to mould infestation and mycotoxin contamination. In developing countries regular mycotoxin presence in the diet will also affect the human populations causing morbidity and premature deaths. Chemistry, detection and biological effects of mycotoxins have been excellently reviewed. Creepy (2002) discussed the incidence and toxic effects of different mycotoxins in Europe. Similarly Bennett (2003) has discussed different aspects of mycotoxins problems.

Mycotoxins are not only diverse in their structure but also in their biosynthetic pathways and they are non-proteinaceous, thermostable and incite variety of health hazards to warm blooded animals and which are called mycotoxicoses. Mycotoxicoses can be defined as "poisoning of host following the entry of toxic substances of fungal origin into body resulting in various pathological conditions. Mycotoxicoses are in general differ from other diseases in the following respects.

1. They are not transmissible.
2. Drug and antibiotic treatment have little or no effect on the disease.
3. In field outbreaks, the friable is often seasonal.
4. The outbreak is usually associated with a specific food or feed stuff.
5. The degree of toxicity is often influenced by the age, sex and nutritional status of the host.
6. The examination of suscepted food or feed reveals signs of fungus activity.

Oldest known mycotoxicoses are associated with various species of parasitic fungus *Claviceps* and their metabolic products. The toxic effect of these compounds has been the cause of mass poisoning in both man and animals all over the world. As a sequal intensive studies have been conducted in this area all over the world and more than 150 moulds are reported to elaborate more than 300 mycotoxins. Probably still many more are awaiting for discovery. Most important and common moulds in foods and feeds are listed in Table 1.1. It is interesting to note that several fungi produce a particular mycotoxin Table 1.2. On the other hand, more than one mycotoxin may be produced by single fungus Table 1.3.

Table 1.1. Fungi producing mycotoxins and their natural substrates

Mycotoxins	Major producing fungi	Typical substrate in nature
Aflatoxins	*Aspergillus flavus, A. parasiticus*	Peanuts, corn, cottonseed, cereals, figs, most tree nuts, milk, sorghum, walnuts
Alternarin (AAL)	*Alternaria alternata*	Cereal grains, tomato, animal feeds
Citrinin(CT)	*Penicillium citrinum*	Barley, corn, rice, walnuts
Cyclochlorotine (CC)	*P. islandicum*	Rice
Cyclopiazonic acid (CPA)	*A. flavus, P. cyclopium*	Peanuts, corn, cheese
Deoxynivalenol (DON)	*Fusarium graminearum*	Wheat, corn
Fumonisins (FM)	*F. moniliforme*	Corn, sorghum, rice
Luteoskyrin (LT)	*P. islandicum, P. rugulosum*	Rice, sorghum
Moniliformin (MN)	*F. moniliforme*	Corn
Ochratoxin A (OA)	*A. ochraceus, P. viridicatum, P. verrucosum.*	Barley, beans, cereals, coffee, feeds, maize, oats, rice, rye, wheat
Patulin (PT)	*P. patulum, P. urticae, A. clavatus.*	Apple, apple juice, beans, wheat
Penicillic acid (PA)	*P. puberulum, A. ochraceus.*	Barley, corn

Penitrem A (PNT)	*P. palitans*	Feed stuffs, corn
Roquefortine (RQF)	*P. roqueforti*	Cheese
Rubratoxin B (RB)	*P.rubrum, P. purpurogenum*	Corn, soybeans
Sterigmatocystin (ST)	*A. versicolor, A. nidulans.*	Corn, grains, cheese
T-2 toxin	*F. sporotrichoides*	Corn, feeds, hay
Zearalenone (ZE)	*F. graminearum.*	Cereals, corn, feeds, rice
12-13, Epoxy trichothecenes (TCTC) other than T-2 & DON	*F. nivale.*	Corn, feeds, hay, peanuts, rice

Table 1.2 Mycotoxins produced by more than one fungus

Mycotoxin	Fungal species*
Aflatoxin B1	*Aspergillus (2)*
Sterigmatocystin	*Aspergillus (5)* *Drechslera* sp. and *Bipolaris* sp.
Patulin	*Penicillium (16)*
Penitrem A	*Penicillium (5)*
Fumitrogen B	*Aspergillus (2), Penicillium lanosum*
Ochratoxin A	*Aspergillus (5), Penicillium (9)*
Citrinin	*Penicillium (16), Aspergillus (3)*
Cyclopiazonic acid	*Penicillium (8), Aspergillus (2)*
Penicillic acid	*Penicillium (15), Aspergillus (4)*
Zearalenone	*Fusarium (7)*
T2 toxin	*Fusarium (5), Trichoderma viride*
Roridin A	*Myrothecium roridum, Stachybotrys atra*
Cytochalasins (14 compounds)	*Metarrhizium anisopollae, Zygosporium (2), Phomopsis paspali, Rosellinia necatnix, Aspergillus clavatus*
Rubratoxin	*Penicillium rubrum*
Skyrins (Hydroxy anthroquinones)	*Penicillium (7), Pachbastum bamatum, Sepedonium ampullosporum, Chaetomium elatum, Cladosporium fulvum, Talanomyces avellaneus, Endothia parasttica and Preussia multispora*
Gliotoxin	*Trichoderma (2), Gliocladium (2), Aspergillus (2)* and *Penicillium (3)*
Secalonic acid	*Aspergillus (2), P. oxalicum, Claviceps purpurea, Phoma terrestris*

*Number of species ( )

Table 1.3. Fungi producing more than one mycotoxin

Aspergillus flavus	Aflatoxin, Aflatrem, CPA
A. terreus	Patulin, Citrioviridin, Citrinin
A. ochraceus	Viomellin, Penicillic acid, Ochratoxins
Penicillium islandicum	Cyclochlorotine, Erythoskyrin, Luicoskyrin
P. viridicatum	Rubratoxic, Viridiotoxin
P. roqueforti	Roquefortine, PR-toxin
Fusarium sporotrichiodes	Zearalenone, T_2 toxin

Mycotoxins are chemically diverse groups of compounds mainly derived from primary metabolites (Fig. 1) ranging from polypeptides, polyketides, diketopiperzines, alkaloids, benzoquinones, xanthones, coumarins, terpenes and steroids derivatives Table 1.4. Mycotoxins are in general low molecular weight compounds, non-antigenic and are capable of eliciting a toxic response in man and warm blooded animals. They include various toxic effects such as acute and chronic toxicity, cytotoxicity, carcinogenicity, antitumor activity, mutagenic, teratogenic, tremorgenic, immunosuppressive, antimicrobial, insecticidal and phytotoxic.

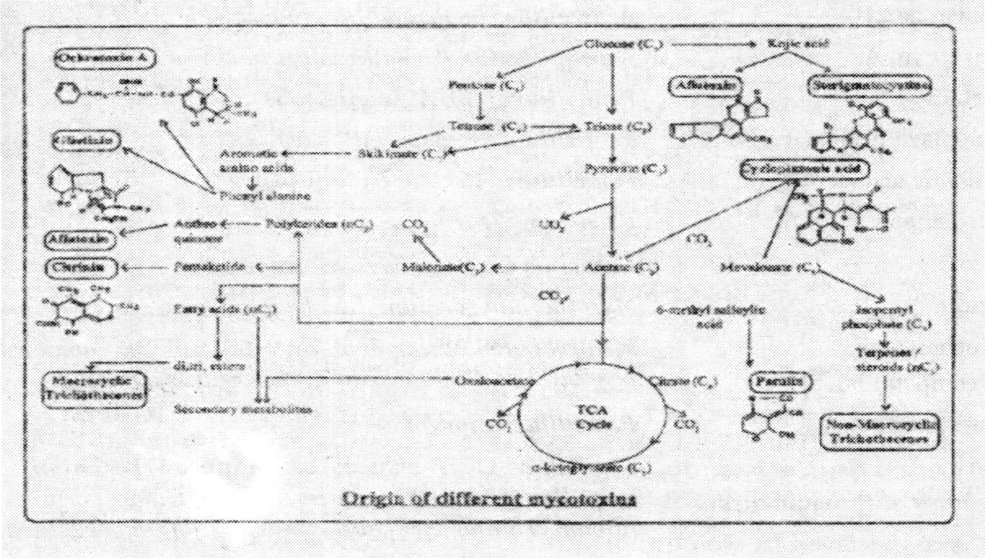

Fig. 1. Origin of different mycotoxins

Toxigenic species are found in all major taxonomic groups of fungi and also in all foods, feeds and fodders. Most of the mycotoxins are the secondary metabolic products of species of *Aspergillus, Penicillium* and *Fusarium*.

Some of the important mould species which elaborate toxic metabolites are *Aspergillus flavus* (aflatoxins, aflatrem, Cyclopiazonic acid and flavotoxin), *A. parasiticus* (aflatoxins), *A. terreus* (patulin, terric acid and citreoviridin), *A. fumigatus* (fuma-gillin, fumitoxin, fumitremorgen, hevolic acid and kojic acid), *A. wentii* (emodin).

Table 1.4. Mycotoxins and their Biosynthetic Categories of Secondary Metabolites

Biosynthetic	Representative mycotoxins
Poly ketides	Zearalenone
Tetra-	Patulin, penicillic acid, chlorflavonin
Penta-	Citrinin, ochratoxins
Hexa-	Maltoryzine
Hepta-	Viriditoxin, cytochalasins, rugulosin.
Octa-	Ergochromes
Nona-	Zearalenone, viridicatum toxin
Deca-	Aflatoxins, austocystins, erythroskyrine
Tetronic/ tertramic acid	Tenuazonic acid, cyclopiazonic acid, cytochalasins
Diketopiperazines Simple	Aspergillic acid, echinulins
Modified	Brevianamides, sporidesmins, fumitremorgons, oxaline
Peptides	Tentoxin, ergotamine, tryptoquivaline
C_6C_3 Products	Chlorflavonin, xanthocillin, terphenyllin
Terpenes Mono-	Viridicatum toxin
Sesqui-	Trichothecenes
Di-	Paspaline

Mycotoxin producing fungi are ubiquitous in nature as favourable conditions for growth and production of mycotoxins are easily achieved. It can be concluded that only few edible substances are considered to be free from mycotoxin contamination. Fungal invasion of foods can occur at any stage of the food production chain i.e., from the time of planting, through harvesting, transportation and storage, during preparation of foods at homes and restaurants and until the food is consumed (Betina, 1984). It is reported that higher levels of exploitation of fertilizers is likely to stimulate the growth of mycotoxin producing fungi (Gambeh *et al.*, 2001). Natural occurance of selected mycotoxins during 1995-1997 along with percentage and range are prescised in Table 1.5.

Mycotoxins elaborated in different agricultural commodities generally are in low concentration and remain undetected in view of extreme low concentration, may get bioamplified in course of time and turnout to be fatal. Therefore, an intensive study have been made to detect these mycotoxins by using simple as well as more modern and sophisticated equipment. Different methods are adopted in mycotoxin analysis including large number of bioassays.

Table 1.5. Natural incidence of mycotoxins in different foods in different parts of the world during 1995-1997

Commodity	Country	Publication Year	No. of samples	Incidence	Range (μg/kg)
Barley	Germany	1997	39	31%	0.1-2.7
Wheat	Denmark	1996	520	32%	0.05-51
Rye	Denmark	1996	616	42%	0.05-121
Oats	Denmark	1996	92	44%	0.05-5.6
Wheat	Switzerland	1996	28	54%	0.1-10.0
Wheat, barley	UK	1995	1061	4%	1.33
Green coffee	Japan	1997	47	30%	0.1-17.4
Green coffee	UK	1996	291	38%	0.2-27.3
Coco powder	UK	1997	20	85%	0.2-1.1
Dried vine fruits	UK	1997	60	88%	0.2-53.6
Wine	Germany	1996	144	42%	0.01-7.0
Spices & herbs	UK	1996	29	86%	1.2-50.4

Mycotoxins may be formed in the field before harvest following infection by any of a variety of fungal species. Members of the genera *Fusarium*, *Alternaria, Penicillium* and *Aspergillus* being the most important. In the same way, crops are susceptible to fungal attack after harvest, during transport or when in storage. There is a close relationship between fungal species and the secondary metabolites produced and it is thus important to have accurate taxonomic information based on correct identification of species according to criteria agreed internationally. Information on the fungus involved has sometimes been contradictory due to mis-identification of fungus at the species level or disagreement over the names applied to specific species. This has been especially true of the genus *Penicillium* (Frisvad and Filtenborg, 1995). Extreme caution is required in attempting to relate fungal infection with the presence of mycotoxins as these substances can only occur if toxigenic strains of fungi are present and conditions suitable for fungal metabolism. Conversely, apparently mould-free products can still be contaminated with mycotoxins. The changing fungal flora throughout food production, together with the chemical stability of many of the mycotoxins, means that it is very unreliable to infer the presence or absence of mycotoxins by the fungi present at any stage in the food chain.

The production of mycotoxins is affected by many factors, including temperature, available moisture, oxygen levels, nutrients, trace elements, genetic variation between strains of specific species or competition with other organisms that may be present. The way in which fungi colonise and grow in crops in the field or infect food commodities during transport or storage, results in a highly heterogeneous distribution of any mycotoxin that subsequently develop. Different probable routes for mycotoxin contamination of foods and feeds are precised in Table 1.1.

This has important implications when samplings: Statistically based protocols are required to ensure that a representative sample is drawn. Several statistically based studies have been carried out to establish the distribution of mycotoxins in specific food commodity. The results of such studies are used to determine the number and size of samples required to be taken. Sometimes this may conflict with the practicalities involved in acquiring samples. Aflatoxins, for example, are controlled in the UK by the Aflatoxins in nuts, nut products, dried figs and dried pig product regulations (Anonymous, 1992). These specify how nuts and figs should be sampled and involves taking a total sample of 10 kg of shelled peanuts comprising 30 sub-samples or 20 kg of whole figs from 20 x 1 kg sub-samples. This would normally be impracticable for application at retail outlets. Processed foods and drinks usually present less of a sampling problem as mycotoxins are likely to be distributed in a much more homogeneous manner.

Detection of g/kg amounts of fungal products within a substratum that is composed of many natural products with similar chemical or physical properties presents a major challenge to the analytical chemist. Hence, specific and precise analytical methods have been developed. Antibodies have then risen to most of the important mycotoxins and a range of immunologically based methods now complements, or has been incorporated into chemically based methods such as thin-layer chromatography (TLC), high performance liquid chromatography (HPLC) and gas chromatography (GC). Antibodies bound to activate sepharose are commercially available as highly selective clean up columns.

Many toxicological studies on the effects of mycotoxins in animals have been reported (Table1.6). However, the administration of a pure crystalline mycotoxin often leads to results at variance with those achieved when similar quantities of the crude product formed under natural conditions. The reason is that secondary fungal metabolism often leads to a cocktail of different metabolites which may act additively or synergistically. The use of animal or human cell lines for toxicological testing is increasingly used. Accumulation of toxicological data and information on occurrence and food consumption, has enabled risk assessment to be carried out for the most important mycotoxins. This has led in some instances to recommend acceptable total daily intake values.

Table 1.6. Toxicological effects of some commonly occurring mycotoxins

Major mycotoxins	Toxicological effect	Mode of action
Aflatoxin	A, C, H, I, M, T	Form DNA adducts and Mutare p53 tumor suppressor gene
Alternaria (AAL) Mycotoxins	A, H, M	Inhibition of ceramide synthesis
Citrinin (CT)	C (?) Nh, M	Inhibitiion of protein synthesis
Cyclopiazonic acid (CPA)	Cv, I, M, Nr	Alteration of Ca^{++} hemostasis & inhibition of Key enzymes in animal cells
Deoxynivalenol (DON)	I, Nr	Inhibition of protein synthesis
Fumonisins (Fm)	A, C, Cv, H, Nr, R	Disruption of sphingolipid metabolism by inhibiting ceramide synthase
Luteoskyrin (LS)	A, C, H, M	Binding with DNA
Moniliformin (MN)	Cv, Nr	Binding with pyruvate dehydrogenase and affecting TCA cycle enzymes
Ochratoxin A (OA)	A, I, Nh, T	Inhibition of protein synthesis and several enzymes; enhancement of lipid peroxi-dation; impairment of Ca^{++} and CAMP homeostasis
Patulin (PT)	C(?), Nr, D, T	Binding with-SH groups of protein and inhibition of –sH enzymes
Penicillic acid (PA)	C(?), Nr, M	Inhibition of-SH enzymes
Rubratoxin B (RB)	A, H, T	Inhibition of protein synthesis
Sterigmatocystin (ST)	H,C, M	Same as aflatoxin
T-2	A, ATA, Cv, D, I, T	Inhibition of protein synthesis, binding with peptidyl transferase, membrane proteins and other enzymes
Tenuazonic acid (Tz)	Hr	Inhibition of protein sythesis
Tremorgenic toxins	Nr	Binding and inhibition of acetylcholinesterase
Zearalenone (ZE)	G,M	Binding with estrogen receptor
12-13, Epoxy trichothecenes (TCTC) other than T-2 & DON	A, D, I, Nr	Same as T-2 toxin

A, apoptosis; ATA alimentary toxic aleukea; C, carcinogenic; C(?), carcinogenic effect is still questionable; Cv, cardiovascular lesion; D, dermatoxin; G, genitotoxin and estrogenic effects; H, hepatotoxic; Hr, hemorrhagic; I, immuno-modulation effects, M, mutagenic, Nh, nephrotoxin; Nr, neurotoxin; R, respiratory; T, teratogenic.

In view of the highly toxic nature of many of the mycotoxins, it is fortunate that their occurrence in food is sporadic and often only in small amounts atleast in developed areas of the world. However, the potential for wide-scale problems exists should the appropriate circumstances arise. Low-level contamination of foods by highly biologically active chemicals presents a major difficulty in assessing their true significance for man. Legislation is usually set on a national basis and is mainly restricted to the aflatoxins, of which aflatoxin B_1 is accepted as a potent liver carcinogen, although some countries have introduced limits for other selected mycotoxins (Food and Agriculture Organization of the United Nations, 1997).

Fungi and mycotoxins intimately associated with growing crops and stored foods are difficult to eliminate, especially when they develop prior to harvest. Attention to good farming and storage practice are important factors in reducing the occurrence and amounts of mycotoxins. It is unfortunate that the less developed nations tend to have climatic conditions that encourage mould growth and mycotoxin formation and hence are faced with much greater problems, while at the same time having fewer resources to detect, control and reduce the extent of contaminated food.

In the production of food crops and consumer food products, the aim should be to reduce levels of mycotoxins to the lowest that can be technologically achieved within the existing economic constraints and with regard to risk assessment, where this has been carried out. The existence of a statutory limit requires instigation of structured sampling and analysis at agreed key points in trading and marketing.

Specific concerns often expressed by the food industry and other bodies include the means or observing legal limits, inspection of appropriate monitoring and control protocols, representative sampling, identification of products likely to be susceptible and the ability to respond to any public or media concern as, and if, the need arises. Study of mycotoxins should be carried out by a multi-disciplinary approach involving, chemists, mycologists, toxicologists, veterinarians and statisticians.

Inspite of advent of green revolution, India continues to be linked with the vagaries of nature. Nearly seventy percent of the total production of food grains in India is retained at farm level where the poverty, unscientific and faulty storage conditions enhance the chances of mould infestation and thereby mycotoxin contamination. Storage moulds are present in a dormant state (usually as spores) in the field and transport on food grains ingrow under favourable storage conditions of food grains (Hussein and Brasel, 2001). During warm, humid conditions, several of the fungi such as species of *Aspergillus penicillium* and *Fusarium* increase rapidly in association with grains (Girish and Goyal, 1986).

Incidence of mycotoxigenic moulds and mycotoxin contamination is reported to be widespread in different parts of the world Table 1.7.

Table 1.7. Mycotoxicosis producing fungus and contamination of foods and feeds

Mycotoxin (Disease)	Fungus	Natural substrates
Aflatoxins	*Aspergillus flavus* *A. parasiticus*	Peanuts, corn, cottonseed meal, Brazil nuts, pecans, rice, sorghum
Akakabi-byo (red mold or scab disease)	*Fusarium* spp.	Rice, barley, wheat, corn, oats
Alimentary toxic aleukia (ATA)	*Fusarium* spp. *Cladosporium* spp.	Grains
Bermudagrass tremors	*Unidentified fungi*	Bermudagrass pastures
Citrinin	*Penicillium citirnum, P. notaum, P. viridicatum P. palitans, P. citreoviride, P. steckii, P. expansum, P. corylophilum, P. fellutanum, P. lividum, P. implicatum P. jensenii, Aspergillus candidus, A. niveus, A. terreus*	Wheat, oats, rye, rice
Ergot alkaloids	*Claviceps purpurea*	Rye and other grains, dallies grass & pasture grasses
Fescue foot	*Fusarium* spp.	Fescue pastures
Ochratoxins	*Aspergillus ochraceus, A. melleus, A. sulphureus, Penicillium viridicatum, P. cyclopium*	Corn, oats, barley, rye, wheat, white beans, peanuts, cottonseed meal, Brazil nuts, citrus fruits, tobacco, coffee
Patulin	*Penicillium expansum, P. leucopus, P. cyclopium, P. melinii, P. claviforme, P. novaezeelondiae, P. divergens, P. urticae, P. patulum P. equinum, P. lanosum, Aspergillus clavatus, A. giganteus, P. griseofulvum, A. terreus, Byssochlamys nivea*	Apple juice, various processed fruits
Penicillic acid	*Penicillium cyclopium, P. roqueforti, P. aurantio-virens, P. simplicissimum, P. baarnense, P. stoloniferum, P. fennelliae, P. viridicatum, P. janthinellum, P. lividum, P. martensii, P.*	Corn, beans, tobacco

	palitans, P. puberulum, Aspergillus ochraceus, A. alliaceus, A. melleus, A. sclerotiorum, A. sulphureus	
Psoralens	*Sclerotinia sclerotiorum*	Celery
Rubratoxin	*Penicillium rubrum, P. purpurogenum*	Corn, various grains
Slaframine	*Rhizoctonia leguminicola*	Red clover
Sporodesmins	*Pithomyces chartarum*	Pasture grasses
Satratoxins	*Stachybotrys atra*	Hay
Sterigmatocystin	*Aspergillus amstelodami, A. chevalieri, A. flavus, A. nidulans, A. versicolor, Bipolaris sorkiniana, Penicillium luteum*	Grains, green coffee, miscellaneous foodstuffs
Tremorgens	*Aspergillus clavatus, P. palitans, A. flavus, P. paxilli, A. fumigatus, P. puberulum, Penicillium crustosum, P. verruculosum, P. cyclopium, Rosellinia necatrix, P. martensii*	Peanuts, various moldy commercial feeds, rice
Trichothecenes	*Fusarium avenaceum, F. moniliforme, F. culmorum, F. oxysporum, F. equisetii, F. poae, F. graminearum, F. roseum, Gibberella zeae, F. solani, F. lateritium, F. tricinctum*	Corn and various grains, contaminated feed
Yellowed rice disease	*Penicillium citreo-viride, P. islandicum, P. rugulosum.*	Rice
Zearalenone	*Fusarium graminearum, F. oxysporum, F. roseum, F. sporotrichioides, F. moniliforme, F. tricinctum.*	Corn and various grains

The activity of fungi may cause decrease in germinability, discoloration of food grains, heating and mustiness, loss in weight, along with biochemical changes and production of toxins (Gajendra Prasad *et al.*, 1993). All these changes may occur before the responsible fungi could be detected on visual examination.

The fungi may produce a variety of toxic substances in food grains, spices and dry fruits (Giridhar and Reddy, 1997a), vegetables, feeds, fodders and their products (Laxma Reddy and Reddy, 1992) either at pre-harvest or post-harvest stage. Natural incidence of mycotoxins in different parts of the world has been recorded (Table 1.8). Because of their potent toxic nature and adverse effect on animals including man (Bilgrami *et al.*, 1986; Bhat *et al.*, 1997) results in illness and economic losses are of fairly high. Smith *et al.* (1994), Dutton (1996) and Moss (1996) are some of the recent reviews appeared on this topic. Smith and Handerson (1991) have excellently reviewed the subject of mycotoxins in animal health. Dirheimer (1998) elegantly discussed the genitoxicity of different mycotoxins. The diseases or physiological abnormalities resulting due to the ingestion of mycotoxins are known as "mycotoxicoses" (Bilgrami and Sinha, 1984a). In acute forms of mycotoxicosis, individual animal will show marked signs of disease and in extreme cases animals may die. Observable acute disease syndromes include hepatitis, haemorrhage, nephritis and necrosis of oral and enteric epithelia. Fink-Gremmels (1999) has discussed different aspects of implication of mycotoxins in human and animal health. In India aflatoxins, ochratoxin A, trichothecenes, zearalenone, tremorgenic toxins and ergot alkaloids are the mycotoxins of great agro-economic importance.

Table 1.8. Natural incidence of mycotoxins in different foods from various parts of the world

Countries	Products	Mycotoxins	Content (μg/kg)
Argentins	Baby food	B_1	0
	Groudnut, maize and by products	B_1	0
		$B_1B_2G_1G_2$	20
	Liquid milk, powdered milk	M_1	0.05
	Milk products	M_1	0.5
Australia	All foods	$B_1B_2G_1G_2$	5
	Peanut butter, nuts and the nut portion of products	$B_1B_2G_1G_2$	15
Austria	All foods	B_1	1
		$B_1B_2G_1G_2$	5
	Milling and shelled products and derived products	B_1	2
		$B_2G_1G_2$	5
	Children foods (in prepared food)	$B_1B_2G_1G_2M_1$	0.02
	Wheat, rye	Ochratoxin A	5
		Deoxynivalenol	500
		Zeralenone	60
	Durum wheat	Ochratoxin A	5
		Deoxynivalenol	750
		Zeralenone	60
	Fruit juice	Patulin	50
	Milk (products)	M_1	0.05
	Whey powder	M_1	0.4
	Whey, liquid whey products	M_1	0.025
	Cheese	M_1	0.25
	Butter	M_1	0.02
	Pasteurized fresh milk for infants/children; children foods	M_1	0.01

	Powdered milk (products), condensed milk, milk concentrates	M_1	0.04
Bahamas	All foods, all grains	$B_1B_2G_1G_2$	20
Barbados	All foods	$B_1B_2G_1G_2$	20
	Milk	M_1	0.05
Belgium	Peanuts	B_1	5
	Milk	M_1	0.05
Belize	Maize, groundnut	$B_1B_2G_1G_2$	20
Bosnia & H	Wheat, maize, rice, cereals	B_1G_1	1
	Beans	B_1G_1	5
Brazil	All food stuffs	B_1	15
		$B_1B_2G_1G_2$	30
	Imported foodstuffs	B_1	5
		$B_1B_2G_1G_2$	10
	Industrially prepared foodstuffs for children from 0-2 years and for school meals	$B_1B_2G_1G_2$	3
	Maize	Zearalenone	200
	Maize and groudnut	B_1G_1	30
	Milk (products)	M_1	0.5
	Imported milk (products)	M_1	0.1
Bulgaria	Peanut (products), kernel (products), cocoa beans, cocoa butter, cocoa powder	$B_1B_2G_1G_2$	5
	Cereals (products)	$B_1B_2G_1G_2$	2.5
	Liquid milk products	M_1	0.5
	Powdered milk	M_1	0.1
	Powdered milk for dietetics and infant feeding	M_1	0
	Cheese and similar products	M_1	0.5
Canada	Nut (products)	$B_1B_2G_1G_2$	15
	Uncleaned soft wheat	Deoxynivalenol	2000
China	Rice and edible oils	B_1	10
	Maize, peanut (products), maize and peanut oil	B_1	20
	What, barley, oats, beans, sorghum, other grains, fermented foodstuffs	B_1	5
	Cow milk, milk products (calculated on the basis of milk)	M_1	0.5
Colombia	Foods	$B_1B_2G_1G_2$	20

	Cereals (sorghum/millet)	$B_1B_2G_1G_2$	30
	Oil seeds	$B_1B_2G_1G_2$	10
Costa Rica	Maize	$B_1B_2G_1G_2$	35
Cuba	Foods	$B_1B_2G_1G_2$	5
	Cereals, groudnuts	$B_1B_2G_1G_2$	5
Cyprus	Cereals, pulses, dried fruit, nuts, sesame and food produced exclusively from these, caraway seeds, poppy seed, seeds used in bakery products and confectionery	$B_1B_2G_1G_2$	10
	Milk, dairy products	All mycotoxins	0.5
Czech Republic	All foods	B_1	5
		$B_1B_2G_1G_2$	10
	Infant/children foods	B_1	1
		$B_1B_2G_1G_2$	2
	All foods	Patulin	50
		Ochratoxin A	20
	Children foods	Patulin	30
		Ochratoxin A	5
	Infant foods	Patulin	20
		Ochratoxin A	1
	All foods	M_1	5
	Milk	M_1	0.5
	Infant/children foods	M_1	1
	Infant foods on milk basis	M_1	0.1
		B_1	0.1
		$B_2G_1G_2$	0.2
Denmark	Peanut (products)	B_1	2
		$B_1B_2G_1G_2$	4
	Brazil nuts	B_1	2
		$B_1B_2G_1G_2$	4
	Dried figs	B_1	2
		$B_1B_2G_1G_2$	4
	Pig kidneys	Ochratoxin A	25
	Pig kidneys	Ochratoxin A	10
	Cereal (products)	Ochratoxin A	5
Dominican Republic	Maize (products), groudnut, soya, tomato (products)	B_1G_1	0
	Imported maize	$B_1B_2G_1G_2$	20

Egypt	Peanut (products), oil seed (products), cereal (products)	$B_1B_2G_1G_2$	10
		B_1	5
	Maize	$B_1B_2G_1G_2$	20
		B_1	10
	Starch and its derivatives	$B_1B_2G_1G_2$	0
		B_1	0
	Milk and dairy products	$G_1G_2M_1M_2$	0
		M_1	0
Finland	All foods	$B_1B_2G_1G_2$	5
		Patulin	50
France	All foods	B_1	10
	Peanuts, pistache nuts, almonds, oil seeds, children foods	B_1	1
	Wheat meal	B_1	3
	Wheat bran	B_1	10
	Vegetal oils, cereals, wheat meal (complete)	B_1	5
	Apple juice	Patulin	50
	Cereals, vegetable oils	Zearalenone	200
	Cereals	Ochratoxin A	5
	Milk, milk powder (calculated on reconstituted product)	M_1	0.05
	Milk, milk powder for infants under 3 years	M_1	0.03
Germany	All foods	B_1	5
		$B_1B_2G_1G_2$	4
	Enzyme (preparation) intended for the production of foodstuffs	$B_1B_2G_1G_2$	0.5
	Foods for infants and young	$B_1B_2G_1G_2$	0.05
	Milk	M_1	0.05
	Foods for infants and young children	M_1	0.01
Greece	Peanuts, hazelnuts, walnuts, cashewnuts, pistaches, almonds, pumkin seeds, sunflower seeds, pine seeds, apricot seeds	B_1	5
		$B_1B_2G_1G_2$	10
	Maize, dried figs, dried apricots, dried prunes, dates and raisins	B_1	5
		$B_1B_2G_1G_2$	10
	Raw coffee beans	Ochratoxin A	20
	Apple juice, apple products	Patulin	50
Guatemala	Maize, kidney beans, rice, sorghum	$B_1B_2G_1G_2$	20
	Groudnuts, groudnut butter	$B_1B_2G_1G_2$	20

Honduras	All foods	$B_2G_1G_2$	1
	Maize (grounded or whole grain)	B_1	1
	Baby food	$B_1B_2G_1G_2$	0.01
		M_1	
	Milk (products)	M_1	0.05
	Cheeses	M_1	0.25
Hungary	All foods	B_1	5
	Groundnut kernels	B_1	30
	Preserved foods	All mycotoxins	0
	Groundnuts	$B_1B_2G_1G_2$	5
India	All foods	B_1	30
Ireland	All foods	B_1	5
		$B_1B_2G_1G_2$	30
IsraeI	Nut (products), peanut (products), maize, flour (jproducts), fig (products)	B_1	5
		$B_1B_2G_1G_2$	15
	Milk, milk powder (calculated on the basis of milk)	M_1	0.05
Italy	All foods	$B_1B_2G_1G_2$	10
		B_1	5
	Dried figs	$B_1B_2G_1G_2$	10
		B_1	5
	Species	$B_1B_2G_1G_2$	40
		B_1	20
Jamaica	Food, grains	$B_1B_2G_1G_2$	20
Japan	All foods	B_1	10
Jordan	Almonds, cereals, maize, peanuts, pistachio nuts, pine nuts, rice	B_1	15
		$B_1B_2G_1G_2$	30
Kenya	Peanut (products), vegetable oils	$B_1B_2G_1G_2$	20
Luxembourg	Peanut (products)	B_1	5
Macedonia	Wheat, maize, rice, cereals	B_1G_1	1
	Beans	B_1G_1	5
Malawi	Peanuts (export)	B_1	5
Malaysia	All foods	$B_1B_2G_1G_2$	35
Mauritius	All foods	B_1	5
		$B_1B_2G_1G_2$	10
	Groudnuts	B_1	15
		$B_1B_2G_1G_2M_1M_2$	

Mexico	Flours	All aflatoxins	20
Morocco	Codex texts followed		
Netherlands (The)	All foods and food ingredients except groud-nuts used for the preparation of peanut oil	B_1	5
	Cereal (products, pulse) (products), legume (products)	All mycotoxins	0
	Milk (products), milk power calcd reconstituted product	M_1	0.05
	Cheese	M_1	0.2
	Butter	M_1	0.02
	Infant foods on milk basis	M_1	0.05
New Zealand	All foods	$B_1B_2G_1G_2$	5
	Peanut butter, shelled nuts, nut portion of products containing nuts	$B_1B_2G_1G_2$	15
Nigeria	All foods	B_1	20
	Infant foods	B_1	
	Fluid milk	M_1	
Norway	All foodstuffs	$B_1B_2G_1G_2$	5
	Brazil nuts, bulk wheat	$B_1B_2G_1G_2$	5
	Apple juice (concentrated)	Patulin	50
Peru	All foodstuffs	$B_1B_2G_1G_2$	10
Philippines	Nut (products)	$B_1B_2G_1G_2$	20
Poland	All foods	B_1	0
Portugal	All foodstuffs	B_1	20
	Peanuts	B_1	25
	Infant foods	B_1	5
Romania	All foods	B_1	0
		Patulin	30
		Ochrtoxin A	5
		Zearalenone	30
	Milk, dairy products	M_1	0
Russia	Animal fats	B_1	0
		M_1	0.5
	Bottles/canned/potted fruits and berries	Patulin	50
	Bottles/potted/canned vegetables	B_1	5
		Patulin	50
	Casein	B_1	0
		M_1	5
	Cereals (wheat of hard and strong types), flour, wheat bran	B_1	5
		Zearalenone	1000
		T2-Toxin	100
		Deoxynivalenol	1000

	Fruits, berries and vegetables (bottled/ canned/potted juices and puree), cacao, cacao powder, chocolate, coffee, eggs, dehydrated egg, meat and poultry (fresh/chilled) Frozen/tinned/potted/bottled). Sausage and culinary products from meat and poultry, subproducts of farming animals and poultry, sweets	B_1	5
	Leguminous, protein isolators and concentrators, vegetable oil	B_1 Zearalenone	5 1000
	Nuts (kernel)	B_1 Zearalenone	? 1000
	Milk, sour dairy products, concentrated milk, cheese and cottage cheese products, cow butter	B_1 M_1	0 0.5
Salvador	Foods	$B_1B_2G_1G_2$	20
Serbia	Wheat, maize, rice, cereals	B_1G_1	1
	Beans	B_1G_1	5
Singapore	All foods	B_1 $B_1B_2G_1G_2$	0 0
South Africa	All foods	B_1 $B_1B_2G_1G_2$	5 10
Spain	All foods	B_1 $B_1B_2G_1G_2$	5 10
Srilanka	Foods	All aflatoxins	30
	Foods intended for children up to 3 years of age	All aflatoxins	1
	Milk (products)	All aflatoxins	1
Surinam	Maize	$B_1B_2G_1G_2$	30
	Groundnut (products), legumes	B1	5
Sweden	All foods	$B_1B_2G_1G_2$	5
	Berries, fruits, juices	Patulin	50
	Liquid milk products	M1	0.05
Switzerland	All foods (except maize/cereals/herbs)	B_1 $B_2G_1G_2$	1 5
	Maize, cereals (granular or ground)	B_1 $B_2G_1G_2$	2 5
	Herbs	B_1 $B_2G_1G_2$	5 5
	Baby/infant foods	$B_1B_2G_1G_2$	0.01
	Cereals products	Ochratoxin A	2
	Maize (products)	Fumonisins B1B2	1000
	Fruit juice	Patulin	50
	Milk (products)	M_1	0.05

	Whey (products)	M_1	0.025
	Cheese	M_1	0.25
	Butter, baby/infant food	M_1	0.02
Taiwan	Cereals	$B_1B_2G_1G_2$	50
Thailand	All foods	$B_1B_2G_1G_2$	20
Trinidad Tobago	Foods	$B_1B_2G_1G_2$	10
United Kingdom	Nut (kproducts), dried fig (products)	$B_1B_2G_1G_2$	20
Uruguay	Foods and spices	$B_1B_2G_1G_2$	20
	Texturized soy protein products; flour, starch, concentrate, isolate	$B_1B_2G_1G_2$	30
	Peanuts, dried fruit (products)	$B_1B_2G_1G_2$	30
	Cocoa beans	$B_1B_2G_1G_2$	10
	Infant foods, produced industrially	$B_1B_2G_1G_2$	3
	Rice, barley, beans, coffee, maize	Ochratoxin A	50
	Maize, barley	Zearalenone	200
	Fruit juice	Patulin	50
	Milk (products)	M1	0.5
USA	All foods	$B_1B_2G_1G_2$	20
	Finished wheat products	Deoxynivalenol	1000
	Whole milk, low fat milk, skim milk	M_1	0.5
Venezuela	Rice flour	$B_1B_2G_1G_2$	5
Zimbabwe	Foods	B_1	5
		G_1	4
	Groundnuts, maize, sorghum	B_1	5
		G_1	4

Though presence of mould produced toxins with food commodities has been known since Biblical times but their role in inciting disease syndromes was realized only when it was discovered that "ergotism" was caused due to the consumption of barley and rye infected with *Claviceps purpurea*. Cardiac beriberi caused by *Penicillium citreoviridin* was recorded in Japan due to consumption of contaminated rice; Stachybotryotoxicosis of horses due to *Stachybotrys atra* infestation in Soviet Union in 1931; Red Mould Diseases or Black Spot Disease caused by *Fusarium* sps. in Japan during 1940-50 and Alimentary Toxic Aleukia (ATA) caused by *Fusarium* and *Cladosporium* species in USSR during 1942-47 are some of the maladies in man and livestock. However, the severity of mycotoxin problem was realized during World War II when Russians eating mouldy over wintered grains suffered with severe dermal necrosis, leukopenia, haemorrhages and destruction of bone marrow. *Fusarium* was found to be the causal organism. World wide scientific recognition of mycotoxin problem was, however, only in 1960 when it was discovered that the aflatoxins were responsible for the mass death of about a lakh turkey poults (Turkey x disease) in England, due to the ingestion of the "poisonous feed" which caused the havoc was due to the feed imported from Brazilian peanuts, contaminated with *Aspergillus flavus* (Sargeant *et al.*, 1961). Another acute and fatal food poisoning of human caused by the consumption of deteriorated sugarcane by species of *Arthrinium saccharicola* in Northern China has also attracted the attention of several scientists (Liu *et al.,* 1992). Discovery of incidence of fumonisins produced by *Fusarium moniliforme* in maize and their association with equiencephalomalacea in 1988 (Gelderblom *et al.,* 1998) is another event which awakened world recognisation to this problem. Marasas (2001) aptly states that if the new era of research on food-borne diseases caused by mycotoxins was introduced by aflatoxins in 1960, the twentieth century closed with world wide attention focused on fumonisins. The next chapter to be written on a food-borne diseases caused by *Fusarium* mycotoxins may well be called by Fumonisins, food-borne fusarium toxins for third millennium impact of mycotoxins on human health in developing countries has been brought out clearly by Shephard (2008). Some of the important mycotoxin induced diseases along with the producing fungus and foods and feeds responsible for such health problems are precised in Table 1.10 and myco-toxicoses along with symptoms and victimized animals are listed in Table 1.10 a.

In India first outbreak of mycotoxicoses has occurred in 1958 as a result of ingesting pearl millet containing 15–26 mg/kg ergotine a clavine-type ergot from *Claviceps fusiformis*. Subsequently several disease outbreaks have been reported due to mycotoxins during last 40 years;

1. The outbreak of aflatoxin hepatitis in 1974-75 in Western India claiming more than 100 tribal lives due to consumption of maize containing aflatoxins (Krishnamachari *et al.*, 1975a).
2. The outbreak of enteroergotism in Rajasthan, Maharashtra and Gujarat in 1976 due to consumption of bajra contaminated with ergot.
3. Occurrence of Degnala disease, a disease of an obscure etiology, with syndromes stimulating chronic ergotism in buffaloes and cattle in 1970's in the rice growing areas of the Punjab and Haryana states (Karla and Bhatia, 1990).
4. Aflatoxin and Indian Childhood cirrhosis (Bhandari and Bhandari, 1980).
5. Endemic familial arthritis in India due to fusarial toxins.
6. "Kodua poisoning" in man due to presence of CPA in Kodo millet (*Papsalum scobiculatatum*) produced by associated fungi (Rao and Hussain, 1985).
7. Deoxynivalenol mycotoxicosis in Kashmir valley in 1987 caused by the consumption of mould-damaged wheat products (Bhat *et al.*, 1989a).

8. Inclusion body hepatitis in poultry in Northern India from 1990-1994 (Singh *et al.*, 1994).
9. A food-borne disease outbreak due to the consumption of moldy sorghum and maize containing fumonisin mycotoxins (Prathap Kumar *et al.,* 1997).
10. Severe aflatoxicosis outbreak in chicken in Himachal Pradesh (Rajesh Chahota *et al.,* 2000).

1.2. NATURAL INCIDENCE OF MYCOTOXINS IN DIFFERENT FOOD COMMODITIES

Cereals constitute the most important food and feed sources, which are affected by different mycotoxigenic fungi. The problem of natural contamination of mycotoxins in cereals aggravated due to rapidly changing agricultural technology. Though in general mycotoxins seem to pose great threat in the tropics than in the temperate regions, but no part of the world can be considered to be mycotoxin-free zone due to the movement of various food stuffs from one part of the globe to the other (Bilgrami and Sinha, 1984b). When some mycotoxins are proven or suscepted carcinogens, mutagens or tremorogens, others have been shown to challenge immune systems of man and animals. These raise the possibility of increased susceptibility to other diseases without suspicion of mycotoxin involvement. The intake of mycotoxins may also occur as a result of their presence in fungal spores in the intramural atmosphere due to humid atmosphere and other similar conditions may prove to be favourable (Johanna Flannigen *et al.*, 1991). Mycotoxins are also reported to carry over to dairy milk from cattle feeds contaminated with mycotoxins (Fink-Gremmels, 2008). Fruit juices can contribute significant levels of certain mycotoxins to the human diet. Some of the important commodities, which have been reported to be naturally contam-inated with one or the other mycotoxins in India, are listed in Table 1.9 and natural toxicants in foods are précised in Table 1.10.

Table 1.9. Incidence of mycotoxigenic fungi and mycotoxins in different food and feeds from India.

Item	Fungi /Mycotoxin	Reference
Apple	AFB_1	Roy and Chourasia (1989), Neelakanthan *et al.* (1984)
Anise fruits	Trichothecenes	Ghosal *et al.* (1982)
Barley	Mycotoxins	Mall *et al.* (1979)
Beetle nut	AFB_1	Raisuddin and Misra (1991)
Black cumin	Mycoflora and Mycotoxins	Giridhar and Reddy (1997a)
Black pepper	Mycoflora and Mycotoxins	Giridhar and Reddy (1997a,e, 2002)
Bread	AFB_1	Gaur and Siradhana (1989)

Cereal grains	*A. flavus*, AFB1	Sashidhar *et al.* (1989)
Chick pea seeds	Mycoflora and Aflatoxins	Ahmad and Singh (1991)
Chilli (dry)	Mycoflora, Mycotoxins	Giridhar and Reddy (1996), Reddy *et al.* (2001)
Coconut copra	Mycoflora and Mycotoxins	Giridhar and Reddy (1996b, 1997a).
Coriander	Mycotoxins	Misra 1987; Tirumala Devi *et al.* (2001).
Date fruit	Mycoflora, Mycotoxins	Giridhar and Reddy (2001b).
Dry fruits	Mycoflora and Mycotoxins	Saxena and Mehrotra (1990), Giridhar and Reddy (1999a, 2001a,b, 2002)
Fennel seeds	Mycoflora, mycotoxins	Misra and Batra (1987)
Fish	AFB_1	Prasad and Sinha (1987)
Feed & fodders	Mycotoxins	Pruthi (1978), Neelakanthan *et al.* (1978), Laxma Reddy and Reddy (1992a), Sinha *et al.* (1999)
Fenugreek	Mycoflora and Mycotoxins	Giridhar and Reddy (1997d, 1999c).
Garlic and onions	Mycoflora and Mycotoxins	Giridhar and Reddy (1999d).
Ginger	Mycoflora and Mycotoxins	Giridhar and Reddy (1997a)
Groundnut cake	AFB_1, T-2 toxin, other Mycotoxins	Subramaniyan and Rao (1974), Rukumini and Bhat (1978), Nagaraj and Kumar (1986), Bhavanishankar and Shantha (1987), Rati and Shanta (1994), Bhat *et al.* (1996), Usha *et al.* (1994), Sashidhar (1993)
Gram flour	Mycotoxins	Sinha and Sinha (1992), Ramakrishna *et al.* (1990)
Herbal medicines	AFB_1	Kumar *et al.* (1993)
Linseed	AFB_1	Sahay *et al.* (1990)
Macuna puriens	AFB_1	Roy and Chourasia (1989)
Makhana	Mycotoxins	Sinha *et al.* (2002)
Maize	AFB_1, Mycotoxins	Bilgrami and Sinha (1982, 1983), Bhavanishankar and Shantha (1987), Sinha and Ranjan (1989), Sinha and Singh (1987); Sinha (1988); Krishna Reddy and Reddy (1990), Vasanthi and Bhat (1997), Bhat *et al.* (1997a), Bilgrami *et al.* (1992), Janardhana *et al.* (1999).
	FB_1	Chatterjee and Mukherjee (1994).
	Cyclopiazonic acid (CPA)	Krishna Reddy and Reddy (1992)
Mango	Aflatoxins, Fumonisins	Ghosal *et al.* (1977)
Macuna	AFB_1	Roy and Chourasia (1989)
Medicinal plant seeds	AFB_1	Roy and Kumar (1993)
Milk and milk products	Mycotoxins	Sinha and Ranjan (1989); Sinha *et al.* (1991)
Mustard	Mycotoxins	Sahay and Prasad (1990), Ranjan *et al.* (1992), Giridhar and Reddy (1997a, b)

Oil seeds	Ochratoxin A, fusarial toxins	Rati *et al.* (1979), Chakrabarti (1987), Rati and Shantha (1994).
Paddy	Mycoflora and Mycotoxins	Chenulu and Dharmvir (1979), Gangopadhyay and Chakraborty (1981), Chary and Reddy (1987a), Waghray *et al.* (1988), Jayaram and Kalyanasundaram (1990), Begum and Samajpathi (2000), Sudhakar Rao and George (2000)
Poultry feeds and foods	Mycotoxins	Gopal (1985), Nagaraj and Kumar (1986), Balachandrana and Ramakrishnan (1987), Surekha and Reddy (1990), Balachandran and Parthasarathy (1996), Prathap Kumar *et al.* (1997), Shetty and Bhat (1997a, b), Jindal *et al.* (1999), Suman Mor and Kishan Singh (2000).
Pearl millet	Mycotoxins	Girisham and Reddy (1988), Misra and Daradhiyar (1991), Reddy and Nusrath (1986), Sinha (1988).
Pulses	Mycoflora and Mycotoxins	Premalatha Singh *et al.* (1990)
Papasalum (Kodua millet)	CPA, Mycotoxins	Rao and Hussein (1985)
Quinces (Cyadonia fruits)	AFB_1	Sharma and Sumbali (1999)
Raisins	Mycoflora and Mycotoxins	Giridhar and Reddy (2001a)
Safflower seeds	Trichothecenes	Ghosal *et al.* (1977)
Scented supari	Patulin	Neelakanthan *et al.* (1982)
Sorghum	Mycoflora, Mycotoxins	Tripathi (1973), Rukmini and Bhat (1978), Reddy and Nusrath (1985), Sashidhar *et al.* (1992a), Mukherjee and Lakshminarasimham (1995), Hareeshvardhan Rao *et al.* (1996 a,b,c, 2001a,b), Shetty and Bhat (1997a).
Stored grains	Mycotoxins	
Sunflower seeds	Mycoflora and Mycotoxins, Toxigenic fungi	Vijaya Lakshmi and Rao (1985)
Turmeric corms	Mycoflora and Mycotoxins	Giridhar and Reddy (1997a)
Tobacco (chewing type)	AFB_1	Sumna Rai and Chaturvedi (1993)
Air borne aflatoxin in grain processing industries	AFB_1	Ghosh *et al.* (1991)

Table 1.10 Natural toxicants in Foods

Mycotoxin	Foods infested	Main fungal species
Aflatoxin B_1G_1	Nuts, figs, dried fruit spices, ricebran, maize	*A. flavus, A. parasiticus*
Aflatoxin M_1,M_2	Milk and dairy	Metobolic productst of aflatoxins B_1 and B_2
Sterigmatocystin	Cereals, Cheese	*A. versicolor, A. nidulans*
Cyclopiazonic acid	Cereals, pulses	*A. flavus, P. communis*
Ochrotoxin A	Cerals, Coffee filed beans, beer, nuts	*P. verrucosum A. ochraceus*
Citrinin	Cereals	*P. verrucosum*
Patulin	Apple juice, Cereals	*P. expansum*
Deoxynivalenol, Nivlenol	Cerals	*F. graminearum* *F. culmorum* *F. crookwellellse*
T-2 toxin	Maize	*F. poae* *F. sporotrichioides*
Fumonisins B_1, B_2	Maize	*F. moniliforme* *F. proliferatum*
Zearalenone	Cereals	*F. graminearum* *F. culmorum* *F. crookwellellse*
Moniliformin	Cereals	*F. moniliforme*
Alternariol	Tomatoes, Oil seeds	*Alternaria alternata*
Alternariol monomethyl ether, tenuzronic acid	Cereals	*A. alternata*

1.2.1. Cereals

Maize is an excellent substratum for mould growth and mycotoxin production. In India, systematic survey of maize grains for mycotoxins contamination has been undertaken in some parts only. Some reports mainly focused on aflatoxin contamination in stored grains have indicated that the problem of aflatoxin contamination is much more serious than usually visualized (Bilgrami and Sinha, 1984a). In an extensive survey (1977-1980) of the important maize growing areas of Bihar State, natural contamination of mycotoxins and particularly of aflatoxins was reported in the grains of maize crop still in the field (Bilgrami *et al.*, 1981). Apart from this out of 615 samples of maize, 316 samples contained aflatoxin B_1 (AFB_1) in concentration ranging from 8 to 1640 ppb. Surveys conducted by the Indian Grain Storage Institute during 1978-79 and 1980-81 indicated mycotoxin contamination. Rati *et al.* (1979) reported the presence aflatoxins and ochratoxin in the range of 30-50 ppm in corn samples collected from Hapur, UP and Mysore, Karnataka State respectively.

1.10a. Incidence of Mycotoxicosis in different parts of the world

Country	Year	No. of subjects affected/ Exposed	Clinical symptoms and signs	Onset of symp-toms/ recovery	Source of exposure	Toxic fungal species	Toxin
USSR	1931-47	--------a	"Alimentary toxic aleukia"		Grain	*Fusarium sporotri-choides*	
			1. Hyperaemina of mucous membranes of oral cavity and pharynx, gastritis, gastroenteritis , abdominal and esophageal pain and diarrhea	A few hours/ 2-3 days		*F. poae*	
			2. Generalized indisposition, vertigo, unpleasant taste in mouth, progressive leukopni, and lymphocytosis.	3-4 weeks			
			3. Haemorrhagic diathesis and angina, petehial rash, catarrhl diphthertitic, gangrenous pharyngitis, ulcerative and gangrenous laryngitis, aphoni, asphyxia	Few days/ 2 weeks			
Japan	1956	25; 0b	"Scabby grain toxicosis"- nausea, vomiting,drowsiness		Rice	*F. roseum* *F. nivele*	
China	1961-1985	7818; 0	"Scabby grain toxicosis"- nausea, vomiting, abdominal pain, diarrhea, dizziness, headache.		Corn Wheat	*F. gramin-earum* sp.	Deoxynivlenol, Zeralenone
China	1984-1985	463/600; 0	Nausea,vomiting,abdominal pain, diarrhea,dizziness, headache	5-30 min.	Corn Wheat	*F. grami-nearum*	Deoxynivlenol, Zeralenone
India	1987	97/224;0	Mild to moderate abdominal pain, feeling of fullness imitation of nausea	15 min- 1 hour / 1-2 days	Wheat	*F. grami-nearum*	Nivalenol Deoxynivlenol T-2 Acetyldeo-xynivalenol
China		97/165 ;0	Nausea	0-30 min.		*F.heteros-porum*	T-2

Under a FAO sponsored Food Contamination Monitoring Project in the year 1980, out of 10 maize samples collected from Western Uttar Pradesh, 3 samples were contaminated with AFB_1 in the range of 20-80 ppb (Anonymous, 1980). Similarly samples of maize collected from traditional storage structures from different parts of India have shown contamination of AFB_1 upto 15,680 ppb (Singh *et al.*, 1982). Aflatoxin production by *Aspergillus parasiticus* in cereals leads to loss in caloric value of maize seeds (Premlatha Singh *et al.,* 1988). A positive correlation of insect damage and aflatoxin contamination of maize crop was observed by Sinha and Ranjan (1989). The incidence of aflatoxins, zearalenone (Bilgrami *et al.*, 1980a), trichothecenes (Krishna Reddy and Reddy, 1989), and cyclopiazonic acid (Krishna Reddy and Reddy, 1992) on standing crop of maize was also reported. Sinha (1990) reported the heavy infestation of maize samples of Kharif crop, which contained more aflatoxins (47%) than the samples of the Rabi crop (17%). Bhat *et al.* (1997a) reported that out of 2074 samples of maize from rural and urban areas of 11 states of the country 26% of maize samples contained aflatoxins beyond the permissible limit (30 mg/kg). Analysis of aflatoxin intake from maize based diets in a rural population in Southern India by Vasanthi and Bhat (1997) revealed that the average aflatoxin intake in house ranged from 0.33 to 1.5 mg/day, while individual intake varied from 0.08 to 2.22 mg/day. They further recorded a considerable variation in fumonisin B_1 (FB_1) contamination between the hybrids and inbreds.

Rice: Natural occurrence of mycotoxins especially aflatoxins and aflatoxin producing fungi in rice has been reported from different parts of the world (Bhat *et al.,* 1978). They identified coastal areas of A.P. and Cauvery delta of Tamilnadu as areas, which are prone to cyclonic weather conditions during paddy harvesting season as high risk areas of mycotoxin contamination and reported aflatoxins, zearalenone and sterigmatocystin in different paddy samples. Joshi *et al.* (1987a) reported that AFB_1 in the range of 40 ppb in paddy harvested from rain affected crop in Punjab and Haryana. Aflatoxin contamination of rice has also been reported from different parts of India (Tyagi *et al.,* 1986). Bilgrami and Sinha (1984a) is of the opinion that though presence of mycotoxins like citrinin, sterigmatocystin and ochratoxin A has been detected as natural contaminants of rice in one or the other regions of the world, only scanty information is available pertaining to these toxins in India. Dehusked rice collected from the flood affected areas of Warangal district of A.P. (Chary and Reddy, 1987a, b) and other parts of A.P. (Nusrath and Ravi, 1984) were contaminated with different mycotoxins.

Wheat: Wheat which contributes 36% to our country's total food grain basket does not appear to be a good substratum for aflatoxin elaboration. Srivastava (1987) reported that out of 223 samples collected from wheat growing belt of Western U.P. only 9 samples were contaminated with aflatoxin B_1 in the range of 8-40 ppb (Vasan, 1980 and Bilgrami and Sinha, 1984a) All the six flood affected wheat samples collected from Punjab contained aflatoxin B_1 in the range of 8-40 ppb (Joshi *et al.*, 1987a). Vora (1978) reported the natural incidence of aflatoxin and sterigmatocystin respectively. Rati *et al.* (1979) reported the presence of 30-50 ppm of ochratoxin A in wheat samples collected from Mysore. The incidence of enterogastric disorder outbreak in Kashmir valley during July-September, 1987 was attributed to trichothecenes contaminated wheat products (Bhat *et al.,* 1989a, b). Most of the wheat samples contained deoxynivalenol in the range of 0.43-4.85 μg/g, followed by acetyl deoxynivalenol (0.6-2.5 µg/g), nivalenol (0.1 μg/g) and T2 (4 μg/g).

Barley, oats, and other millets: Micromillets also do not favour natural contamination of mycotoxins. However, bajra has been reported to be suitable substratum for aflatoxin production. Presence of cyclopiazonic acid (CPA) in Kodo millet (*Papsalum scobiculatatum*) causing "Kodua poisoning" in man was reported by Rao and Hussain (1985). David Miller (2008) discussed mycotoxins in small grain millets.

Sorghum: Sorghum is prone to head moulds infestation and mycotoxin contamination. Bhat (1988), Ramakrishna *et al.* (1990) recorded to the extent of 24.5 mg/ kg of T2. Rukmini and Bhat (1978) recorded T-2 toxin in *Fusarium incarnatum* infested sorghum. Similarly the production of DON by *Fusarium* isolates of wheat and sorghum was reported to be associated with human mycotoxicoses outbreak (Ramakrishna *et al.,* 1989). Reddy and Nusrath (1985) reported the infestation of different varieties of sorghum seeds collected from A.P. by mycotoxigenic fungi and mycotoxin contamination. Rati *et al.* (1979) reported the presence of 50–70 ppm of ochratoxin A in sorghum samples collected from Mysore. In Western India sorghum samples were contaminated with AFB_1, depending on seed moisture (Mall *et al.,* 1986). An extensive study undertaken by NIN, Hyderabad in A.P. revealed that 26 isolates of sorghum belonging to different species of *Fusarium* such as *F. proliferatum* and *F. subglutinans* produced fumonisin B_1 upto 2 μg/g which is significantly low compared to maize (93-218 μg/g). The natural incidence of *Alternaria* mycotoxins i.e. alternariol (AOH), alternariol methyl ether (AME), alternuene (ALT), altertoxin-1 (ATX-1) and tenuazonic acid (TA) in sorghum has also been recorded by Ansari and Srivastava (1991).

1.2.2. Pulses

Pulses being proteinaceous in nature constitute a good substratum for mycotoxin producing fungi. However, there are only a few reports about natural occurrence of mycotoxins in these substrates. Association of toxigenic fungi with some samples of pulses has been confirmed by Vora (1978). A survey under FAO sponsored by Food Contamination Monitoring Project in

Western Uttar Pradesh revealed that few samples of pulses including green gram, black gram, lentil and few samples of cotton seed were contaminated with AFB_1 in the range of 4-80 ppb and 35-200 ppb respectively. Similarly, ten percent of stored soybean samples from Uttar Pradesh were contaminated with aflatoxin B_1 in the range of 4-40 ppb (Joshi *et al.*, 1986). The incidence of mycotoxin producing fungi in black gram, red gram and green gram collected form different regions of Andhra Pradesh State (A.P.) was dominated by species of *Aspergillus* and *Penicillium* sp.

Premalatha Singh *et al* (1990) reported nutritional deterioration and presence of aflatoxins in some pulse cultivars under the influence of *Aspergillus flavus*. Premalatha Singh and Ahmed (1989) recorded aflatoxin in seed samples collected during the development of pod in the range of 30- 110 μg/kg.

1.2.3. Oil Seeds

Groundnut: Groundnut is one of the most vulnerable commodity for *A. flavus* infestation and AFB1 contamination. Rati *et al.* (1979) recorded 50–2000 ppb of ochratoxin A in ground nut samples collected from Mysore region. Bhavanishankar and Shantha (1987) reported the natural incidence of *Fusarium* toxins in peanut samples collected from Mysore. Of the total number of peanut samples analyzed, 6.9% were positive for the trichothecenes (T-2 up to 38.89 mg/kg^{-1}; DAS up to 2.03 mg/kg^{-1}). Relative humidity during storage is reported to play pivotal role in mycotoxin contamination of ground nut (Gambhir and Khairnar, 1986). A multicentric study carried out by the FAO under the Food Contamination Monitoring Project (FCMP) in India indicated that groundnut is most prone to aflatoxin contamination. Various surveys carried out in different parts of India also point to ground nut as a high-risk commodity for aflatoxins (Bhat *et al.*, 1989a). The levels of aflatoxin in unrefined oil in India varied, the highest recorded level being 5000 $\mu g\ kg^{-1}$. Assuming a realistic average toxin level of 80 $\mu g/kg^{-1}$, use of 50 g of unrefined groundnut oil per day has been calculated to contribute 4 μg of aflatoxin to the diet per day (Acharya, 1986). A multicentric study for aflatoxin B_1 contamination in groundnut samples from different geographical regions of India revealed the presence of aflatoxins B_1 in the range of less than 5 to 833 mg/kg (Bhat *et al.,* 1996). The Codex Alimentarius Commission Working Group–recommended aflatoxin limit of 5 mg/kg in nuts was exceeded in 44.9% of samples. The trichothencenes and zearalenone were reported to contaminate groundnut oil cakes (Rajakumari Kaushal *et al.,* 1994) and sesamum (Reddy and Reddy, 1983). The levels of aflatoxins were about 31% in sweet and spiced peanut products collected from shops (Rati and Shantha, 1994). However, aflatoxin levels in the toffees were, lower (2-400 ppb). In 1994-95 a large quantity of groundnut was rejected as unfit for export due to the presence of aflatoxins. Rejection of groundnut for export was attributed to aflatoxin B_1 contamination and in appropriate sampling techniques.

Mustard: Mustard is also susceptible to mycotoxigenic fungal infestation (Giridhar and Reddy, 1997a). Survey of Incidence of mycotoxin producing fungi on mustard in Bihar by Sinha *et al.* (1988) and Sahay and Prasad (1990) and Bilgrami *et al.* (1991) indicated the aflatoxin contamination in the range of 750, 870 and 1420 μg/kg respectively in 80% of samples. Mycotoxins producing fungi can damage mustard seedlings (Giridhar and Reddy, 1996a, 1997a). The incidence of aflatoxin in mustard seeds, cakes and oil was also reported from other parts of India (Sahay, 1988).

1.2.4. Dry fruits and spices

Aflatoxins have been reported as the main contaminant of coconut and other dry fruits viz. areca nut, almond, cashew nut, walnut, raisin, makhana and emblic (Chaturvedi and Chaturvedi, 1992; Giridhar and Reddy, 1999a,b). In some samples of these dry fruits, presence of zearalenone, citrinin and ochratoxin A has also been reported (Giridhar and Reddy, 1996a, 1999a, 2001a, b). Natural contamination of aflatoxin has been reported in different samples of spices (Madhyasth and Bhat, 1984), chilli (Jyothi Saxena *et al.*, 1988; Giridhar and Reddy, 1999), ginger (Giridhar and Reddy, 1996), cardamom and turmeric (Giridhar and Reddy, 1997) from different parts of India. Presence of ochratoxin A, citrinin and zearalenone has also been reported in a few samples of spices (Anjana Singh, 1983; Bilgrami, 1984). In a survey in Uttar Pradesh conducted by the Indian Grain Storage Institute, Hapur in 1984, out of 99 dry fruit samples only 19 samples contaminated with aflatoxin B_1 in the range of trace to 50 ppb. Production of aflatoxin in mustard by isolates of *Aspergillus flavus* was reported by Anisa Athas *et al.* (1988). Aflatoxin was one of the contaminants of copra (Krishna Kumari *et al.*, 1984 and Giridhar and Reddy, 1997b) of A.P. and Maharashtra State (Kulkarni *et al.*, 1986) and other parts of country (Narasimhan, 1968). Natural incidence of mycotoxigenic fungi and mycotoxins has been reported on bitter gourd seeds (Rajendra Singh *et al.*, 1991). Survey for the incidence of myctoxigenic fungi in different spices indicated that 48.07% of black pepper, 31% of caraway, 30% of dry chilli, 28% of methi, 60% of ginger 35% of turmeric samples contained one or other mycotoxins produced naturally. These were found to contain aflatoxins, sterigmatocystin, patulin, citrinin and cyclopiazonic acid (Giridhar and Reddy, 1997a). Date fruits, areca nut, raisins, coconut copra, cardamum, cashewnut etc were good substrates for mycotoxingenic fungi and toxin production under faulty storage conditions (Giridhar and Reddy, 1996a,b, 1997a). Variation in the incidence of mycotoxigenic fungi in different agricultural commodities may be attributed to suitability of substratum nature, climatic conditions, storage practices, temperature and high humidity (Giridhar and Reddy, 2002).

1.2.5. Milk and Milk Products

Available evidence indicate that aflatoxin B_1 present in groundnut meal, one of the major ingredients of cattle feeds, not only causes liver damage in milk cattle but could also be excreted in milk as such or in the form of aflatoxin M_1 (Lewis *et al.*, 1994). Yadagiri and Tulpule (1974) reported aflatoxin M_1 upto 8 mg/l in the buffalo's milk. Outbreaks of aflatoxicosis in dairy cattle from Mysore region and in Murrah buffaloes from Andhra Pradesh region emphasize the need for a constant awareness and monitoring.

1.2.6. Feeds

Mycotoxins in the feed are exceptionally important in animals reared under intensive husbandry conditions. Aflatoxin in feeds is still the primary concern because of wide range of host susceptibility, ubiquitous presence of aflatoxigenic fungi and conducive growth conditions, immuno-suppressive properties and heat stable properties of aflatoxins (Lillehoj, 1973). Usually, the food which is declared unfit for human consumption finds its way as feed for animals and poultry birds (Bilgrami and Sinha, 1984). Incidence of mycotoxicosis in poultry due to the ingestion of contaminated feed has been reported from different parts of India (Neelakanthan *et al.*, 1978; Sivaswamy *et al.*, 1991; Singh *et al.*, 1994; Balachandran and Parthasarathy, 1996; Jindal *et al.*, 1999). Still many times they go unnoticed, as proper reporting is not being done. Similarly the incidence of toxigenic fungi associated with poultry feed have been reported very frequently (Neelakanthan *et al.*, 1978; Surekha and Reddy, 1990, 1991; Laxma Reddy and Reddy, 1992). Survey on the presence of fusarial toxins, sterigmatocystin etc in animal feeds in Tamil Nadu State of India was reported by Department of Food Technology, Tamil Nadu Agricultural University in 1980. Out of 52 strains of *Fusarium* screened, 36 were positive for one or other trichothecene group of fungi in cattle feed samples collected from different regions of Andhra Pradesh (Surekha and Reddy, 1990).

1.2.7. Fodders

Straw of rice, stover of maize and sorghum left after harvesting of the grains are the main bulk of fodders for livestock throughout India. Tropical conditions, harvesting practices, unseasonal rains, high moisture content of straw during baling and open storage not only promote fungal proliferation but also encourage to elaborate many mycotoxins.

1.2.8. Sea foods

Gayatri Murjani (2000) recorded remarkable gross changes in different internal organs of fishes (Indian major carp, *Labeo rohita*) that received AFB_1. Verma and Pandey (1987) have studied the effect of aflatoxin B_1 on the liver of an air breathing teleost, *Channa punctatus*. Studies proved that aquatic animals are sensitive to carcinogenic stimuli and can serve as indicators of carcinogenic hazards to man (Shank *et al.*, 1972). It has also been shown that livestock products like meat, milk and eggs do contain aflatoxin in the form of residue (Rajan, 1985).

1.3. HIGH RISK AREAS FOR MYCOTOXINS IN INDIA

Climatic conditions in India are most conducive for mould invasion, proliferation and elaboration of mycotoxins. Moisture content of the grain or the relative humidity surrounding the substrate are the most important factors governing the aflatoxin production by *A. flavus*. The high risk areas identified in India are Kerala, Western India, Gangetic plains, north eastern as well as coastal areas of Andhra Pradesh, Karnataka and Tamil Nadu (Bilgrami, 1984). Unseasonal rains and flash floods which are very common in India enhance the moisture content of the grain making them more vulnerable for fungal attack (Srivastava, 1987).

1.4. WEATHER CONDITIONS AND MYCOTOXINS

Warm humid climate provides congenial atmosphere for the growth of fungi and production of mycotoxins. Food grains are normally harvested at higher moisture content and then dried to bring down the moisture content up to safe level (less than 10%) before storage. Delay in drying to safe moisture levels increases risks of mould growth and mycotoxin production (Srivastava, 1987). Natural calamities like floods or torrential and unseasonable rains during pre, mid or post-harvest stages may render the crops vulnerable to microbial attack.

Faulty storage conditions may also enhance the chances of microbial attack and production of mycotoxins. Thus, starting from harvesting of the crop till the food or food products are consumed by the consumers, fungus may invade at any stage/step. Two important environmental factors that govern mould infestation are substrate water activity and temperature. The minimum a_w at which molds are capable of growing is approximately above 0.61. Further the a_w and temperature limits for growth and mycotoxin production are some times markedly different. Giridhar and Reddy (1997b) have discussed elegantly the role of water activity and temperature in production of mycotoxins on mustard, maize etc. Apart from this, the insects that attack the crops at pre- and post-harvest conditions are responsible for *A. flavus* infestation and aflatoxin production (Sinha and Sinha, 1992).

The increased contamination of maize with fumonisins, AFB_1 and wheat samples with AFB_1, DON, Zearalenone etc. in India may also attributed to draught conditions. Further, the demand for increased food grain production has paved the way for changing farming pattern may also contribute to the severity of mycotoxin problem.

1.5. MANAGEMENT OF MYCOTOXINS

Increased awareness of mycotoxin induced health hazards, resulted in the intensive and extensive studies on the management of these biocides. The method of preventing or reducing levels of mycotoxin contamination can be broadly classified as preventive or curative. Under this there have been five approaches.

1. Prevention of the initial growth of moulds and subsequent contamination of mycotoxin (Anonymous, 1969).
2. Detection of mycotoxins in foods and selective removal of contaminated portion which however, reduces the toxin only to a certain level (Goldblatt, 1971).
3. In activation or destruction of toxin by physical, chemical and biological means (Cucculu *et al.*, 1976).
4. To develop effective economic methods for controlling the mould growth and subsequent production of mycotoxins.
5. Utilization of genetic resistance towards toxigenicity of fungus. Scott *et al.* (1990) have excellently reviewed the different strategies adopted in the management of mycotoxin problem.

Several methods of control of different mycotoxins such as heating (Davies and Diener, 1987), boiling (Mann *et al.*, 1976), roasting (Goldblatt, 1971), radiation (Ellis *et al.*, 1991), light (Cole *et al.*, 1989), chemical (Muller, 1989) and biological (Well *et al.*, 1976; Hao *et al.*, 1988; Wilson 1988; Park *et al.*, 1989; Karunaratne *et al.*, 1990; Brown *et al.*, 1991; Cotty, 1992 and 1994; Cole *et al.*, 1994 and Dorner *et al.*, 1994) and raising of resistant crops (Zuber, 1977; Zuber *et al.*, 1973) have been suggested from time to time. However, these methods are beset with one or other problem and no method proved to be safe and effective.

Mycotoxins are one of the important factors contributing for food loss. Based on the Global Environment Monitoring System – Food Contamination Monitoring and Assessment Programme (GEMS/FCMAP) data, mycotoxin contamination is a widespread problem of the food supplies in most countries. As a result, many countries have enacted regulations to control level of mycotoxins in the national food supply as well as in food moving in international trade. Eighteen countries all over the world have guidelines or regulations, which prescribe maximum acceptable limits for mycotoxins in foods and feeds. The limits prescribed for aflatoxins vary from 0 to 50 ppb in foods and from 0 to 1000 ppb in feeds in different parts of the world (Srivastava, 1987). In India the permitted levels of AFB_1 in food commodities is 30 ppb but no such regulations are prescribed for the other mycotoxins. It is time to realize the problem and enact wisely by making regulations for these mycotoxins also to avoid health risks of man. In general all the agricultural commodities before export are subjected to stringent analysis of mycotoxins such as aflatoxins. Though health hazards from aflatoxin in developing countries such as India is more from maize than groundnut, the economic significance of aflatoxin contamination in terms of reduced export earnings and loss at different levels of the trade chain such as shelled groundnuts, edible oils and oil cakes is more in grounduts than in maize (Bhat, 1989). Hence, the Governmental agencies should procure food grains confirming to prescribed specifications there by minimizing the chances of contamination and thus ensure the supply of quality food grains to the consumers. At farm level also quality consciousness should be created amongst the farmers through a network of expert teams of Save Grain Campaign.

There is an urgent need to form a National Guidelines for prevention and control of mycotoxins contamination specially aflatoxin in agricultural commodities, which should include: harvesting of fully matured grains/cobs/ears and drying to safe moisture level (less than 10%), separating the healthy, sound grains/cobs/ears from damaged or rotted or infected, drying the grains/cobs/ears on raised bamboo floors along with other appropriate drying methods as required for the merchants (for eg. Dry shelled maize on concrete drying floors), using a water proof sheet (plastic not advisable) to cover drying produce for rain protection, even mechanical drying if required, storage at proper dry conditions along with good ventilation, efficient sanitation practices to destroy or discarded infested ears/cobs/etc, and effective fumigation for control of insect problem.

Prevention of the growth of moulds, separation of infected grains, detoxification and growing resistant varieties are some of the other strategies for preventing mycotoxin contamination (Bilgrami and Sinha, 1984b).

1.5.1. Prevention of mould growth

Mould damage in stored grain, may be prevented mainly by three kinds of treatments viz. drying of grain, controlled storage atmosphere and chemical treatment.

Chemical control of fungal deterioration of stored grain is restricted to the treatment of grain for seed purposes only but not for food and feed. Experiments conducted by Indian Grain Storage Institute, Hapur, U.P. showed that "Grain treat" (Mixture of propionic acid, acetic acid and benzoic acid) against species of *Aspergillus* and *Penicillium* failed to give effective protection. Phenol and 2,4-D inhibited the cyclopiazonic acid production by *Penicillium griseofulvum* (Krishna Reddy and Reddy, 1989). Similarly Laxma Reddy and Reddy (1992) reported that p-nitrophenol and alpha-naphthol as effective inhibitors of satratoxin production by *Stachybotrys atra*. Casadio and Niola (1993) reported the destabilization of patulin in fruit juices with microwaves. Clavero *et al.* (1993) could destroy aflatoxin with the help of hydrogen peroxide. Seed treatment with Bavistin and TMTD (Trimethyl thiuram disulphide) at 0.25% concentration gave 100% protection to wheat grain during one year storage against species of *Aspergillus* and *Penicillium* (Srivastava, 1986). Blasticideen A showed strong inhibitory activity towards aflatoxins production by *A. parasiticus* (Sakuda *et al.,* 2000). Jayashree and Subramanyam (1999) reported the inhibition of aflatoxin production by *A. parasiticus* in a dose dependent manner up to a concentration of 0.75 mmol/L without inhibiting the growth of the fungus. Acetosyringore a phenolic compound inhibited aflatoxin biosynthesis (Hua *et al.,* 1999). Certain mild phenolics like ferulic acid and O-Vanillin have also been reported to prevent aflatoxin production on rice, wheat, maize, groundnut and mustard seeds (Bilgrami and Sinha, 1984a). Oxalic acid (3%) application on ground nut kernels was effective in inhibiting AFB_1 production. Volatile compounds which have no residues and highly active at low concentrations are being advocated in the protection of food grains at post-harvest stage. High moisture grain has been protected from aflatoxin contamination by using volatile organic acids such as propionic acid and acetic acid. By decreasing level of oxygen and creating anaerobic conditions by increasing level of CO_2 or N_2 in its place discourages mould growth and mycotoxins contamination. Modified Atmospheric Packing (MAP) and Controlled Atmosphere Storage (CAS) are ideal to keep away mycotoxigenic moulds from stored food grains for 6-9 months.

1.5.2. Detoxification

Cooking at atmospheric pressure can destroy about 50% of mycotoxins. Dry roasting and oil roasting of groundnut reduces aflatoxins to a significant degree. Cooking rice under 15 lbs. pressure for 5 minutes gave maximum destruction of aflatoxins (72%) as compared to ordinary cooking or cooking with excess water. Detoxification by various cooking methods is highly effective in detoxification of AFB_1 (Fasiha Rehana and Basappa, 1990). Light has also been employed successfully to destroy aflatoxin in crude groundnut oil. Studies have shown that visible light is more effective than either ultraviolet or infra-red light (Pruthi, 1978). Similarly oven drying and autoclaving detoxified AFB_1 by more than 65% (Bilgrami *et al.*, 1984). Detoxification also could be possible by bulb and tube light treatments to the extent of 51-55% in black cumin seeds and almond respectively (Bilgrami *et al.*, 1984a). It was observed that on cooking 17-56% reduction in AFB_1 concentration in maize based rotis was recorded (NIN, 1992). However, the above said drying treatment efficiency varied with mycotoxins. Some of the major cooked products such as roti and gutka (Poridge) prepared from contaminated maize and sorghum flour respectively showed about 29% reduction in extractable fumonisins B_1 in maize and no reduction in sorghum rotis. NaCl (30%) is reported to remove FB_1 from the contaminated maize (86% removal). Even the buoyant fraction of water itself removes upto 74% of FB_1 when contaminated samples of maize soaked in water (NIN, 1992). Kamimura *et al.* (1979) reported the decomposition of trichothecenes in noodles cooked under alkaline conditions. Roasting of coffee beans resulted in the decrease of ochratoxin A level (Vander Stegen *et al.*, 2001). Extrusion cooking appears to be effective for detoxifying deoxynivalenol but not aflatoxin (Cazzaniga *et al.*, 2001).

Although several chemical detoxification methods including acids, alkalies, aldehydes and oxidizing agents and gases like chlorine, sulphur dioxide sodium nitrite, ozone and ammonia have been tested for several mycotoxins, only the ammonium process is an effective and practical method for detoxification of aflatoxins (Rustom, 1997; Lamke *et al.*, 2001). Other than ammoniation, many of the techniques proposed to remove mycotoxins are currently perceived to be impractical, ineffective or potentially unsafe for large scale utilization (Marguardt, 1997). Though solvent extractions have been shown to be effective but are not economically feasible.

De Vries *et al.* (2002) have excellently discussed the effect of food processing on different mycotoxins. Loss of FB_1 occurs during extrusion and baking of corn based foods with sugars (Castelo *et al.*, 2001) and nixtamalization (alkaline cooking) and rinsing in the preparation of tortilla chips and masa (Dombrink Kurtzman, 2000; Voss *et al.*, 2001). Patulin can be removed from apple juice by treatment with certain types of active carbons (Leggott *et al.*,, 2001).

1.5.3 Dietary modifications

Dietary modifications greatly affect the absorption, distribution and metabolism of mycotoxins and this can subsequently affect their toxicity (Atroshi *et al.*, 2002). Diets containing chemoprotective agents, antioxidants such as ascorbic acid, BHA, BHT, ethoxyquin, oltipraz, pentaacetyl geniposide, kolaviran, bioflavanoids and even greater have been found to inhibit carcinogenesis caused by aflatoxins. Dietary administration of naturally occurring chemopreventive agents like ellagic and coumarin or α-angelicalcatone caused an increase in glutamate, cystein ligase activity which inturn suppressed carcinogenic activity of aflatoxins, anthraquinones, coumarins and flavone type flavonoids present naturally in fruits, vegetables and spices were found to be potential inhibitors for the formation of aflatoxins B_1 (Lee *et al.*, 2001). Diets containing Nova SL and related absorbents have been found to be effective in preventing aflatoxins and other mycotoxins in test animals. Several other absorbants such as zeolite, bentonite and superactive charcoal have been found to be effective in decreasing the toxicity of other mycotoxins such as T_2 toxin.

1.5.4. Resistant varieties

In recent times, efforts are being made to develop mould resistant varieties which will be mould free not only in the field on standing crop but also during storage and restrict development of moulds and mycotoxin contamination. However, evaluation and selection for resistance to mycotoxins at post-harvest stage may be more difficult than it has been for plant resistance to diseases. Substantial efforts have been made to identify maize genotypes that resist infection by *A. flavus* (Brown *et al.*, 1995). Genetically modified crops resistant to insects infestation (eg. BT cotton) have shown some effect in minimizing fungal propagation and aflatoxins, fumonisin and deoxynivalenol formation (Bacon and Hinton, 2002; Mag *et al.*, 2002). Naturally occurring *Aspergillus* strains non-toxic to peanut, corn and cotton have resulted in great reduction of aflatoxin formation (Cotty and Bhatnagar, 1994; Horn and Dorner, 2002). Current research is aimed at identifying and characterizing naturally occurring antifungal agents in commodities like corn (Nencere and Clevelano, 1997; Gambeh *et al.*, 2001). Abbas *et al.* (2002) and Windham and Williams (2002) have developed germplasm lines resistant to aflatoxins. Tubajika and Danman (2001) found that drought tolerant maize genotypes might be used as potential source of resistance to *Aspergillus* ear rot and aflatoxins production. Munkvold (2003) suggested different cultural and genetic approaches in the management of mycotoxins in maize.

Integrated approach to contain mycotoxins in food and feeds is desirable. This approach includes efforts both to prevent contamination and to detoxify contaminated grain especially where food resources are limited. Analytical quality assurance in agricultural export produce is required (Sashidhar *et al.*, 1992a). Biological detoxification methods such as fermentation can eliminate some classes of mycotoxins.

1.5.6. Biological control

Significant research has been done on biological control of mycotoxins. Fungal degradation of aflatoxins B_1 was reported by Shantha (1999). Cell free extracts of *Phoma* sp. were able to degrade AFB_1 upto 99% (50 ug/100 ml of culture media). Similarly Singh and Singh (1998) reported enzymatic degradation of aflatoxins in shake cultures of *A. flavus* CMI 102566 and highlighted the possible role of perxoidase in biological control mechanism. Karunarathre *et al.* (1990) reported that *Lactobacillus acidophilus* could prevent the growth of the moulds which in turn may pave the way for mycotoxin non production. *Marganella morganil* an intestinal organism can degrade aflatoxin (Hoh and Aibana, 1982). Tuason (1993) reported that *Flavobacterium aurantiacum* metabolizes aflatoxin. *Trichoderma viride* and *Aspergillus niger* could completely check the production of cyclopiazonic acid by *Penicillium griseofulvum* (Krishna Reddy and Reddy, 1989). Aflastatin A, a noval inhibitor of aflatoxin production has been isolated from a *Streptomyces* sp. (Sakuda *et al.,* 1996; Ono *et al.,* 1997). Schisler *et al.* (2002) reported successful control of deoxynivalenol in wheat through microbial antagonist.

1.6. LIMITS AND REGULATIONS FOR MYCOTOXINS

The contamination of agricultural commodities with mycotoxins may prevent a serious health hazards to both human beings and animals. Although world wide regulations exist to control aflatoxins in animal feed. Similar legislations are not available for any of the *Fusarium* mycotoxins including carcinogenic fumonisins (Placinta *et al.,* 1999). However, many countries have established some measures to control the contamination of these mycotoxins in foods and animal feed stuff. These limits are imposed to protect animal health as well as the health of potential consumers of contaminated animal products and to limit financial losses due to adverse effects of mycotoxins on animal productivity (Marquardt, 1996). In USA advisary directive has been set at 10 mg/kg of deoxynivalenol in grains and by products offered to cattle and chicken and 5mg/kg for pigs (Trucksess, 1995).

1.7. CURRENT RESEARCH SCENARIO

Development of validated new PCR assays for trichothecene and fumonisins producing *Fusarium* species is the current trend in mycotoxins research. Though earlier investigations in India were mainly focused on incidence of mycotoxigenic fungi on various food commodities and management of mycotoxins problem, finding of suitable molecular markers for mycotoxigenic fungi is the need of the day. Similarly the regulation of mycotoxin production at cellular levels needs to be investigated with special reference to AFB_1. Jayashree and Subramanyan (1999) reported regulation of aflatoxins production by Ca^{2+}/calmodulin-dependent protein phosphorylation and dephosphorylation. The recent discussions on the *Fusarium* mycotoxins in Food Chain Project and Ochratoxin A (OTA) prevention project of The

European Mycotoxin Prevention Cluster, held in June 2003 at Uppsala, Sweden warrants the necessity of looking the mycotoxins problem in a different way with a main focus on expanding the data base on *Fusarium* toxins in cereals and aflatoxins in groundnuts. Identifying critical control points in the production of maize based products and groundnut, using natural antagonists as pre-harvest biocontrol agents against head blight and other *Fusarium* diseases. Similarly a membrane based Molecular Imprinted Polymer (MIP) biosensors, ELISA assay and the competitive Lateral Flow Device (LFD) are the promising tools for analysis of grain extracts for mycotoxins contamination in grain as there is a lot of significance about these mycotoxins contamination in various food commodities.

Acute mycotoxicoses is can cause serious and some times fatal diseases. The possibility of mycotoxin intoxication should be considered when a acute disease occurs in several persons, when there is no evidence for infection with a known etiological agent and no improvement in the clinical picture following treatment. Most of the outbreaks of mycotoxicoses described are a consequence of the ingestion of food that is contaminated with mycotoxins. The strict control of food quality in both industrialized and developing countries is therefore, necessary to avoid such breaks. New tools and more knowledge are needed and will greatly help in the on going struggle to keep mycotoxins out of the food supply and off the minds of consumers.

There should be extention of research into the analysis of human body fluids eg: blood, urine and milk by way of adducts specific to the perceived most harmful mycotoxins. There should also continued examination of noval ways of determining toxic effects on the immune system. The use of human cell lines for toxicity studies should be further developed especially using cells that have been transfected with human cytochromes.

Though there is a wealth of knowledge concerning toxic effects of mycotoxins on animals, there is still insufficient knowledge concerning the health effects of mycotoxins in humans. There is an urgent need to gain better understanding of mycotoxins in the epidemiology of human disease and to further firm growing evidence of relationship between mycotoxin exposure and immunotoxicity. The insidious longterm effects of low concentrations of certain mycotoxins on the immune system identify mycotoxins as serious human environmentally derived poisons. Firm establishment of these relationship must be central feature of future research.

CHAPTER-2

TAXONOMY OF MYCOTOXIGENIC FUNGI

2.1. INTRODUCTION

As a sequel to the report of aflatoxin production by *Aspergillus flavus* in 1960 (Sargeant *et al.*, 1961) a great interest has been arose to know the harmful effects of fungi associated with stored foods and feeds. Large assembly of fungi are now known to elaborate wide variety of metabolites which exhibit varying toxicity ranging from chronic to acute toxicity. This fungus mainly exists in anamorphic state and has ascomycetes connection. These fungi which are comparatively xerophilic and potential to produce conidial mass in abundance and need warmer conditions to eloborate these toxic metabolites on nutritionally rich substrates making them unsuitable and hazardous to health of a man and animal are popularly called moulds. More than fifty genera of fungi are known to contain mycotoxigenic species, most of them remain in asexual stage abscuring their phylogenetic relationships and makes prediction of toxigenic potential difficult. Most of these moulds produce their sexual states from groups of ascomycetous orders eurotiales and hypocreales with a smaller number distributed among other orders of ascomycetes. The genera which are reported to contain mycotoxigenic species are precised in Table 2.1.

Perusal of Table 2.1 leads to the conclusion that toxigenic ability is sporadically distributed that any fungus is as likely as any other to be mycotoxigenic. However, it is logical to suggest that these fungi are phylogenically related as these mycotoxins are synthesized at the end of long enzyme mediated reaction sequences. Since most of these fungi do not reproduce sexually or produce rarely, their taxonomic identification is based on anamorphic state which makes determination of their phylogeny difficult and consequently there is little correlation between the pattern of toxin production by particular fungus and its classificatory position. However, connections between teleomorphs and their anamorphs are now known (Tubaki, 1981) to enable us to predict the likely teleomorphic genus (or groups or teleomorph genera) for most anamorph genera and only a few genera with toxigenic species cannot be linked to teleomorphs.

Table 2.1. Genera of fungi containing Mycotoxigenic species

Acremonium	*Dichotomomyces**	*Myrothecium*	*Rosellinia**
Aspergillus	*Diplodia*	*Microdochium*	*Sclerotinia**
Bipolaris	*Drechslera*	*Monographella* *	*Sphacelia*
Lasiodiplodia	*Epichloe**	*Nigrosabulum**	*Stachybotrys*
*Byssochlamys**	*Epicoccum*	*Nigrospora*	*Talaromyces**
*Ceratocystis**	*Fusarium*	*Paecilomyces*	*Thielavia**
*Cheatomium**	*Gibberella**	*Penicillum*	*Trichoderma*
Cladosporium	*Gliocladium*	*Periconia*	*Trichothecium*
*Claviceps**	*Gloeotinia**	*Phoma*	*Verticillium*
Colletotrichum	*Khuskia**	*Phomopsis*	*Verticimonosporium*
Curvularia	*Metarrihizium*	*Pithomyces*	*Zygosporium*

* - Teleomorphic stage

Majority and most significant mycotoxigenic species can be assigned to two orders, eurotiales and hypocreales. Probably the toxins produced by these fungi share same basic biosynthetic pathways but specific toxins are not shared between the orders.

Species of other orders of ascomycetes produce various toxins but these toxins are also produced by species of either eurotiales or hypocreales. However, claviceptales only produce ergot alkaloids.

Members of eurotiales dominate toxigenic species represented by *Aspergillus* and *Penicillium*. Other toxigenic members of this group are species of *Byssochlamys* and *Paecilomyces*.

Toxins produced by members of this order are diverse representing atleast six of the main biosynthetic categories. These include among others, (1) polyketides-patulin, citrinin, ochratoxins, maltoryzine, viriditoxin cytochalasins, rugulosin, ergochromes, aflatoxins and sterigmatocystins, (2) tetramic acid - cyclopiazonic acid; (3) diketopiperzines - aspergillic acid, fumitrogens and oxaline (4) neurotrophic peptide-tryptoquivaliene; the C_6 C_3 product - xanthocilin and (6) nonadride - rubratoxin a compound of mixed fatty acid – TCA cycle origin. However, a few of them are produced by non-eurotialean genera such as a ergochromes are also formed by *Claviceps, phoma* and lichen. Cyclopiazonic acid is produced by species of *Alternaria, pyricularia* and *Phoma*. Similarly, sterigmatocystins are produced by species of *Bipolaris* and *Chaetomium*.

Toxigenic species are found in all of the six subdivisions of *Penicillium* (Pitt, 1979) and out of 18 groups of *Penicillia,* 10 contained toxigenic. *Penicillia* are found in four subgenera but the production of specific toxins is not localized to any subgenus. For example, patulin is produced by teleomorphic genera *Byssochlamys* and *Eupenicillium* and by two sections each of *Penicillium* and *Aspergillus*. On the other hand, aflatoxins are produced by only two species of *A. flavus* group, but structurally similar sterigmatocystins are produced by species belonging to four groups including the *A. flavus* group, one species of *Penicillium* and other non-eurotiales fungi. The ability to produce particular toxin by species is not necessarily to produce toxins by other closely related species. Variation in the ability to produce a particular toxin also rests among different strains of the same species. For instance *Aspergillus flavus, A. parasiticus, A. sojae* and *A. oryzae* are difficult to distinguish morphologically but *A. flavus* and *A. parasiticus* produce aflatoxins, while no strain of *A. oryzae* and *A. sojae* which are used in production of food are not known to produce toxins. Wicklow and Cole (1984) feels that these species evolved from *A. flavus* and *A. parasiticus* and lost toxigenicity under domestication.

Over 40 species of anamorphs of hypocrealean fungi are toxigenic and belong to *Acremonium, Fusarium, Gliocladium, Myrothecium, Trichoderma* and *Verticillium*. Teleomorphs are not known for *Metarrhizium* or *Verticimonosporium* but their morphology and biology suggests their hypocrealean nature. These fungi are known to produce comparatively less number of toxins which include zearalenone (polyketide), trichothecenes (sesquiterpenes) and butenolides whose biosynthetic pathway is as yet unknown when butenolides are produced by only these fungi, zearalenone is also produced by species of *Curvularia*, a genus of pleosporales. On the other hand, trichothecenes are produced exclusively by species of *Fusarium, Myrothecium, Acremonium, Trichothecium* and *Trichoderma*. However; *Stachybotrys* which has affiliation to sordariales also produce macrocyclic trichothecenes. Similarly *Microdochium nivale* which was formerly known as *Fusarium nivale* now it is incuded in hyphonectriaceae of polystigmales and by the flowering plant genus *Baccharis*.

Claviceps with anomorphic state *Sphacelia* is the only genus of Claviceptales positively identified as toxigenic. The pharmacologically most active components of ergots are alkaloids which are amides of D-lysergic acid. No lysergic acid is found outside the *Claviceps*. The other two alkaloids produced by *Claviceps* are clavine alkaloids and ergolines. Clavine alkaloids are also formed by species of *Aspergillus* and *Penicillium*, while ergolines are found in *Aspergillus, Penicillium* and *Phoma* which is a genus of pleosporiales and lichens.

Acremonium typhium, the anomorph of *Epichloe typhina,* a pasture grass endophyte, has been linked to the fescue foot toxic syndrome. Butanolides produced by species of *Fusarium* have also been implicated in this condition.

Some of the anamorphic genera like *Phoma, Curvularia, Bipolaris, Drechslera, Alternaria, Diplodia* and *Periconia* are reported to have their teleomorphs in bitunicate ascomycetes belonging to pleosporales. Of these fungi, *Alternaria* is pharmacologically important and produces polyketides alternarioles and related altertoxins by its one of the species, *A. alternata*. This species also produces the tetramic acid, tenuazonic acid, in common with *Pyricularia* and *Phoma*. The other tetramic acid, the cyclopiazonic acid is elaborated by species of *Aspergillus* and *Penicillium*.

The other toxigenic fungi belong to four unrelated orders of ascomycetes and represented by one or two species. *Chaetomium* belonging to order sordariales is reported to produce cytochalasins and sterigmatocystins with several other unrelated fungi. Several species of *Cladosporium* whose teleomorph stage is in the dothideales, produces epi- and fagicladosporic acids, fatty acid derivatives which can induce alimentary toxic aleukia in common with *Fusarium*. *Stachybotrys* an anamorph of the sordariales produces trichothecenes in common with *Monographella* and flowering plant, *Baccharis*.

Some anamorphic fungi such as *Pithomyces, Zygosporium* and *Epicoccum* are reported to be toxigenic. Teleomorphs are not known and it is not possible to speculate their affinities. Though conidial ontogeny in *Pithomyces* and *Epicoccum* are similar, they are not related. *P. chartarum* produces sporidesmins, the cause of photosensitivity, liver damage and death in sheep, cattle and deer. These toxins include gliotoxins produced by hypocrealean genus *Gliocladium* and flavipine produced by *Epicoccum purpurescens* and species of *Aspergillus* and *Zygosporium* in common with many unrelated fungi produce cytochalasins.

Most of the known mycotoxins are derived either from polyketides or from sesquiterpenes, two basically different categories of secondary metabolites (Bulouc, 1980) both of which are common in ascomycetes and fungi imperfecti (Mc Corkindale, 1976). Thus, the basic pathways to most mycotoxins are generally resident within the ascomycetes and their anamorphs.

Key for the identification of these mycotoxigenic fungi is proposed as given below.

2.2. KEY TO THE IDENTIFICATION OF MYCOTOXIN PRODUCING FUNGI

1. Mycelium present in life cycle 2

1A. Mycelium unknown, asexual reproduction by budding (Fig. 1) .... *Torulopsis* (1)

2. Ascospores borne in a ascus, the ascus naked or enclosed within disk-shaped, cup-shaped, or more or less globose structure. The asexual state may have a separate name 39

2A. Ascospres not present 3

3. Conidia formed and borne on simple or branched conidiophores or arising directly from vegetative mycelium 7

3A. Conidia absent, sclerotia present 4

4. Sclerotia slender-fusiform, cylindrical with tapering ends, spherical, or shaped like a banana fruit hard, brown, black, or purple in the mature inflorescence of grasses (Fig 30) .... *Claviceps* (2)

4A. Sclerotia not in the inflorescence of grasses 5

5. Sclerotia homogeneous in structure, i.e., no specialized tissues 6

5A. Sclerotia not homogeneous; organized into tissues; rind, cortex, medulla; globose to concave, pinkish to dark purple-brown, clamp connections present (Fig. 3) .... *Sclerotium* (3)
Similar to *S. rolfsii* and included in the latter species by some authors.

6. Sclerotia black, 1.0 to 2.5 mm diameter, mycelium dark-green, becoming black, plant parasitic on cloves and other legumes (Fig. 4) ....*Rhizoctonia* (4)

6A. Sclerotia not black, very small, consisting of regular to irregular groups of cells. Sclerotia referred to as bulbils (Fig. 5) .... *Papulospora* (5)

7. Conidia formed in pycnidia 8

7A. Conidia not formed in pycnidia 12

8. Pycnidia separate, wall distinct, yellowish-brown to dark-brown with a circular short protruding opening, globose to subglobose. Immersed in host tissue 9

8A. Pycnidia without a distinct wall, but reduced to cavities in a stroma. Spores, hyaline, fusiform, large in bark (Fig 6) ....*Fusicoccum* (6)

9. Conidia one-celled 10

9A. Conidia two-celled 11

10. Conidia elliptical, oval, short, cylindrical, hayline (Fig. 7) ....*Phoma* [7]

10A. Pycnidia separate, immersed and scattered in tissues; subglobose, with short protruding ostiole. Conidia oval to lanceolate, hyaline. Conidiophores long, tapering toward apex. Beta conidia (curved, needle like) present (Fig. 8) .... *Phomopsis* (8)

11. Conidia brown, 2-celled, cylindrical or slightly curved, with rounded apex (Fig. 9). On roots, stalk, ear of corn....................*Diplodia* (9)

11A. Conidia 2-celled with longitudinal structure (Fig. 10) ....*Lasidiplodia* (10)

12. Conidia formed in an acervulus 13

12A. Conidia not formed in an acervulus 14

13. Conidia hyaline, one-celled; elliptical, cylindrical, fusiform, or sickle-shaped. Dark setae within and around the acervulus present or absent (Fig. 11) .... *Colletotrichum* (11)

13A. Conidia with several cells. Central cells dark, end cells hyaline. Hyaline appendages arising from end cells hyaline (Fig. 12)....................*Pestalotiopsis* (12)

14. Conidia formed in sporodochia or naked conidiophore 15

14A. Conidia not formed in sporodochia; conidiophores free on the host or substrate 18

15. Sporodochia disk-, cup-, or cushion-shaped....................16

15A. Sporodochia not disk-, cup-, or cushion-shaped but thin, irregular, and consisting mainly of discrete aggregations of conidiophores 17

16. Sporodochia cushion-shaped to irregular, at first white, then with a black or olive-black layer of conidia. Conidiophores tightly packed, irregularly branched. Conidia fusiform, pale-green. (Fig 13) .... *Dendrodochium* (13)

16A. Sporodochia disk- or cup-shaped, with or without setae, sessile, or with short stalk, forming a conidial mass, at first green, then black. Conidiophore branched, terminated by an apical whorl of phialides. Conidia hyaline, one-celled, olive-brown (Fig 14) .... *Myrothecium* (14)

17. Macroconida hyaline, three to many celled, fusoid, generally curved atleast to some degree. Basal cell somewhat resembling a foot (foot cell), the latter often poorly developed.Conidiophores simple or branched, aggregated into slimy sporodochia (Fig 15) .... *Fusarium* (15)

17A. Conidia dark, large, globose, septa in various planes (Fig 16) .... *Epicoccum* (16)

18. Conidiophore simple (i.e., unbranched) 19

18A. Conidiophore branched or unbranched 30

19. Conidia hyaline or light-colored 20

19A. Conidia dark 25

20. Conidia one-celled 21

20A. Conidia two-celled 24

21. Conidiophore mostly with straight parallel walls throughout or tapering gradually towards the apex 22

21A. Conidiophore not with straight parallel walls throughout but modified by a swelling or crooks 23

22. Conidia aggregated in a slimy mass at the tip of the conidiophore, globose, oval to short cylindric (Fig 17) .... *Acremonium* (17)

22A. Conidia arising from a convoluted mass of compact hyphae (stroma) in the young ovaries of certain grasses. Conidia hyaline, ovoid, on simple conidophores in slime. Conidial state of *Claviceps* (Fig. 18) .... *Sphacelia* (18)

23. Conidiophores (=conidiogenous cell) somewhat swollen at the base, borne directly on the hyphae, tapering to a long slender tube, conidia in chains (Fig. 19) .... *Paecilomyces* (19)

23A. Conidiophore terminated by a swelling (vesicle). Sterigmata arising in one or two series from upper part or all of the surface of the vesicle (Fig. 20) .... *Aspergillus* (20)

24. Conidia with long thin rod like appendages arising from apical cell (Fig 21) .... *Clavaria* (21)

24A. Conidia without appendages, formed in chains; the latter composed of conidia oriented sideways to the long axis of the conidiophore. Colony pink. (Fig 22) .... *Trichothecium* (22)

25. Conidia one-celled, brown to black, subglobose or ovoid. Conidiophore inflated at base, tapering toward apex (Fig 23) .... *Nigrospora* (23)

25A. Conidia two-to many-celled, condiophore not inflated

26. Conidia only with transverse septa 27

26A. Conidia with septa in two or more planes (muriform) 29

27. Conidia cylindrical (some species taper in apical part), smooth; pigmented in light shades of yellow, green, or brown, or combinations of these; formed through a pore, germinating laterally, conidiophore straight, bent, or wavy in spore-producing zone. Mostly parasitic on grasses, *Pyrenophora* perfect state. (Fig 24) .... *Drechslera* (24)

27A. Conidia not cylindric, perfect state not in .... *Pyrenophora* (28)

28. Conidia fusiform or ellipsoid, straight or curved. If predominantly curved, end cells pigmented. Germination only from end cells. Conidia pigmented in various shades of brown, green or yellow, or combinations there of, formed through a pore, perfect states in *Trichometasphaeria* and *Cochliobolus* (Fig 25) .... *Bipolaris* (25)

28A. Conidia fusiform, dark, curved, with light-colored end cells. Central ells somewhat larger and darker. Perfect state where known as *Cochliobolus*. (Fig 26) .... *Curvularia* (26)

29. Conidia ovoid, obclavate, solitary or in chains, with a beak young conida rounded at the base formed through a pore. (Fig 27) .... *Alternaria* (27)

29A. Conidia oval, dark-brown, usually with three transverse and two longitudinal septa, roughened; each conidium formed as a blown-out end of the conidiophore (holoblastic) on dead plant parts (Fig. 28) .... *Pithomyces* (28)

30. Condiophore unbranched 38

30A. Condiophore branched 31

31. Conidia formed in chains 32

31A. Conidia not in chains 36

32. Conidophore with vesicle at the apex. One or two layers of cells develop over surface of vesicle; the outer cells (phialides) produce conidia in chains (Fig 20) .... *Aspergillus* (29)

32A. Conidiophore without a swelling at the apex 33

33. Conidiophore brush like, composed of a stipe terminated by few or many sterile branches and conidiogenous cells (phialides) or only by phialides produce condia...... 34

33A. Conidiophore not brush like 35

34. Conidia ovoid, globose, cylindrical, hyaline in most species, smooth or rough. Long chains of dry conidia arise from phialides. Colonies mostly in shades of green or blue-green.Conidiophore with a single cluster of phialides at apex, or complexly branched (Fig. 29) .... *Penicillium* (30)

34A. Conidiogenous cell swollen at the base tapering to a long, slender, widely divergent tube (Fig. 19).Conidia hyaline, in chains. .... *Paecilomyces* (31)

35. Conidia developing blastically (budding) from the conidiophore branches or mostly from other conidia to form a complex of branched, conidial chains (a tree like system). Conidiophores densely crowded. Conidia dark, elliptical or lemon-shaped, one- or two-celled (Fig. 2). .... *Cladosporium* (32)

35A. Conidiophore dark with few, very short, apical branches; curved downward at the apex. Conidia dark, globose, rough, developing successively in chains from globose conidiogenous cells (Fig. 31) .... *Periconia* (33)

36. Ultimate branches of the conidiophore (conidiogenous cells) in whorls; other branching patterns may be present. Conidia one-celled, ovoid, aggregating in slimy

balls at the tip of slender conidiogenous cells. Irregular branching sometimes present (Fig. 32) .... *Verticillium* (34)

36A. Conidiophore branches not in whorls 37

37. Conidiophore profusely and irregularly branched, forming a tree like complex. The ultimate branches, the conidiogenous cells, bottle-shaped, disposed irregularly, opposite or alternate. Irregular or false whorls of phialides common. Conidia mostly subglobose to short obovoid, in loose heads, green, finely roughened (oil immersion lens required) (Fig. 33). Culture green .... *Trichoderma (*35)

37A. Conidiophore branched irregular or simple. Each stipe or branch terminated by a cluster of club-shaped, lightly pigmented phialides. Conidia one-celled, roughened, subglobose, cylindrical, ovate, and other shapes, formed in a slimy mass around the phialides (Fig. 34) .... *Stachybotrys (*35)

38. Conidiophore irregularly branched or simple (Fig. 35). Ultimate branches of the conidiophore (conidiogenous cell) flask-shaped or globose, terminated by a thread like zigzag apex. .... *Beauveria* (37) (See entry 36B)

38A. Conidiophore penicillately branched.Conidia light-colored or hyaline, ovoid, elliptical, or reniform. Conidia aggregate at the tip of phialides in slimy masses (Fig. 36). Cultures may be rose, yellow, orange, or green ....*Gliocladium* (38)

Perfect states in *Nectria, Hypocrea, Hypomyces, Thuemenella* and *Lilliputia.*

39. Ascocarp present 40

39A. Ascocarp lacking, asci globose, arising in clusters on the mycelium. Ascospores hyaline, ellipsoidal, smooth, eight per ascus (Fig 37) ...... *Byssochlamys (*39)

40. Ascocarp a perithecium, Perithecia free or on a stroma 41

40A. Ascocarp not a perithecium, Perithecia immersed within a stroma 44

41. Ascospores one-celled 42

41A. Ascospores two- to many-celled 43

42. Ascospores small, hyaline, hat-shaped, eight per ascus. Asci globose, borne irregularly within the perithecium, the ascus wall breaking down at maturity. The spores collect in slime at the tip of the long perithecial neck (Fig. 38). Perithecium superficial. Conidia present .... *Ceratocystis* (40)

42A. Perithecia superficial, globose, oval, vase-shaped, variously shaped hairlike hyphae of diagnostic value arise from the upper part. Spores mostly lemon-shaped and olive-brown but shape and color may vary. Asci clavate, disappering early (Fig 39) .... *Chaetomium* (41)

43. Ascospores with three or more cells, hyaline, elliptical to fusiform. Asci elliptical to clavate, perithecia superficial or on an erumpent stroma, ovoid to globose-papillate, dark-purple or blue (Fig 40) .... *Gibberella* (42)

43A. Stroma globose; terminating in a stout or slender stipe which arises from a hard sclerotium. Asci very narrow with a thick apex transversed with a pore. Ascospores needle like (Fig. 30) .... *Claviceps* (43)

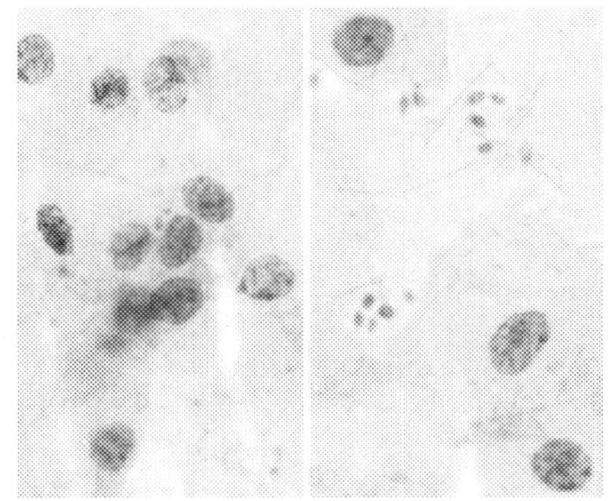

Fig 1 *Torulopsis*

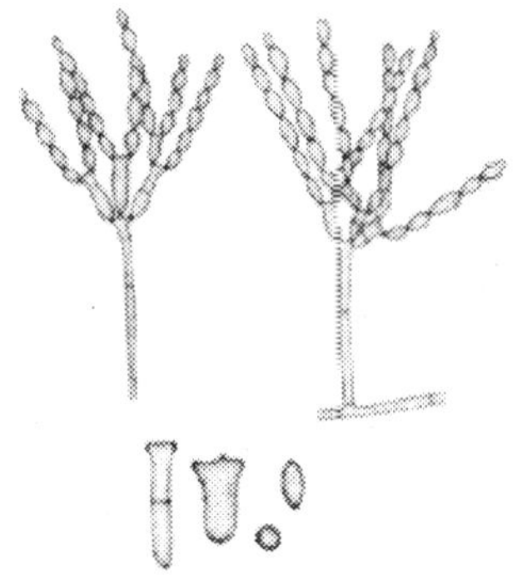

Fig 2 *Cladosporium*

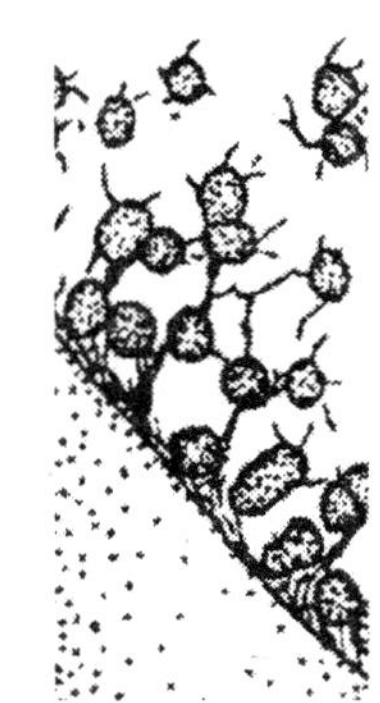

Fig 3 *Sclerotium*

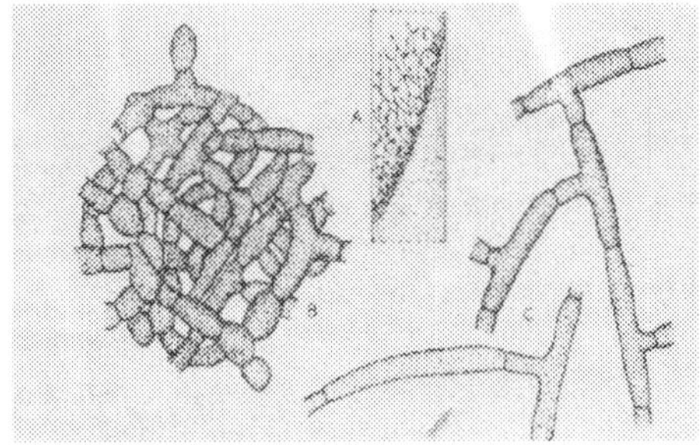

Fig 6 *Fusicoccum*

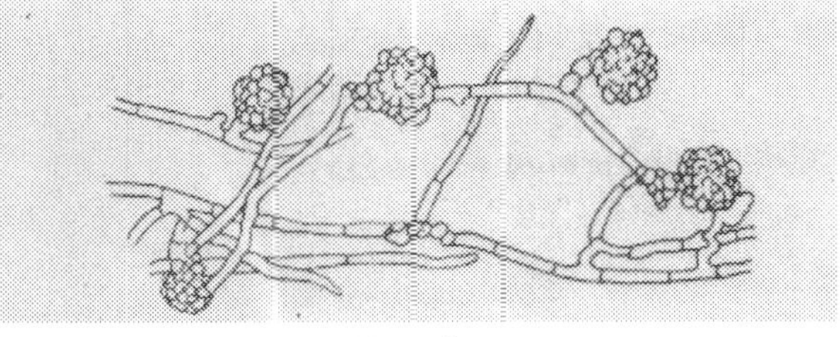

Fig 5 *Papulaspora*

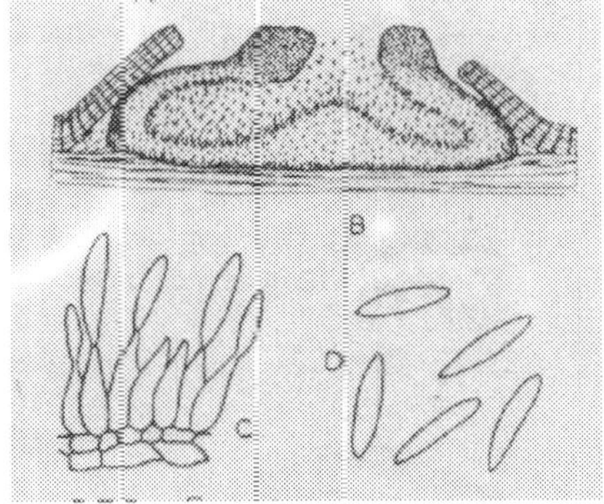

Fig 4 *Rhizoctonia*

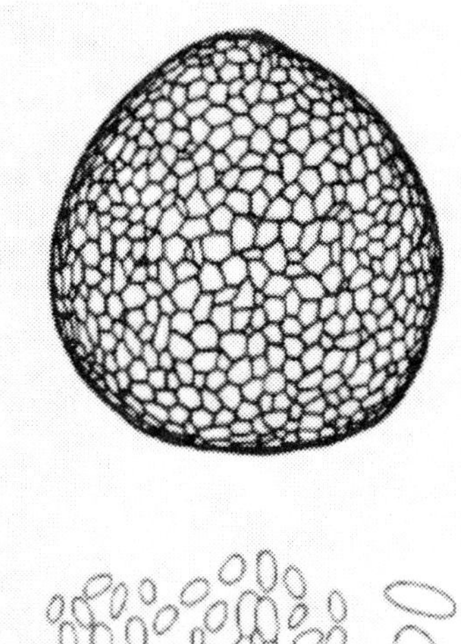

Fig 7 *Phoma*

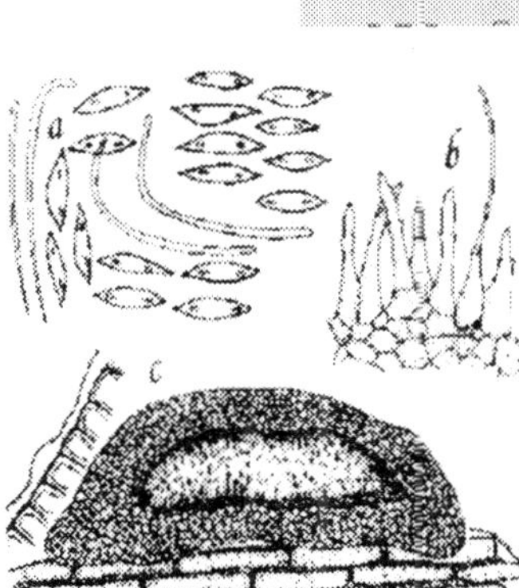

Fig 8 *Phomopsis*

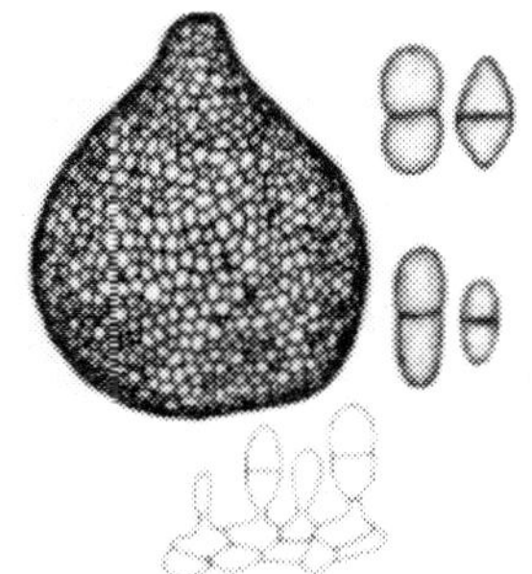

Fig 9 *Diplodia*

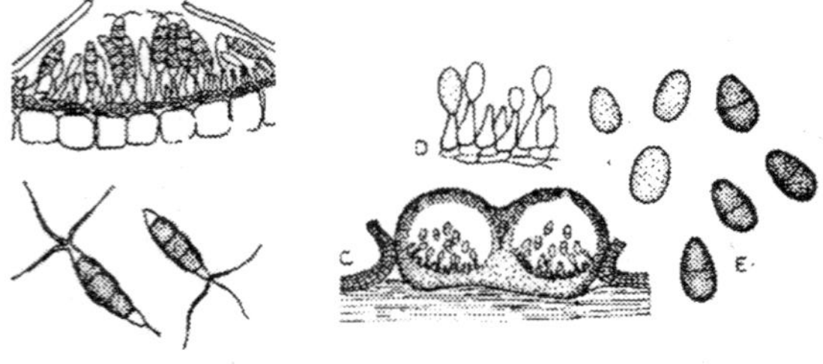

Fig 12 *Pestalotiaopsis* Fig 10 *Lasidiplodia* Fig 11 *Colletrotricum*

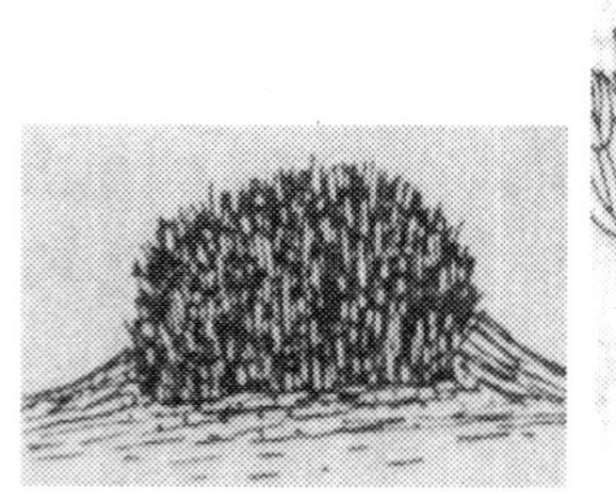

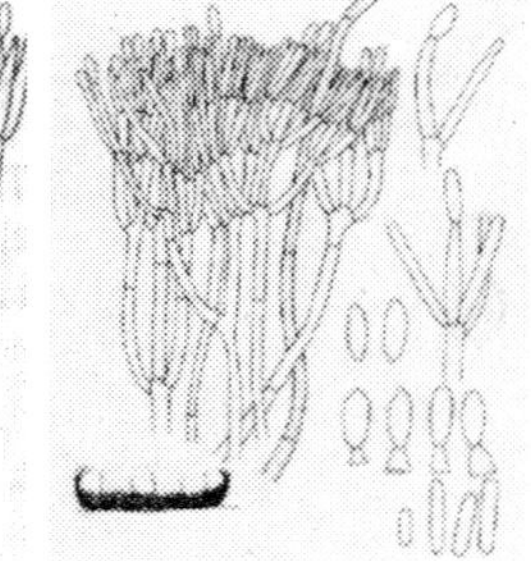

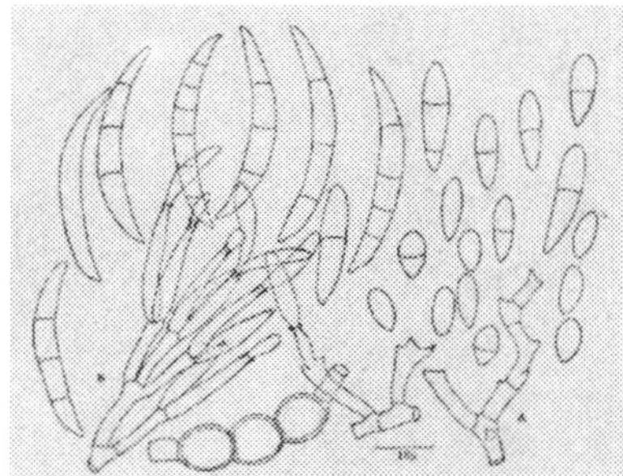

Fig 13 *Dendrodochium* Fig 14 *Myrothecium* Fig 15 *Fusarium*

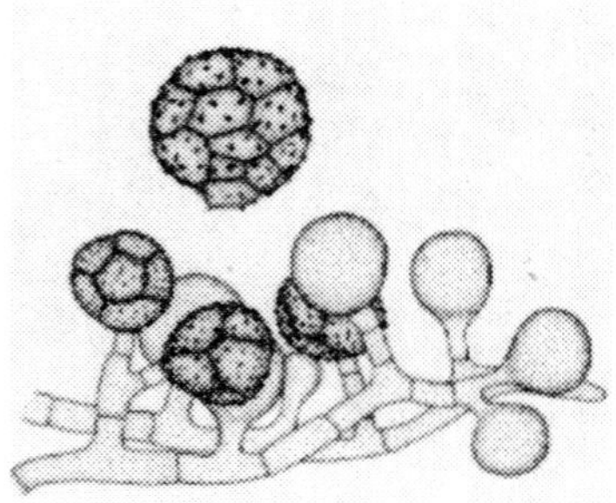

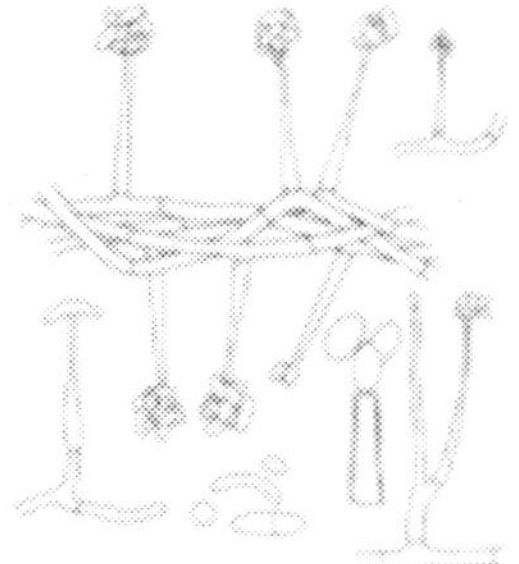

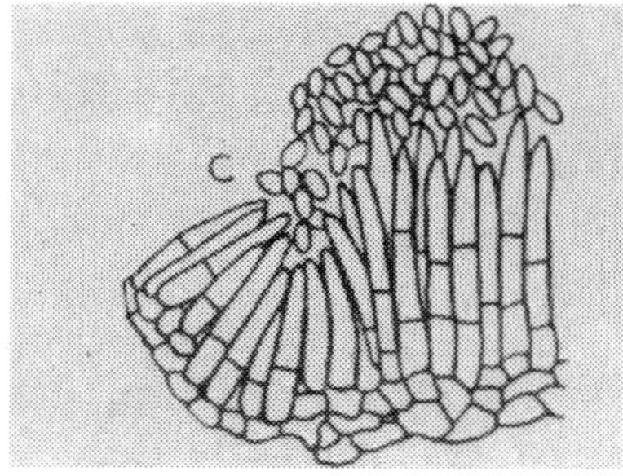

Fig 16 *Epicoccum* Fig 17 *Acremonium* Fig 18 *Sphacelia*

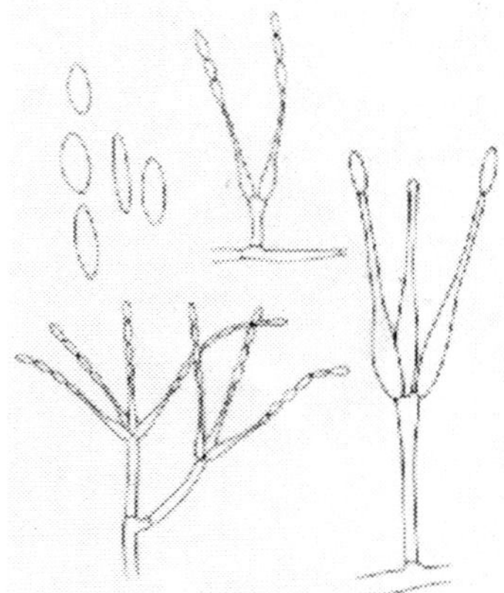

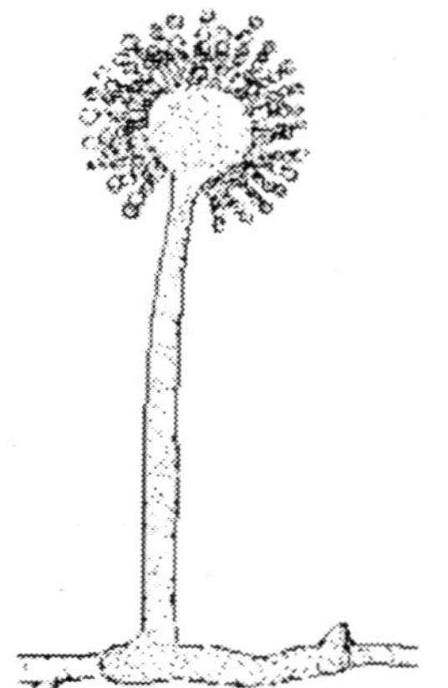

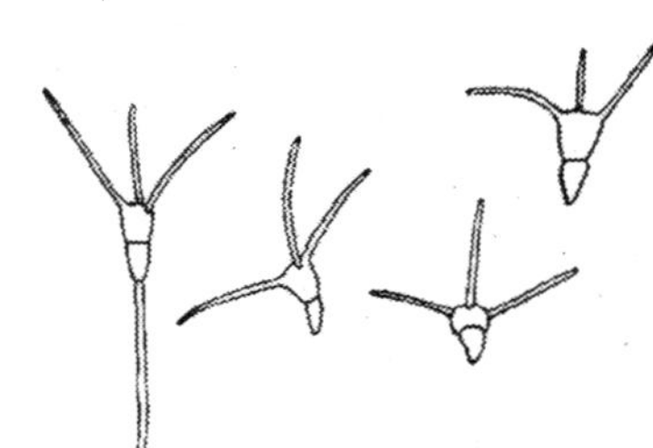

Fig 19 *Paecilomyces* Fig 20 *Aspergillus* Fig 21 *Clavaria*

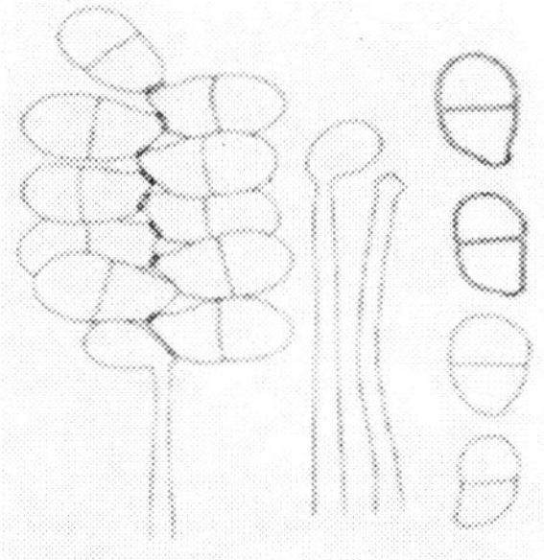

Fig 22 *Trichothecium*

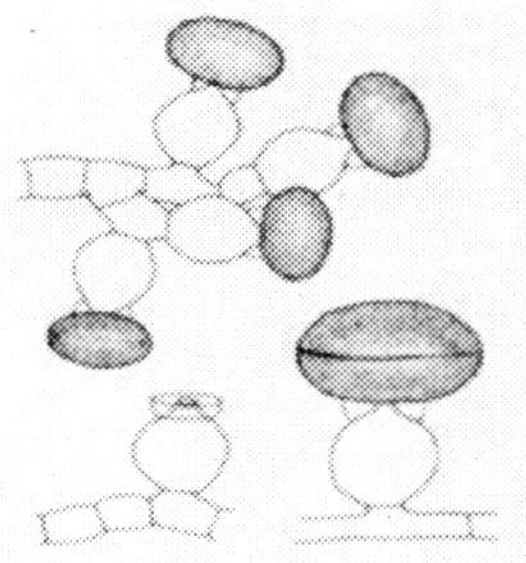

Fig 23 *Nigrospora*

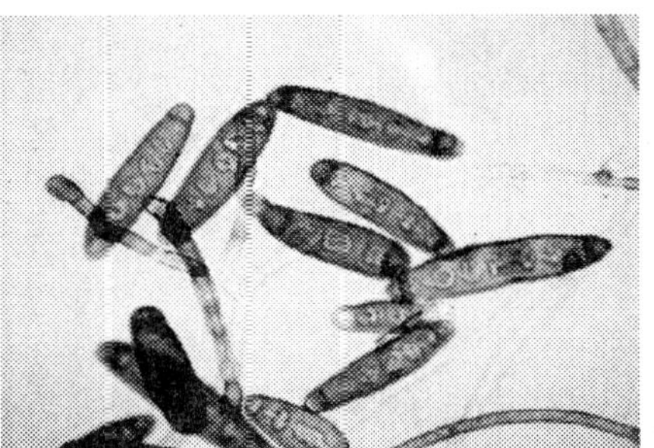

Fig 24 *Drechslera*

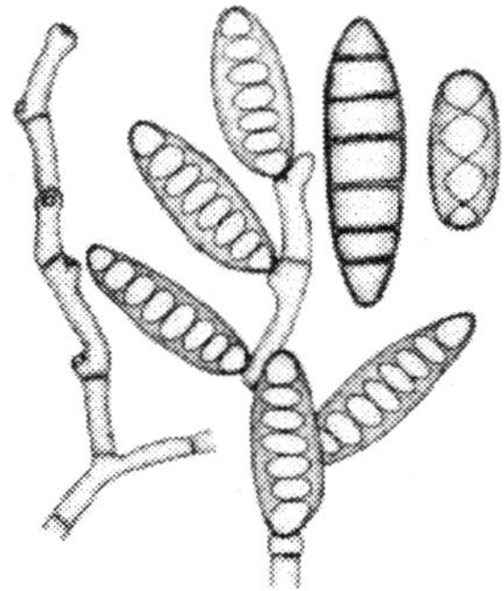

Fig 25 *Bipolaris*

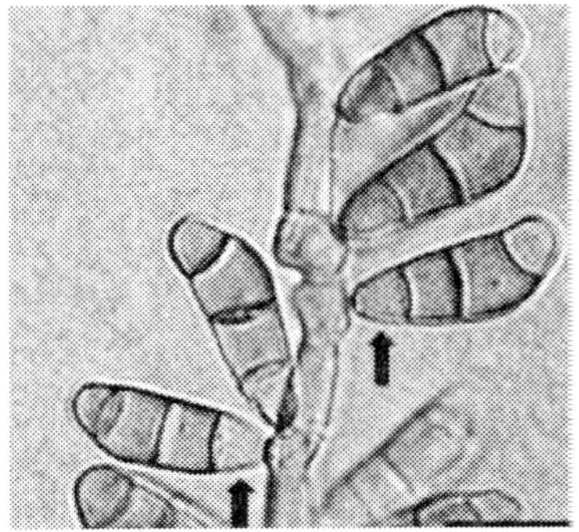

Fig 26 *Curvularia*

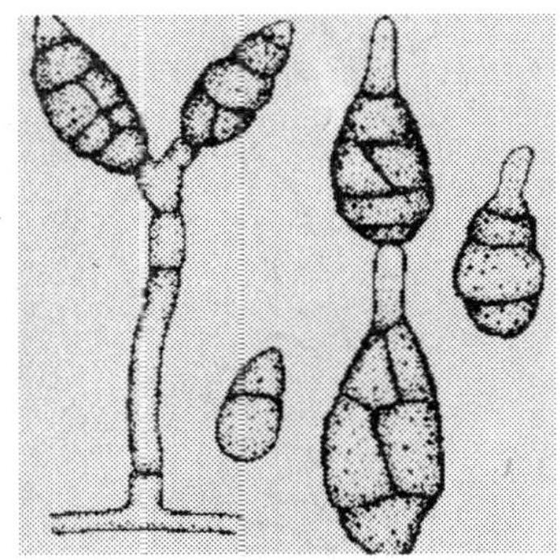

Fig 27 *Alternaria*

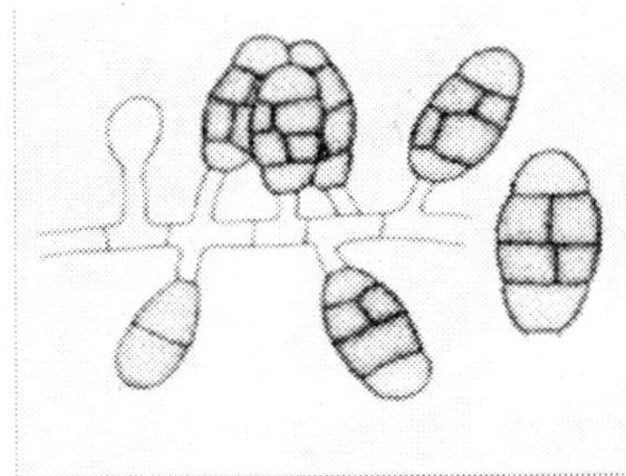

Fig 28 *Pithomyces*

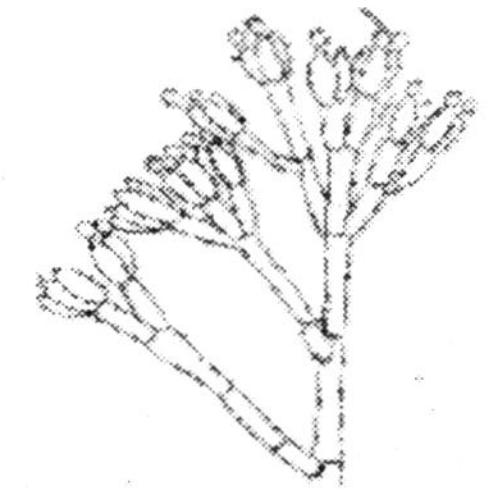

Fig 29 *Penicillium*

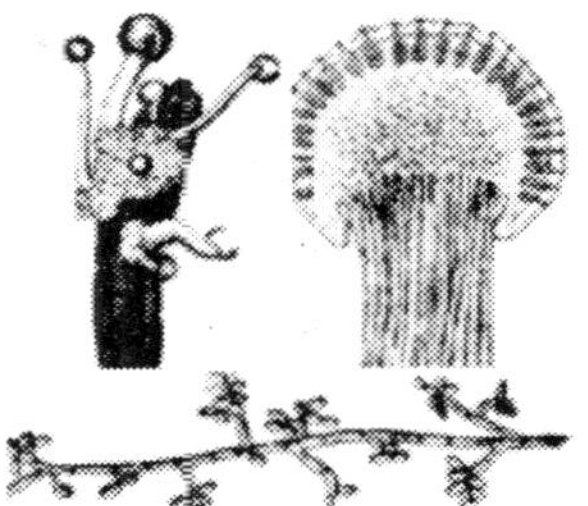

Fig 30 *Claviceps*

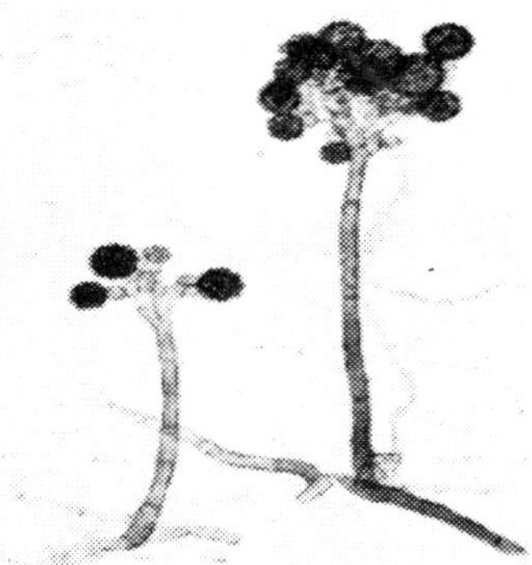

Fig 31 *Periconia*

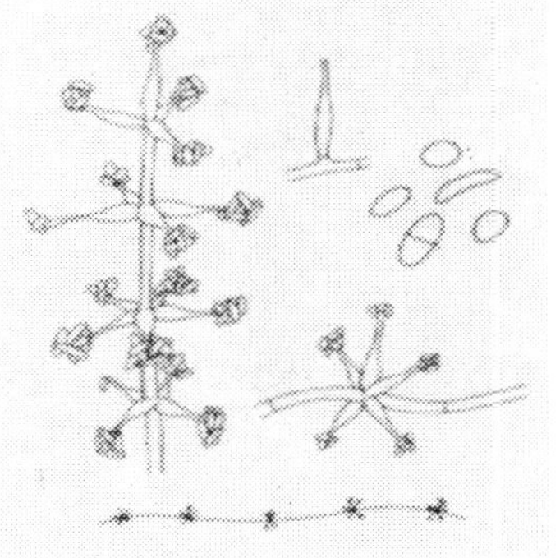

Fig 32 *Verticillium*

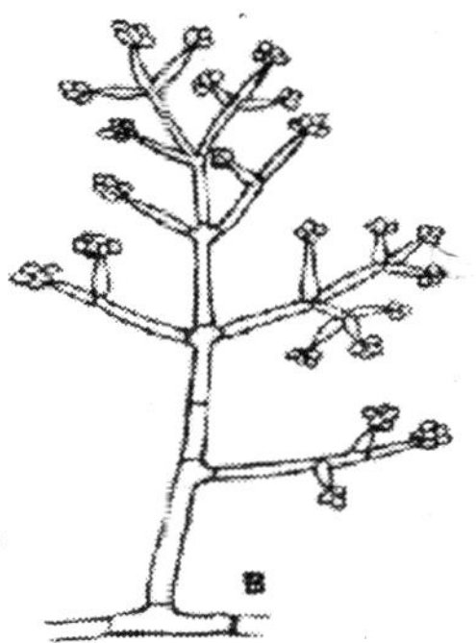

Fig 33 *Trichoderma*

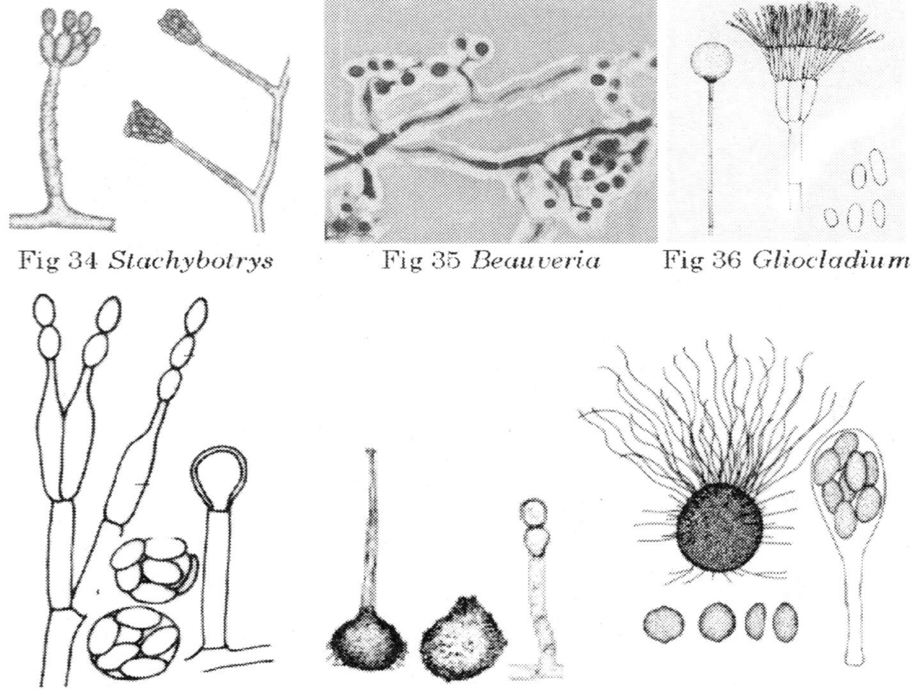

Fig 34 *Stachybotrys* Fig 35 *Beauveria* Fig 36 *Gliocladium*

Fig 37 *Byssochlamys* Fig 38 *Ceratocystis* Fig 39 *Chaetomium*

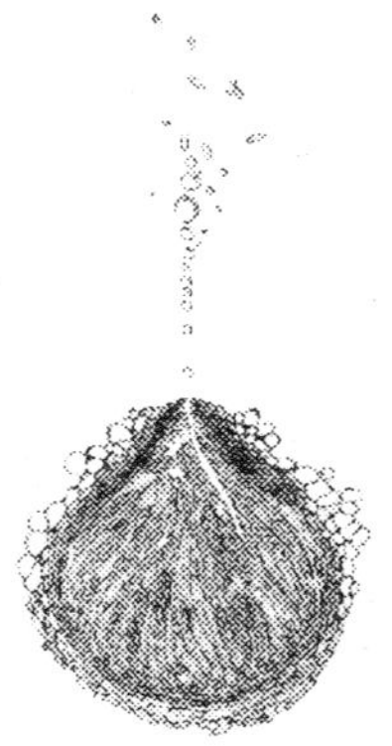

Fig 40 *Gibberella*

2.3. MORPHOLOGY OF MYCOTOXIGENIC FUNGI

2.3.1. *Aspergillus*

Ever since the reports of carcinogenic aflatoxins produced by *A. flavus*, the genus *Aspergillus* has gained importance. Extensive and intensive search for toxigenic moulds associated with foods and feeds revealed quite a large number of species of *Aspergillus* are mycotoxigenic and are important from the health point of view (Table 2.2). Their presence leads to deterioration and spoilage of many food stuffs, while exposure to their spores frequently results in allergic reactions often of a serious nature. Correct identification of the species is, therefore, of paramount importance.

Table 2.2. Important mycotoxins producing species of *Aspergillus*

Mycotoxin	Species
Aflatoxin	*A. flavus, A. parasiticus, A. nomius*
Citrinin	*A. terreus*
Cyclopiazonic acid	*A. flavus, A. tamorii*
Fumitoxins	*A. fumigatus*
Kojic acid	*A. flavus, A. nomius, A. parasiticus, A. tamarii*
Nidulotoxin	*A. versicolor, Emericella nidulans*
Ochratoxin A	*A. ochraceus*
Patulin	*A. clavatus, A. terreus*
Physcions	*Eurotium amestelodani, E. chevalieri, E. repens, E. rubrum*
Echinulins	*E. chevalieri, E. amestelodai*
Sterigmatocystin	*A. versicolor, E. nidulans*
Terretrems	*A. terreus*
Tryptoqualins	*A. clavatus*
Cytochalasins	*A. clavatus*

Aspergillus was first described almost 300 years ago and is an important genus in foods. They are extremely common in stored commodities such as grains, nuts and spices and occur more frequently in tropical and subtropical than in temperate climates. It is a large genus containing more than 100 species. There are number of teleomorphic genera which have *Aspergillus* conidial states but only two genera *Eurotium* and *Neosartoria* produce heat resistant ascospores and are mycotoxigenic and sporulate heavily in stored products.

Raper and Fennel (1965) monograph is the most widely used monograph for identification of species of *Aspergillus*. However, concepts have changed due to the addition of some more species (Samson *et al.*, 1990). Gams *et al.* (1985) proposed nomenclaturally correct classification for species of genus *Aspergillus* by grouping into six sub-genera which are subclubbed with sections. Klich and Pitt (1988) have also provided a key to species of *Aspergillus* which are important in foods. In addition to traditional morphological taxonomic characters, chemical characteristics such as isozyme patterns

(Cruickshank and Pitt, 1990), secondary metabolites (Frisvad, 1985) ubiquinone system (Kuraishi *et al.*, 1990) and molecular characterstics (Mullaney and Klich, 1990) have been used to clarify relationships within the genus *Aspergillus*.

Raper and Fennel (1965) assigned species into four groups based on conidiogeneous apparatus (Table 2.3). They are

1. Those with one series of conidiogenous cells (uniseriate)
2. Those with two series (conidiogenous cells plus supporting cells) (biseriate).
3. Those in which both conditions are found. However, recent changes in the taxonomy of *Aspergillus* have divided the genus into six subgenera (Gams *et al.*, 1985), each containing one or more sections. Sections correspond to the groups of Raper and Fennel (1965). Only those sections that have mycotoxiological importance will be discussed here.

Table 2.3. *Aspergillus* sections according to structure of conidiogenous cells

Uniseriate section	Biseriate section	Both sections
Clavati	*Nidulans*	*Flavi*
Restricti	*Versicolores*	*Wentii*
Fumigati	*Flavipedes*	*Nigri*
Aspergillus	*Terrei*	*Candidi*
(A. glaucus group)		*Circumdati*
		(*A. ochraceus* group)

For identification of species of *Aspergillus* primary emphasis was on arrangement of conidiogenous cells, secondary emphasis on shape and pigmentation of conidial heads and finally on the presence or absence of ascomata, Hulle cells (thick-walled cells of no known function) and pigmentation of the conidiophore.

A key to 14 sections and mycotoxigenic species is given below.

Key to sections and species of *Aspergillus*

1	Conidiogenous cells strictly uniseriate	2
	Conidiogenous cells biseriate or uniseriate	5
2(1)	Conidial heads clavate	Clavati
	Conidial structures not exceeding 4.0mm length	*A. clavatus (1)*
	Conidial heads not clavate	3
3 (2)	Conidial heads radiate, ascomata present	*Aspergillus*
	Conidial heads large, radiate to loosely columnar, borne above the surface layer of cleistothecium and enveloping hyphae, ascospores	

	lenticular, 6 μ or less in long axis, equatorial ridges facing furrow absent or showing only as a trace	*A. ripens* (2)
	Equatorial ridges low or rounded, furrow broad and shallow, colonies strong red, orange red pigment	*A.rubrum* (3)
	Equatorial ridges thin, and flexuous, crest like ascospores resembling a pulley	*A.chevalieri* (4)
	Conidial heads not radiate, ascomata absent	
4 (3)	Conidial heads columnar, conidia echinulate	Restricti
	Conidial heads columnar, conidia spinose	
	Cultures strictly conidial, varying from velvety to floccose	*A. fumigatus* (5)
	Cultures producing claistothecia conidial development generally limited	*A. fischeri* (6)
	Vesicles often borne at an angle to the condiphore, thin walled, vesicle, hyaline, often strongly noded conidia in pale-blue green shades	*A. viridi-nutans* (7)
5 (1)	Conidigenous cells strictly biseriate	6
	Condiogenous cells uniseriate and biseriate	10
6 (5)	Ascomata present	
	Ascospores present, orange-red in colour equtorial crests two in number, convex wall smooth, conidial stage dark yellow-green, crest entire	*A. nidulans (8)*
	Ascomata absent	7
7 (6)	Conidial heads columnar	8
	Conidial heads not columnar	9
8 (7)	Conidial heads white to brown, conidiophores brown	
	Conidiophores definitely pigmented in yellow to light-brown shades, conidial heads usually white to very pale-buff	*A. flavipes (9)*
	Conidial heads cinnamon, conidiophores not-brown	terrei
	Colonies velvety, conidial heads long compactly columnar in Cinnamon to orange-brown to brown shades borne on short conidiophares sclerotium like masses of swollen, relatively heavy waled cells lacking	*A. terreus (10)*
9 (7)	Conidial heads radiate, dull-brown, conidiophore brown	Usti
	Conidial heads olive gray to drab or red-brawn, hulle cells scattered, thoughout the colony or forming irregular masses not associated with pigmented mycelium.	Versicolores *(11)* A. ustus *(12)*
	Conidial head radiate, green conidiophores not brown	
	Vesicles globose to some what elongate, fertile over most of the vesicular surface, globose to sub-globose hulle cells often present	
	Conidial heads variable in colour, light yellow-green buff orange-yellow	*A. versicolor* (13)
	Conidial heads always blue-green when young	*A. sydowi (14)*
10(5)	Conidial heads yellow-green	Flavi
	Sterigmata single or double, heads lossely columnar	*A. flavus (15)*
	Heads radiate, stergmata uniseriate	

	Colonies shifting to light-brownish-green, conidia large elliptical to globose to subglobose, irregularly	*A. oryzae (16)*
	Conidial heads not yellow-green	11
11(10)	Conidial heads ginger-brown	Wentii
	Conidial heads large upto 500 μm on smooth or slightly granular stalks which may reach several mm in length	*A. wentii (17)*
	Conidial heads not ginger-brown	12
12(11)	Conidial heads globose, black	Nigri
	Sterigmata biseriate, conidia globose, roughened	*A. niger (18)*
	Sterigmata uniseriate, conspicuously echinulate	*A. japonicus (19)*
	Conidial heads globose, not black	13
13(12)	Conidial heads globose, white	Candidi
	Conidial heads globose, yellow to ochraceus	Circumdati
	Sclerotia abundant, small, pure-yellow then brown conidia globose to sub-globose to ovate	*A. melleus (20)*
	Sclerotia large, pink to venaceous, purple, when mature ovate to cylindrical, conidia globose to sub-globose	*A. ochraceus (21)*

Diagnostic features of some of the common mycotoxigenic species are given below.

Section *Aspergillus* (*A. glaucus* group)

Species in *A. glaucus* group possess a teleomorph named *Eurotium*. *Eurotium* is a well defined genus characterized by the formation of yellow to orange ascomata. The *Aspergillus,* anamorphs possess bluish-green, radiating conidial heads producing uniseriate conidiogenous cells which inturn produce large barrel shaped spinose conidia.

Although about 20 species belong to this group, only four species are common: *E. amstelodami* (Fig. 41), *E. rubrum, E. repens* and *E. chevalieri*. Species differentiation is based on ascospore ornamentation, that is, smooth or rough convex walls, presence or absence of a longitudinal furrow and equatorial crest or flange. Both *E. repens* and *E. rubrum* have smooth, convex walls, no equatorial crest, but the former has a hint of a furrow and the latter a definite furrow. *E. amstelodami* has rough convex walls, a prominent furrow, and rounded crests and *E. chevalieri* has smooth walls and thin prominent crests resembling a pulley.

Eurotium species have been reported as producing such carcinogenic mycotoxins as aflatoxin, sterigmatocystin and ochratoxin A (Frisvad and Samson, 1991).

All species of *Eurotium* are xerotolerant (Pitt and Hocking, 1997) and will, therefore, be found on foods containing little free water (eg., stored cereals, flour, nuts, jams, jellies, fruits, cakes, cheese cake, filled chocolates, smoked bacon, dried salted fish and dried meats).

Section fumigati (*A. fumigatus* Fresen group)

Diagnostic features. Species of *A. fumigatus* (Fig. 42) group are recognized by rapidly growing, dark-smoky-green colonies, bearing uniseriate, short, columnar conidial heads. Conidiophores often are pigmented green, as are the swollen apex and conidiogenous cells. Conidia are small, globose, and delicately roughened. Some species in this group have a teleomorph *Neosartoriara*, of which *N. fischeri* is the most common. It is characterized by rapidly growing colonies, sparsely sporulating columnar heads with soft-walled white to cream-colored ascomata or containing rough walled ascospores with two sinuous equatorial crests or flanges.

A. fumigatus causes human lung infections. Colonies are blue-green and produces columns of conidia similar to *A. nidulans*. However, conidiophores have but one set of sterigmata, condiophore vesicles and sterigmata are pigmented blue. Colony reverse is also blue to black.

A. fumigatus produces a large number of mycotoxins: fumiclavines, fumigatin, fumitoxins, fumitremorgins, verruculogen, gliotoxin and tryptoquivalins (Fisvad and Samson, 1991). Gliotoxin is the most important of these, causing numerous mycotoxicoses (Moreau, 1979). Evidence suggests that the toxin is important in the invasion of animal and bird lungs *N. fischeri* also produces fumitremorgen and verruculogen (Pitt and Hocking, 1997).

Section Clavati (*A. clavatus* Desm Group)

Diagnostic features: The *Clavati* section is a small group, of which *A. clavatus* (Fig. 43) is the commonest. Growth is rapid, velvety, with grayish-green, uniseriate heads producing smooth, elliptical conidia. The group is characterized by its clavate-shaped swollen apex.

A. clavatus produces patulin, ascladiol, tryptoquivaline and a derivative of patulin. Cytochalasin E. The last two are known as potent mycotoxins (Cole and Cox, 1981).

Section Nidulantes (*A. nidulans* Wint group)

Diagnostic features: Species in the *A. nidulans* (Fig. 44) group possess a teleomorph, *Emericella,* which is distinguished by hulle cells surrounding black ascomata (white when young), bearing red-purple, lenticular, and smooth-walled ascospores with two equatorial plated but entire crests. The *Aspergillus* anamorph bears biseriate, deep-green short columnar heads with a distinct short, sinuous, brown-pigmented conidiophore. Conidia are globose and rough. The colony reverse is usually deep red-purple. There are about 20 species in this group of which only *E. nidulans* is common in food. It can be distinguished by the ornamentation of the ascospores, which are red, smooth walled and bear two equatorial crests or flanges.

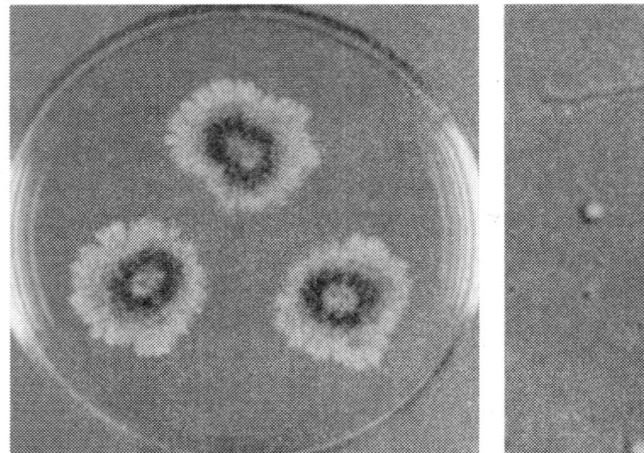

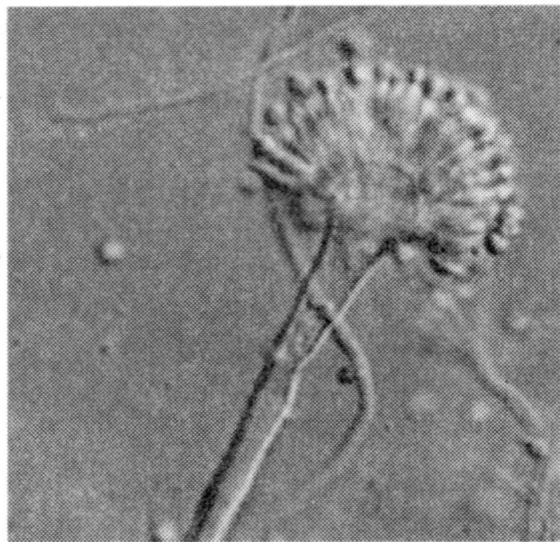

Fig 41. *Aspergillus amstelodami*

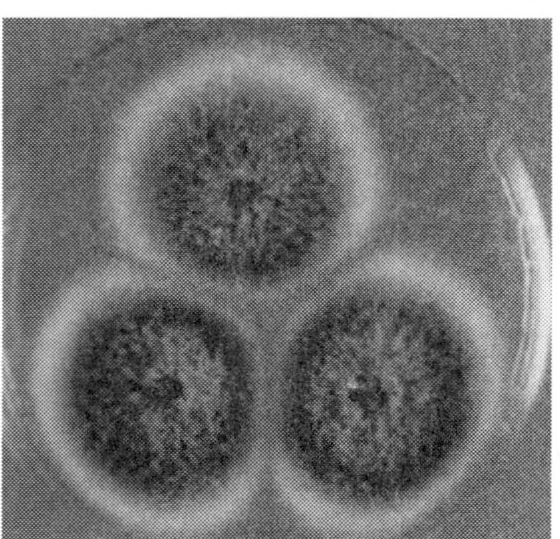

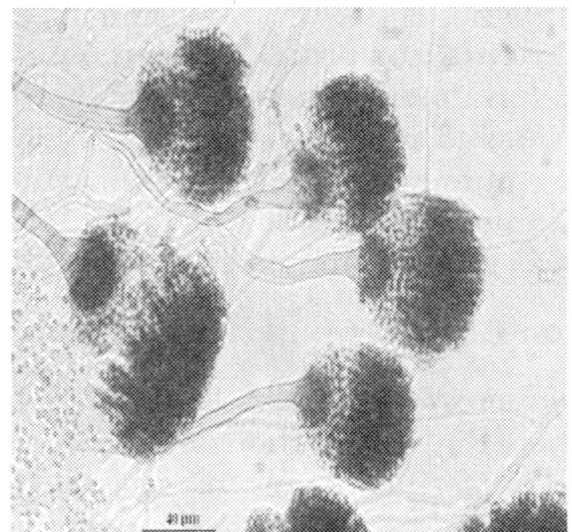

Fig 42. *Aspergillus fumigatus*

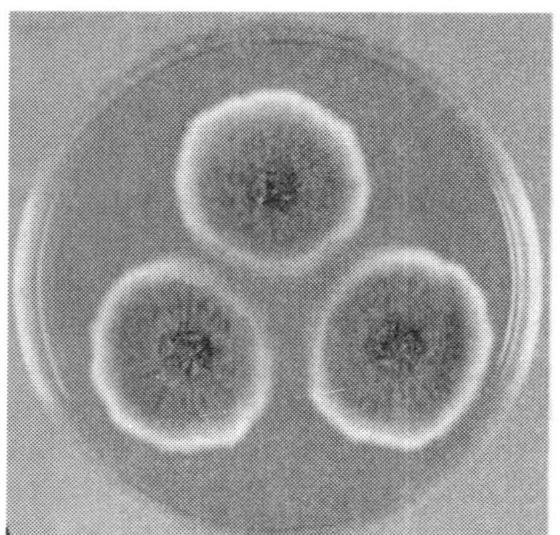

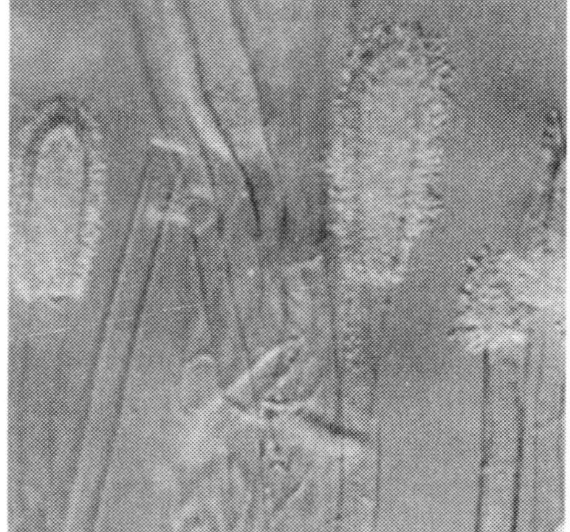

Fig 43. *A.clavatus*

A. nidulans colonies dark-green and produce long columns of conidia that resemble sticks or cigarettes. Conidia and condiophores are similar to *A. versicolor* (with 2 sets of sterigmata, smooth and colourless stack and small vesicles) but stack and vesicle are pigmented brown. The species has a sexual stage and produces cleistothecia surrounded by hulle cells plus bright red ascospores.

E. nidulans produces sterigmatocystin (a precursor of aflatoxin), nidulotoxin, and emestrin (Pitt and Hocking, 1997).

Section Versicolores (*A. versicolor* TiraboschiGroup)

Diagnostic features: species are slow growing, with dull-green, radiate, biseriate heads bearing small, globose, spinose conidia. Conidiophores are colorless or sometimes pigmented brown and hulle cells are occasionally present. Although there are over 20 species in the group, only *A. versicolor* (Fig. 45) and *A. sydowii* are common. *A. versicolor* can show a range of colony color from pale-green to yellow-green, gray-green, buff or even pink; reverse is deep-red to plum, and the spinose conidia turn emerald-green when mounted in lactophenol. The turquoise colony color of *A. sydowii* easily distinguishes it from *A. versicolor*. It is mistaken easily for a *Penicillium* at first sight, until examined with a stereomicroscope. Dwarfed heads can also be found in nearly all isolates. Colony reverse is very deep-red.

A. versicolor colonies dark-green but with often highly and variously coloured mycelial sectors. Colony reverse is purple. Conidiophore with 2 sets of sterigmata, smooth and colourless stack and small vesicles, conidia slightly roughened. *A. versicolor*, like *E. nidulans*, produces sterigmatocystin and nidulotoxin (Frisvad and Samson, 1990). *A. sydowii* is not known to produce any mycotoxin.

Section usti (*A. ustus* Thom and church Group)

Diagnostic features: Ustus is a small group, of which only *A. ustus* (Fig. 46) is important. Distingishing features are floccose, moderately fast growth, brown-gray to dull-gray colonies, bearing small, biseriate and radiate to loosely columnar heads. Conidiophores are short, smooth and brown with small, globose, echinulate conidia. The reverse is yellow with irregular to elongate (sausage-shaped). Hulle cells are sometimes present.

Some isolates of *A. ustus* can produce austamid, ausadiol, austins and austocystins (Frisvad and Filtenborg, 1995).

Section Terrei (*A. terreus* Thom Group)

Diagnostic features: The *Terrei* have velvety to floccose growth and biseriate, cinnamon to tan, compactly columnar conidial heads, bearing very small, smooth-walled conidia. Two varieties have been described: *A. terreus* var. *africanus* and *A. terreus* var. *aureus* (Raper and Fennel, 1965). The former possesses yellow to buff-colored sclerotial-like masses on MEA and the latter bright-yellow mycelium with conidiophores over 500 μm long.

A. terreus (Fig. 47) produces long columns of conidia. However, colony colour is tan to beige. Conidiophores are colourless and smooth have a small vesicle and 2 sets of sterigmata.

Isolates of *A. terreus* are able to produce citrinin, patulin, terrain, terricacid territrems, and citreoviridin. *A. terreus* commonly can be isolated from air tight stored cereals (Ceynowa, 1986) and a variety of nuts (hazelnuts, peanuts, walnuts, pistachios and pecans). It also has been isolated from barely, flour, pasta, and flour products, pulses, soybeans, meat and meat products (Pitt and Hocking, 1997).

Section Flavipedes (*A. flavipes* Thom and Church Group)

Diagnostic features: The *flavipedes* have velutinous to slightly granular growth, with small, white, biseriate columnar heads. Conidiophores are brown, bearing small, globose, smooth-walled conidia. A teleomorph rarely develops, but according to Wiley and Simmons (1973), the ascomata are yellow, surrounded by hulle cells and produce smooth walled ascospores with an indistinct longitudinal furrow. The teleomorph is *Fennellia flavipes*. Two other anamorphic species found in this group are *A. niveus* and *A. carneus*. The former bears white, columnar conidial heads, but has an unpigmented conidiophore, while the latter bears fawn-colored columnar heads also with unpigmented conidiophores.

A. flavipes (Fig. 48) is reported to produce flavipin and flavipucins, while *A. carneus* produces citrinin (Frisvad and Samson, 1991).

Section Wentii (*A. wentii* Group)

Diagnostic features: Section Wentii is a small group of which *A. wentii* is the most common. Distinguishing features is floccose growth bearing sparsely produced, large, globose, ginger-colored, uniseriate and biseriate conidial heads, the biseriate condition is predominating. Conidiophores can be several millimeters long. Conidia are globose to elliptical, smooth to very rough.

A. wentii colonies deep chocolate-brown but produce much floccose mycelium and conidiophores are quite long with large vesicles. Conidia with single wall with colour bars rather than spines.

Emodin, lwentilacton, physicons, β-nitropropionic acid and numerous other metabolites are produced by *A. wentii* (Frisvad and Filtenborg, 1995). Only emodin is of mycotoxic significance.

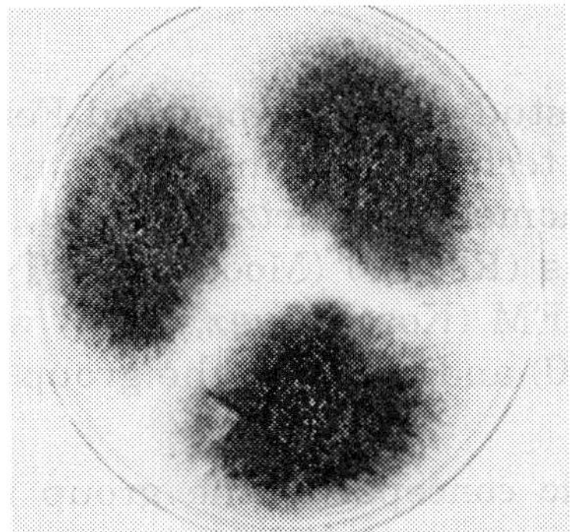
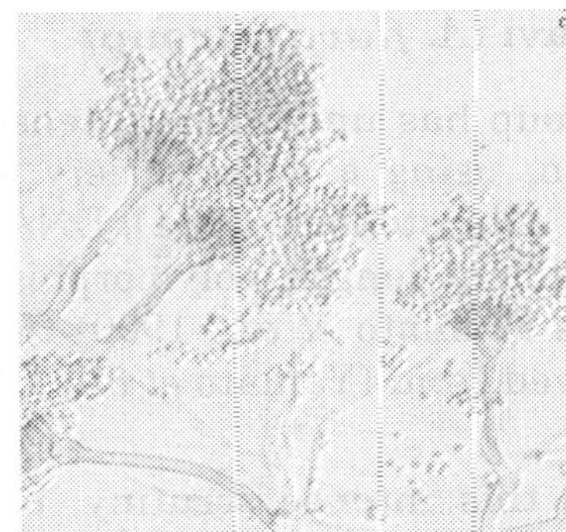

Fig 44 *Aspergillus nidulans*

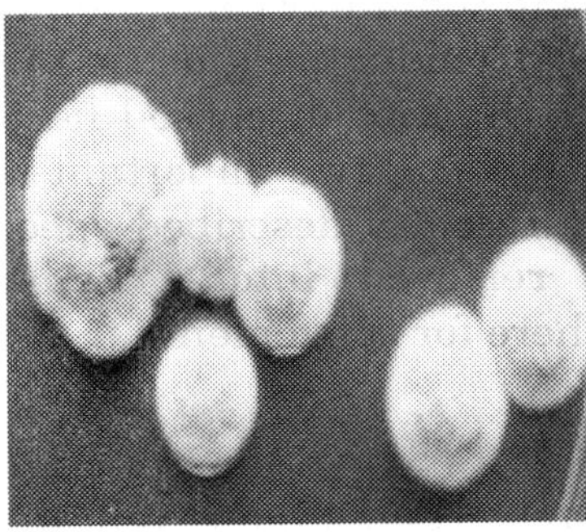
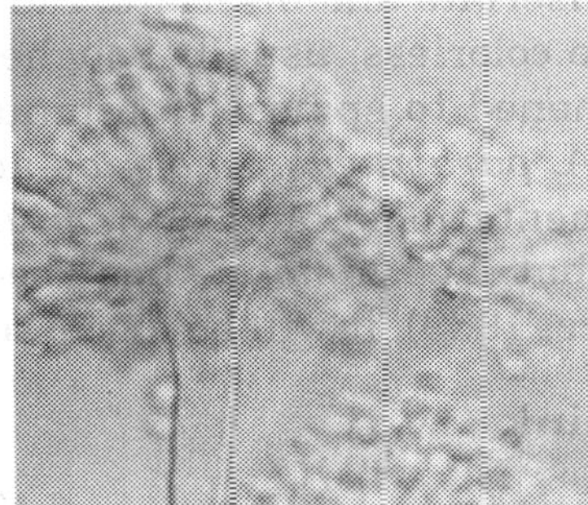

Fig 45 *Aspergillus versicolor*

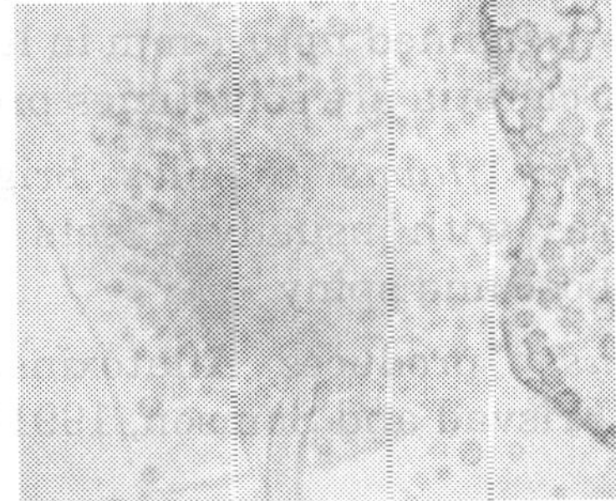

Fig 46 *Aspergillus ustus*

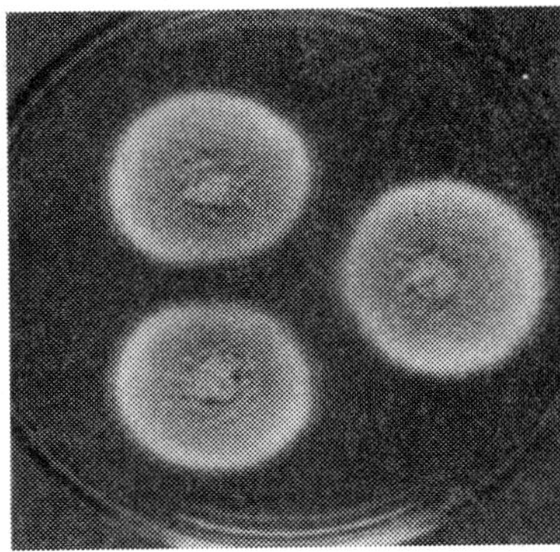

Fig 47 *Aspergillus terreus*

Section Flavi (*A. flavus* Group)

This group has undergone intense study since Raper and Fennel (1965) classification, using a wide variety of techniques ranging from traditional morphology (Christensen, 1981), DNA homology (Kurtzman *et al.*, 1986), restriction fragment length polymorphisms (RFLPs) (Moody and Tyler, 1990), ubiquinones (Yamato *et al.,* 1990) to SEM (Kozakiewez, 1986) with species being removed from (Yamato *et al.*, 1990) and added to the group (Kurtzman *et al.*, 1987).

Despite this intense scrutiny, basic concepts of the group remain the same. Distinguishing features are yellow-green to jade-green to brown colonies, with uniseriate and biseriate, radiate to columnar conditial heads. Conidiophores are colorless, usually roughened. Conidia are globose to subglobose, finely roughened to spinose to echinulate. Species discussed here include *A. flavus* and *A. parasiticus*, *A. oryzae* and *A. sojae* (two important species used in food fermentations), and *A. nomius* is recently described species. Despite the fact that *A. tamorii* has been removed into a separate group (Kozakiewicz, 1989), it will be discussed here for convenience.

A. flavus Link

Diagnostic features: Colonies of *A. flavus* (Fig. 49) are yellow-green, conidial heads are predominatly biseriate, radiate or columnar. Conidiophores are coarsely roughned upto 1 mm in length. Conidia are globose to subglobose or spinose. Sclerotia are sometimes produced.

Aspergillus flavus colonies bright-green often with black sclerotia. Conidiophores with usually 2 sets of sterigmata and with spiny stalks, Conidia globose and spiny.

A flavus produces Aflatoxins, Cyclopiozonic acid Kojic acid and Aflatream (Frisvad and Samson, 1991).

A. parasiticus spears

Diagnostic features: Colonies of *A. parasiticus* are velutinous, jade-green (a much richer green, deeper in shade than in *A. flavus*). Conidial heads are predominantly uniseriate (although about 10% can be biseriate) and radiate; Conidiophores are roughened, usually less than 400 μm in length. Conidia are globose and echinulate (Fig. 50).

Aspergillus parasiticus: Colony colour yellow-green, just one set of sterigmata with smooth conidiophore stalks. Conidia are usually smooth elliptical and larger than those of *A. flavus*.

Aflatoxin B_1 and G_1, but not cyclopiazoic acid are produced consistently (Bryden, 1982).

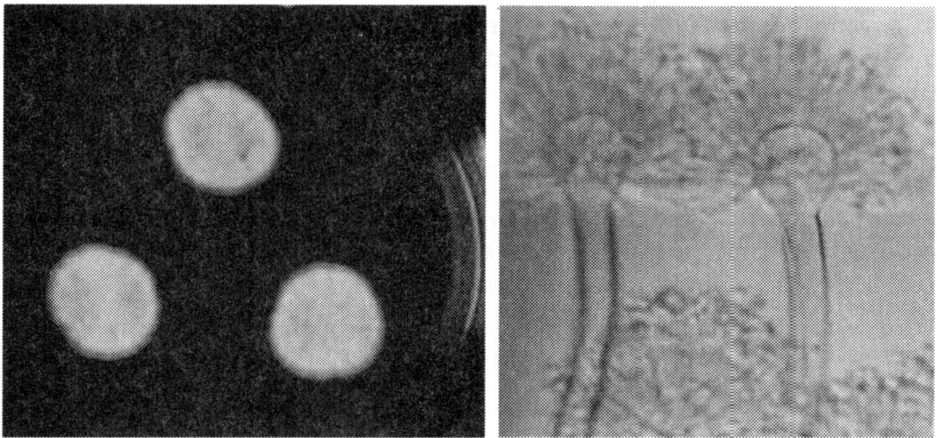

Fig 48 *Aspergillus flavipes*

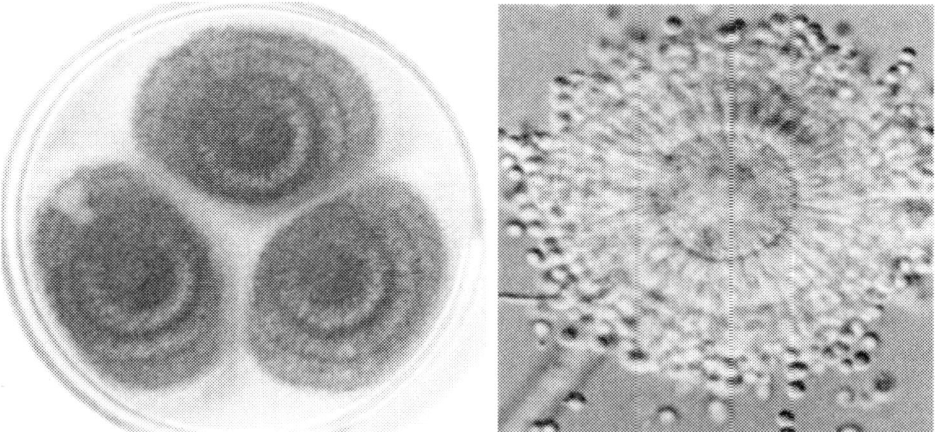

Fig 49 *Aspergillus flavus*

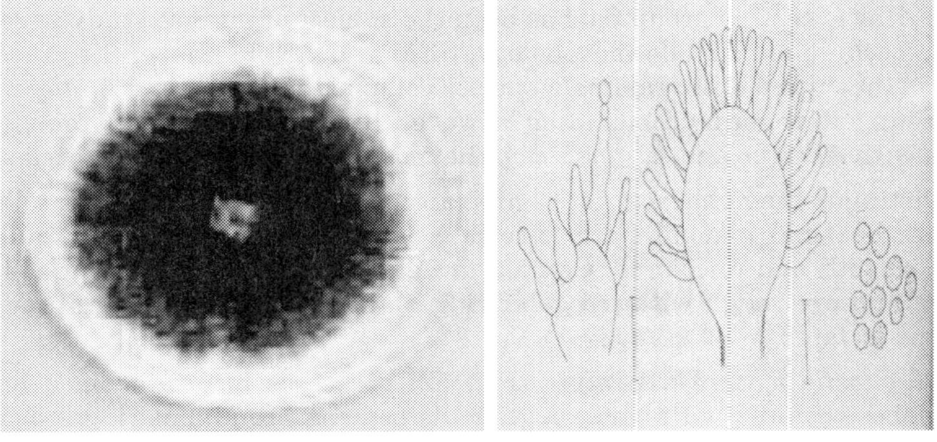

Fig 50 *A. parasiticus*

A. oryzae colony yellow-green just one set of sterigmata with smooth conidiophore stalks. Conidia are usually smooth and elliptical and larger than those of *A. flavus.*

A. nomius

Diagnostic features: The *A. nomius* species resembles *A. flavus* except that sclerotia (when produced) are smaller and vertically elongate, unlike those of *A. flavus*, which are larger and spherical.

A. nomius produces aflatoxin B_1 and G_1 and is the only member of the section *Flavi* to produce nominine (Bryden, 1982).

A. tamorii Kita

Diagnostic features: Colonies of *A. tamorii* are brown-green to brown. Conidial heads are predominently biseriate, large and radiate. Conidiophores are roughened. Conidia are large, globose and echinulate with a distinct double wall (Fig. 51).

A. tamorii colonies soon become deep chocolate-brown. Conidiophores are the same as *A. flavus* but condia are unique having double wall and colour bars than spines.

A. tamorii produces cyclopiazonic acid (Frisvad and Samson, 1991). It has also been implicated in "kodua poisoning", where high levels of CPA and fumiclavine produced by *A. tamorii* were found in kodo millet seed (Pitt and Hocking, 1997).

Section Nigri (*A. niger* Tiegh Group)

Raper and Fennel (1965) recognized approximately 12 species in this group. However, this group too has been the subject of intense study in recent years. Techniques include morphology, SEM and RFLPs. Several species are distinct, but there still remains a species complex based around *A. niger,* and the group will remain problematic. However, basic group concepts remain the same, with *A. niger* and *A. japonicus,* the most commonly isolated species.

Distinguishing characteristics are rapidly growing, purple-black to black colonies bearing predominently biseriate, globose, conidial heads. Conidiophores smooth and uncolored. Conidia are globose to elliptical to horizontally flatten and are rough with irregular ridges and bars. Sclerotia are produced by some isolates.

Diagnostic features: *A. niger* (Fig. 52) possesses biseriate black heads, and globose conidia with very rough walls bearing prominent ridges or color bars.

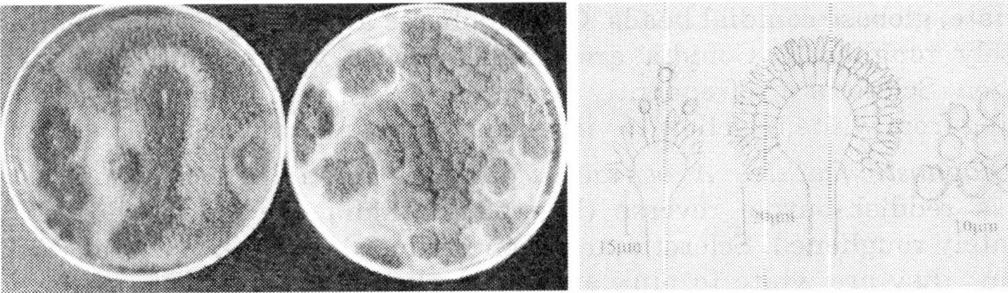

Fig 51 *A. tamorii*

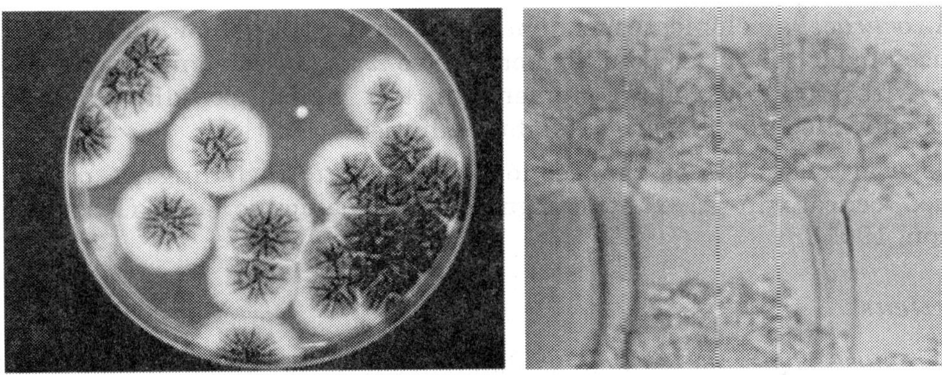

Fig 52 *Aspergillus niger*

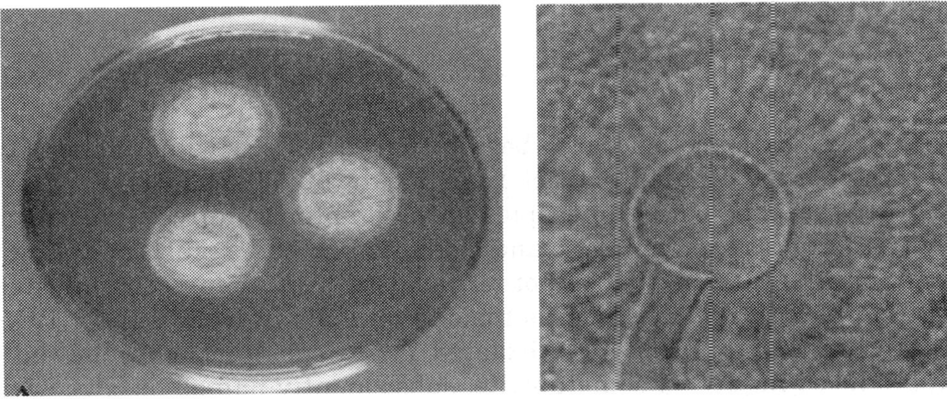

Fig 53 *Aspergillus ochraceus*

A. niger produces malformins and naphthopyrones (Frisvad and Samson, 1991). In addition, recently certain isolates have been shown to produce ochratoxin A (Pitt and Hocking, 1997).

Section Circumdati (*A. ochraceus* wilh Group)

Circumdati is a large group comprising 15 species of which only *A. ochraceus* and *A. melleus* are common. Distinguishing features are pale-yellow to orange-yellow to ochraceus or buff-colored colonies, bearing large, biseriate, globose conidial heads. Conidiophores are faintly colored yellow and coarsely roughened. Conidia are small, globose and smooth to delicately roughen. Sclerotia are frequently produced and vary in size, shape and color, ranging from white to yellow to vinaceous to brown and black.

Diagnostic features: *A. ochraceus* produces spreading ochraceus colonies, with a reddish-purple reverse (Fig. 53). Conidia are small, globose, and delicately roughened. Sclerotia are often produced, but take several weeks to appear; they are white to pink at first, darkening to vinaceous or lavender with age.

A. ochraceus colonies consistently remain bright-yellow and sometimes produce purple sclerotia. Conidiophores have 2 sets of sterigmata; the stalk is spiny with definite yellow pigment in the walls. Conidia are usually smoothed walled and globose to slightly elliptical.

Several members of the section *Circumdati*, including *A. ochraceus*, produce ochratoxin A, xanthomegnins, and penicillic acid (Frisvad and Samson, 1991).

A. melleus

Diagnostic features: *A. melleus* differs from *A. ochraceus* in that it produces smaller, paler conidial heads, and numerous small, yellow sclerotia, 400-500 μm in diameter.

2.3.2. *Penicillium*

The study of mycotoxicology began 100 years ago when Sakki (1988) demonstrated that ethanol extract from moldy unpolished yellow rice was fatal to dogs, rabbits and guinea pigs with symptons indirectly paralysis of the central nervous system (Ueno and Ueno, 1972). As a result the yellow rice was banned in Japan in 1910. Another incident that moldy corn infested by *Penicillium puberulum* extract was toxic to animals in Nebraska in 1913. In 1940. Miyake *et al.* (1940) have isolated *P. toxicarium* from yellowed rice which produced highly toxic metabolite, subsequently named as citreoviridin. Cole and Cox (1980) have recorded no less than 85 species of Penicillia which are known to be mycotoxigenic and about 42 mycotoxins are known to be produced by one or other species of Pencillia. The literature on this aspect has accumulated mainly on the chemistry and toxicology and most of the literature lacks specificity with misidentification of species of *Penicillium*. For instance citrinin was reported to be produced by atleast 22 species which may be attributed to misidentification. Misidentifications even by experts have also been common. Of 12 *Penicillium* species reported in the literature to produce tremorgenic toxin only two were found to be accurate (Pitt, 1979). Pitt (1979) have developed a classification of species of *Penicillium* based on broader range of standardized characters that used previously. Classification based on secondary metabolites (Fisvad, 1986) electrophoretic finger printing of certain isozymes (Cruickshank and Pitt, 1987) was also proposed. Pitt (1989) and Samson *et al.* (1990) have developed more convenient guide to the identification of toxigenic pencicillia.

Penicillium is a large genus with 150 species recognized which probably is the most conservative. The *Penicillium* is divided into sub-genera based on number and arrangement of phialides (elements producing conidia) and metulae and rami (elements supporting phialides) on the main stalk cells (stipes). Pitt (1989) classified *Penicillium* into four subgenera Aspergilloides in which phallides are borne directly on the stipe without intervening supporting elements, while *furcatum* and *biverticillatum* in which phialides are supported by metulaed. On the other hand, *Penicillium* in which both metulae and rami are usually present. Majority of important mycotoxigenic and food spoilage species are found in the subgenus *Penicillium.*

General procedures for enumeration of food-borne fungi are effective for enumerating all common *Penicillium* species. However, some *Penicillium* species grow weakly on media like PDA. Dichloran-rose Bengal – chloromophenicol (DRBC) agar or Dichloran + 18% glycerol (DG 18) agar is generally recommended (Pitt and Hocking, 1985).

Pitt (1979) gave comprehensive account of taxonomy of genus *Penicillium.* Good computer assisted key for identification of species of *Penicillium* is provided by Pitt (1988), while Samson *et al.* (1995) and Pitt (1996) classified food-borne species of *Penicillium* based on secondary metabolites. Correct identification of species of *Penicillium* can be made by observations under standard conditions such as media, incubation period and temperature, in addition to microscopic morphology, gross physiological features including colony diameter, nature of conidiophore metulae and sterigmata, colour of conidia and colony pigments. Culture media, colony characters (colours), exudates, odour and growth rate are some other characters which will be taken into consideration in the identification of species of *Penicillium* characterstic features of some common species of *Penicillium* are précised in Table 2.4.

About 14 species of *Penicillium* are reported to be principle mycotoxigenic. Majority of them produce single toxin, while some species such as *P. expansum* (citrinin and patulin), *P. crustosum* (CPA, penetrem A and roqusifortin), *P. roqueforti* (Patulin, PR toxin, Roquefortine) and *P. aurantiogrisem* (penicillic acid, vermcosidin, viomellein and xanthomegnin) produces more than one mycotoxins. Similarly toxins like penetrem A and patulin are elaborated by four species, while CPA by seven species. Toxins like citrinin and patulin are simple molecules, while others like penetrems are very complex. Thus *Penicillium* toxins are of great complex and diverse. Only ochratoxin A is carcinogenic, while citreovirdin, citrinin, patulin and penetrem A exhibit wide range of toxicity from acute to chronic. The toxins produced by *Penicillium* can be broadly divided into 2 groups. Those affect liver and kidney are asymptomatic or cause generalized debility in humans or animals, while others are neurotoxins which are characterized by sustained trembling. With the low levels of *Penicillium* in nature animal health is affected resulting reduced fertility, increased susceptability to infectious diseases, reduced feed conversion and ill thrift. Many species produce several compounds which are known to be toxic and the possibility of synergy also cannot be ignored. The potential involvement of *Penicillium* toxins in low level diseases in humans, undoubtedly warrants continued research.

In assessing the claim of toxigenicity of *Penicillium,* several points need to be kept in mind.

1. Many of the identifications reported in the literature are incorrect and thus has resulted in confusion.
2. A number of genuinely toxic compounds are produced by species not usually occurring in foods or feeds. Such toxins are of little practical importance.
3. Some compounds often cited as mycotoxins in the literature are of low toxicity.
4. Growth of moulds does not always mean production of toxin.

A key to the identification of some common mycotoxin producing species of *Penicillium* is given below:

1. Conidiophore heads symmetrical, with sterigmata, usually metulae, but no branches 2

1a Conidiophore asymmetrical with sterigmata, metulae and branches 7

2. Heads, with sterigmata only .... *P. citreoviride* (1)

2a. Heads, with sterigmata and metulae 3

3. Sterigmata flask-shaped, metulae loosely packed .... *P. citrinum* (2)

3a. Sterigmata lanceolate, metulae closely packed 4

4. Colony reverse on CZA deep red, pigment diffusing into surrounding agar. 5

4a. Colony reverse on CZA, yellow-brown to orange-brown, pigment not diffusing into surrounding agar 6

5. Conidia rough-walled, colony producing a fragrant, fruity odor .... *P. purpurogenum* (3)

5a. Conidia smooth-walled, colony odorless .... *P. rubrum* (4)

6. Conidia smooth-walled, colony diameter exceeding 2 cm in 2 weeks .... *P. islandicum* (5)

6a. Conidia rough-walled, colony diameter not exceeding 2 cm in 2 weeks .... *P. rugulosum* (6)

Table 2.4. Characteristic features of Mycotoxigenic Penicillia species

Species	Colonies			Conidiophores			Conidia
	CYE	MEA	G25N	Stipes	Metulae	Phialides	
P. aurant-iogriseum	30-35 mm radially sulfates moderate deep surface texture. Smooth to granular marginal hyphae. Entire, deep white, *Mycelium* white	25-35 mm plane rarely radially sucate. Marginal hyphae – Sub surface bearing fasicles of conidiophres. *Mycelium*: Subsurface, bright yellow condiation – Moderate to heavy dull green or near glaceum green. *Exudate* – Absent S. pigment – brown reddish	20-24 mm radically sulfate moderately deep absent to heavy deep gray green.	Longer or indefinite length	Verticels of 3-4	5-8 per metulae ampulli	Subspherical to ellipsoid smooth long in well defined columns.
P. chrysog-enum	*Colonies*: 35-45 mm, radially sulcate to plicate central umb-onate moderately deep and floccose. Margins usually deep 3-5 mm wide. *Mycelium*-marginal white, centrally white to yellowish, condiation-light to moderate grayish turquotise to dull green. *Exudate*-produce yellow brown, droplets, soluble pigment bright yellow.	*Colonies*: 25-35 mm diamradially sulcate / centrally umbonate rarely with floccose. Margins: low, subsurface. *Mycelium*:Inconspicuous, white or centrally pale orange. *Conidiatian* moderate to heavy, grayish to dull green. *Exudate*: absent soluble pigment absent.	*Colonies*: 18-24 mm radially sulcate dense with floccose. *Mycelium*: White to buff. *Conidiation*: light to heavy. *Exudate* : absent Soluble pigment – absent.	Surface 3-4 mm long	Verticels of 3-5 cylindrical	4-7 perven-ticite ampull-iforms.	Ellipsoidls penicidal smooth borne in long irregular columns.
P. citrinum	*Colonies*: 25-30 mm, smaller, radially sulcate, marginal velutinous, centrally floccose.	Colonies: 14-18 mm, rarely, 22 mm, plane, radially sulcate, deep and floccus.	*Colonies:* 13-18 mm radically sulcate centrall umbonate centrally floccose.	Smooth verticels 3-5	Uniform lengths spathulate terminally	Compact verticals ampulliform	Spheroidal to sulospheroi-dal.

	Margins: Low narrow. Mycelium: White in peripheral areas centers white to grayish orange to apricot. *Conidiation*: Moderate turquo-ise grey to grayish turquoise arglancous sky blue. *Exudate*: Clear, pale yellow to reddish brown. Soluble pigment : Bright yellow or absent.	*Mycelium:* White to grayish orange. *Margins*: Very narrow and typically irregular. *Conidiatian*: Moderate to heavy, grey-blue. *Exudate:* Absent soluble pigment.	*Margins :* Narrow *Mycelium :* White *Conidiation :* Moderate to heavy, dull-green		vesiculate		
P. commune	*Colonies :* 25-30 mm radially sulcate centrally umbonate surface texture. *Margins :* low, narrow. *Mycelium :* Inconspi-cuous, white. *Conidiation :* Greenish grey. *Conidia :* Green, grey. *Exudate :* Clear, small amount soluble / pigment absent.	*Colonies :* 20-28 mm plane sulcate lightly floccose. Light to moderate exudates – absent. Soluble pigment absent *Colonies :* 22-28 mm plane, radially sulcate, lightly floccose. *Margin :* low narrow, *Mycelium :* White *Conidiation :* Dull-green *Exudate* – Absent soluble pigment absent.	*Colonies :* 20-22 mm plane, sulcate lightly floccose. *Conidiation :* Light to moderate. *Exudate :* absent *Soluble pigment:* Orange-brown	Smooth walls	Verticels	4-8 ampulli form	Spher-oidal smooth 3-4 mm diam borne in disordered chain.
P. islan-dicum	*Colonies :* 17-22 mm umbonate, lightly floccose. *Margins :* low, narrow *Mycelium :* Coloured deep-orange to brownish orange. *Conidiation*: Moderate	*Colonies :* 17-22 mm umbonate, floccose. *Margins :* Low 2-3 mm wide. *Mycelium :* Colour *Conidiation :* Heavy grayish-turquoise	10-15 mm, plane sulcate, deep, floccose. *Mycelium:* White / coloured. *Conidiation:* Dark-green. Exudate-absent *Soluble pigment* : Absent	Smooth walled non-vesiculated	Verticels	5-8 per metulae acerose	Broadly ellipsoidal to subsphe-roidal.

	grayish-green. *Exudate* : Pale-yellow *Soluble pigment* : Absent.	*Exudate* : Absent *Soluble pigment* : Absent					
P. oxalicum	*Colonies* : 35-60 mm, plane / sulcate at center. *Margins* : low 2-4 mm wide. *Mycelium* : Usually inconspicuous. *Condition* : Heavy visible under low magnification of the microscope grayishgreen. *Exudate* : Limited *soluble pigment*: absent.	*Colonies* : 20-50 mm lightly radially sulcate, velutinuous. *Margins* : subsurface. *Mycelium* : Usually inconspicuous white. *Condiation* : Very heavy *Exudate* : Absent *Soluble pigment* : Absent	*Colonies:* 14-16 mm diameter plane, radially sulcate. *Mycelium* : White *Condiation* Absent *Exudate* : Absent *Soluble pigment* : Absent	Verticil of 2-4	Verticels	Verticels	Ellipsoidal, walls smooth borne in long chains.
P. purpu-rogeum	*Colonies* : 15-25 mm, plane radially sulcate. *Margins* : Narrow *Mycelium* : White, bright yellow. *Conidi ation* : Moderate to heavy dark-green. *Exudate* : Occasio-nally produced orange to red. *Soluble pigment* : Chara-cteristically produced, vicid red.	*Colonies* : 25-35 mm, plane, dense, slightly flo-ccose funiculose. *Margins* : Low and narrow, entire, irregular. *Mycelium:* Usually incon-spicuous white to-bright yellow. *Conidiation* : Heavy, dark green, rarely lighter. *Exudate* : Absent *soluble pigment*: absent.	*Colonies:* Microcolo-nies, to 6 mm, plane radially sulcate. *Conidiation:* Moderate. *Exudate* : Absent *Soluble pigment:* absent.	10 mg, smooth wells.	Verticels	5-7 acerose	Ellipsoidal, walls thick
P. simplic-issium	Colonies : 40-50 mm. Radially sulcate Margins:2-3 mm more wide Mycelium: dense, white	*Colonies* 40-50 mm plane, velutinouste floccose *Margines*: sub surface, *Mycelium*: white	*Colonies* 14-18 mm plane, sulcate velutinous to floccose, *Mycelium*: white	Very long, rough walled	Rarely short	Ampulliform	Ellipsoidal, borne in short to mode-ratly long disor-derd chains.

	Conidiation: absent typically coloured blue green. *Exudate:* clear *Soluble pigment:* not produced	*Conidiation*: Modrate heavy *Exudate*: obsent soluble pigment absent	*Conidiation*: light to heavy *Exudate*: absent *Soluble pigment*: absent				
P. variable	*Colony*: 15-20 irregular sulcate strictly velutinous *Margins*: low, narrow *Mycelium*: white, sulphur yellow, *Margins*: limited extent near colony centers. *Conidiation*:Moderate to heavy *Exudate*: absent *soluble pigment*: absent	*Colony*: 15-22. irregular sulcate, *Margins*: low narrow *Mycelin*: white sulphur yellow *Margins*: limited *Exudate*: occasional formation *Soluble pigment*: Absent	*Colonies*: Commonly no growth, rowly upto 5 mm, plane *canidiation*: light	Shorter acrial hyphae, smooth walled	Quite divergent verticels	Acerose ampulliform	Ellipsoidal smooth disorderd chains
P. viridicatum	*Colonies*: 28-32 mm, radially sulcate under low power fasciculate. *Margins*: typically narrow, 1-2mm wide, *Mycelium*: white, pale-brown, *Conidiation*: typically moderate to heavy, yellow green *Exudates*: clear pale yellow, pale brown, pale pink *Soluble pigment*: orange to reddish.	*Colony*: 25-30 plane, centrally floccose, *Margins*: typically by deep white, *Mycelium*: inconspicuous white, brown *Conidiation*: yellow-green soluble pigments absent	*Colony*: 20-22 mm plane, surface granular *Margins*: low deep, 2-3 mm wide *Mycelium*: white *Conidiation:* Moderate. Blue-grey *Exudates*: absent *Soluble pigment*: absent	Roughened walls. Penicillicterivally	Vertical	5-8 ampulliform	Ellipsoidal irregular chains.
P. cyclopium	*Colonies*: Weakly fasciculate *Mycelium*: yellow to orange *Exudate*: droplet	*Mycelium*: strangly yellow		Rough-walled	Cylindrical	Cylindrical	Smooth walled globose to subglobase

P. crustasum			Rough-walled	Cylindrical	Cylindrical	Smooth walled globose, to sub globose
P. expa-nsum	*Colonies*: Floccose to weakly fasiciculate, *Exudates*: droplets	Cream yellow	Smooth walled	Cylindrical	Cylindrical	Smooth walled ellipsoidal blue green
P. griseo-fulvum	*Colony*:Fas-cidulate, *Exudate*; droplets		Smooth walled	Cylindrical	Cylindrical	Smooth globose sub globose
P. rayue	*Colony*: Velutinous, green, *Exudates*: droplet.		Rough walled	Cylindrical	Cylindrical with short coloum	Smooth walled
P. virilitum	Velutionous to weakly fasiculate green, *Exudate*: droplets yellow		Rough walled	Cylindrical	Flask shaped	Rough walled globose, to sub globose

7. Conidia strongly elliptical and greater than 5μ in diameter ....*P. oxalicum* (7)
7a Conidia not strongly elliptical and less than 5 μ in diameter 8
8 Colonies blue 9
8a. Colonies other than blue
9. Fasicles regularly produced on MEA ....*P. expansum* (8)
9a Fasciles not regulary produced on MEA 10
10. Colony growth and conidium production restricted on MEA 11
10a. Colony growth and conidium production not restricted on MEA
11. Colony reverse on CZ, reddish-maroon, pigment diffusing into surrounding agar ... . *P. martensii* (9)
11a. Colony reverse on CZ, colourless to purple-brown pigment surrounding agar 12
12. Fascicles regularly produced on CZB ... .*P. cyclopium* (10)
12a Fascicles rarely produced on CZB ....*P. puberulum* (11)
13. Colony light gray and producing a fragrant, fruity odor ....*P. urticae* (12)
13a. Colony green, producing a strong musty odour 14
14. Colony granular to fasciculate, bright green to yellow-green ....*P. viridicatum* (13)
14a Colony velvety, dull-green to grey-green 15
15. Conidia delicately rough-walled, rarely exceeding 4 μm in diameter ....*P. olivio-viride* (14)
15a. Conidia smooth-walled, regularly exceeding 4 μm in diameter ....*P. palitans* (15)

CZB = CZ plus 20 ppm Botran.

CZA = Czapek solution agar.

MEA = Malt Extract Agar.

1. *P. citreo-viride* Biourage

It belongs to sub-genus Aspergilloides. Colony on Czapek agar (CZ) bright-yellow to yellow-gray, never definitely green or blue. Growth restricted colony diameter (Fig. 54) rarely exceeding 3 cm in 10 to 14 days at 23 to 25°C. Surface velvety or woolly, the colony leather like and tears with difficulty. Radial furrows regularly produced. Sporulation restricted. Surface droplets, when produced clear to light-yellow. Colony reverse bright-yellow, pigment usually diffusing into surrounding. Colony on Malt extract agar (MEA) gray to yellow-gray, not leathery, neither growth nor sporulation restricted. Surface plane not furrowed. Colony reverse yellow to brown, the pigment not diffusing into the surrounding agar.

Conidiophore heads are symmetrical consisting of only flask-shaped sterigmata. Conidia smooth-walled, globose, small rarely exceeding 3 μm in diameter. In slide mounts, conidia often adhere in long chains.

Diagnositc features. (1) Colony on CZ yellow to yellow-gray, leathery, radially furrowed. (2) Growth rate and sporulation restricted on CZ. (3) Colony reverse on CZ bright yellow, the pigment diffusing into the surrounding agar. Conidiophore heads symmetrical, consisting only of sterigmata.

2. *P. citrinum* Thom

Colony on CZ blue-green to blue-gray, the color often restricted to colony margins because of poor sporulation elsewhere, the poorly sporulating areas colorless to faint yellow, growth restricted and colony diameter rarely exceeding 3 cm in 10 to 14 days at 23 to 25°C. Surface velvety to woolly, never granular or fasciculate. Surface droplets, clear to faint yellow, usually produced and concentrated in central areas of the colony. Colony reverse, flesh to yellow to sometimes pink, the pigment occasionally diffusing into the surrounding agar. Growth occurs at 35°C, but not below 8°C. Colonies on Malt extract agar (MEA) blue-green to blue-gray over the entire surface due to abundant sporulation. Growth, however, more restricted than on CZA, the diameter rarely exceeding 2 cm in 10 to 14 days at 23 to 25°C.

Conidiophore heads are symmetrical, consisting of sterigmata and metulae, but rarely branches, all parts being smooth-walled (Fig. 55). Metulae loosely packed and divergent, not closely packed and parallel. Sterigmata flask-shaped, not lanceolate or elongate. Conidia smooth-walled, globose to subglobose and small, 2.5 to 3.5 μm in diameter. Conidia produced in long, adherent columns and characteristic that is readily detected when sporulating areas, especially at colony margins are examined under low power magnification.

Diagnostic features: (1) Colony growth more restricted on Malt extract agar (MEA) than on CZ. (2) Conidiophore heads symmetrical, consisting of sterigmata and metulae, but no branches. (3) Metulae loosely packed and divergent, not closely packed and parallel. (4) Sterigmata flask-shaped, not lanceolate or elongate. (5) Conidia produced in long, adherent columns.

3. *P. purpurogenum* Stoll

Colony on CZ dark-green to greenish-black except for yellow to orange-red margins. Growth restricted, colony diameter rarely exceeding 2cm in 10 to 14 days at 23 to 25°C. Surface velvety, rarely woolly, never granular or fasciculate. Small, beadlike, orange to red surface droplets are usually produced. No detectable odor is produced. Colony reverse, highly colored, deep-red, the pigment diffusing into the surrounding agar. Colonies on Malt extract agar (MEA) grow more rapidly, 3 to 5 cm in diameter in 10 to 14 days at 23 to 25°C, but produce no red coloration in surface mycelium or in colony reverse. A pungent odor, suggestive of apples or walnuts is usually produced on Malt extract agar (MEA). Conidiophore heads are symmetrical, biverticellate consisting of sterigmata and metulae, but no branches, all parts being smooth-walled. Sterigmata are lanceolate or elongate not flask-shaped. Metulae are closely packed and parallel, not loosely packed and divergent. Conidia are elliptical, globose often pointed at the tips and spiny rough-walled, 3 to 4 μm in length.

Diagnostic features: (1) Colony growth restricted on CZA, (2) Colony reverse on CZA deep-red, with pigment diffusing into the surrounding agar (3) A pungent odor suggestive of apples or walnuts produced on Malt extract agar (MEA) (4) Conidia elliptical, rough-walled, sometimes pointed at the tips (5) Isolates grow at 35°C, but not below 8°C.

4. *P. rubrum* Stoll

P. rubrum in general, indistinguishable from isolates of *P. purpurogenum*, except for the following: (1) Colony on CZA dull gray-green rather than dark-green to greenish-black (Fig. 56). (2) Colony on MEA odorless (3) Colony reverse on MEA is red, especially under colony center, but the pigment does not diffuse into the surrounding agar (4) Conidia elliptical, sometimes pointed at the tips, smooth-walled, not rough-walled.

5. *P. islandicum* Sopp

Colony on CZ is bright orange-brown because of the presence of abundant, highly colored mycelium. Areas of sporulation, blue to blue-gray are usually restricted to marginal areas of the colony, often zonate, that is produced in concentric circles, the circles interspaced by bright-orange to orange-brown mycelium (Fig. 57). Growth is fairly restricted, colony diameter rarely exceeding 3 cm in 10 to 14 days at 23 to 25°C. Surface is velvety to woolly, sometimes granular, but never fasciculate. Surface droplets, when produced, are small, enmeshed in the mycelium, clear to straw-colored. No detectable odor is produced. Colony reverse is highly colored, orange-brown to reddish-brown, never distinctly red, the pigment not diffusing into the surrounding agar. Colony on MEA is likewise restricted at 23 to 25°C, blue to blue-gray due to heavier sporulation, but with abundant patches of yellow to faint orange mycelium is evident.

Conidiophore heads are symmetrical, consisting of sterigmata and metulae, but no branches, all parts being smooth-walled. Sterigmata are lanceolate or elongate, not flask shaped. Metulae are closely packed and parallel. Conidia are elliptical, without pointed tips, smooth-walled, 3 to 4 μm in length.

Diagnostic features: (1) Colony on CZA bright orange to orange-brown, the blue to blue-gray areas of sporulation produced in concentric circles in marginal areas. (2) Colony reverse on CZA orange-brown, never distinctly red, the pigment not diffusing into the surrounding agar. (3) Colony on MEA likewise restricted, the surface containing abundant patches of yellow to orange mycelium. (4) Conidia elliptical, not pointed at the tips, smooth-walled. Isolates grow at 35°C, but not below 8°C.

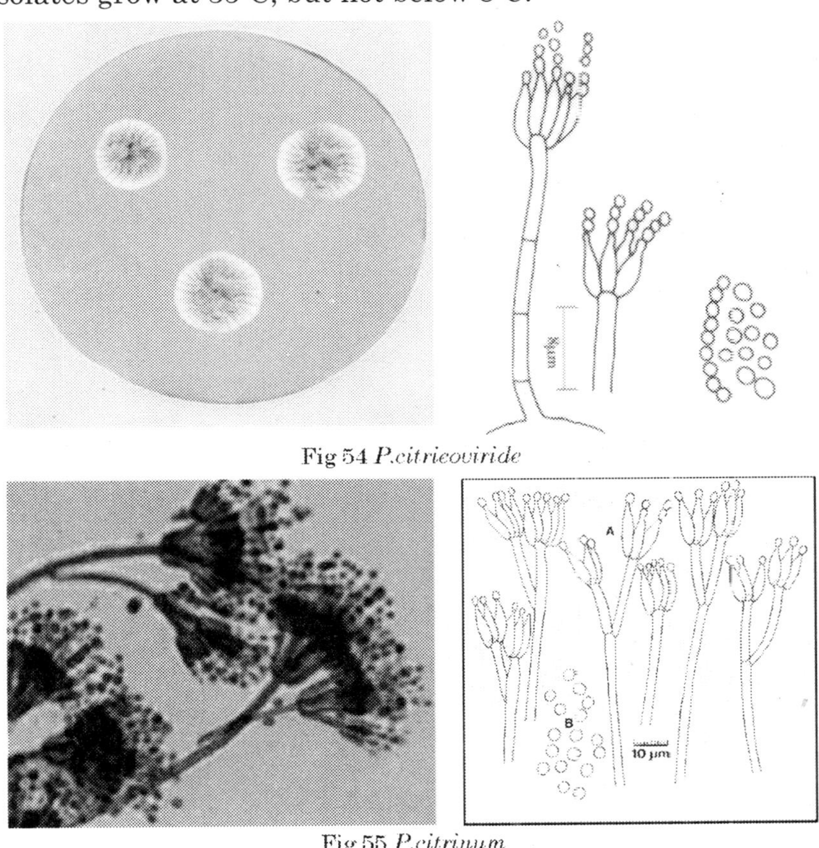

Fig 54 *P.citrieoviride*

Fig 55 *P.citrinum*

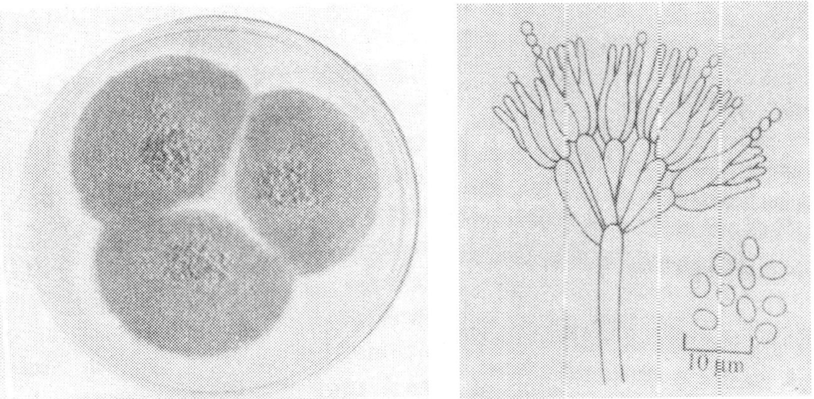

Fig 56 *P.rubrum*

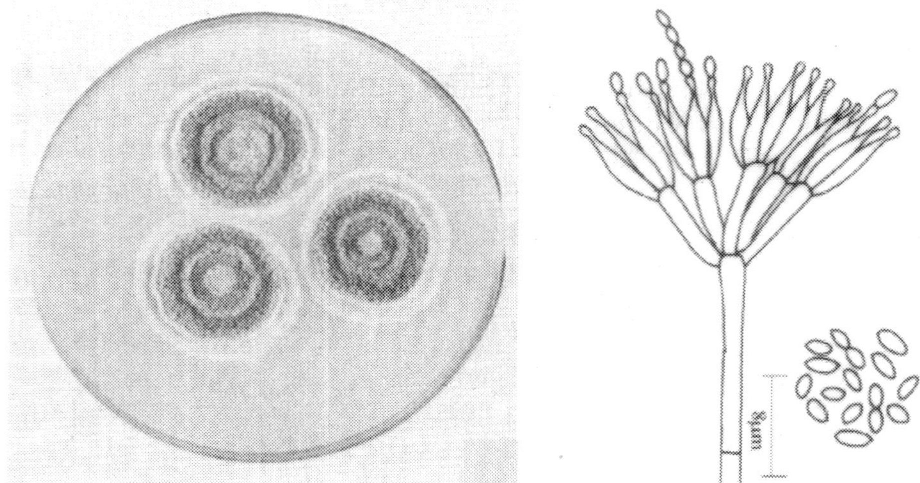

Fig 57 *P. islandicum*

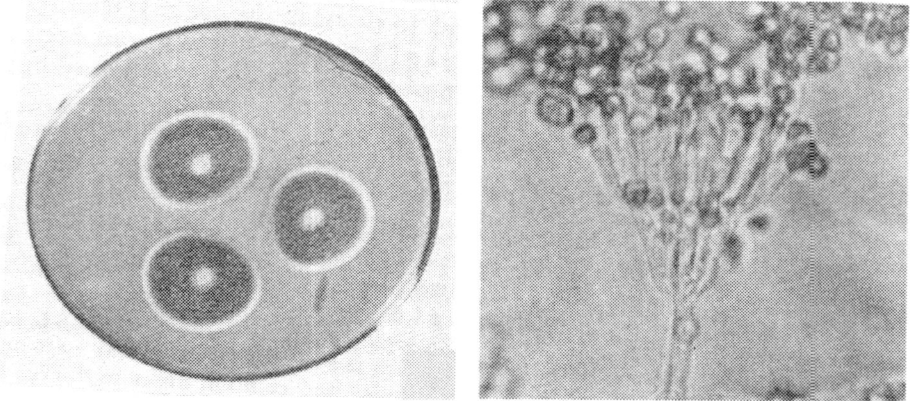

Fig 58 *P. rugulosum*

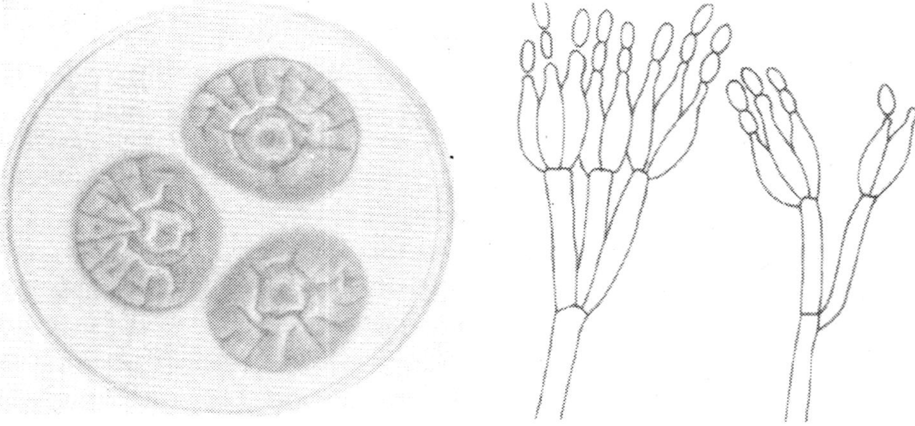

Fig 59 *P. oxalicum*

6. *P. rugulosum* Thom

Isolates of *P. rugulosum* are in general, indistinguishable from isolates of *P. purpurogenum* or *P. islandicum,* except for the following: (1) Colony on CZA yellow-green to gray-green rather than dark-green to greenish-black or bright -orange to orange-brown (Fig. 58) (2) Growth on CZA extremely restricted, colony diameter rarely exceeding 1.0 to 1.5 cm in 10 to 14 days at 23 to 25°C, the growth velvety but extremely wrinkled, radially furrowed and leathery in consistency. (3) Colony reverse on CZA colorless, or with patches of yellow or orange-red, the pigment not diffusing into the surrounding agar. (4) No detectable odor produced on MEA. (5) Conidia elliptical, but extremely rough-walled or rugulose.

7. *P. oxalicum* Currie and Thom

It belongs to sub-genus *furcatum* and section *furcatum*. Colony on CZ is dull-green to gray-green, never distinctly blue, sporulating throughout except for a 2- to 5-mm band at the colony margin which, when present, is white and often raised and woolly (Fig. 59). Growth is rapid, colony diameter exceeding 5 cm in 10 days at 23 to 25°C. A few, if any, species of *Penicillium* grow more rapidly on this medium than *P. oxalicum*. Surface is velvety, sometimes woolly in marginal areas, never granular or fasciculate. Conidia are abundantly produced, often forming crusts which break away from the colony surface when the petri dish is tapped; this characteristic feature is more apparent on MEA. Neither surface droplets nor detectable odor is produced. Colony reverse is colorless to faint yellow or pink. It can be enumerated either on DRBC or DG 18, biverticellate, minimum a_w requirment is 0.86.

Conidiophore heads are asymmetrical, consisting of sterigmata, metulae, and branches, all parts being smooth-walled. Sterigmata are long flask-shaped, not lanceolate or elongate. Conidia are strongly elliptical and large, ranging from 5.0 to 6.5 μm in length.

Diagnostic features: (1) Growth extremely rapid on CZA (2) Conidia form crusts which break away from the colony surface when the petri dish is tapped (3) Conidia strongly elliptical, smooth-walled and large. (4) Growth excellent at 35°C but rare at 8°C.

8. *P. expansum* Link

Colony on CZA is blue to blue-gray. Growth is rapid, colony diameter approaching 5 cm in 10 to 14 days at 23 to 25°C. Colony surface is granular, especially in marginal areas with distinct fascicles often detectable. Fasicicles are routinely encountered on MEA and On CZB (Fig. 60). A strong, obnoxious odor, suggestive of rotting apples is produced regularly. Colony reverse is colorless or light to deep-brown, the pigment not diffusing into the surrounding agar. Growth occurs at 4°C, but rarely at 35°C or above. All isolates of *P. expansum* cause rapid rot of apples. The minimum a_w is 0.82-0.83.

Conidiophore heads are asymmetrical, tetraverticellate with smooth stipe consisting of sterigmata, metulae and branches, all parts being smooth walled. Sterigmata are flask-shaped, not lanceolate or elongate. Conidia are globose to slightly elliptical, smooth-walled, rarely exceeding 3.5 µm in diameter.

Diagnostic features: (1) Colony on CZA granular to fasciculate, fascicle production enhanced on MEA and CZB. (2) Obnoxious odor, suggestive of rotting apples or, sometimes of acetic acid, regularly produced. (3) Conidiophores asymmetrical. (4) Rapid rot of apples produced. (5) Growth occurs at 4°C, but rarely at 35°C.

9. *P. aurentiovirens* Biourge

Isolates of this species are, in general, indistinguishable from isolates of *P. cyclopium*, except for the following: (1) Colony reverse on CZ is red to maroon, the surrounding agar likewise colored (similar to *P. martensii*) and usually producing a halo effect in the agar (unlike *P. martensii*) (Fig. 61). (2) Growth on MEA is more restricted than that on CZ with sporulation greatly reduced. Few species of *Penicillium* are more restricted in growth and sporulation on MEA than on CZ.

10. *P. martensii* Biourge

This species is indistinguishable from isolates of *P. cyclopium*, except for the following: (1) Colony reverse on CZA is red to maroon, the surrounding agar likewise colored (similar to *P. aurentiovirens)* but without producing any halo effect (unlike *P. aurentiovirens*). (2) Conidia are rather large, approaching 5.0 to 5.5 µm in diameter.

11. *P. cyclopium* Westling

Colonies on CZ are bright-blue to blue-gray, colony diameter approaching 4 to 5 cm in 10 to 14 days at 23 to 25°C. Surface is velvety to granular, rarely fasciculate. Detectable fascicles are readily produced on CZB. Surface droplets, clear to faint yellow, are usually produced (Fig. 62). A strong, obnoxious, musty odor is regularly produced. Colony reverse is colorless to purple-brown, the color not diffusing into the surrounding agar. On MEA, colony reverse is yellow, the surrounding agar usually likewise colored.

Conidiophore heads are asymmetrical, consisting of sterigmata, metulae, and branches. The branches and metulae often being rough walled (Fig. 5). Sterigmata are flask-shaped, not lanceolate or elongate. Conidia are smooth-walled, globose, up to 4.5 µm in diameter.

Diagnostic features: (1) Colony blue to blue-gray (2) Fascicles produced on CZB, this characteristic feature usually not evident on the other media. (3) Colony reverse on MEA yellow, the surrounding agar usually likewise colored. (4) A strong, musty odor regularly produced on all substrates. (5) Branches and metulae usually rough-walled. (6) Growth occurs at 4°C, but rarely at 35°C or above.

12. *P. puberulum* Bainier

This species is indistinguishable from isolates of *P. cyclopium*, except for the following: (1) Colony more consistently blue-gray rather than bright-blue. (2) Growth on CZA restricted, rarely exceeding 3.5 μm in diameter in 10 to 14 days at 23 to 25°C (Fig. 63). (3) Fascicles rarely encountered, not even on CZB.

13. *P. urticae* Bainier

Synonyms: *Penicillium patulum* Bainier, *Penicillium griseofulvum* Dierckx

Colony on CZ is light-gray to gray-blue. Growth is restricted, colony diameter rarely exceeding 3 cm in diameter in 10 to 14 days at 23 to 25°C. Surface is regularly granular to fasciculate, especially near colony margin, where fascicles are often produced in concentric circles. Fascicles are more apparent on M and CZB (Fig. 64). Surface droplets, clear to dull-red, are regularly produced. A distinct, fruity odor, suggestive of strawberries, is regularly produced on all substrates. Colony reverse is orange to reddish-brown, the pigment sometimes diffusing into the surrounding agar.

Conidiophore heads are asymmetrical, consisting of sterigmata, metulae, and branches, all parts being smooth-walled. Metulae are unusually short and stalky, sometimes wedge-shaped, rarely exceeding 10 μm in length. Sterigmata are flask-shaped, not lanceolate or elongate, and unusually short, rarely exceeding 6 μm in length. Conidia are globose to slightly elliptical and small, rarely exceeding 3 μm in diameter.

Diagnostic features: (1) Colony definitely light-gray to gray-blue and regularly granular to fasciculate. (2) Fascicles often produced in concentric circles in marginal areas. (3) A fruity, strawberry like odor regularly produced. (4) Conidiophores asymmetrical with unusually small metulae and sterigmata. (5) Growth occurs at 4°C, but rarely at 35°C or above.

14. *P. viridicatum* Westling

Colony on CZ is green to yellow-green. Growth is restricted, colony diameter rarely exceeding 3 cm in 10 to 14 days at 23 to 25°C. Surface is velvety to granular, rarely fasciculate (Fig. 65). Detectable fascicles are produced on CZB. Surface droplets, clear to faint yellow, are usually produced. A strong, obnoxious, musty odor is regularly produced. Colony reverse is colorless to purple-brown, the pigment not diffusing into the surrounding agar.

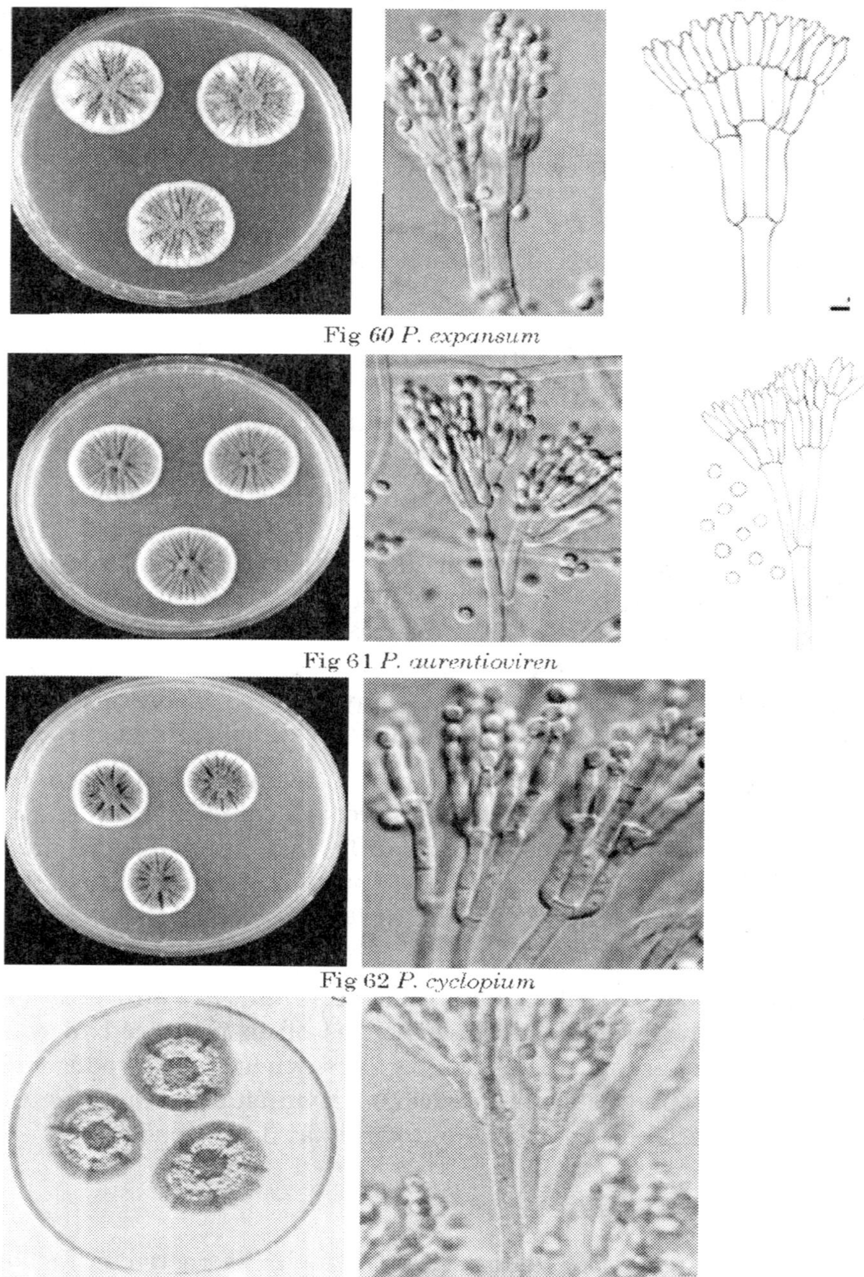

Fig 60 *P. expansum*

Fig 61 *P. aurentioviren*

Fig 62 *P. cyclopium*

Fig 63 *P. puberulum*

Conidiophore heads are asymmetrical, consisting of sterigmata, metulae, and branches, the branches and metulae often rough-walled. Sterigmata are flask-shaped, not lanceolate or elongate. Conidia are subglobose to elliptical and rough-walled, up to 4.5 µm in diameter.

Diagnostic features: (1) Colony green to yellow-green, never distinctly blue or light gray. (2) Granular, rarely fasciculate, fascicles readily produced on CZB. (3) Strong, musty odor produced on all substrates. (4) Conidiophore heads asymmetrical, the branches and metulae usually rough-walled. (5) Growth occurs at 4°C through 35°C, but rarely higher.

15. *P. olivino-viride* Biourge

Isolates of this species are, in general, indistinguishable from isolates of *P. viridicatum*, except for the following: (1) Colony on CZ dull-green to gray-green rather than green to yellow-green. (2) Fascicles rare, even on CZB. (3) Conidia globose and smooth-walled.

16. *P. palitans* Westling

Isolates of this species are, in general, indistinguishable from isolates of *P. viridicatum*, except for the following: (1) Colony on CZ dark gray-green rather than green to yellow-green. (2) Growth on CZ more rapid, colony diameter approaching 4 cm in 10 to 14 days at 23 to 25°C (Fig. 66). (3) Fascicles rare, even on CZB. (4) Conidia sub-globose, smooth-walled, and relatively large, attaining a diameter of 5 µm or more.

17. *P. verrucosum*

It belongs to sub-genus *Penicillium* section *Penicillium* is the fungus which produces ochratoxin A. *P. verrucosum* is distinguished by slow growth on Czapek yeast extract agar (CYA) and malt extract (MEA) at 25°C. Bright-green conidia, clear to pale-yellow exudates and rough stipes are the characteristic features of this fungus (Fig. 67). It differs from *P. viridicatum* by its slow growth, while few *P. solitum* in having green rather than blue conidia. On dichloron – rose bengal – yeast extract sucrose agar (DCRBY SA), a selective medium for enumeration of *P. verrucosum* and *P. viridicatum,* it produces violet-brown reverse colouration. It grows most strongly at relatively low temperature down to 0°C with a maximum near 31°C. It is xerophilic growing down to a_w 0.81. It produces maximum ochratoxin A at 20°C.

18. *P. funiculosum*

Colonies are grey and produce definite ropes or funicles or sporulating mycelium on the surface (Fig. 68). No other *Penicillium* species does this.

Reverse is usually purple. Conidiophores biverticellate and identical with those of *P. variable.*

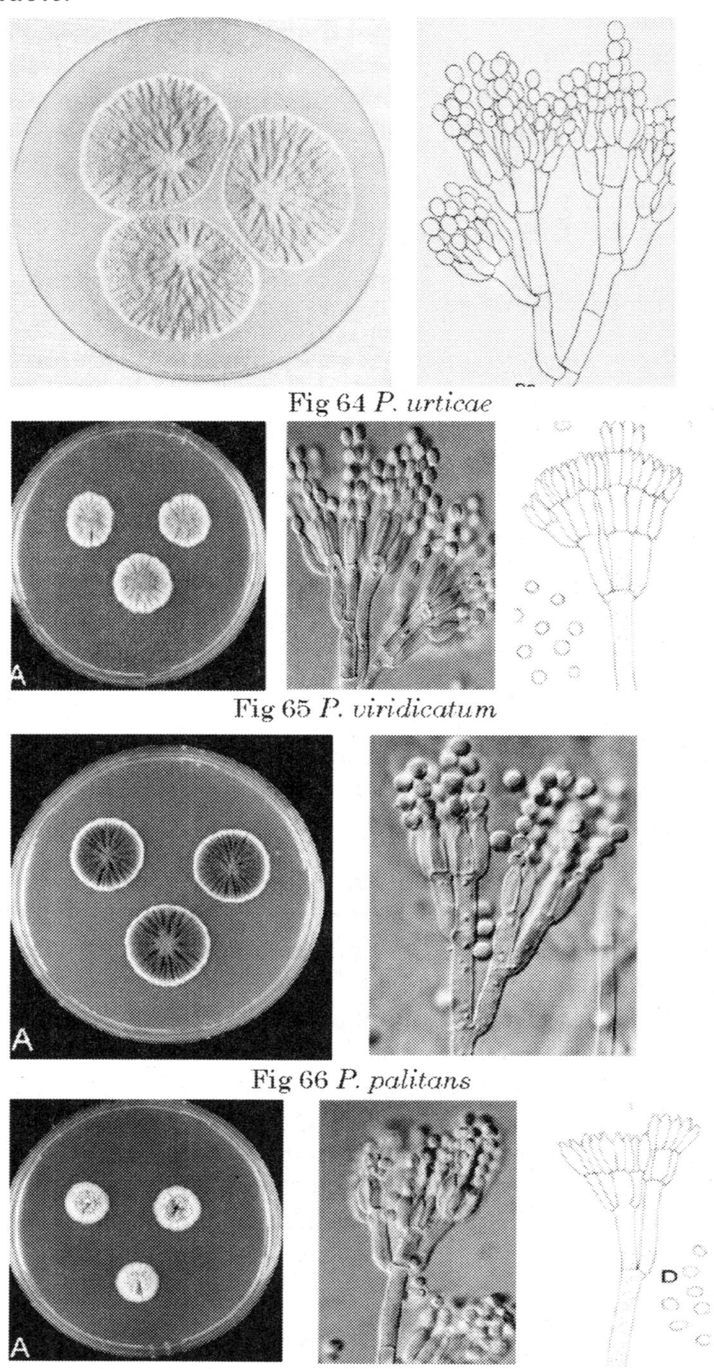

Fig 64 *P. urticae*

Fig 65 *P. viridicatum*

Fig 66 *P. palitans*

Fig 67 *P. verrucosum*

19. *P. commune* Thom

It is the most common source of CPA which probably misidentified as *P. cyclopium*. It is a wild ancestor of *P. combrelinm*. It belongs to *Penicillium* sub-group and is a common cause of cheese spoilage and incapable of growing at 0°C optima around 25°C (Fig. 69).

20. *P. crustosum* Thom

A member of *Penicillium* subgenus, produces large penicillin with rami, metulae and phialides characteristic of this sub-genus. Colonies dull-green colonies with granular texture on both CYA and MEA and grows fast (Fig. 70). Microscopically *P. crustosum* is characterized by large rough walled stipes and smooth walled usually spherical conidia. Production of enormous number of conidia on MEA which became detached from the colony when the petridish is jarred. DRBC medium inhibits spreading of the colony. It is a ubiqutous spoilage mould having been found in majority of cereals examined such as corn, processed meats, nuts, cheese and fruit juice and a weak pathogen of pomaceous fruits. It does not grow at 37°C. Penetrem A is produced at high a_w.

21. *P. roqueforti* Thom

It belongs to subgenus *Penicillium* sect *Penicillium*. It comprises two distinct varieties *P. roqueforti V. roqueforti* widely used mould in ripening roqueforti cheese, while *P. roqueiforti* var. *varneum* occurs in meats and silage. The former produces PR toxin, while the latter produces patulin and both varieties produce roquefortine. *P. roqueforti* grows very rapidly on CYA and MEA. It could produce green reverse colour on one or both media. Stipes are often very rough, conidia are large spherical and smooth walled (Fig. 71). DRBC and DG18 media are used in its enumeration. On malt acetic agar (MEA + 0.5% glacial acetic acid) it grows well in the atmosphere with less than 0.5% O_2 and even in the presence of upto 20% CO_2.

2.3.3. *Fusarium*

Fusarium is one of the most important genera of plant pathogenic fungi on earth with record of devastating infections causing wilts, blights, root-rots and cankers in legumes, coffee, pine trees, wheat, corn, carnations and grass-es. Importance of *Fusarium* species with reference to mycotoxins infections may sometimes occur in developing seeds especially in cereals and also in maturing fruits and vegetables having potential to produce mycotoxins in foods.

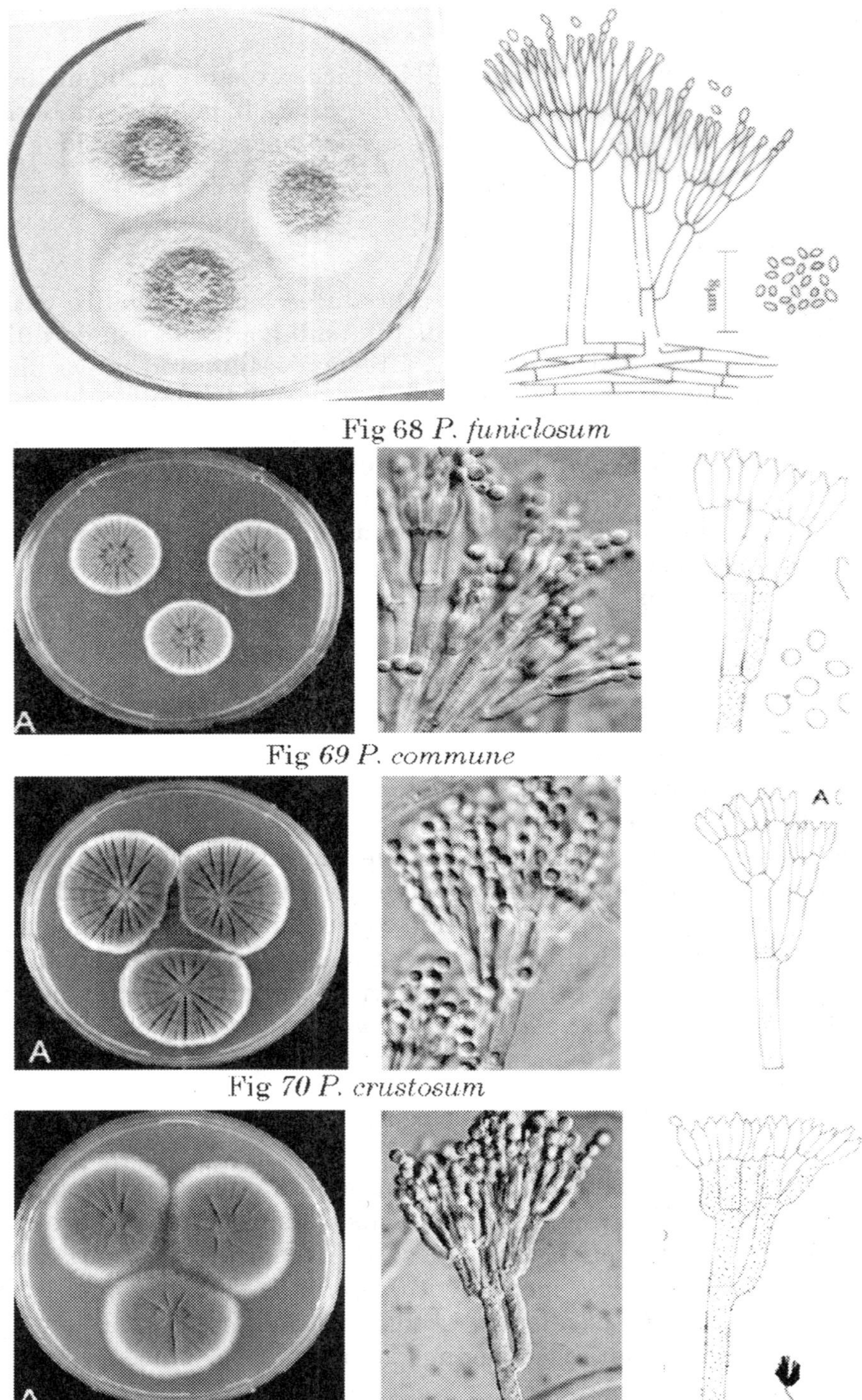

Fig 68 *P. funiclosum*

Fig 69 *P. commune*

Fig 70 *P. crustosum*

Fig 71 *P.roqueforti*

At one time the taxonomy of *Fusarium* species may have been a subject of purely academic interest. However, ever since it was proved that species of this genus incited grave disorders in man (Joffe, 1977) and animals (Ueno *et al.*, 1972; Joffe, 1978) by the production of toxins, determination of the exact identity of these species and strain has become a matter of acute importance. The taxonomy of the sporotrichella, section is a matter of some dispute. The various species in this section are bunched by Snyder and Toussom (1965) under *F. tricinctum*. This section includes species which are not of particular phytopathological importance and therefore, the Snyder-Hansen system is not acceptable.

The fusaria was reported to be the cause of alimentary toxic aleukea (ATA), human mycotoxicoses epedimic in the USSR which killed about one lakh people between 1942 and 1948 (Joffe, 1978). It also occurred in Russia during 1913, 1932 and also earlier. According to Marasas *etal.* (1981), ATA also occurred in other countries including England during 16th to 18th Centuries. Recent researches have shown that *Fusarium* species are capable of producing a bewildering array of mycotoxins. Formost among these are atleast about 140 trichlothecenes are known to be produced by fusaria and suspected to be responsible for acute and chronic diseases of man and animals.

The taxonomy of *Fusarium* is complex because many of species, even when grown from monospore cultures, exhibit extreme variability in their morphological characteristics. This led early taxonomists, foremost among them Wollenweber (1935) and Wollenweber and Reinking (1935), to distinguish between hundreds of so-called species and varieties. Later classifications by Gordon (1960), Subramanian (1952), Raillo (1959), Bieai (1955), Seemuller (1968), Gerlach (1970), Booth (1971), Joffe *et al.* (1973), Joffe (1974) and Joffe and Palti (1975), were largely based on Wollenweber and Reinking's earlier system.

The taxonomy of has undergone a phenomenal change in species concept due to multitude problems during last 60 years. Wollenweber and Reinking (1935) recognized over 100 species, while Snyder and Hansen (1945) and Tousoun and Nelson (1968) have accepted only nine species each with many formae species. On the other hand, Booth (1971) and Gerlach and Nirenberg (1982) have recognized about 75 species. Identification of *Fusarium* species is ideally carried from the growth on carnation leaf agar, an effective medium for macroconidium production (Nelson *et al.*, 1985).

Determined attack on the taxonomy of these fungi by International collaborative group led by Nelson has resolved most of the conflicts and developed most acceptable classification with 30 valid species (Nelson *et al.*, 1985). An attempt is made to present precisely the characteristics of those *Fusarium* species that are known to produce mycotoxins.

The Snyder and Hansen classification of *Fusarium* chiefly in regard to the distinction between toxigenic *Fusarium* sections and species, the most important being the *Sporotrichiella* section is quite inadequate and leads to wrong identification. The species of *Sprotrichiella* have attracted world wide attention because of their capacity to produce toxins dangerous to man and animals. The most prominent of these are *F. poae* and *F. sporotrichioides*, which grow in the Soviet Union on over wintered cereals and cause the often fatal disease, alimentary toxic aleukia in man (Joffe, 1977) and *F. tricinctum*, which grows in the United States on mouldy corn, wheat and fescue hay and produces toxic symptoms in many animals (Ueno *et al.*, 1972). Metabolites of *Fusarium* have received a great deal of attention from research scientists because of their mycotoxic potential (Smalley and Strong, 1974).

A direct consequence of confusion in taxonomy is the confusion in species mycotoxin associations. *Fusarium* isolates producing a particular toxin have been given different names as a result of the different taxonomic systems used or simply as a result of misidentification. Marasas *et al.* (1984) after intensive study of about 200 toxigenic *Fusarium* isolates provided accurate information for species identification and toxins produced by them. They listed 24 species of which they considered *F. sporotrichioides, F. equiseti, F. graminearum* and *F. moniliforme* to be most important from human health point of view.

Of the sixteen sections of *Fusarium* accepted by Gerlach and Nirenbreg (1982), 12 have one or more toxigenic species. But three sections Sporotrichiella, Liseola and Roseae are rich in toxigenic species. Species belonging to section Sporotrichiella produce trichothecene and butanelides. Species of section Liseola produce zearalenone, while species placed in Rosea produce both trichothecenes and zearalenone.

The taxonomy of species of the Sporotrichiella section is a matter of some dispute. The various species in this section are bunched together by Snyder and Toussoun (1965) under *F. tricinctum*. This section includes species which are not of particular phytopathological importance and therefore, the Snyder and Hansen system is not acceptable.

Most significant change in the species composition is of *F. moniliforme* along with two other species which formed a Liseola section of Wollenweber is reduced to one species *F. moniliforme* and one sub-species *F. moniliforme* var. *subglutinans* by Booth (1971). Another interesting nomenclatural change is that of *F. roseum* which was used by Snyder and Hansen (1945) and commonly used in phytopathological literature has been applied to five species and one variety in three sections of *Fusarium* (Gerlach and Nirenberg, 1982). Nirenberg (1976) replaced *F. moniliforme* with the older name *F. verticilloides* and expanded the old concept of *F. moniliforme* to include in addition to *F. verticilloides, F. proliferatum* and *F. lactus*. She divided the old *F. moniliforme* var *subglutinans* among *F. sacchari, F. sacchari* var. *subglutinans*, *F. neoceras* and *F. anthophilum.*

Another problem with regard to *Fusarium nivale* is reported to have teleomorph in *Micronectriella nivalis* and *Calenectria nivalis*. The teleomorph of this is also reported to be *Monographella* and anamorph superficially resembles *Fusarium* and now it is placed in *Gerlachia* (Gams and Muller, 1980) and subsequently in *Microdochia* (Samuels and Hallett, 1983) which produces trichothecenes.

Another interesting nomenclatural change is that of *F. roseum* which was used by Snyder and Hansen (1945) and commonly used in phytopathological literature has been known to be applied to five species and one variety in three sections of *Fusarium* (Booth, 1971a; Gerlach and Nirenberg, 1982).

Gams and Muller veth (1980) feels that *Fusarium* species be typified by material producing only microconidia, while other argue that macroconidia should be taken as a character for classifying species.

Another group of *Fusarium* of considerable toxicological interest is the *Discolor* section, comprising such species as *F. graminearum* (*Gibberella zeae*), *F. sambucinum (G. pulicaris)* and *F. culmorum*. Snyder and Hansen (1965) devoted much attention to these species, which cause well-known cereal diseases. Largely on the basis that they all attack cereals, species as disparate as these *Discolor* species were grouped together with *F. equiseti (G. intricans)* of the *Gibbosum* section and *F. avenaceum (G. avenancea)* of the *Roseum* section. All these species belonged to different sections (Discolor, Gibbosum, Roseum), but were included by Snyder and Hansen (1945) in their Roseum section. In view of this, a grouping that disregards important morphological and physiological features is completely unacceptable. On the other hand, Snyder and Hansen's treatment of other sections of interest to toxicologists, such as the Arachnites *(F. nivale)*, Liseola *(F. moniliforme)*, and Martiella *(F. solani)* sections, deviates only in less important details from the generally accepted schemes based to some degree on Wollenweber and Reinking (1935). A toxicologist or other scientist who wishes to identify a species of *Fusarium* for toxicological purposes may have difficulty utilizing Snyder and Hansen's (1945) system, since it is oversimplified and tailored for use in plant pathology.

According to Nelson *et al.* (1985) there are 12 sections or groups in *Fusarium*. Only four sections contain most common toxic species. They are: (1) Sporotrichella (*F. sporotrichoides, F. poae*), (2) Gibbosum (*F. equiseti*) (3) Discolor (*F. graminearum, F. culmorum*) and (4) Liseola (*F. moniliforme, F. proliferatum* and *F. subglutinans*). Samson *et al.* (1995) have provided excellent keys for identification of species of *Fusarium*.

The toxigenicity of *F. graminearum, F. sambucinum, F. culmorum, F. equiseti,* and *F. avenaceum* has been studied fairly extensively, especially with regard to the estrogenic and emetic effects observed in swine and some other animals in the United States and toxicoses in humans, animals, and plants. *F. nivale,* which is highly lethal to animals, induced pathological changes, degenerations, and necroses in various organs and produced cytotoxic and strongly irritant effects on rabbit skin. *F. moniliforme* is the most prevalent fungus on corn kernels and causes acute toxicity in different animals. *F. moniliforme* produced a cytotoxic compound with antitumor activity. Neurotoxic syndrome (Wilson and Maronpot, 1971) and necrosis in the brain (leukoencephamalacia) in horses in South Africa (Kellerman *et al.*, 1981). *F. solani* a soil fungus causes numerous destructive plant diseases. *F. solani* also cause intoxication of horses in Japan, characterized by disturbances of the central nervous system, and hemorrhages and crust formation on rabbit skin (Ueno *et al.*, 1972). It was responsible in some cases for corneal disease (keratitis) in man in Florida (Jones *et al.*, 1969). Characteristic features of different sections are given below.

1. Section Arachnites Wr. (*F. nivale, F. larvarum*) Macroconidia on aerial mycelium, rarely in sporodochia or pionnotes, curved, apedicellate. Chlamydospores and sclerotia absent.
2. Section Sporotrichiella Wr. Joffe *(F. poae, F. sporotrichioides, F. sporotrichioides* var. *tricinctum, F. sporotrichioides* var. *chlamydosporum)*: The basis for classification in this section is (a) shape of microconidia, whether lemon or pear shaped, globose, ellipsoid, or elongate, dispersed in aerial mycelium or formed in false heads; (b) the relative frequency of micro – and macroconidia. Macroconidia sparse, small, oblong, narrowly fusoid to falcate, pedicellate, formed in aerial mycelium or in sporodochia. Chlamy-dospores intercalary, terminal, in chains or knots, occasionally with plectenchymatous sclerotia. Perithecial states absent.
3. Section Roseum Wr. (*F. avenaceum, F. arthrosporioides*): Cultures yellow, ochre, carmine, purple or red. Microconidia absent or sparse, Chlamydospores absent, Macroconidia in sporodochia, pionnotes, or on aerial mycelium, sublunate, slender almost filiform, falcate with thin walls, narrowing at both ends, pedicellate. Sclerotia white, yellow, purple to brown. Stroma yellow-red.
4. Section Liseola Wr. (*F. moniliforme, F. moniliforme* var. *subglutinans, F. moniliforme* var. *anthophilum):* Cultures white-cream-brown, organge, violet. Microconidia on aerial mycelium, usually in long chains or small false heads, oval, fusiform, oblong, rarely pyriform. Macroconidia thin walled in sporodochia, pionnotes, or on aerial mycelium, sublunate, spindle shaped to cylindrical, straight or

curved with narrow apex and base cells, typically 3-septate. Chalmydospores absent.

5. Section Gibbosum Wr. em. Joffe (*F. equiseti* and *F. equiseti* var. *acuminatum, F. equiseti* var. *compactum, F. equiseti* var. *caudatum, F. equiseti* var. *longipes*): Culture white-pale pink, pale ocher, olive, carmine-red. Microconidia absent or sparse in aerial mycelium. Macroconidia in pionnotes and sporodochia, falcate, narrowing at both ends with elongated apical cell and well-developed pedicellate foot cell, dorsiventral, parabolic or hyperbolic, typically 5- rarely 3-septate. Chlamydospores intercalary, abundant, smooth or rough walled, single or in chains and knots, yellow-brown.
6. Section Discolor Wr. (*F. heterosporum, F. graminearum, F. sambucinum, F. sambucinum* var. *coeruleum, F. sambucinum* var. *trichothecioides, F. culmorum, F. tumidum)* : Cultures white-rose, peach, grayish, rose, red to brown. Microconidia absent; Macroconidia thick-walled, in aerial mycelium, sporodochia and pionnotes, either broad, falcate with short apical cell, and well developed foot cell, spindle or sickle-shaped with elongated gradually narrowing apical cell, well-marked basal cell, typically 5-septate. Chlamydospores intercalary, sometimes terminal, often in knots and chains. Sclerotia purple-blue, brown to dark. Stroma yellow.
7. Section Martiella Wr. em. Joffe Palti (*F. solani, F. solani* var. *coeruleum, F. solani* var. *ventricosum, F. javanicum*): Cultures white, cream, orange-blue to brown. Microconidia abundant, oval or oblong, hyaline. Macro-conidia in aerial mycelium, sporodochia or pionnotes, fusoid, cylindrical, curved or elongate, with thick walls and short, rounded apical cells and foot cells. Chlamydospores globose, oval, smooth or rough-walled, terminal and intercalary, single or in pairs, short chains or knots.

Key to mycotoxigenic species of *Fusarium*

1. Microconidia very heterogeneous, singly or in false heads or in short chains. 2

1a. Microconidia abundant, broadly oval, elongated, singly or in false heads. 7

2. Microconidia elongated in chains usually with a broad and rounded apex 8

2b. Microconida absent or rare..11

2. Microconidia globose, spherical with basal papilla, rarely pear-shaped 3

2a. Microconidia pyriform to ellipsoid, globose or elongate..4

2b. Microconidia narrowly clavate with rounded apex, pyriform to fusiform, not spherical or globose 6

3. Microconidia abundant. Macroconidia very sparse, small, curved, falcate, usually without foot cell, typically 3-septate, 19 to 38 x 3.5 to 6.8 μm. Sporodochia and pionnotes absent (Fig. 72) .... *F. poae* (1)

3a. Microconidia often as numerous as macroconidia. Macroconidia formed in sporodochia and in aerial mycelium, curved, falcate, elongated, and narrowly fusoid. Typically 3-to 5-septate, 22 to 45 x 3.6 to5.2 μm (Fig. 73).......... *F. sporotrichioides* (2)

4. Microconidia more abundant than macroconidia. The latter formed in sporodochia, rarely in aerial mycelium, curved, spindly ellipsoid, falcate, 3-to 5-septate, 23 to 50 x 3.3 to 4.4 μm. Typically 3-septate (Fig. 74) .... *F. tricinectum* (3)

4a. Microconidia more or less numerous than macroconidia, the latter formed in aerial mycelium, curved, fusoid with narrowly pointed apex and marked foot cell. Sporodochia absent or very rare, 3- to 5-septate, 28 to 42 x 3.2 to 4.5 μm. Typically 3-septate (Fig. 75).......... *F. chlamydosporum* (4)

7. Macroconidia elliptical, sausage-shaped, elongated, slightly curved, with round apical and round basal cells or marked foot cell, 3 to 5-septate, 27 to 52 x 4.4 to 6.8 μm. Typically 3-septate (Fig .76).......... *F. solani* (5)

7a.

8. Culture peach to violet. Macroconidia spindly, elongated, cylindrical, straight, or slightly falcate with narrow curved apical cell and marked foot cell, 3-to 5-septate, 29 to 67 x 2.6 to 4.0 μm. Typically 3-septate Microconidia monophialidic (Fig.77).......... *F. moniliforme* 6

8a. Microconidia are produced from polyphialidic conidiophores .... *F. proliferatum* (7)

9. Culture light-cream, dark-yellow to brown. Microconidia rare, short, oval, spindle-shaped or clavate, 0 to 2-septate. Macroconidia falcate to parabolic with more or less gradually or sharply narrowing elongated, curved or non-curved apical cell and well-marked foot cell, 3 to 7-septate, 27 to 62 x 3.5 to 5.5 μm. Typically 5-septate (Fig. 78) .... *F. equiseti* (8)

9a. Cultures from below sulphurous conidia 3-4 septate, 23 – 32 x 3.5 – 4.0 μm (Fig. 79) .... *F. sulphureum* (9)

10. Culture sometimes whitish, mostly strongly colored, rose, red, dark red, or crimson, Microconidia rare, appear initially in young aerial mycelium, short elliptical to spindle-shaped. Macroconidia elongated up to 90 μm, long narrow, filiform or slightly hyperbolic, curved, with elongated narrowing apical cell and well-marked foot cell, 3 to 7-septate, 31 to 92 x 2.5 to 4.1 μm. Typically 5-septate (Fig. 80) .... *F. avenaceum* (10)

Microconidia absent. Macroconidia usually in aerial mycelium, small, often sparse, 1 to 3-septate, curved, sometimes aseptate

10a. Microconida with an ovate or hooked apical cell macroconidia curved, beaked or elongated apical cells (Fig. 81).......... *F. lateritium* (11)

11. Culture light-colored, white, light-rose, or yellow to place, slow-growing. Macroconidia short, elliptical, curved, sickle-shaped with pointed apex and wedge-shaped or round basal cell, 0 to 3-septate, 8 to 32 x 2.2 to 4.8 μm (Fig .82) .... *F. nivale* (12)

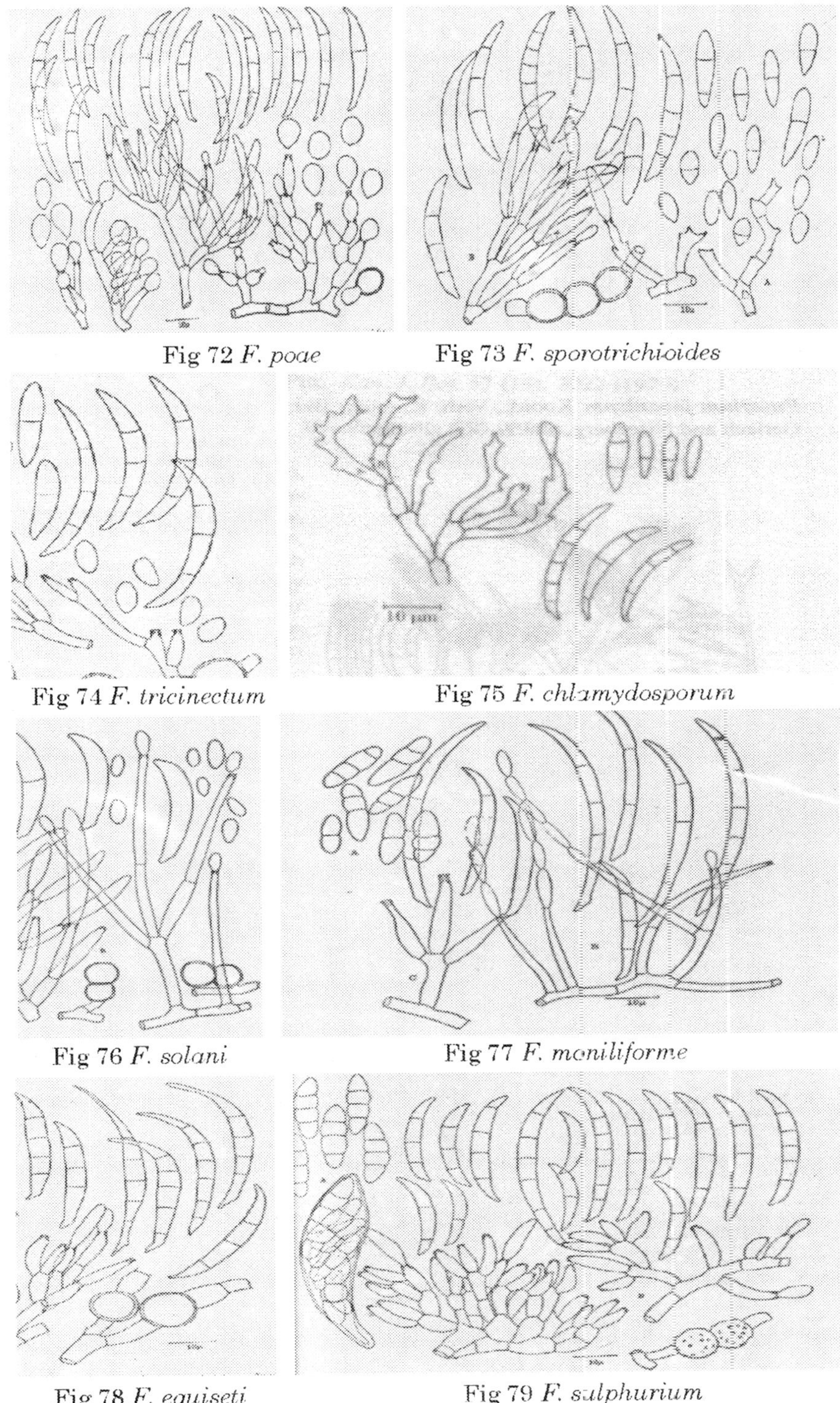

Fig 72 *F. poae*

Fig 73 *F. sporotrichioides*

Fig 74 *F. tricinectum*

Fig 75 *F. chlamydosporum*

Fig 76 *F. solani*

Fig 77 *F. moniliforme*

Fig 78 *F. equiseti*

Fig 79 *F. sulphurium*

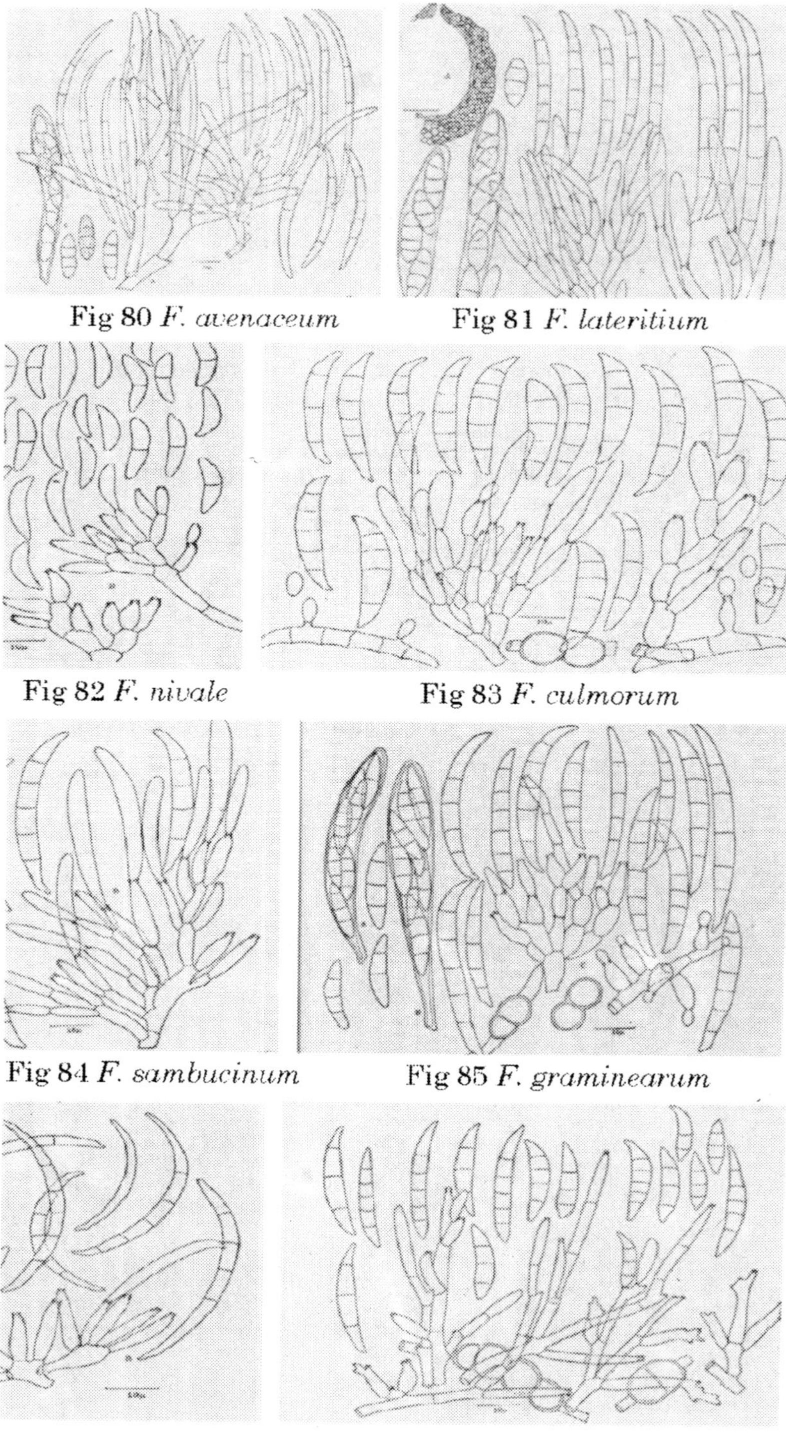

Fig 80 *F. avenaceum*

Fig 81 *F. lateritium*

Fig 82 *F. nivale*

Fig 83 *F. culmorum*

Fig 84 *F. sambucinum*

Fig 85 *F. graminearum*

Fig 86 *F. acumiatum*

Fig 87 *F. semitectum*

11a. Microconidia absent. Macroconidia typically 5-septate, mostly abundant, slightly curved, strongly dorsiventral, with a short and wide pointed apex and well-marked foot cell

12. Macroconidia 3-to 7-septate, broad, 25 to 68 x 4.8 to 82 µm. (Fig. 83) .... *F. culmorum* (13)

12a. Macroconidia less broad, cylindrical, curved, strongly dorsiventral, falcate, elliptic or sickle-shaped with short beaked apical cell or slightly curved and gradually narrowing, elongated apical cell, and well-marked foot cell.

13. Macroconidia cylindrical, curved, strongly dorsiventral, fusoid with short beaked apical cell, 3-to 7-septate, 25 to 53 x 3.8 to 5.9 µm. (Fig. 84).......... *F. sambucinum* (14)

13a. Macroconidia elongated up to 78 µm, spindle-or sickle-shaped, slightly dorsiventral, curved with elongated apical cell, 3 to 7-septate, 28 to 78 x 3.2 to 5.0 µm (Fig. 85) .... *F. graminearum* (15)

Apical cell of macroconidia elongated narrowing eventy to a point with chlamydospores formed in knots or chains. (Fig .86) .... *F. acuminatum* (16)

Microconidia present or absent macroconidiate usually present macroconidia with wedge shaped foot cell, cultures from below beige brown becoming dark-brown (Fig. 87) .... *F. semitectum* (17)

Cultural and morphological characters of some of the mycotoxigenic species of *Fusrium* are précised in Table. 2.5.

Table 2.5. Cultural and morphological characters of mycotoxigenic species of *Fusarium*

Name of the species	Growth rate (in mm)	Culture Pigmentation	Microconidia	Macroconidia	Chlamydospores
Fusarium avenaceum	54	Rose, Red with White	1-3 Septate 8-50 x 3.5 – 4.5 µm	Primary macroconidia polyblastic, secondary macroconidia from phialides, 4-7 septate, 40-80 x 3.5 – 4.5 µm	Absent, rarely formed in chains
F. acuminatum	45	Saffron to by to carmine to red		3-7 septate, 30-70 x 3.5 - 5 µm curved elongation of the apical cell produced from phialids.	Globose, intercalary in knots or chains
F. culmorum	85	Red becoming reddish- brown		Uniform in shape, Borne on simple phialides, borne on complex, loosely branched conidiophores, 3-septate.	Oval to globose intercalary, singly or in chains, smooth to rough wall.
F. equiseti	58	Peach usually changing to vellaneous and finally becoming buff-brown.		Conidia produced from single solitary or grouped phialides, conidia 4-7 septate, 20-60 x 3.5-9 µm. Often with elongated apical cell.	Globose intercalary in solitary chains or clumps

F. graminearum	89	Greyish, yellow or white, rose, coral becoming vinaceous with brown tinge.		Single, lateral, sickle-shaped or falcate, 3-7 septate, simple lateral phialides may or may not be grouped on branched conidiophores.	Absent or rare if present intercalary.
F. later-itium	28	Variable peach to deep orange, vinaceous to reddish-brown, greenish-yellow to bluiseblack	Absent	Straight, beaked at the apex produced in sporodocnia, phialides simple 7-10 x 7-8 μm	Sparsley oval to globose, interclary
F. moniliforme	46	Vinaceous, purple to violet, peach solmon	Fusiform to clavate, 0-1 septate, conidia from subulate, lateral phialides, in chain, 5-12 x 1.5-2.5 μm	Elongated, sharply curved, equilateraly fusoid, 25-60 x 2.5-4 μm	Absent but globose stromatic initial cells may be present.
F. nivale	13	White to pale-peach	Absent	Curved broadly falcate 1 to 3 septate, 10-30 x 2.5-5μm	Absent
F. oxysporum	45	White, peach, salmon, vinaceous grey to purple, violet	Cylindrical to curved, oval ellipsoid, produced from simple, short lateral phialides, 5-12 x 2.2-3.5μm.	Hooked apex & pedicellate base, 3-7 septate 27-66 x 3.5μm	Globose, solitary or in pair. Intercalary formed on short lateral branch.
F. poae	76	White to yellow, salmon to livid-red, vinacous	Globose, single spore or small globes to obclavate phiaides, 8-12 x 7-10 um, O-septate.	Single spore curved, 3- septate 20-40 x 3-4.5 μm	Globose, formed sparsely, single or in chains in old cultures.
F. sambucinum	52	Peach to orange or ochreous, white in some isolates vinaceous to bay.	Absent	Strongly dorsiventral with beaked apical cell, curved pointed apex, marked foot cell, 3-5 septate 30-55 x 4-5.5 μm	Formed sparsely globose, single or chains or knots intercalary.
F. semitectum	61	Peach changing to buff-brown		Two types. 1) Primary conidia 0-5 septate, blastospores, polyblastic. 2) Secondary conidia 3-7 septate, typically heeled foot cell.	Sparse, globose, intercalary, single or in chains.

F. solani	32	Greyish-white to blue or bluish-brown	Single, cylindrical to ovate, 1-septate, borne on lateral phialides.	Curved, 1-9 septate inequilaterally fusoid.	Globose, singly or in pairs on short lateral branches or intercalary.
F. sporotrichloides	35	Surface mycelium white and red agarlivid-red becoming with brown	Clavate to cymbiform, 0-septate, blastic spores from pegs.	An incurved, 3-7 septate 3-0-7.0 x 3.5-5μm	Globose intercalary chains
F. sulphureum	64	Cream or light-brown or reddish-brown		Curved pointed apical cell epedicellate foot cell, 3-5 septate, 14-15 x 3.5 μm	Globose short chain or single formed sparsely.
F. proliferatum		Heavy white aerial mycelcium that may become tinded with purple	Single celled with flattend base in short or varying length change or in false head for polyphialides	Abundantly, 3-5 septa slightly curve to almost straight basal cell foot shaped teleomorph G fuzikuroi	No Chlamydospores

Description of mycotoxigenic species of *Fusarium*

1. *F. sporotrichioides* Sherb: It is responsible for ATA in USSR, scabby grain intoxication, bean hull poisoning in Japan, mouldy corn toxicosis in USA and fescue foot disease in USA, Australia and New Zealand (Marasas *et al.*, 1984). It belongs to section *sporotrichiella*. Marasas *et al.* (1984) have given a clear cut distinction between *F. sporotrichoides* and *F. tricinctum*:

Colonies of *F. sporotrichioides* grow rapidly on CYA, MEA, DDA or DCPA are deep and floccose with mycelium coloured pale-pink or salmon and reverse on PDA grayish-rose to burgundy. Conidiophores are monophiladic, macroconidia, microconidia are abundant on DCPA. Microconidia both fusiform and pear shaped are borne on phialides with more than one fertile pore telemorphic stage is reported. Produces abundant chlamydspores in singly and in chains and bunches.

Besides T_2 toxin, it produces butenolide, fusarinon X, neosolanol and nivalenol Zearalenone, deoxynevalenol and some less well characterized trichothecenes occur less frequently. *F. tricinctum* and *F. chlamydosporum* are other members of this section.

2. *Fusarium moniliforme* Sheldone: also known as *F. verticilloides* was described on corn more than a century ago. Its possible involvement in human or animal disease also dates back to a century (Marasas *et al.*, 1984).

Its involvement in equineleukoencephalamalacea was also known as early as 1850.

F. moniliforme has been recognized as a distinct entity for many years. Disagreement still exists over the correct name, *F. verticilloides* has nomenclature priority. Nelson *et al.* (1985) disagree on the ground that the provisions of the ICBN are difficult to apply to the *Fusarium* in the absence of type material. Eventual conservation of *F. moniliforme* seems likely. Its teleomorph is *Gibberella fuzikuroi*.

Colonies are 3-4 cm in diameter, mycelium floccose powdery. Colonies of *F. moniliforme* on PDA are white sometimes tinged with purple. Monophialidic conidiophores. Macroconidia vary from slightly sickle shaped to almost straight, 3-5-7 septate. Microconidia are abundant and are usually single celled, ellipsoid to clavate with a flattened base and formed in long chains. No chlamydospores are produced. *F. moniliforme* is primarily a plant pathogen causing both stalk and cob-rot of corn and disease in rice, sorghum, sugarcane and other gramineae. It is much more common in tropics than in temperate climates.

3. *F. poae* Wollenw: is widespread in soils of temperate climate and on grains such as wheat, corn and barley. It is mainly a saprophyte but may be a weak parasite. It is also reported in warmer regions such as India, Australia, Iraq and South Africa. It is reported to be associated with ATA, a hemorrhage syndrome, urov or kashin beck disease and human esophageal cancer in Northern China.

F. poae produces rapid, profuse white to pink mycelial growth on PDA with a red to very deep carmine-red reverse coloration. The most distinctive feature of *F. poae* is its production of abundant globose to oval, almost papiform (Pear shaped) microconidia with a few macroconidia conidia produced on branched or unbranched monophialides. Chlamydospores are produced infrequently. No teleomorphic stage has been observed.

4. *F. proliferatum* is closely related to *F. moniliforme* and less known possibly because of its frequent misidentification as *F. moniliforme*. Though it produces fumonisins, it has not so far been reported to be associated with animal or human diseases. It is also reported to produce moniliformin and fusaric acid.

On PDA it grows rapidly to produce heavy white aerial mycelium that become tinged with purple. Macroconidia are produced abundantly, long and thin, have 3-speta and are only slightly curved to almost straight. The basal cell is foot shaped. Microconidia are single celled with a flattened base and are produced in short to varying length chains or in false heads from polyphialides (more than one opening) which distinguishes *F. proliferatum* from *F. moniliforme*. Chlamydospores are not produced. Teleomorphic state is not reported.

5. *F. subglutinans* is similar to both *F. moniliforme* and *F. proliferatum*. Very little information is available about this fungus which may be again attributed to misidentification. Culture of *F. subglutinans* has been shown to be acutely toxic to ducklings and rats and to be dermatoxic to rabbit skin. It does not produce fumonisins and its toxicity may be due to other metabolites such as moniliformin.

F. subglutinans grows rapidly on PDA forming whie aereal mycelium that may be tinged with purple. The reverse of the colony may range from colorless to dark-purple. Macroconidia are long, slender and almost straight to slightly curved with foot shaped basal cell and septa 3-5, microconidia are oval and usually single celled (but may have 2-3 septa) and are produced only in false heads. The teleomorphic stage is *Gibberella subglutinans* (synonys *F. fuzilcurai* var. *subglutinans*). *F. anthophilum* is another species in this section.

6. *F. equiseti* Saccardo: a broad spectrum plant pathogen and wide spread soil sapryolyte, has a long history of association with animal disease and a possible implication in human leukaemia. Nelson *et al.* (1985) classified *F. equseti* in Gibbosum sections. It has a teteomorph called *Gibberella intricans.* Earlier it used to be called *F. roseum* sensu Snyder and Hansen (1945), *F. roseum* 'gibbosum' *F. roseum-avenaceum*, *F. roseum-culmorum or F. roseum-equiseti* and several other distinct species as well. Hence, it is difficult to determine which of the species earlier referred to *F. equiseti* (Marasas *et al.*, 1984).

Colonies of *F. equiseti* on CYA, MEA and PDA usually cover the whole petri dish, while those on DCPA are smaller. Mycelium and colony reverses are not strongly coloured but white or pale-salmon to brown. Macroconidia are distinctly curved (hunch backed), microconidia are not produced.

Mycotoxins produced by *F. equseti* include nivalenol, fusarenon X_1 T-2, diacetoxyscirpenol, butenolide, zearalenone and several others (Marasas *et al.*, 1984). Several of these may be produced simultaneously.

Other species of the section are *F. scirpi, F. acuminatum* and *F. longipes, F. graminearum* and cause estrogenic syndromes, feed refusal and emetic syndromes in pigs and other animals and is very likely to be the cause of human Akakabi-byo (Scabby grown intoxication) in Japan (Yoshizawa, 1983).

7. *F. culmorum* Sacc: Causes Bean hull poisoning in Japan. Mycotoxins reported to be produced by *F. culmorum* include deoxynivalenol, zearalenone and acetyl deoxynivalenol. Growth of *F. culmorum* on PDA is rapid with the formation of dense aerial mycelia that are white with some tinges of yellow and brown. The reverse of the colony is a deep carmine-red colour.

Sporodochia that are orange to red -brown may be produced in old cultures, while *F. culmorum* may resemble *F. graminearum* in macroscopic appearance when growing on PDA. The macroconidia are quite different. Macroconidia are very short and stout and have 3-5 speta. The macroconidia have curved dorsal and ventral surfaces and the basal cell varies from being foot shaped to having notched appearance. Chlamydospores are formed readily in the macroconidia and mycelia. Microconidia are absent. No teleomorphic state has also been observed. *F. heterosporum, F. reticulatum, F. sambucinum* and *F. crookswellense* are other species in section discolor. Abnormal bone development or rickens in chickens and pigs in different countries and a toxicosis due to mouldy weed potatoes in USA (Marasas *et al.,* 1984) are due to this fungus.

F. culmorum is in many ways similar to *F. graminearum* and in the part has been with *F. graminearum* is the non-existant species. *F. roseum* and *F. culmorum, F. roseum* var. *culmorum* are other species included in this section.

8. *F. graminearum* Schwabe: is placed in the section Dicolor and has a well recognized teleomorph *Gibberella zeae*.

Colonies of the fungus fills the petri dish when grown on CYA, MEA or PDA for 7 d, while colonies of DCPA are smaller. On CYA and MEA colonies are floccose in muted or pastel shades of grayish-red or yellow. On PDA colonies are usually highly coloured with dense to floccose grayish-rose to golden-brown mycelium and a dark-ruby reverse, while on DCPA colony appearance is dominated by salmon to orange clusters of macroconidia often in concentric rings. Macroconidia are relatively straight and thick walled with a foot-shaped basal cell. Monophialidic conidiophores, microconidia are not produced. Chlamydospores are formed abundantly singly or in pairs, chains or clumps.

F. graminearum is primarily a pathogen of graminaceous plants particularly wheat causing crown-rot at the base of the stem and head scab in developing grain. It also causes cob-rot of corn in many countries (Marasas *et al.*, 1984) and uncommon in other foods (Pitt and Hocking, 1985).

2.3.4. Alternaria (Deuteromycotina, Hyphomyctes, Dematiaceae)

Colonies effuse usually grey, dark blackish-brown or black. Conidiophores mcaronematous, mononematous, simple or irregualry and loosely branched, pale-brown or brown, solitary or in conidiogenous cell, integrate terminal, becoming, intecalary, polytretic, sympoidal sometimes monotretic cicatrized. Conidia catenate or solitary dry typically ovoid or obclavate, often rostrate, pale to mid olivaceous brown or brown smooth or verrucose with trans verrucose and frequently also oblique and long longitudinal septa (Fig.

26). Species of *Alternaria* produce Alteraniol, Alternarion methylester, Alternuene, isoaltneuene, ATX-1, ATX-II and Tenuazonic acid.

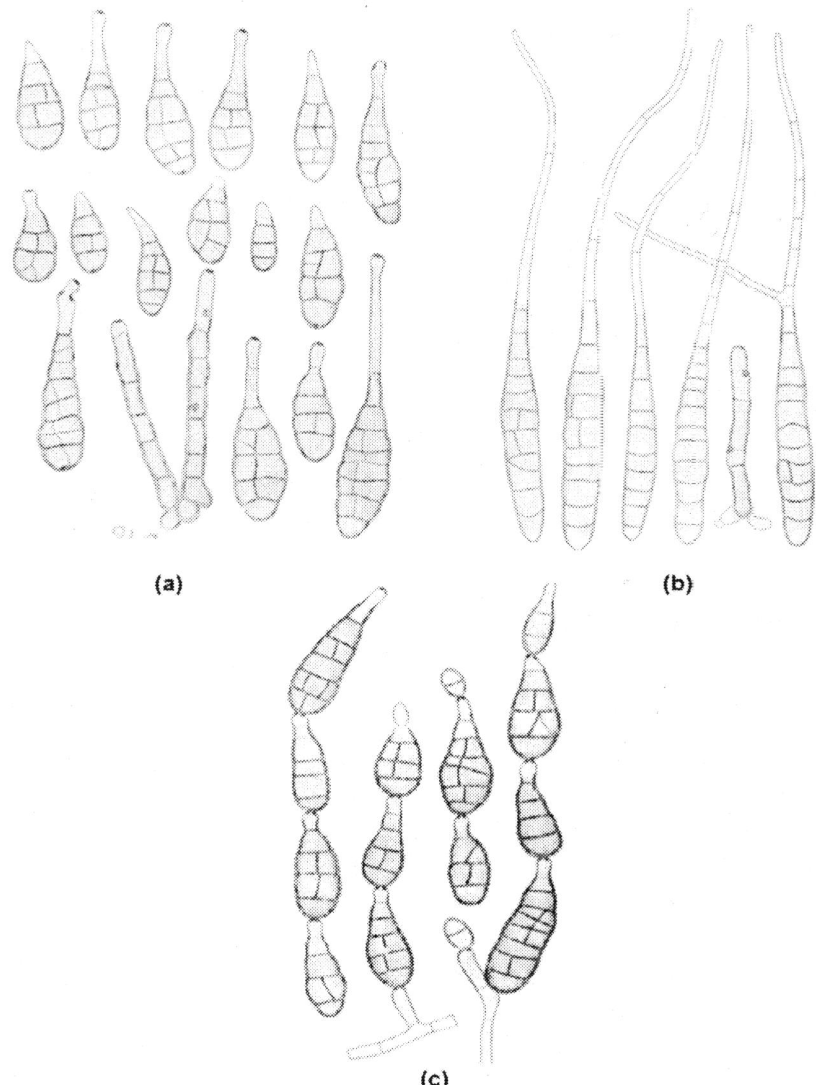

Fig. 26 (a) *Alternaria tenuissima* (b) *Alternaria solani* (c) *Alternaria alternata*

Key to species

1. Conidia solitary or in short chains 2

1a Conidia in long chain 3

2. Conidia solitary or in short chains obclavate, beak of the length of the conidium, golden brown 20-95mM (Fig. 26a) .... *A. tenuissima* (1)

2a. Conidia in short chain shape variable (Fig. 26b) .... *A. solani* (2)

3. Conidia mostly in chains often or more chains branched, Conidia irregularly shaped, beak short not exceeding 1/3 of conidial length, medium golden brown (Fig. 26c) .... *A. alternata* (3)

3a. *A. kukichiana* .... (4)

2.3.5. STACHYBOTRYS (Deuteromycotina, Hyphomycetes, Dematiaceae)

Stachybotrys is a saprophytic and world wide in distribution. Soviet scientists have described this fungus under the name *S. alternans* Bonorden var. *jateli* with two different variants, one toxic and other nontoxic, while in the literature of other parts of the words used the name *S. atra* Cords and in a few cases as *S. charatarum* (Ehrenberg ex Link) Hughes. *S. atra* is a typical cellulose-decomposing fungus that grows well on cellulose-rich substrates such as straw and hay, but it has also been isolated from cereal grains, plant debris, soil, peas, cotton, and sugar cane roots.

The aerial mycelium is colorless. The colonies on agar become black on standing, except for the peripheral zone, which remains grayish-white with abundant aerial mycelium. *S. atra* can grow at pH levels of 3.0 to 9.8, although the optimum pH values are 5.6 to 6.6. The optimum temperature for growth is 20 to 25°C, although the fungus grows within the range of 5 to 40°C.

Young hyphae are hyaline, become olive gray-brown, and 2 to 4 μm in thickness. Conidiophores are olive-brown or olive-gray, 30 to 70 μm long, and 3 to 5 μm wide with a few septa, either simple or sympodially branched. The upper, usually darker part of the conidiophore bears a group of phialides (sterigmata). The phialides are hyaline or dark-colored, 7 to 12 μm long, and 3 to 5 μm wide, varying from 6 to 10 in number (Fig. 34). The conidia, which are produced at the small tips of the phialides, differ greatly in size: 6 to 14 μm x 4 to 9 μm. Conidia are oval or rounded hyaline or dark-colored, smooth-walled or occasionally showing short spiny structures.

Key to the species of *Stachybotrys*

1. Conidia mostly more than 20 μm long (Fig.34.1a) .... *S. theobromae* (1)

1a. Conidia always less than 17 μm.......... 2

2. Conidia reniform.......... 3

2a. Conidia not reniform 4

3. Phialides brone in 1 or 2 rings around the swollenapex of the Conidiophore .... *S. oenanthes* (2)

3a. Phialides borne in one ring, simple, dark-green and verrucose apex of the conidiophore (Fig. 34 b) .... *S. nephrospora* (3)

4. Conidia with dark longitudinal striations (Fig. 34.1c) .... *S. cylindrospora* (4)

4a. Conidia without striations 5

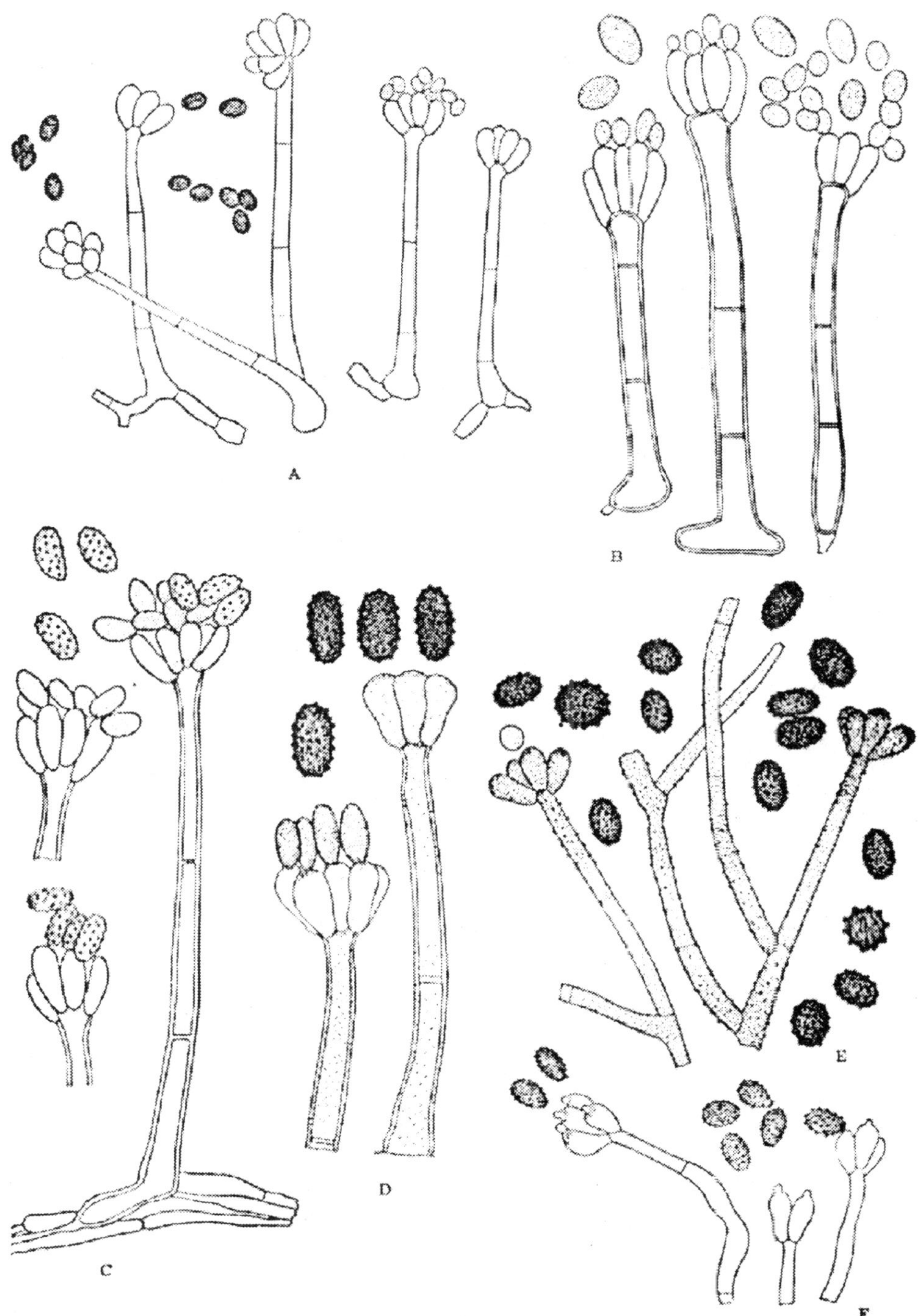

Fig. 34 /1 (a) *S. parvispora* (b) *S. Melanopsamma pomiformis*; (c) *S. dichroa* (d) *S. kampalensis* (e) *S. atra* var. *microspora*

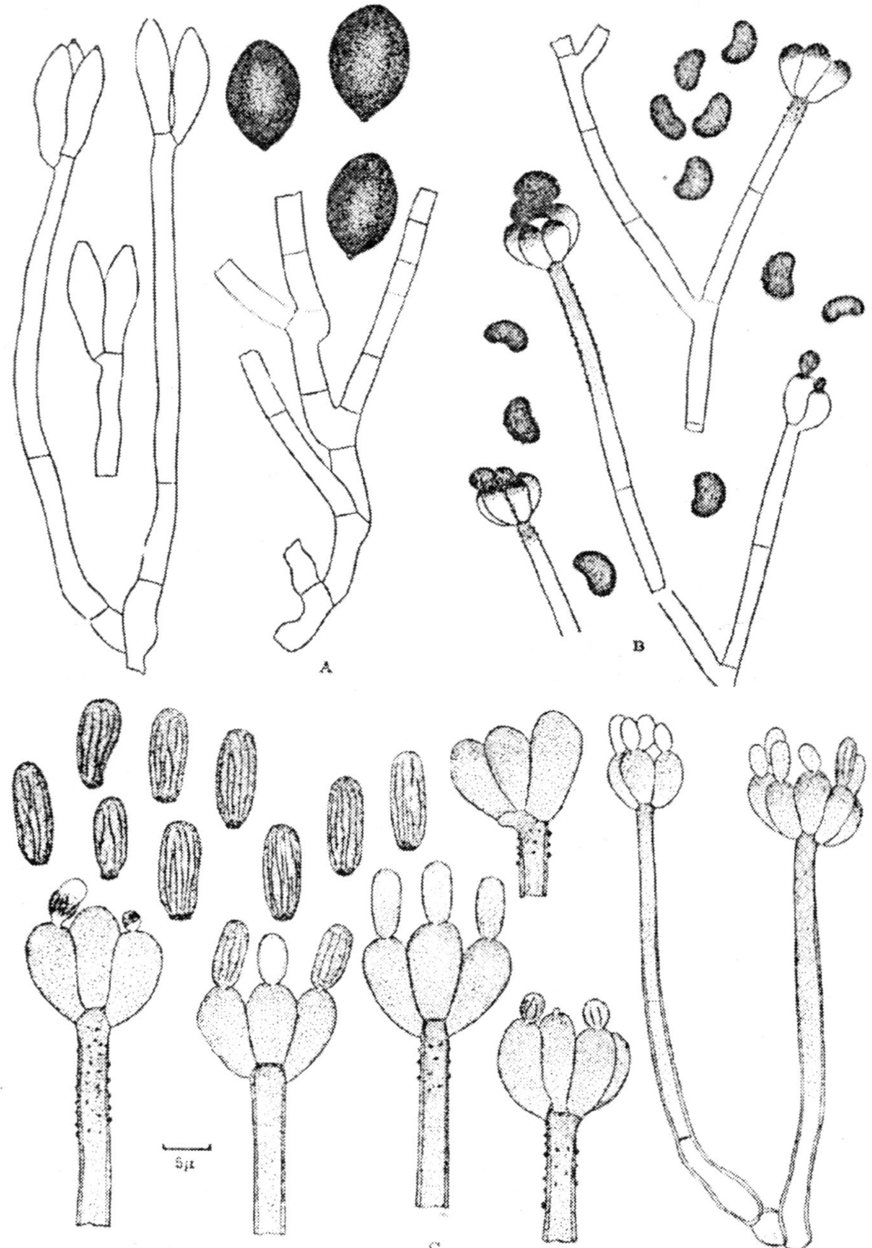

Fig. 34/2 (a) *S.theobromae* (b) *S. nephrospora* (c) *S. cylindrospora*

5. Conidia when mature smooth 6

5a. Conidia when mature verrucose 8

6. Conidia ellipsoidal or boat shaped truncate at the base .... *S. sanseveriae* (5)

6a. Conidia ellipsoidal not truncate at the base.. 7

7. Conidia dark-brown, 3.0-5.5 μm long.(Fig. 34.2 d) .... *S. parvispora* (6)

7a. Conidia greenish or grayish-brown 6.0-11.0 μm long (Fig 34.2 e) .... *Melanopsamma pomiformis* .... (7)

8. Conidia often obliquely attenuated at the base, 8-12 x 4-6 μm (Fig. 34.2 f) *S. dichroa* .... (8)

8a. Conidia not obliquely attenuated at the base 9

9. Conidia 11-15 x 6-8 μm (Fig.34.2 g) .... *S. kampalensis* (9)

9a. Conidia 8-11 x 5-11 μm (Fig.34.2 h) .... *S. atra* (10)

9b. Conidia 6-8 x 4-5 μm (Fig.34.2 i) .... *S. atra* var. *microspora* (11)

2.3.6. *Myrothecium* (Deuteromycotina, Hyphomycetes, Dematiaceae)

The genus *Myrothecium* includes those imperfect molds in which the fructification is a sporodochium of hyaline, compacted, irregularly branched conidiophores bearing unicellular slimy phialospores which are green to black in mass. Two other genera, *Metarrhizium* (Metsch.) Sorokin and *Dendrodochium* Bonorden, include species in which unicellular spores, green to black in mass, are borne on branched conidiophores grouped in sporodochia. *Metarrhizium* isolates can be distinguished from those of *Myrothecium* by their dry spores which adhere in tall columns, but green-spored *Dendrodochium* spp., including *D. toxicum* Pid. and Bilai, are very similar to *Myrothecium,* and Tulloch (1972) considers that it is necessary to extend the present concept of *Myrothecium* to include them.

Three of the 13 species included in *Myrothecium* by Tulloch (1972) have been reported to produce material toxic to mammals. *M. roridum M. verrucaria* and *M. leucotrichum* (synonyms: *M. jollymannii* preston, *M. indicum* Rama Rao). *M. leucotrichum* is distinguished from the other two toxin-producing species by the hyaline, thin-walled, several-celled setae which surround its sporodochia. *M. verrucaria* by fusiform spores, 6.5 to 8.0 μm in length, and *M. roridum* by rod-shaped or narrowly ellipsoid spores, 5.5 to 7.0 μm in length (Fig. 14).

Key to the species of *Myrothecium*

1. Conidiophores in synnemata or sporodochium .... *M. masonii* (1)

1a. Conidiophores single...2

2. Conidia not more than 4 μm long on agarics (Fig. 14a) .... *M. inundatum* (2)

2a. Conidia more than 4 μm...3

3. Setae present conspicuous mostly non-septate (Fig. 14b) .... *M. gramineum* 4

3a. Setae absent or rarely seen..5

4. Conidia with longitudinal striations .... *M. cinctum* (4)

4a. Conidia without longitudinal striations...5

5. Conidia slightly curved, sporodochia stalked (Fig. 14c) ....*Nectria bacteridoides* (5)
5a. Conidia straight, sporodochia usually sessile 6
6. Conidia cylindrical or narrowly ellipsoidal 7
6a. Conidia navicular, limoniform or broadly ellipsoidal 8
7. Conidia 5.7 x 1.5-2.5µm, usually on Coffea (Fig. 14d) ....*M. advena* (6)
7a. Conidia mostly 6-8 x 1.5-2.5 µm (Fig. 14e) ....*M. roridum* (7)
7b. Conidia mostly 10-11 x 1.3 µm ....*M. carmichaelii* (8)
8. Conidia 6-10 x 1.5-2 µm (Fig. 14f) .... *M. verrucaria* (9)
8a. Conidia 12-17 x 7-9 µm ....*Nectria ralfsii* (10)

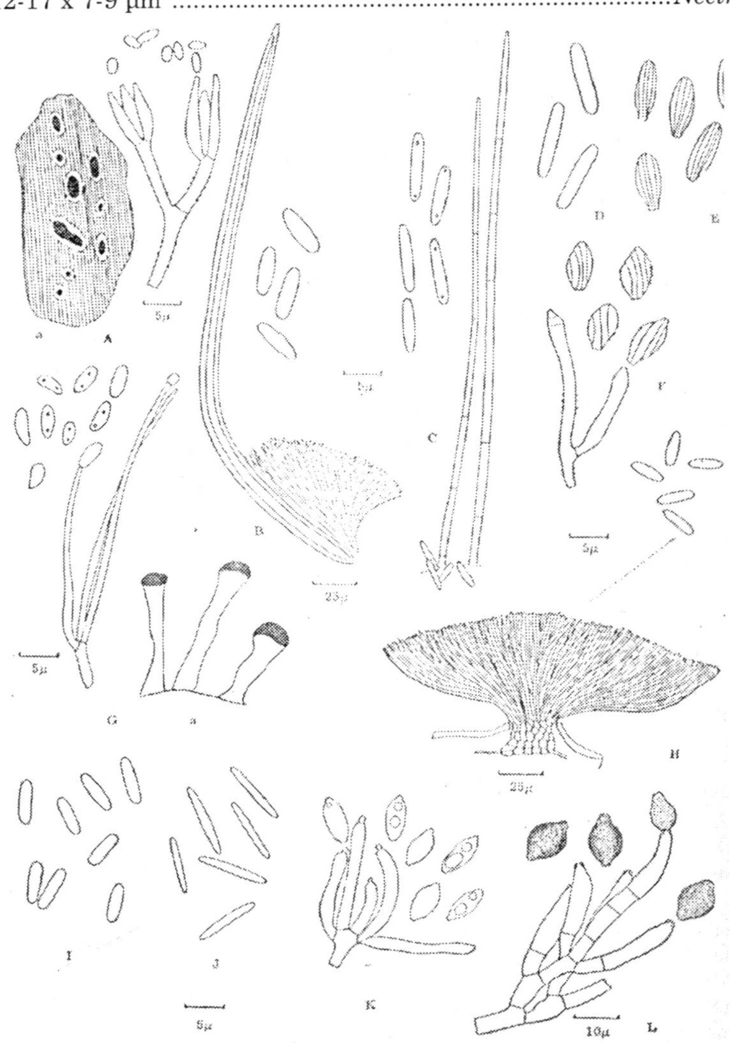

Fig 14. a. *M. inundatum*; b. *M. gramineum*; c. *M. indicum*; d. *M. jollymannii*; e. *M. striatisporum*; f. *M brachysporum*; g. Annomorph of *Necteria bactridiodes*; h. *M. advena*; i. *M. roridum*; j. *M. camichaelii*; k. *M. verrucaria*; l. anamorph of *Nectria ralfsii*

2.3.7. *Trichoderma* (Deuteromycotina, Hyphomycetes, Moniliaceae)

Colonies spread, repeated, branched, hyphae hyaline, septate, conidiophore in tufts with divergent often irregularly branched, and flask shaped phialides. Conidiophores may end in a sterile appendage with phialides borne on lateral branches in some species. Conidia hyaline or usually green, o-septate, globose to spherical (Fig. 33). Hyaline chlamydospores are usually present in the mycelium of older cultures.

Ishikawa *et al.* (1973) have reported production of trichodermin and trichoderminol by *T. hamatum*, *T. harzianum*. *Hypocrea* is the perfect stage of some species belonging to *Trichoderma,* and *Trichothecium roseum*.

Key to the species of Trichoderma

1. Conidiophore and branches short and stout often with sterile hyphal elongation bearing crowded phialides Colonies white, whitish-green or green..............................2

1a. Conidiophore and its brnahces strong and slender without sterile hyphae elongation, phialides not crowded. Colonies yellowish, bright-dull to dark-green 5

2. Sterile hyphal elongation absent, conidia lobose, hyaline..................*T. piluliferum* (1)

2b. Sterile hyphal elangation present or modified or rarely absent, conidia not globose..3

3. Conidia green, short, ellipsoidal, surrounded by a wide irregular veil ....*T. saturnisporum* (2)

3a. Conidia smooth walled or finally punctuated 4

4. Conidia hyaline, small, 2.4-3.8 x 1.8-2.2 µm .... *T. polysporum* (3)

4b. Conidia green, small to large, 3.8-6.0 x 2.2-2.8 µm..............................*T. homatum* (4)

5. Conidia rough, 3.6-4.8 x 3.5-4.5 µm ....*T. viride* (6)

5a. Conidia smooth walled 6

6. Conidiophore with complicated dendroid branching, Phialides regularly disposed in no. f 3 or more..7

6a. Conidiophore with simple branching system, phialides irregularly laterally disposed often arising singly..9

7. Conidia ellipsoidal or oblong often appearing angular 3.0-4.8 x 1.9-2.8 µm ....*T. koningii* (6)

7a. Conidia shorter with a length: width ratio of less than 1.5 µm 8

8. Conidia obovoid with truncate ....*T. aureoviride* (7)

8a. Conidia shorter with a length: width ratio of less short obovoid .... *T. harzianum (8)*

9. Conidia subglobose to ovoid ....*T. reesei* (9)

9a. Conidia ellipsoidal 10

10. Phialides usually distinctly attenuate at the base, conidia smaller, pale green 2.8-4.8 µm, mostly oblong ellipsoidal ....*T. pseudokoningii (10)*

10a. Phialides usually only slightly attenuate at the base, conidia large, partly dark-green upto 7.0 µm long mostly ellipsoidal............*T. longibrachiatum (11)*

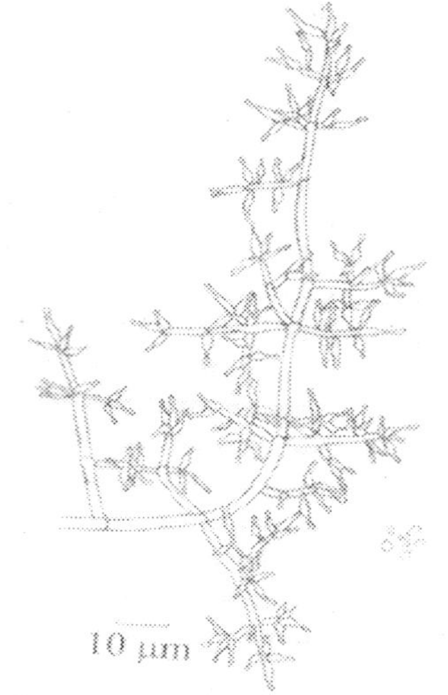

a. *T. aureoviride*

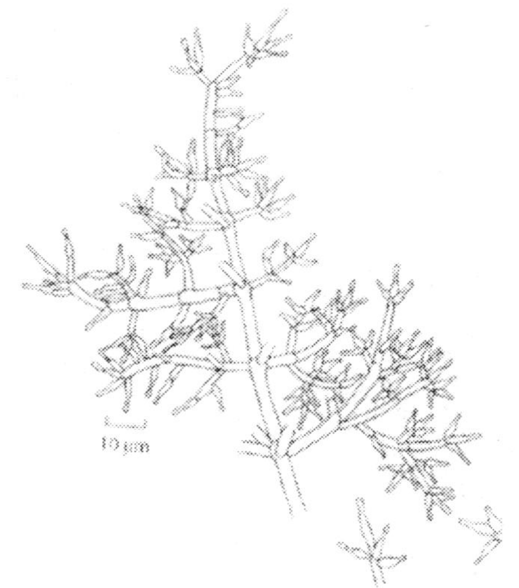

b. *T. harzianum*

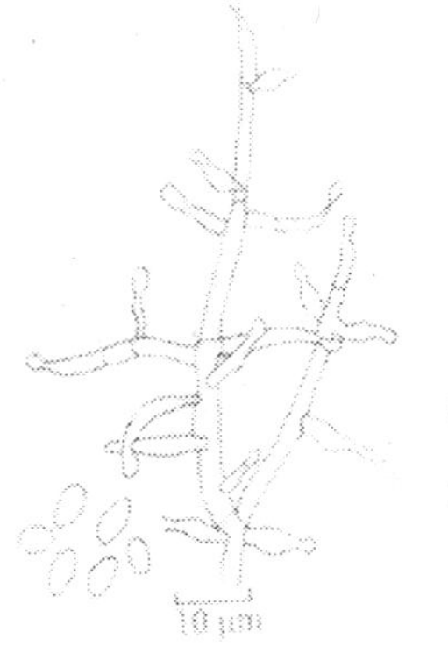

c. *T. longibrachiatum*

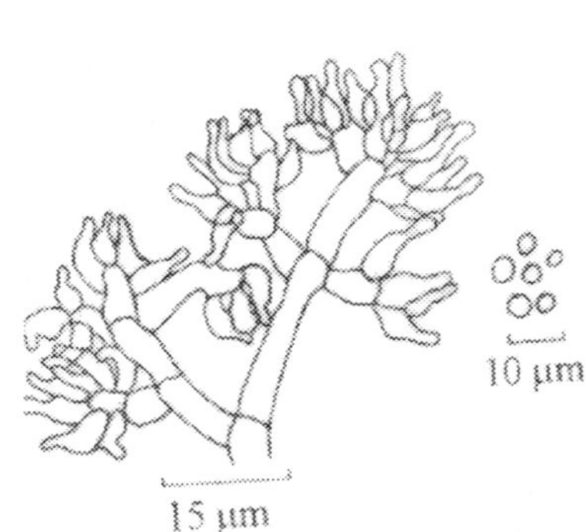

d. *T. piluliferum*

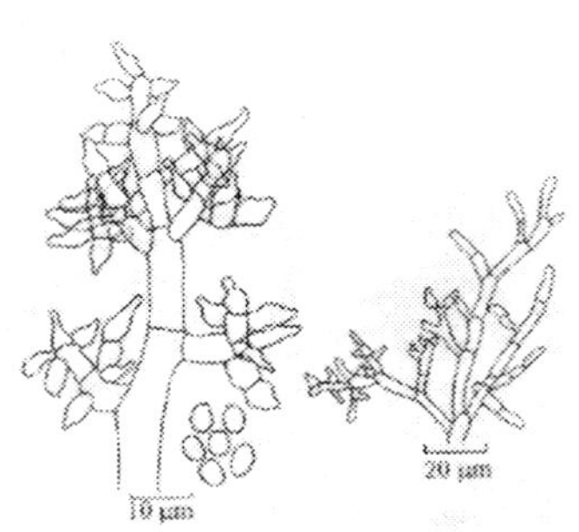

e. *T. polysporum*

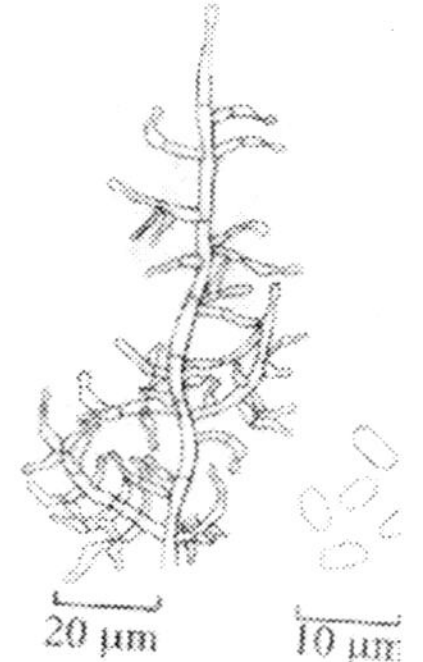

f. *T. pseudokoningi*

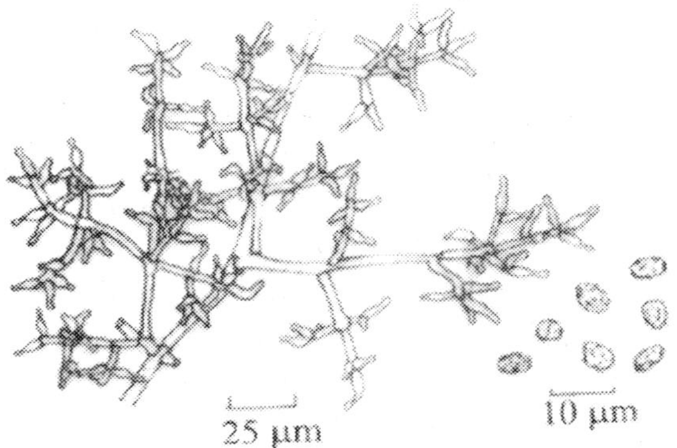

g. *T. viride*

23.8 *Trichotheicum* (Deuteromycotina, Hyphomyctes, Moniliaceae)

Colonies fast growing at first whit later pinkish, culminate velvety, powdery form conidial formation zonate diurnal rhythm conidiophores erect of sub erect, single or in grooves, unbranched or sparingly branched, hyaline septate, the tip bear basipetal imbricate, zigzag chains of conidia held together by mucilage conidia ellipsoidal two celled thick walled hyaline with well marked truncate attachment point (Fig. 22). This is monotypic genus. it produce trichothecene mycotoxin.

2.3.9. *Pithomyces* (Deuteromycotina, Hyphomycetes, Dematiaceae)

Pithomyces chartarum is the causative agent of pithomycotoxicosis, commonly called facial eczema, which occurs in sheep and cattle grazing pasture in summer and fall in New Zealand and less frequently in Australia and South Africa. *P. chartarum grows* saprophytically on dead vegetable matter and is cosmopolitan in distribution. It has been reported more often from the tropical warm temperate zones than from the cool temperate zones. It occurs in sufficient numbers to cause animal disease virtually only in pasture and under rather restricted regimes of moisture and temperature.

P. chartarum is an imperfect mold which forms rough, dark, several-celled, barrel-shaped aleurlospores, 8 to 20 x 10 to 30 μm, borne singly on short simple conidiophores. A small piece of denticle remains attached to the base of free spores (Fig. 28). Septa in the spores are both transverse and longitudinal and their number, none to five (usually three) transverse, non to three (usually two) longitudinal, distinguishes the spores of *P. chartarum* from other species.

2.3.10 *Phoma* (Deuteromycotina, Coelomycetes, Sphaeropsidales, Sphaeropsidaceae)

Freshly isolated strains grow rapidly on standard nutrient media at 20 to 25°C, producing dark mycelium and forming pycnidia within a few days of inoculation; old cultures may form pycnidia less freely. The black pycnidia are globose or ovoid with ostioles at the tips of small papillae (Fig. 7). Conidiophores are inconspicuous. The very numerous pycnospores are ovoid, colorless, guttulate, and either unicellular or with a single septum. The presence of both one-and two-celled pycnospores has caused the mold to be assigned to both *Phoma* and *Ascochyta*. Macroscopically, cultures of *P. herbarum* var. *medicaginis* are brown-black, moist and somewhat wrinkled with little aerial mycelium.

2.3.11. *Phomopsis* (Deuteromycotina, Coelomycetes, Sphaeropsidales, Sphaeropsidaceae)

Lupin plants are infected by *P. leptostromiformis* and cause field outbreaks of a mycotoxicosis known as lupinosis in sheep, cattle, horses, and pigs (Marasass, 1974).

The fungus produces black stromatic pycnidia on the stems, pods, and seeds of infected lupin plants. The first symptoms of infection are susally evident on the lower part of the stems, where yellow-brown, sunken lesions are formed. Black, stromatic pycnidia are produced in rows or irregularly scattered on these stem lesions. Dark-brown, water-soaked spots develop on infected pods; these pods then become dark-brown to black with pycnidia frequently produced in concentric rings, A conspicuous, coarse, white mycelium is visible in such discolored pods and infected seeds become shriveled and brown. The interior of the pycnidial stroma consists of white pseudoparenchymatous cells surrounded by a black sclerotial plectenchyma. A flat fertile locule eventually develops in the upper part of the stroma under the black outermost layer (Fig. 8). Along the inner wall of the locule, a dense pallisade of slender, simple or branched, tapering conidiophores are produced. Conidia are hyaline, one-celled, biguttulate, cylindrical with tapered ends, straight or very slightly curved, 5 to 12 x 1.5 to 2.5 μm (Fig. 8). β-conidia are produced in separate pycnidia on overwintered stems, and these are hyaline, filiform, and curved to strongly hooked

Conidia are extruded through the ostioles of mature pycnidia in cream-colored to salmon-pink droplets after 14 to 21 days. Only α-conidia have been observed on artificial media and they are similar in shape and size to those produced on the host. The ascigerous state of *P. leptostromiformis* is presumed to be *Diaporthe lupini* Harkness, but attempts by several investigators to induce the formation of the perfect state have been unsuccessful.

2.3.12. *Diplodia* (Deuteromycotina, Coelomycetes, Sphaeropsidales, Sphaeropsidaceae)

Diplodia maydis (Berk.) Sacc. [=*D. zeae* (Schw.) Lev.] occurs in nature as an important pathogen of maize (*Zea mays* L.). Maize ears infected by *D. maydis* cause field outbreaks of a mycotoxicosis known as diplodiosis in cattle and sheep.

The fungus causes a seedling blight, stalk-rot and ear-rot of maize wherever the crop is grown intensively. The ear rot usually progresses upward from the base of the ear and is visible as a conspicuous, coarse, and white to grayish-brown mycelial growth over the husk and kernels. The husks are often glued to the kernels by the white mycelium and infected kernels have a lusterless appearance and a dull-gray to light-brown color.

Black pycnidia are produced late in the season on the invaded husks and kernels as well as on the rotted stalks. Pycnidia are subglobose or flask-shaped, 150 to 350 µm in diameter, completely immersed with a distinct, slightly protruding ostiole. Conidiophores are phialidic and formed from the cells of the inner pycnidial wall and are elongate, straight or curved, hyaline, simple, aseptate, 10 to 20 x 2 to 3 µm. Conidia are formed from the apex of the conidiophores and are cylindrical or ellipsoidal, narrowing at both ends to a rounded apex and truncate base, straight or curved, brown, smooth-walled, 0 to 2 but usually 1-septate, slightly or not constricted, 15 to 45 x 3 to 8 µm

The fungus grows easily on media such as cornmeal, potato dextrose, or malt extract agars, producing a thick, coarse, white, woolly aerial mycelium. The reverse of the colony becomes yellowish-brown with age, and some strains produce a greenish pigment. Dark-brown to black globose pycnidia containing conidia are gradually produced on the surface and become immersed in the agar. Considerable variation occurs in conidial shape and size but conidia are typically light-brown, 1-septate, and somewhat smaller than conidia formed on the host. Certain strains in culture produce slender, thread like, hyline scolecospores, 21 to 45 µm long, in separate pycnidia or sometimes in the same pycnidium as brown conidia.

In addition to *D. maydis,* at least two other species of *Diplodia* are known to cause ear rot of maize, i.e., *D. macrospora* and *D. frumenti*. The three species of *Diplodia* on *Zea mays* can be identified with the help of following key.

Key to species of mycotoxigenic species of *Diplodia*

1. Conidia cylindrical or ellipsoidal, 0- to 3-septate, light-brown, smooth-walled, mycelium white 2

1A. Conidia ellipsoidal, 0- to 1-septate, dark-brown, longitudinally striate, 15 to 31 x 10 to 15 µm, mycelium becoming brown .... *D. frumenti*

2. Conidia cylindrical or ellipsoidal, 0- to 2-septate, 15 to 45 x 6 to 8 µm .... *D. maydis*

2a. Conidia cylindrical, 0- to 3-septate, 43 to 98 x 6 to 14 µm .... *D. macrospora*

2.3.13. *Chaetomium* (Ascomycotina, Pyrenomycetes, Sphaerial Chaetomiaceae)

Chaetomium is one of the familiar genra of ascomycetes associated with agricultural commodities due to their cellulolytic activity. Species of this genus are characterized by dark-coloured perithecia with short neck which are covered with irregularly coiled or tightly spiraled hairs or with tuff simple or branched hairs which are of taxonomic importance (Fig. 39). However, recent *Chaetomium* taxonomists have selected ascospore morphology as a major taxonomic character. These are world wide in occurrence and are found as contaminants of straw, roots, tubers, seeds, grains and may cause detereioration of foods and food stuffs.

The toxicity of *C. globosum* to roots was first demonstrated by Christenisen *et al.* (1966). Subsequently two toxins cochliodinol and chaetomin were isolated from cultures of *Chaetomium* (Bsewer *et al.*, 1978; Brewer *et al.*, 1972). Umeda *et al.* (1975) and Shekita *et al.* (1977) have isolated group of new cytochalasins and chaetoglobosins from rice cultures of *C. globosum*. Oosporin was isolated from cultures of *C. aurum* and *C. trilaterale* by Cole *et al.* (1974). Bsewer and Taylor (1978) reported chaetoglobins production by 3 strains of *C. globosum* and chaetomin production by *C. cochliodes, C. fimicola, C. globosum* and *C. umbonatum*. Mollicellins A-H are reported to be produced by *C. mollicellum* (Stark *et al.*, 1978). Carter (1981) divided the genus into four subgenera *Chaetomium, Murora, Bomerella* and *Achaetomium*. The subgenus *Chaetomium* consists of six sections (*Chaetomium, Bostrychodes, Torulosa, Brasiliensia, Crispata* and *Indica*), while *Murora* has five sections (Murora, Aura, Atrobruno, Spiralia and Cuniculosa). The mycotoxins produced by some of the *Chaemium* species are précised in Table 2.6.

At present much more work is still required to indicate how frequently toxigenic *Chaetomium* infested material are used as foods and feeds and more information on different factors contributing to human and animal mycotoxicoses. In *Chaetomium,* a very well defined correlation between particular species and the mycotoxins they produce so that secondary metabolism could readily find a place as an adjunct to taxonomy. Further, identification of *Chaetomium* to species level will provide a good basis for recognization of the toxins produced.

Table 2.6. Mycotoxins produced by different species of *Chaetomium*

Species	Chaetochromin	Chaetoglobosins	Chaetomin	Cochliodin	Sterigmatocystin
C. abuense	-	-	-	+	-
C. coarctum	-	±	-	-	-
C. cochliodes	-	+	+	+	-
C. elatum	-	-	-	+	-
C. fimeti (Chaetomidium)	-	-	-	+	-
C. fumicola	-	-	+	-	-
C. globosum	-	+	+	+	-
C. globosum var. *rectum*	-	+	-	-	-
C. megalocarpum	-	+	-	-	-
C. ochraconum	-	-	-	-	-
C. olivaceum	-	-	-	-	-

C. subuffine	-	+	-	-	-
C. subglobosum	-	-	+	-	-
C. terissimum	+	-	+	-	-
C. tetraspomum	+	-	-	-	-
C. umbonatum	+	-	+	-	-
C. thielavioidarum	+	-	+	-	+
Achaetomella virescens	+	-	-	-	+
A. udgawae	+	-	-	-	+
C. caprinum	+	-	-	-	-

CHAPTER – 3

ANALYTICAL TECHNIQUES

Mycotoxins are chemicals produced by fungi (moulds) under certain conditions. They are not essential for fungal growth or reproduction but are toxic to animals or humans. They are naturally present in feed and food and cannot always be avoided. Analysis of feeds and foods for the presence of mycotoxins can help grain, feed and animal producers to minimize, if not eliminate, or avoid adverse health effects of mycotoxins in animals. Krska *et al.* (2008) discussed elegantly use of different methods of mycotoxins analysis. Spanjer *et al.* (2008) devised LC-MS/MS multi methods for detection of aflatoxins, ochrotoxin A, DON, zearalenone, T-2, HT-2, sterigmatocystin cyclopiazonic acid, penicillic acid and fuminosins in fast 30 min run. Sampling feed or grain for mycotoxins analysis is critical for their detection. Analysis of an inappropriate sample is a waste of resources. General recommendations for specimen collection are given below.

Mycotoxins are not evenly distributed in feeds or grains. Grain from individual plants or from individual seeds or kernels from one plant, may be highly contaminated with mycotoxins, while grains from other plants, or other seeds or kernels from same plant may be mycotoxin free. If the specimen collected for analysis includes contaminated grain, the mycotoxin will likely be detected. If not, the mycotoxin will not be detected. The objective of sample collection is to collect a specimen from as many parts of the grain or feed pile as possible, that way the chance of collecting some of the contaminated grain is increased. Correct sampling is especially critical for mycotoxin analysis. The basic steps involved in sampling and detection are précised in Table 3.1.

The size of the collected specimen depends on the size of the load or storage bin in which the grain or feed is located. To estimate how much feed or grain should be collected for analysis. For a typical truck- or wagon-load size of grain or feed, collect at least 1 kg. For a train-car load size of grain or feed, collect atleast 4 kg.

Table 3.1. Basic steps in chemical analysis of mycotoxins

Step	Description	Purpose
Sampling	Probe of automatic sampler	Representative sample
Sample preparation	Grinding, mixing, sub sampling	Representative sample
Extraction	Shaker or blender	Separate the toxin from compounds insoluble extraction solution
Clean-up	Liquid-liquid partitioning (Separatory funnel)	Separate the toxin from groups of compounds of sample extract
Final separation	Thin layer chromatography (TLC)	Separate the toxin from remaining compounds.
	Gas-liquid chromatography (GLC)	The sample extract might interfere with toxin
	Liquid-Liquid chromatography (LC), Minicolumn chromatography	
Detection and quantitation	Fluorescence on TLC plate Fluorescence in solution UV absorption in solution GLC flame detector	Detection and measurement of response
Confirmation	TLC separation and detection of derivative of mycotoxin, Biological test, Mass spectrometry	Identification of chemical compounds

Samples are most conveniently collected when the feed is being moved from one container to another, such as when feed is being augured from truck to storage bin. Collect small amounts of sample at periodic intervals from the moving stream of feed. Space the sampling so that it occurs from start to finish of movement. Combine and mix all of the sub-samples well. Submit all of the collected samples to the laboratory for analysis.

Sampling feed or grain when it is not moving is more difficult. For example when it is loaded on a truck, or already in a storage bin, it is best to collect small samples from as many different places in the load or storage bin as possible. Probes can be used to collect samples deep into the grain. Use the same guidelines for sample size suggested above. Mix the collected samples well and submit it all for analysis. Dry feed is better for shipment to the laboratory than wet grain because fungi may begin to grow on wet grain during shipment. If possible, oven dry the feed to less than 13 percent moisture before shipment. Known mycotoxins are not significantly degraded if feed is dried at moderate temperatures (eg. 60 to 70°C). If wet feed cannot be dried, freeze and ship it so it remains frozen until it arrives at the laboratory where analysis will be performed.

It is better to ship feed in cloth or paper bags than in plastic bags to help prevent fungal growth during shipment. If fungi begin to grow during shipment, it may be difficult to determine if any mycotoxin detected was present before the sample was shipped.

3.1. SAMPLING STRATEGIES

The presence or absence of visible mold growth in feed are not good predictors of presence or absence of mycotoxins. If ill health due to mycotoxin exposure is suspected, it is best to collect specimens of feed from the bunks or feeders that best represent the feed the animals consumed just prior to or shortly after the problem was noticed. Samples taken from feed not yet offered to the animals may not contain any mycotoxins, even though they were present in the feed already consumed.

If you wish to compare results of mycotoxin analysis between laboratories, then the sample provided to each laboratory must contain the same amount of mycotoxin. Thoroughly mix a large sample, mill it, mix it again, and then split the sample, sending part of it to each laboratory. The objective of the sample preparation is to produce a sample as uniform with respect to mycotoxin concentration as possible. If a uniform sample is not provided to each laboratory and the laboratories find different concentrations, all results may be correct for the samples each laboratory was provided but likely to give a wrong message.

3.2. METHODS OF ANALYSIS

1. Different methods of mycotoxins analysis are available and the one(s) chosen by a laboratory will vary. Some laboratories may use more than one method, depending upon its circumstances. Cost of mycotoxin analysis depends partly upon the method used to detect their presence. When comparing cost between laboratories, it is important to know if the methods used by the laboratories are similar or different.
2. Some methods of mycotoxin analysis are able to detect several different mycotoxins with one test, while others detect only one mycotoxin. Some are able to determine the concentration of mycotox- ins present, while others cannot. Different methods adapted for the detection and estimation of different mycotoxins are précised in Fig. 3.1.
3. Some laboratories may screen samples for the presence of mycotoxins, using a relatively fast, easy and cheap method of analysis. If that analysis finds a mycotoxin to be present, then the sample is re-analyzed using another method that confirm the presence of the mycotoxins and determine the amount present.

Two methods of analysis commonly used to detect aflatoxins at grain elevators are black light and enzyme-linked immunosorbent assay (ELISA). The black light test for aflatoxin is based on the glow (fluorescence) of aflatoxin under ultraviolet light, commonly called black light. Research has demonstrated that results obtained using this method are rarely falsely negative, but often falsely positive. In other words, if glowers are observed in an appropriately collected sample, aflatoxin may be present but one cannot say that with certainty. Other chemicals present in the grain may also cause it to glow. If no glowers are found, then it is very likely that aflatoxin is not present in significant concentration.

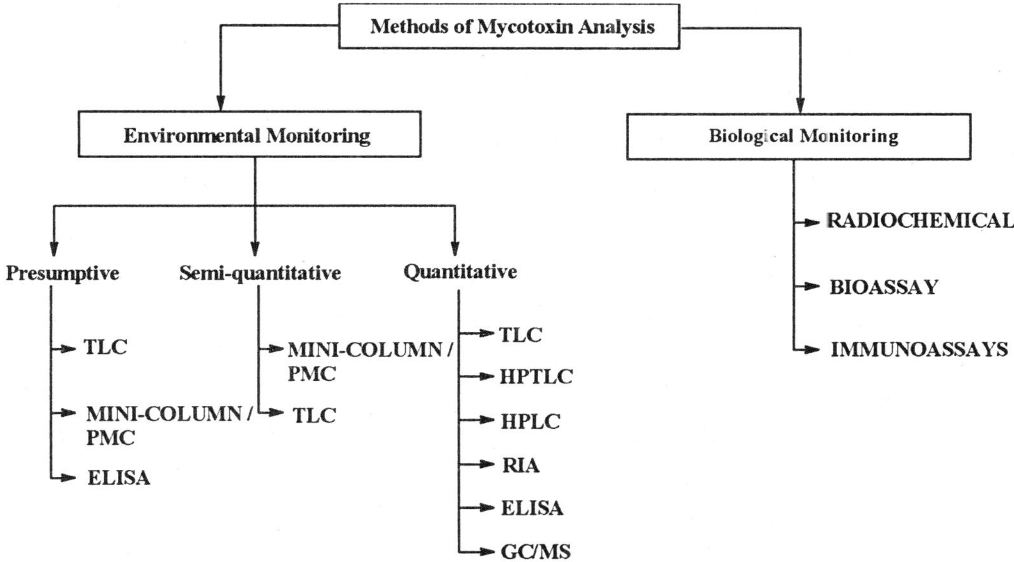

Fig. 3.1. Schematic presentation of different methods of mycotoxin analysis

If glowers are found to be present, the concentration of aflatoxin present in a sample cannot be reliably determined by counting the number of glowers present under black light. Samples judged to be positive for aflatoxins by black light should be analyzed by a more reliable method to confirm the presence of aflatoxin and determine the concentration present.

Enzyme-linked immunosorbent test kits are available for aflatoxin, fumonisin, vomitoxin and zearalenone. The kits test for only one mycotoxin, so a separate kit is necessary for each. They are based upon the interaction between the mycotoxin and antibodies produced against the mycotoxin. Antibodies against the mycotoxin are bound to the sides and bottom of a plastic well. Mycotoxins in a sample of grain are dissolved into a solution, which is placed in the plastic well. If the mycotoxin for which the kit is designed to detect is present in the solution, the mycotoxin binds to the antibody in the well. After a series of solutions are added and removed from the well, the presence of the mycotoxin is detected based on a color change in the final step. Recently number of enzyme linked immunosorbent assays (ELISA) have been developed (Bullerman and Gourama, 2001) and are commercially available (Table 3.5). Some ELISA kits are designed to determine the concentration of mycotoxin in the grain, others are designed to detect the presence of the mycotoxin above a specified concentration, but cannot be used to determine how much above that concentration is present.

More sophisticated methods of mycotoxin analysis may be used in laboratories for several reasons. They can detect mycotoxins in many different kinds of feeds, detect them more reliably and allow the concentrations of mycotoxins present to be more precisely determined. Analytical services using such methods usually cost more and take longer than ELISA tests. Each method has advantages and disadvantages, which are summarized in (Table 3.2). One has to use discretion for employing different techniques based on precision required. Bioassay methods also proved to be more useful and precise in detection of different mycotoxins (Table 3.3).

Table 3.2. Advantages and disadvantages of different methods for detection of mycotoxins in foods/feeds.

	Detection level (μg/g)	Advantages	Disadvantages
TLC	10	Versatile, Compact, Rapid	Tedious
HPLC	1	High precision, Good recovery, High sensitivity, Automated	Expensive
Minicolumn	5	Fast, Easy to handle, Good for routine analysis	Low precision
RIA	5	High specific, More sensitive, Fast	Expensive

Table 3.3. Detection of mycotoxins by bioassay

	Type of assay	Mycotoxin
Microbial Assays		
	Bacillus megaterium	MT
	B. stearothermophilus	MT
	B. thuringiensis	MT
	B. subtilis	AF
	Streptomyces and Nocardia spp.	AF
	Lactic acid bacteria	AF
	Rumen bacteria	AF
	Salmonella typhimurium	MT
	Staphylococcus aureus	CPA

Cytological assays		
	Tracheal organ cultures	MT
	Calf kidney cells	AF
	Cultured embryonic lung cells	AF
	Cultured liver cells	AF
	Cultured mouse cells	AF
Immuno assays		
	Immunochemical assays	MT
	Radioimmunoassay	AF
Invertebrate assays		
	Brine shrimp	MT
	Mollusc eggs	AF
	Planaria	AF
	Insects	MT
	Parameocia	AF
	Amoeba	AF
Vertebrate assays		
	Chick embryo	MT
	Day-old chick	MT
	Ducklings	MT
	Rabbit skin test	TR
	Mouse water refusal	*TRE*
	Fish	MT
	Mice	MT
	Bull Spermatozoas	AF
	Amphibians	MT
	Rats	MT

AF = Aflatoxins; MT = Multitoxins; TR = Trichothecenes

It is debatable whether culturing and identification of molds present in feed provides useful information for the assessment of mycotoxin contamination. The presence or absences of molds do not necessarily indicate the mycotoxin contamination.

The identities of molds found in feeds may be of some value in assessing the possibility of mycotoxin contamination. Identification of the genus and species of a mold present in feed is of most important. However, it is difficult to determine if a particular mold present in feed is capable of producing mycotoxins. Molds capable of producing mycotoxins are called toxigenic. Even if the genus and species of the mould are identified, it may not be possible to determine if that particular mold belongs to a toxigenic strain. If only the genus is identified, it is even more difficult to predict its toxigenicity. If an identified mold is believed to be toxigenic, it is not possible to determine if the mold has or is actively producing mycotoxins in the feed. The presence or absence of mycotoxins can be determined only by chemical analysis for mycotoxins.

3.3 UNITS OF CONCENTRATION

Quantitative results provided by chemical analysis consist of two parts, both are equally important: a number and a unit of concentration. The numerical part, which may be a number or a range of numbers, tells how many "units of measure" are present. Simply providing the numerical part can lead to confusion. For example, if aflatoxin is said to be present at a concentration of "30" it makes a big difference if the result is 30 parts per billion (ppb) or 30 parts per million (ppm). Aflatoxin at a concentration of 30 ppm is present at a concentration 1,000 times higher than at 30 ppb. Such a difference in concentration can make a big difference in whether a feed is safe to use or not. When seeking advice how to use mycotoxin-contaminated feed, always provide both parts of quantitative results. Units of concentration commonly used to report mycotoxin results are listed in Table 3.4

Table 3.4. Units of concentrations used to report mycotoxin concentrations in foods and feeds.

Unit of concentration	Definition	Remarks
Parts per billion (ppb); May appear as μg/g or mg/kg	1 part of the mycotoxin in 1,000,000,000 parts of feed.	1 ppb equals 1 billionth of a pound of mycotoxin in a pound of feed.
Parts per million (ppm); May appear as μg/g or mg/g	1 part of the mycotoxin in 1,000,000 parts of feed.	1 ppm equals to 1 millionth of a pound of mycotoxin in a pound of feed.

3.4. NO DETECTABLE AMOUNTS (NDA)

No analytical method can determine if a sample is completely free of mycotoxins. Every method has a limit of detection. Concentrations present in the sample below the detection limit cannot be detected. Result reported as no detectable amount with a detection limit of 1 ppm means that the mycotoxin is not present at concentrations greater than 1 ppm, but it may be present at a concentration less than 1 ppm.

When no mycotoxin is detected in a sample that may be reported as "none detected", "no detectable amounts" or "negative to test". An estimate of the detection limit should be provided. Any results reported as "0" should not be taken literally.

Failure to detect mycotoxins in the sample provided for analysis may be the result of sample selection such results do not prove mycotoxins are not present in other parts of the feed from which the sample was collected.

3.5. CHEMICAL ANALYSIS

3.5.1. Extraction

The organic solvents most commonly used for extraction of mycotoxins are: chloroform, acetonitrile, methanol and acetone. These solvents are mixed with a given ratio of a more polar solvent (water, dilute acid, aqueous solution of salts) to aid the breaking of weak electrostatic bonds which bind some mycotoxins to other substrate molecules (eg. proteins).

The ground sample, or preferably an aqueous slurry, is either shaken with the extraction solvent for 30-45 minutes or blended at high speed for about 3 minutes. It should be noted that an explosion proof blender is recommended for use with inflammable solvents such as acetone and methanol.

3.5.2. Clean-Up

Mycotoxins are such a diverse group of chemical compounds that it is difficult to find a simple procedure which specifically removes non-mycotoxin interfering compounds, whilst leaving the mycotoxins in the extract. For this reason it is difficult to find a good method for screening a wide-range of mycotoxins simultaneously. It is possible, however, to devise procedures which remove interfering non-mycotoxin compounds from the extract of a particular commodity and leave a particular mycotoxin or group of mycotoxins in the extract. Some of common clean-up techniques which have been used are:

(i) ***Defatting***: This is carried out prior to the toxin extraction step and uses petroleum ether or hexane to extract lipids from the sample using a Soxhlet extractor. This step is only required when the subsequent clean-up step is not capable of removing the lipids.

(ii) ***Column–chromatography***: This technique has very wide applications, and is used in a number of regulatory or officially approved methods. A glass column is packed with one or more adsorbent materials and the crude extract is added to the top of the column. The column is then eluted with a series of solvents or solvent mixtures which are designed to first wash off interfering compounds and then elute the desired mycotoxins, whilst other interfering compounds remain strongly bound on the column. A miniature column, called a mini-column is used in many rapid aflatoxin assay methods to remove interfering compounds and to qualitatively detect aflatoxin down to a few parts per billion (ppb).

(iii) ***Precipitation*** : This is an extremely useful technique whereby certain chemicals, sometimes in colloidal form are added to the crude extract and these adsorb certain pigments, proteins and other interfering compounds onto their surface. The complex so formed precipitates out of solution and can be filtered off, leaving a cleaned-up solution. Useful precipitating agents include cupric carbonate, ammonium sulphate, lead acetate and ferric gel.

(iv) ***Liquid-liquid partition*** : It is commonly used, often in conjunction with one or the other clean-up procedures, to provide additional clean-up and also to transfer toxins from one solvent system to another, whilst at the same time effecting a considerable increase in concentration of the toxins. The partition is carried out in a separating funnel which contains the two immissible solvents. The funnel is shaken for a few minutes to allow the dissolved compounds, including the mycotoxins, to partition between the two phases. The solvents are selected so that the mycotoxins are preferentially partitioned into one of the solvents. Care must be taken in the choice of solvents in order to minimize the risk of emulsion formation.

(v) ***Thin layer chromatography (TLC)*** : Although this technique is used mainly for detection and quantification of toxins it is also useful for separating the mycotoxins from interfering compounds in the extract. Sometimes it is necessary to try a range of developing solvents in order to separate the toxins of interest from the interfering compounds. If this proves impossible, or too time consuming, then two dimensional TLC may be employed.

(vi) ***Other clean-up procedures*** : The clean-up procedures outlined above are those employed in officially approved methods for aflatoxin analysis and they are also used in analytical methods for many other mycotoxins. Other clean-up procedures such as reverse dialysis and base extraction are also used for the detection of different mycotoxins.

3.5.3. Work-Up

After the clean-up step, the extract must be "worked-up" in order to prepare it for the detection and/or quantification step. Following clean-up the extract is often dissolved in a large volume of aqueous solvent, so it must be transferred into a small volume (10-50 ml) of a volatile solvent, such as chloroform. The chloroform solution often requires drying and this is achieved by passing it through a bed of anhydrous sodium sulphate. The solvent is evaporated off, as described below to near dryness. Care must be taken at this stage as some mycotoxins can break-down if the dry extract is heated at 100°C. To avoid this, the evaporation is best carried out using a rotary evaporator at 30°C to reduce the volume to a few ml, followed by evaporation to dryness in an inert atmosphere using a sample concentrator. If such items of equipment are not available, then evaporation should be carried out using a steam-bath, preferably under a stream of nitrogen. Great care should be taken to ensure that flasks are removed from the steam-bath just before they go dry and that vials are removed on the point of dryness. The extracts are now ready for detection and quantification.

3.5.4. Detection and Quantification

It is fortunate that the aflatoxins and many of the other important mycotoxins are fluorescent under ultraviolet light. This enables them to be detected at very low levels (parts per billion) and this method of detection is used in the majority of analytical methods for mycotoxins.

Detection or "qualitative assay" is usually by TLC or mini-column and only requires the use of a qualitative (not accurately determined) standard. Such a standard is also sufficient to enable a semi-quantitative assessment to be made by TLC, using a "dilution-to extinction" principle. Fully quantitative determinations, which require the use of a standard of known concentration, can be carried out by a variety of techniques including TLC, high performance liquid chromatography (HPLC), high performance thin layer chromatography (HPTLC) and the fluorotoxinmeter (FTM). Though HPLC and GLC techniques have excellent sensitivity and reproducibility and can be adopted to automation, they do have certain disadvantages such as

i) Samples require to be highly processed in relation to clean-up to achieve sensitive detection limits.

ii) Only one sample can be analysed at any given time.

iii) Capital output is high

iv) Running costs are high

v) Highly trained persons are normally required to operate the equipment and interpret the results.

vi) They are time consuming.

Analytical methods which are simple and can be handled by in experienced operator and institutions where good laboratory facilities are not available (Table 3.5).

Table 3.5. Comparison of Analytical methods for detection of mycotoxins in foods and feeds

Service	Precision of the Method	Advantages	Disadvantages	Cost (Approx.)
Mold culture and identification	Microbiological culturing; microscopic or biochemical identification tests.	Determines presence of molds, but not necessarily toxigenic strains.	Does not determine if molds found to be present. It may take a relatively long time to obtain results	Rs. 500-1000
Chemical analysis	Enzyme-Linked Immunosorbent Assay (ELISA)	Rapid, relatively inexpensive	Only detects one mycotoxin at a time. Proof of mycotoxin presence usually requires additional testing.	Rs. 500-1000
	Thin layer chromatography (TLC)	Can detect more than one kind of mycotoxin	Slower than ELISA. Proof of mycotoxin presence may require additional testing. May not work well for feeds other than grains.	Rs. 200-500
	High performance liquid chromatography (HPLC)	Easier to provide quantitative results. May be able to detect more than one kind of mycotoxin.	May be more expensive. Proof of mycotoxin presence may require additional testing	
	Gas chromatography/mass spectroscopy (GC/MS)	Can identify the mycotoxin without additional testing. May be able to detect more than one kind of mycotoxin.	More expensive	Rs. 1000

3.5.5. Confirmation

It is essential that confirmatory tests be carried out if a mycotoxin is thought to have been detected, especially when a new commodity or new source of commodity is being analyzed. Failure to do this could easily lead to false-positive results (i.e. an interfering compound is mistaken for a known mycotoxin).

Most confirmatory tests involve the formation of a derivative which has different properties (eg. colour of fluorescence or polarity), than the presumptive mycotoxin. When detection and/or quantification is by TLC, then derivatives can be formed "on the plate" either by spotting a reagent on the plate before development or by spraying on a reagent after development. An example of the former approach is the use of trifluoroacetic acid (TFA) to form the hemi-acetal derivatives of aflatoxin B, G and M which are identified, if present, as slower running fluorescent spots. A commonly used spray reagent for the aflatoxins is 50% sulphuric acid which reacts with the blue and green-blue fluorescent aflatoxins to give yellow fluorescent derivatives.

Table 3.6. Spectrophotometer chatacters of different Mycotoxins

Mycotoxin	Molecular weight	Solvent	Absorbtivity (E)	Max. (nm)
Aflatoxin B	312	Benzene : acetonitrile (98:2 v/v)	19,800	353
Aflatoxin B_1	312	Chloroform	22,300	353
Aflatoxin B_2	314	Benzene : acetonitrile (98:2 v/v)	20,900	355
Aflatoxin G_1	328	Benzene:acetonitrile (98:2 v/v)	17,100	355
Aflatoxin G_2	330	Benzene:acetonitrile (98:2 v/v)		357
Aflatoxin M_1	328	Chloroform	19,950	357
Ochratoxin A	403	Benzene:acetic acid (99:1 v/v)	5,550	333
Ochratoxin B	369	Benzene:acetic acid (99:1 v/v)	6,000	320
Ochratoxin A	431	Benene:acetic acid (99:1 v/v)	6,200	333
Ochratoxin B ethyl ester	397	Benzene:acetic acid (99:1 v/v)		320
Patulin	154	Absolute ethanol	14,540	276
Patulin	154	Methanol	12,880	275
Sterigmatocystin	324	Benzene	15,200	325
Citrinin	259	Chloroform	16,100	322
Zearalenone	318	Ethanol	29,700	236
Zearalenone	318	Ethanol	13,909	274
Zearalenone	318	Ethanol	6,020	316

3.5.6. Determination of the Purity and Concentration of Mycotoxin Standards

Pure mycotoxins are available from several manufacturers (listed in Appendix: 2) and these are used to prepare standard solutions for use in mycotoxin analytical procedures. It is essential that the exact concentration of the standard solutions are known these may be determined by UV absorbance measurements on a spectrophotometer. Details of spectophotometric parameters of different mycotoxins are given in Table 3.6 and concentration of mycotoxin standard solution are given in Table 3.7. Some of the simple methods and suitable to Indian conditions devised by different workers are précised in Table 3.8.

Some of the precautions required in rapid mycotoxins analytical methodology are

1. Minimum sample preparation prior to analysis.
2. Good sensitivity and specificity for the appropriate mycotoxins (s) under test.
3. Applicability for wide range of natural substrates.
4. Easy adaptability to in-field situations using untrained personnel or at least using the minimum of laboratory equipment and technical assistance.
5. Laboratory oriented prototype to assess the full capabilities of the system for toxin determination.
6. Availability of final results within a short period of time from sample preparation (5-30 min).

Table 3.7. Concentrations of Mycotoxin Standard Solutions

Mycotoxin	Stock (μg/ml)	UV (μg/ml)	TLC (μg/ml)
Aflatoxin B_1	100	10	1
Aflatoxin B_2	100	10	1
Aflatoxin G_1	100	10	1
Aflatoxin G_2	100	10	1
Aflatoxin M_1	10	10	1
Ochratoxin	25	25	5
Patulin	40	10	40
Sterigmatocystin	100	100	100
Citrinin	40	20	40
Zearalenone	100	10	50
T-2 toxin	5000	----	100
Diacetoxyscirpenol	5000	----	100

Although the list in Table 3.7 below does not necessarily correspond with the concentrations quoted in the literature, we have found them to be most useful.

a. **Stock solutions :** These are made directly from the pure crystals or dry film.

b. **UV solutions :** These are direct dilutions from the stock solution and are those which are used for determining the correct concentration of standard by UV spectrophotometric measurement.

c. **TLC working solutions :** These are made by dilution from either the stock or the UV solutions.

Table 3.8. Analytical methods developed for mycotoxins analysis from India

Area of research	References
Modified Pon's method for estimation of AFB_1	Rati *et al.* (1987)
Single test tube screening method for AFB_1 detection	Pallavi *et al.* (1997)
Screening method for detecting ergosterol in cereals	Sashidhar *et al.* (1989, 1991)
A method for detection of DON	Ramakrishna *et al.* (1989)
Evaluation kit for AFB_1 at field level	Sudershan *et al.* (1992)
p-Tolualdehyde test for penicillic acid detection	Neelakanthan *et al.* (1978a)
Chemical changes in maize seeds	Bilgrami *et al.* (1981a)
Chemical changes in apples infected with *A. flavus*	Sinha and Singh (1987)
Colorimetric method for citrinin	Damodaran *et al.* (1973)
Colorimetric method for terreic acid	Subramanian *et al.* (1978)
Colorimetric method for patulin	Subramaniyan (1982)
Spray reagent for trichothecenes	Rao *et al.* (1985)
Colorimetric method for CPA	Rathinavelu and Shanmuga-sundaram (1984)
Mini-column for the detection of sterigmatocystin	Ramakrishna and Bhat (1990)
Detection of aflatoxins by pressure mini-column technique	Sashidhar *et al.* (1988)

3.6. IMMUNOLOGICAL TECHNIQUES

Intensive studies to develop new sensitive, specific and simple methods have led to many improved and innovative analytical methods for mycotoxins analysis in the last a few years (Truksers and Pohland, 2001 and De Vries *et al.*, 2002). Simplified and efficient clean up procedures such as solid phase extraction cartridges have been established. Better analysis and quality control has been established by using certified reference materials (Boenke, 1997). More sensitive TLC, HPLC and GC techniques are now available. Sensitive and versatile high resolution MS and GC/tandem MS/MS are coming to the market. MS methods have also been incorporated into HPLC systems. New Chemical methods including capillary electrophoresis, fluorescence polarization immunoassay and biosensors are emerging and have gained application for mycotoxin analysis (Chu, 2000; Maragos *et al.*, 2001; Maragos and Plattner, 2002; Nasir and Jolly, 2002). Recent progress in nanotechnology has lead to new innovative approaches for monitoring of mycotoxins in grains as well as new diagnostic tools for mycotoxicosis.

Immunoassays are gaining wide acceptance as an analytical tool for mycotoxins. Antibodies against almost all of the mycotoxins are now available (Chu, 2000). New immunoassay systems are emerging (Maragos and Plattner, 2002). Immunoscreening methods have been widely accepted as a simple approach to screening for mycotoxins in several commodities. Many immunoscreening kits which require less than 15 min to complete are commercially available. These assays eliminate many of the above steps. Automation of mycotoxin analysis has become a reality by combining these newer techniques. Instead of using antibodies, molecularly imprinted polymers are being used for solid phase extraction of mycotoxins such as ochratoxin A (Jodlbauer *et al.*, 2002). Rather than analysis of toxin, PCR methods based on primers of key enzymes involved in the biosynthesis of mycotoxins have been introduced for the determination of toxigenic fungi present in food (Schanerr *et al.*, 2002).

Although several biological methods are available, most of these methods are nonspecific and relatively insensitive. To overcome the difficulties encountered with the chemical and biological methods, new immunochemical methods have been developed (Chu, 1996). Many immunoassay kits for mycotoxin analysis are also commercially available (Beier and Stanker, 1996). The application of mycotoxin immunoassay is not limited to foods and feeds, but also used as a sensitive approach to monitor mycotoxins in body fluids and tissues or organs of humans and animals that have been exposed to the mycotoxins (Pestka, 1994).

Immunochemical methods are based on the specific interaction between antibodies and the toxin. Like most other natural products, mycotoxins must first be conjugated to a protein/polypetide carrier before subsequent use in immunization for antibody production (Chu, 1995). Newer innovative approaches for the coupling of mycotoxin to marcomolecule have been developed in recent years (Chu, 1996). With the advances in hybridoma technology monoclonal antibodies against many mycotoxins were also made. Antibodies against almost all the important mycotoxins have been made available. Many types of immunoassays, including radioimmunoassay (RIA) and enzyme-linked immunosorbent assays (ELISA), as well as several novel immunochemical screening tests, have been developed. Most of these methods are very sensitive, specific and simple to operate. Specific antibodies have also been used as immunohistochemical reagents and to arm affinity columns that are used as a clean-up tool for analysis of mycotoxins by other methods. Antibodies against different mycotoxins are precised in Table 3.9.

3.6.1. Radioimmunoassay (RIA)

The RIA procedure involves incubation of specific antibody simultaneously with a solution of unknown sample or known standard, and a constant amount of labeled toxin. After separation of the free and bound toxin, the radioactivity in those fractions is then determined. The toxin concentration of the unknown sample is determined by comparing the results to a standard curve that is established by plotting the ratio of radio activities in the bound fraction and free fraction versus log concentration of unlabelled standard toxin.

Table 3.9. Antibodies against Mycotoxins

Mycotoxins	Type of antibody	References
AAL toxin	pab	Szurdoki *et al.* (1996)
Aflatoxins: B_1, B_2, G_1, G_2	mab, pab	Hsu and Chu (1992)
Aflatoxin metabolities: B2a, Q1, M_1aflatoxical, DNA adducts	mab, pab	Yu and Chu (1998)
Citrinin	pab	Usleber *et al.* (1996)
Cyclopiazonic acid	mab, pab	Haung and Chu (1993)
Ergot alkaloids	mab, pab	Shelby (1996)
Fusarochromanone	pab	Yu *et al.* (1990)
Fumonisin	mab, pab	Yu and Chu (1999)
Kojic acid	pab	
Ochratoxin A	mab, pab	Clarke *et al.* (1994)
Patulin	pab	Mc Elory and Weiss (1993)
Paxilline related	mab, pab	Garthwaite *et al.* (1994)
PR-toxin	pab	
Rubratoxin B	pab	
Secalonic acid	pab	Brigys *et al.* (1994)
Sporidesmin	mab, pab	Li *et al.* (1996)
Sterigmatocystin	pab	Schmitt *et al.* (1996)
Trichothecenes: DAS, DON, FX, DOVE, AeDON, NIV, Rgridin A, T-2 toxin, and T-2 toxin metabolites (HT-2, T-2-tetraacetate 3'-OH-T-2, dep-T-2)	mab, pab	Wang and Chu (1991); Barna-vetro *et al.* (1994)
Versicolorin A	pab	Reynolds and Pestka (1991)
Zearalenone	mab, pab	Berna Vertro *et al.* (1994).

pab- polyclonal antibody mab - manoclonal antibody.

3.6.2. Enzyme Immunoassay

Enzyme immunoassay (EIA) is a general term for the immunoassays involving use of an enzyme as a marker for the detection of immunocomplex formation. The general principle of EIA is similar to that of RIA, there is an amplification system present in this assay, and thus, EIA is more sensitive. Since no radioactive substances are used, this assay avoids the problems encountered in handling radioactivity. Enzyme labeling can be done by conjugation of enzyme to Ag or Ab via the periodate oxidation and subsequent reductive alkylation method or by cross-linking using glutaraldehyde. Some of the methods used in conjugation of hapten to proteins can also be used. Although Horse Radish Perodxidase (HRP) and alkaline phosphatase are the two enzymes most commonly used, others, such as glucose-6-phosphate dehydrogenase coupled with oxidoreductase and luciferase, glucose oxidase, beta-galactosidase, and urease, have also been used. ELISA is further divided into two major types. One type is competitive ELISA (c-ELISA), which can be used for the analysis of both hapten and macromolecule, while the other is noncompetitive sandwich ELISA, which is used for divalent and multivalent Ag. c-ELISA is used most frequently for mycotoxin analysis.

(a) Direct competitive ELISA (dc-ELISA)

In this assay, specific *antibodies* against mycotoxins are coated on the ELISA plate. The sample or mycotoxin standard solution is generally incubated simultaneously with enzyme-conjugate or incubated separately in two steps. The amount of enzyme bound to the plate is then determined by incubation with a chromogenic substrate solution. The resulting color/ fluorescence, which is inversely proportional to the mycotoxin concentration present in the sample, is then measured instrumentally or by visual comparison with the standards. In this assay, the mycotoxin enzyme conjugate (marker) and free mycotoxin compete for the same binding site on the solid-phase antibody. Although HRP is most commonly used as the enzyme for conjugation, other enzymes such as alkaline phosphatase and beta-galactosidase also have been used. Excluding the time for sample preparation, dc-ELISA itself generally can be completed in 0.5-2 hr. In general, dc-ELISA is approximately 10-100 times more sensitive than RIA when purified standard is used and as little as 2.5 pg of pure mycotoxin can be measured (Table 3.10).

Since a clean-up step is usually not necessary, many samples can be analyzed within a relatively short period. Dc-ELISA can detect 0.05-50 ppb of mycotoxins in foods and feeds. Like RIA, the sensitivity of ELISA is improved when a clean-up treatment is included in the assay protocol.

(b) Indirect Competitive ELISA (idc-ELISA or Double-Antibody ELISA)

In the indirect competitive (idc-ELISA), a mycotoxin-protein (or polypeptide) conjugate is first prepared and then coated to the microtiter plate before assay. The plate is then incubated with specific rabbit (or other type) antibody in the presence or absence of the homologous mycotoxin. The amount of antibody bound to the plate coated with mycotoxin-protein conjugate is then determined by reaction with goat anti-rabbit (or anti-other type) IgG-enzyme complex (which is commercially available) and by subsequent reaction with the substrate. Thus, toxin in the samples and toxin in the solid phase compete for the same binding site with the specific antibody in the solution. The idc-ELISA has also been widely used for analysis of a number of mycotoxins (Table 3.11) with a sensitivity that is comparable to or slightly better than that of the direct ELISA in some cases. This type of ELISA requires less antibody (100 times less) and does not require preparation of a toxin-enzyme conjugate. However, it takes more analytical time (2 hr). To optimize the assay, selection of secondary antibody-enzyme conjugate in the idc-ELISA is important.

Table 3.10. Sensitivity of Direct Competitive ELISA for Selected Mycotoxins

Mycotoxins	Foods/feeds	Standard range (ng/assay)	Detection limits (μg/kg) or (μg/l)	References
AAL	C			Yu and Chu (1998)
AFB/Afs	C, Wh, P, Pb	0.0025-1	1-10 (1)	Barna Vetro *et al.*, (1994)
AFM,	M, Ch	0.0012-1	0.10 (0.01)	Biancardi (1997)
CPA	C, P, MF	0.1-0.6	(0.05-0.1)	Yu and Chu (1998)
DON	C, Wh	0.012-125	1000 (10)	Sinha and Savard (1997)
3-AcDON	B	0.0002-0.4	16	Sinha *et al.* (1995)
15-Ac-DON	Wh	0.002-0.025	50-100	
Fm	C, MF, M, BR	0.0005-50	10-500	Scott *et al.* (1997)
H-Fm	C	0.1-15	(5-10)	
OA	Wh, B, K	0.025-0.5	30 (1-2)	Tapai *et al.* (1997)
T-2	C, Wh	0.0025-0.2	2.5-50 (1)	
Type A TCTC	C	0.0025-0.2	50-100	
ZE	C	0.02-2.5	50 (10)	Tanaka *et al.* (1995)

*B-Barley, BR-Beer, C-Corn, Ch-Cheese, Fm-Fumanisis, H-Fm- Hydrolyzed Fm, K - Kidney, M - milk, mF - mixed feed, P - Peanuts, Wh - Wheat, OA-Ochratoxin A, Tete-Trichothearnes, Other abbreviations see table 3.9.

3.6.3. Immunoscreening methods

By shortening the incubation time and adjusting the antibody and enzyme concentrations in the dc- or modified idc-ELISA assay system, it is possible to do a quick screening test toxin levels in less than 30 min (e.g., 20 ppb) (Schneider *et al.*, 1991). Based on the same principle as the dc-ELISA, several other types of immunoscreening tests with sensitivity similar to ELISA were also developed. Rather than coating the antibody onto the microtiter plate, the antibody is immobilized on a paper disk or other membrane (Usleber *et al.*, 1993) which is used directly as a strip (Smith *et al.*,

1994) or mounted either on a plastic card (card screening test), a plastic strip as dipstick (Edrington *et al.*, 1997), in a plastic cup (Cup test and Cite), or in a syringe (Cite probe and Idexx probe) (Sarr *et al.*, 1995). Antibody has also been coated on polystyrene beads (Mayura *et al.*, 1998). The reaction is carried out on the wetted membrane disk. Thus, after reaction, the absence of color (or decrease in color), generally blue, at the sample spot indicates the presence of toxin in the sample. The reaction is generally very rapid and takes less than 10-15 min to complete. Like the above formats, dipstick-type enzyme immunoassay has been developed for quick screening of *Penicillium islandicum* in rice grain and 3-acetyl-DON (Chu, 1995) and T-2 toxin (De Saeger and Van Peteghem, 1999) in wheat. Abouzied and Pestka (1994) immobilized different mabs against AFB_1, FmB_1 and Ze.

Table 3.11. Senistitivity of Indirect Competitive ELISA for selected Mycotoxins

Mycotoxins	Foods/Feeds	Standard range (μg/assay)	Detection limits (ug/kg) or (ug/L)
Idc -ELISA			
Aflatoxins	C,P, Peanut, Buffer (Pb), F, Rs, S	0.0002-1	0.25-5
AFM	Milk (M)	0.0001-0.005	0.005
Diocetoxyscirpenol	C, Wheat (Wh)	0.005-5	0.005
Dioxynivalenol	C, Wh	0.010-100	300(1)h
3-Ac-DON	Rye (R)	0.005-1	1000(10)
Fumonisins	Corn (C)	1-100	1
Nivalenol	B	0.05-5	200(50)
Ochratoxin A	B, K, mixed feed (M F), Wh	0.004-10	(30)
Sabatoxin	S, U, W	0.0001-5	0.06-50(1)
T-2	M, S, U	0.002-0.2	5(0.2-1)
IIT-2	Urine (U)	0.005-0.5	(0.5)
Zearalenone	C,W, M, K, F	0.05-2.5	1-60
M-dc-ELISA			
AFB	B, P, Pb, C	0.01-1.5	10
T-2	Barley (B)	0.05-5	(50)

Rs-Raisins, S - Serum, Md- modified, F- Figs.

Another screening test is the immunoaffinity method, which was originally designed for mycotoxins such as AF, OA, and ZE that fluoresce (Trucksess *et al.*, 1997). In this assay, sample extracts diluted in phosphate buffer are applied to the affinity a column in which specific antibody was covalently bound to the solid matrix. After washing to remove the unbound materials, specific mycotoxin is then eluted from the column with the appropriate solvent system and then subjected to other chemical analyses. For mycotoxins with native fluorescence such as AF, OA and ZE, the toxin level is then equate could be directly determined fluorometrically or be determined after derivatization to enhance the fluorescence (Hansen, 1990). For fumonisin and DON screening, it is necessary to introduce a fluorophore to the materials eluted from the immunoaffinity column (IAC). The sensitivity to the IAC screening tests for AFBI, FmBI, OA, and ZE is 2 ppb, 1 ppm, 5 ppb and 0.2 ppm, respectively.

The application of various immunoscreening tests to mycotoxins has been summarized by Pestka (1994) and Chu (1996), and most of the screening tests are commercially available as kits. All of the rapid-screening test kits permit monitoring of mycotoxins semi quantitatively and have been found to be effective in screening mycotoxins in the field by FSIGS (Pestka, 1994).

3.6.4. Complementation of immunoassays with chemical

(a) Immunoaffinity Chromatography

With the availability of antibodies against various mycotoxins, Immuno-Affinity Columns (IACs) were made by conjugating antibodies to a solid-phase matrix. These columns are then used either in a screening test as discussed above or as a cleanup column for subsequent chemical analysis (Chu, 1996). IAC was first used in the RIA and later for recovery of AFM from urine and milk samples (Wu *et al.*, 1983) for subsequent analysis. Although earlier applications of this technique were primarily aimed at biological fluids (Gropman *et al.*, 1992), the IAC has gained wide application as a cleanup tool for a number of mycotoxins and is not limited to fluid samples (Scott and Trucksess, 1997). IACs for a number of mycotoxins are also commercially available (Table 3.12). It is apparent that once the contaminants are removed by specific IACs, the solution can be directly subjected to liquid chromatographic (LC) quantitation either off-line in an automated system, or by fluorometry. IAC not only serves as a cleanup tool but also concentrates the mycotoxin from a large amount of sample. Thus, it lowers the detection limits sometimes as low as parts-per-trillion of mycotoxins can be measured.

(b) Combination with HPLC and TLC

With the availability of sensitive ELISA methods, this technique has proved effective as a post column monitoring system for HPLC (Kramer *et al.*, 1994). This is especially useful for the analysis of compounds with no specific absorption, such as TCTCs and Fms. For example, in the analysis of various type A TCTC mycotoxins, the sample extract with no cleanup treatment was first subjected to HPLC with a C-18 reverse-phase column. Individual fractions eluted from the column were analyzed by ELISA using "generic" antibodies against group A TCTCs. This approach cannot only identify each individual group A TCTC, but can also determine its concentration quantitatively. As little as 2 ng of T-2 toxin and related TCTCs as well as their metabolites can be monitored by this method. A combination of HPLC and ELISA technology proved to be an efficient, sensitive, and specific method for the analysis of TCTC (Park and Chu, 1996) and other mycotoxins (Chu, 1991).

Table 3.12. Immunoaffinity Chromatography of Mycotoxins

Mycotoxins	Commodities	Analysis	Selected ref.
AFM	M[a]	F1 h	Fremy *et al.* (1992)
AFB	F, P	TLC	Candlish *et al.* (1989)
AFB	Cff, IP, nuts, figs, etc	HPLC/PsCD	Barmark *et al.* (1994)
	C, P, Ct, F, Pb, BR	F1/Br; HPLC/PCD	Groopman *et al.* (1988)
AFL	S,U	HPLC	Kussak *et al.* (1998)
AFQ 1	U	HPLC	Kussak *et al.* (1994)
AF-adducts	T,S,U	HPLC, ELISA	Groopman *et al.* (1992)
AF-albumin	S	HPLC, ELISA	Sabbioni *et al.* (1998)
CPA	C, P, feed	ELISA	Yu *et al.* (1998)
DON	C, W	HPLC	Yu *et al.* (1999)
FmB, HFms	C, starch	HPLC; LC/MS	Ware *et al.* (1994)
OA	Cff, T, W, C, So	HPLC	Nakajima *et al.* (1994)
OA	Cff, C, So	CE	Corneli *et al.* (1998)
ZE	M, U, C	Elisa, HPLC, MS	Azcona *et al.* (1990)

Cff- Coffee, T - animal tissues, So- Sorghum.

3.6.5. Other newly developed immunochemical methods

(a) Anti-idiotype Antibody-Based Immunoassays

The development of immunochemical methods for mycotoxin detection has led to a great demand for specific antibodies and related immunochemical reagents for the assay. An alternate approach to preparing immunochemical reagents is through generating anti-idiotype (anti-ID) antibodies (Nisonoff, 1991). Anti-ID (Ab2) for large molecules have been well developed and have been applied to clinical diagnosis and immunotherapy. Recent studies have succeeded in generating Ab2 against a number of small-molecular-weight haptens including mycotoxins Afs (Krska *et al.*, 1997), FmB_1 (Chu, *et al.*, 1995), and T-2 toxin (Chanh *et al.*, 1992).

(b) Immunoassay for Mycotoxin-Producing Fungi

Other than mycotoxins, antibodies against specific fungi (Lu *et al.*, 1995) as well as several enzymes involved in the biosynthesis of aflatoxin (Lu *et al.;* 1997) and trichothecenes (Park and Chu, 1996) have also been generated. Five mabs against *Aspergillus flavus were* recently produced by Candlish *et al.* (1997) and those mAbs were used in the immunoassays for identification of *A. flavus*.

(c) Development of Biosensors

Although development of Ag/Ab-based biosensors for mycotoxin detection was initiated in the late 1980s, application of this technology only emerged in recent years. Nevertheless, some of the principles and methodology are useful for designing the biosensors. In the so-called "hit-and-run" assay (Warden *et al.*, 1987) for T-2 toxin column was equilibrated with a fluorescein isothiocyanate (FITC)-labelled Fab fragment of IgG (anti-T-2 toxin). Samples containing T-2 toxin were injected into the column. The FITC-Fab that eluted together with the samples containing T-2 toxin was then determined in a standard flow through fluorometer. Whereas these methods were not as sensitive as the regular ELISAs, several biosensors with some modifications such as line resolved chromatography (Bacigaupo *et al.*, 1994), fiberoptic immunosensor (Strachen *et al.*, 1997) miniatured ELISA (Yacoub-Geroge *et al.*, 1996) and surface plasma resonance biosensor (Mullett *et al.*, 1998) can be used for rapid screening of mycotoxins.

It is apparent that immunoassay techniques have gained wide acceptance as analytical tools for mycotoxins. Antibodies against almost all the important mycotoxins are currently available. Sensitive, simple and specific immunoassays have been established for the analysis of various mycotoxins in foods, feed, animal tissue and body fluids.

CHAPTER – 4

TOXICOLOGY

Mycotoxins which are widespread as a natural contaminants of foods and feeds enter food chain and reach human beings. Biological effects of these mycotoxins are varied from acute to chronic toxicity. Cytotoxicity, carcinogenicity, antitumor properties, mutagenicity, teratogenicity, nephrotoxicity, hepatotoxicity, estrogenic, tremorgenic, haemorrhage immunosuppressive activity, antimicrobial activity, gastro intestinal disorders, insecticidal activity and phototoxicity are some of the major biological effects (Table 4.1 and Fig. 4.1). Fung *et al.* (2004) elegantly discussed various aspects of health effects of mycotoxins. The inhalation of dust containing mycotoxins can cause a variety of toxic effects in human including acute and chronic respiratory disease (Bungar *et al.*, 2004).

4.1. ACUTE AND CHRONIC TOXICITY

Toxicity of mycotoxins varies greatly depending on both their structure and animal species involved. Both these aspects can be well illustrated by aflatoxins. Aflatoxin B_1 is most acutely toxic followed by G_1, B_2, G_2 in order of toxicity. Some mycotoxins are target oriented and act on specific organs of the body. For instance aflatoxins act on liver, citrinin on kidney, penicillic acid on heart, penitrem A and other tremorgenic toxins on central nervous system, vomitoxin and some trichothecenes on digestive tract. *Artemesia salina* larvae used as model organism for comparison of toxicity of 42 fungal metabolites (Durackova *et al.,* 1977).

4.2. CYTOTOXICITY

Growth inhibition and mitotic suppression was observed when heteroploid human embryonic lung cells were treated with a mixture of four aflatoxins. Aflatoxins also inhibit mitosis in cultures of human leucocytes. The characteristic effect of cytochalasins is the inhibition of cytokinesis.

Table 4.1. Toxic effects of some mycotoxins

Mycotoxins Cell line/species	Apoptotic effects	Reference
Fusarenon-X (FS-X) Human promyelotic leukemia HL-60°	Releases cytochrome C and activates caspase –3, -8, and –9	Miura *et al.* (2002)
Satratoxin and other TCTCs U937 human leukemia cells and murine macrophage RAW 264.7	Activation of MAPKs, extra cellular single-regulated protein kinase (ERK), p38 MAPK, and stress-activated protein kinase/ c-JNK	Yang *et al.* (2000)
T-2 toxin and other TCTCs HL-60	Activates caspase-3 and 9 and DFF/ CAD. Elevates Ca^{++} signal by the activation of endonuclease and protease	Yshino *et al.* (2000), Nagare *et al.* (2001), Shinozuka *et al.* (1997)
Dorsal skin of hypotrichotic WBN/ILA-Ht rats Thymocytes/mice	Depresses basal cell proliferation and elevates TGF-beta q mRA level independent of Fas/Fas	Alam *et al.* (2000)
Fumonisin B_1		
	Increases expression of TNFα, interleukin (IL)-1β and TNFα signaling molecules, TNF (Tum or necrosis factor) receptor 55 and receptor protein with no alteration of Fas related signal	Bhandari and Sharma (2000 a & b); Lemner *et al.* (1999); Dugyala *et al.* (1998)
Mouse/liver and kidney	Increases expression of IL-1α and IL-1 receptor antagonist (IL-1Ra) and caspase 8	
	Increases expression of oncogenic transcription factors B-Myc, c-Myc, Max and Mad	
	Increases apoptotic genes Bcl-2, Bax and Bad	
Renal epithelial LLC-PK1	Involves serine palmitoyltransferase and induces expression of calmodulin	Riley *et al.* (1999), Kim *et al.* (2001)
Tomato suspension	Involves L-pirin gene which encoding a protein 53% identical with human protein PIRIN that interacts with oncogene Bc1-3	Orzaez *et al.* (2001)
CV-1 and MEF (Mouse embryo fibroblasts)	Involves the TNF signal transduction pathway and cleavage of caspase; the tumor suppressor gene p53 was not involved	Jones *et al.* (2001)
African green monkey fibroblasts (CV1)	Repressed specific isoforms of protein kinase C and cyclin-dependent kinase 2 (CDK2)	Zhang *et al.* (2001); Ciacci and Jones (1999)

	Induced exprssion of CDK inhibitors, p21 Waf1/Cip1, p27 Kip 1 and p57 Kip2	
	The baculovirus IAP gene (blocks TNF induced APT) was involved	
	Inhibition of interleukin converting enzymes proteases or caspases by the baculovirus gene p35	
	The tumour suppressor gene p53 and Bcl-2 were not involved	
Simian virus 40	Resistant to the antiproliferative or apoptotic	Zhang *et al.* (1999).
Transformed African green monkey kidney fibroblasts CV-1	Effects of FB_1 Sp1 or Sp1-related proteins mediate FB_1 induced activation p21 promoter	

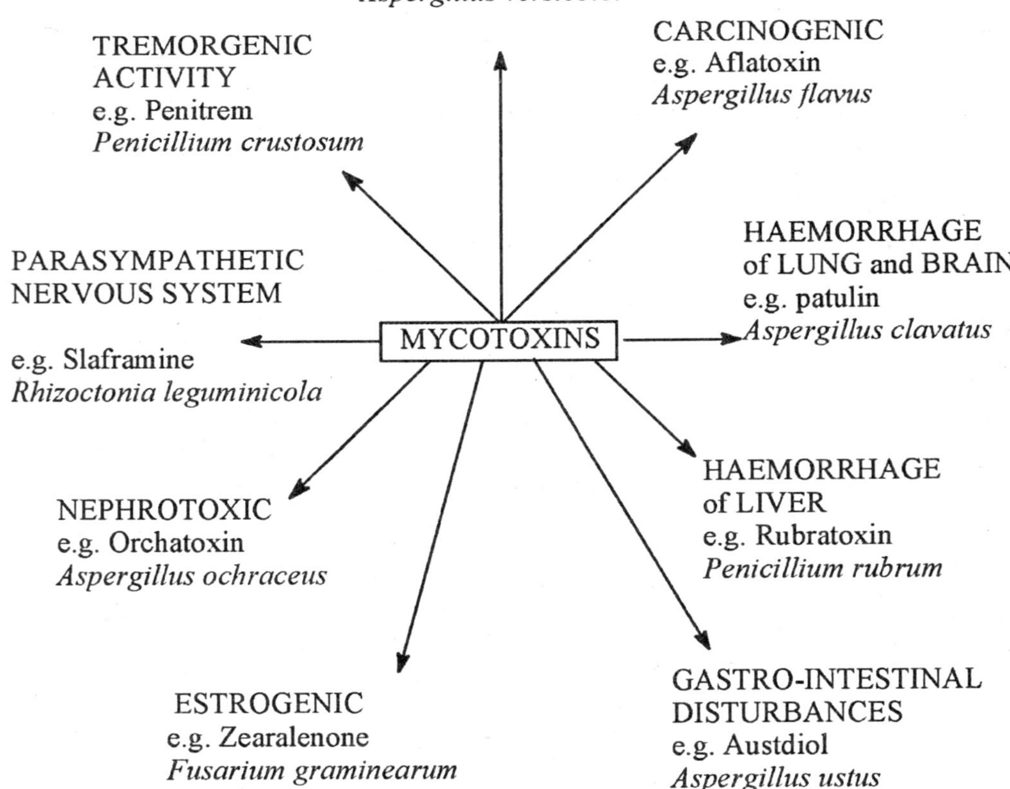

Fig. 4.1. Toxicity of different mycotoxins

4.3. CARCINOGENICITY

Carcinogenic potential of number of mycotoxins is well known, in experimental animals. Aflatoxin B_1 and sterigmatocystin are considered as most potent carcinogens.

Sterigmatocystin which is structurally related to aflatoxin B_1 is a potent hepatocarcinogen. It is believed that interaction with DNA is responsible for carcinogenicity and the mutagenicity of these mycotoxins. Carcinogenicity *in vitro* has also been confirmed with other mycotoxins such as citrinin, penicillic acid, patulin, luteoskyrin and rugulosin.

4.4. MUTAGENICITY

Mycotoxins were demonstrated to interfere with DNA and cause mutations. Aflatoxins B_1, strigmatocystin, versicolorin A, austocystins A and D and kojic acid are some of the established mutagenic mycotoxins. A wide variety of toxins such as rubric skyrin, rugulosin and crysophenol were also reported to be positive for mutagenicity.

4.5. TERATOGENICITY

Citrinin and aflatoxin B_1 cause teratogenic effects in chick embryos with malformations primarily in the extremities. However, other changes also occurred. Ochratoxin A was also reported to be teratogenic in pregnant rats.

4.6. TREMORGENICITY

Penetrim, aflatrim etc. effect central nervous system and cause nerve disorder. Pentrems are reported to be most wide spread.

4.7. IMMUNOTOXICITY

The sensitivity of the immune system to mycotoxin induced immunosuppression arises from the vulnerability of the continually proliferating and differentiating cells that participate in immune mediated activities and regulate the complex communication network between cellular and humoral components. Mycotoxicity is manifested by depressed T or B lymphocyte activity, suppressed antibody production and impaired macrophage/neutrophil – effector functions.

4.8. HAEMATOTOXICITY

Haematotoxicity of different mycotoxins can be classified into 3 categories.

1. Those for which haematoxic effects after intoxication have not been described in the literature.
2. Those able to promote adverse haematological effects yet the principal toxicity responsible for ensuing pathology is non-hematologic in nature.
3. Those principally cause adverse haematology effects associated or not with other minor symptoms of trichothecenes.

4.9. INDUCTION OF APOPTOSIS

Apoptosis (APT) is a regular programmed cell death induced by natural toxicants like mycotoxins. APT is characterized morphologically by cell shrinkage, membrane blebbing and chromatin condensation. Fragmentation of DNA and nuclear conditions in cells are used to detect the early phase of APT. Induction of APT by aflatoxins, beauvericin (Logrieco, 1998), ochratoxin A (Seegers *et al.*, 1994), several trichothecenes, leuteoskyrin (Ueno, 1995) rubratoxin (Nagashima, 1996), fumonisin B and gliotoxin (Sutton *et al.*, 1995) has been observed in number of cell lines as well as cell lines of target organs after animal received the toxin. The concentration causing such induction varies with the mycotoxin. Some mycotoxins may induce APT in one system but not in others. For examples several trichothecenes induce APT in HL-60 at significant low levels with minimum effective dose ranging from 0.001 – 0.2 μg/ml, where as aflatoxins, ochratoxin A induce the same at concentration range of 1-20 μg/ml. No induction of APT was observed with cytochalasin A, fumonisins (Ueno, 1995) and fusaric acid (Voss *et al.*, 1999).

Wang *et al.* (1996) have studied the effect of fumonisin B and AAL toxin on APT in many cell lines and concluded that kidney and liver cells of animals or of tomato plants exhibited typical apoptotic characters which varied with the dose and time of application.

Similar to other toxicants modification of signal transduction pathway by mycotoxins plays a key role for APT induction. Gliotoxin induced apoptosis in spleen cells was found to be mediated by camp (Sutton *et al.*, 1995). Much effort in elucidating the sites involved in induction APT by mycotoxins in the signal transduction pathway was focused on FMB_1 and trichothecenes. It is clear from the data presented in Table 4.1 that mitogen activated protein kinase plays a significant role in induction of APT and toxic effects of trichothecenes.

Alteration of membrane structures by trichothecenes and several other mycotoxins may directly affect cellular components. Modulation of signal transduction of protein kinase by mycotoxins has been suggested as another mechanism. Oxidative DNA damage has been found to be involved in the genitoxicity of aflatoxins (Shen *et al.*, 1996). Lipid peroxidation was considered as a possible secondary mechanism for sterigmatocystin. Hoehler and Marquardt (1996) feel that enhancement of lipid peroxidation is considered to be one of the manifestation of cellular damage in ochratoxin A toxicity.

4.10. ANTITUMOR PROPERTIES

Several mycotoxins possess antitumor properties. Trichothecenes, verrucarins and roridins were cytotoxic and inhibited the tumor in mice and walker carcinoma in rats. Inhibitory effect of diacetoxyscirpenol on the growth and experimental tumors in mice and rats was also reported. Penicillic acid and mycophenolic acid also exhibit antitumor effects.

4.11. ANTIMICROBIAL ACTIVITY

A variety of fungal metabolites presently known as mycotoxins were originally discovered as antibiotics. Mycophenolic acid, penicillic acid, citrinin, patulin and trichothecenes are some of such examples.

The antiviral activity of mycophenolic acid was demonstrated in tissue culture against vaccinia, measles, herpes simplex and new castle diseases. Patulin inhibits the replication of mycovirus in *Penicillium stoloniferum* (Detroy and Still, 1975). *Bacillus thuringiensis* is sensitive to kojic acid, patulin, penicillic acid, DAS, T-2 toxin and rubratoxin B (Boutibones, 1979). Roquefortine selectively inhibits only gram +ve bacteria (Koppe and Rehm, 1979). Morphological changes in *Botrytis cinerea* and *Paecilomyces variotii-extra* were induced by aspergillic acid, citrinin, griseofulvin, patulin, rugulosin and trichothecenes.

4.12 INSECTICIDAL PROPERTIES

Several mycotoxins such as aflatoxins, fusaric acid, kojic acid, citrinin and rugulosin are known to possess insecticidal properties. Insecticidal and larvicidal activities of aflatoxins B_1, rubratoxin, patulin and diacetoxyscirpenol towards *Drasophila melanogaster* was also reported (Reiss, 1975).

4.13. PHYTOTOXICITY

Aflatoxins inhibited the germination of grass seeds and later at lower concentrations induced chlorophyll deficiency in plants. Cytochalasin B acts as a thiol reagent influencing tip growth, cytoplasmic streaming, organelle movement, cell wall synthesis, auxin transport etc. Permeability of plant membranes is damaged by patulin, citrinin and fusaric acid. T_2 toxin and diacetoxyscirpenol inhibited the germination of *Nicotiana sylvestris* pollen.

4.14. MECHANISM OF ACTION OF MYCOTOXINS

Biochemical effects of mycotoxins include the reactions of mycotoxins or their reactive metabolites with critical molecules in animal cells, resulting in impairment of vital biochemical processes involving these molecules. The susceptible critical molecules are known as molecular receptors of the acting toxins. The structurally and functionally altered molecular receptors are known as molecular or biochemical lesions. Modern toxicology has subscribed to the theory that the various biological effects, or toxic end points, caused by a toxin are initiated by certain biochemical lesions induced by the toxin. In this chapter discussion will be limited to those biochemical injuries that are related to human mycotoxicoses of demonstrated significance.

Due to diversity of their chemical structures, mycotoxins may exhibit a number of biological effects including both acute and chronic toxic effects. Toxic effects of most mycotoxins are very organ specific and their mode of action vary considerably. Once mycotoxins have been ingested the toxins may both react directly with the target organs to exert toxic effects or be metabolized to an active intermediate (activated) and exert their toxic effects. The toxins or their metabolites may also be detoxified and then excreted from the body. The interaction of mycotoxins either may get activated or as intact molecules with cellular macromolecules plays a dominant role in exerting their toxicity (Chu, 2003). Mycotoxins can also be either immunosuppressable or immunostimulatory (Bondy and Pestka, 2000).

4.14.1. Receptors

Mycotoxins, with their diverse chemical structures, are capable of reacting with various types of molecular receptors, including DNA and RNA, functional proteins, enzyme cofactors and membrane constituents. The relative significance of a molecular receptor of a mycotoxin is determined by:

(a) Its affinity to the active form of the toxin

(b) Its role in vital biochemical processes

(c) The persistence of the lesion it bears

(d) The severity of the consequences of the lesion.

Based on these considerations, lesions occurring in DNA and RNA are of particular significance because they not only destroy the structural integrity of these informational molecules, but also interrupt their biosynthesis of proteins and other cellular constituents essential to the vitality of the cell.

The reactions between mycotoxins and their molecular receptors may be covalent-irreversible or non-covalent-reversible. In the non-covalent-reversible reactions, mycotoxins form complexes with receptors in a specific fashion comparable to the complex formation between the receptors and their endogenous substrates. In this type of reaction, mycotoxins usually fit the three-dimensional shape of the specific active sites in receptor molecules, rendering the receptors unavailable for endogenous substrates. The non-covalent interactions are reversed by dissociation of receptor-mycotoxin complexes as metabolic processes remove the toxins from the receptor sites.

In the covalent irreversible reactions, a reactive form of mycotoxins conjugate with receptors to form adducts. The most significant reactions are nonspecific electrophilic attacks on the nucleophilic centres of the receptor molecules, usually nitrogen, oxygen and sulphur heteroatoms in proteins and nucleic acids. When the adduction occurs at the active site of a critical receptor, or results in a sufficient conformational change of the receptor, it produces chemical injury and impairs the function of the receptor. Binding to heteroatoms away from the active site, however, could represent a significant means by which the toxins are sequestered and deactivated. The receptor-mycotoxin adducts can be removed from the cell by a number of processes, including spontaneous decomposition of the covalent bonds, receptor turnover and specific repair processes.

In some cases, a mycotoxin would initially form a complex with a receptor due to structural compatibility, and adduction would then take place at the binding site to transform the complex to a covalently bound adduct. When a mycotoxin can interact with a critical receptor in this enhanced manner, it is usually very potent.

4.14.2 Metabolic activation and detoxification

Metabolism or biotransformation by drug-metabolizing enzymes is a central mechanism by which foreign compounds such as mycotoxins are converted to more polar, water-soluble forms that can be excreted and removed from the body. There are two types of biotransformation reactions: phase 1 and phase 2. Phase 1 reactions, which are generally oxidative, reductive or hydrolytic processes, provide the necessary chemical structure for the phase 2 reactions, which are generally conjugation reactions. A mycotoxin which possesses the necessary chemical structure for phase 2 reactions will undergo these reactions without prior phase 1 transformations. Phase 1 reactions may result in activation as well as detoxification of a compound. Phase 2 reactions depending upon the conjugated cellular constituents may lead either to detoxification or to the formation of biochemical lesions.

4.15. NUCLEIC ACIDS AS RECEPTORS

Nucleic acids, in their vital roles in cellular heredity, survival, growth and division, are crucial targets of genitoxic mycotoxins. Nucleophilic heteroatoms in the organic bases of nucleic acids are vulnerable to attack by electrophilic metabolites of mycotoxins, forming covalent adducts.

Alterations in the precise structures of DNA and RNA molecules by these adductions will impair the template activity of these molecules and result both in inhibition of the synthesis of DNA, RNA and proteins and in point mutations leading to synthesis of erroneous macromolecules. The consequences of adduct formation *in vivo* may be cellular death or cellular transformation to cancerous conditions. The relatively well-studied genotoxic mycotoxins are aflatoxins, their biosynthetic precursors (sterigmatocystin and versicolorin A). The *Penicillium* mycotoxins involved in the yellow rice disease (luteoskyrin, rugulosin), and emodin. Of these, AFB_1 has been extensively studied as a model carcinogen for the elucidation of the mechanism of chemical carcinogenesis.

4.16 LESIONS ON PROTEINS

Proteins are the most common receptors in the cell because of their abundance as cellular constituents and the presence of nucleophilic N, O and S heteroatoms in their functional groups. Both nonspecific-irreversible-covalent and specific-reversible-non covalent binding by mycotoxins are important in altering the structures and activities of proteins. As mentioned earlier, the former type of binding may result in conformational changes which denature or block the active sites for binding endogenous substrates, while the latter type of binding allows the mycotoxin to compete with endogenous substrates for the active sites. Proteins which bind mycotoxins reversibly and non-covalently may also serve as carriers to transport reactive metabolites, or as reservoirs of the toxins to stabilize and prolong cellular exposure to the toxins.

Non-covalent, reversible interactions between proteins and mycotoxins have been observed. In numerous examples of competitive inhibition of metabolic enzymes by various mycotoxins and of the complex formation between mycotoxins and cellular constituents.

4.17. TARGET SITES AND BIOLOGICAL EFFECTS

Mycotoxins usually enter the human body through ingestion. From its absorption to its excretion, the toxin encounters various types of molecules characteristic of the organs involved in its transmission and transformation. These molecules are all potential receptors.

The prominent sites of interaction following oral administration of a mycotoxin include gastrointestinal microflora, intestinal epithelial cells, portal blood, liver, bile, circulatory blood and extrahepatic organs. The fate of processes that most significantly influences the toxicity of mycotoxins include the metabolism by enzymes present in these sites, especially liver, the interactions with blood constituents, the enterohepatic cycle and the accumulation in target organs.

The liver is obviously the primary target organ of orally administered mycotoxins. However, due to their structural diversity, mycotoxins are capable of causing harm to other organs as well. The organs that have been observed as major targets are liver, kidney, intestine, blood, the reproductive system, the central nervous system and the immune system. With the implication of mycotoxins in chemical warfare, the skin and lung may become the primary target organs of some mycotoxins.

Within the target organs, mycotoxins or their reactive metabolites react with critical molecular receptors to result in a spectrum of effects:

(a) Specific and overt disorders

(b) Nonspecific impairment of normal functions

(c) Reduction in native mechanisms of resistance and immunity

(d) Induction of delayed, irreversible cellular transformation such as mutagenesis, teratogenesis and carcinogenesis.

These effects have been observed as the end points of certain critical biochemical processes being impaired or altered by some prototype mycotoxins. Elucidation of these biochemical effects offers a mechanistic explanation of the toxicity of these mycotoxins. The most important biochemical effects underlying the toxic end points are inhibition of the activity and the biosynthesis of functional macromolecules, and inhibition of the energy production in cells. These biochemical effects will be elaborated in the subsequent sections.

4.18. INHIBITION OF ENERGY PRODUCTION

Inhibition of cellular energy production is a major metabolic effect of mycotoxins at acute levels of exposure. The molecular receptors in this biochemical effect are the enzymes that catalyse certain key reactions in the tricarboxylic acid (TCA) cycle, the electron transport to oxygen, and the oxidative phosphorylation for ATP production. Both nonspecific-irreversible-covalent and specific-reversible-noncovalent bondings have been observed to occur between these enzymes and toxins. An inadequate supply of ATP resulting from the biochemical effects of mycotoxins will reduce cellular functions and biosynthesis and eventually cause cell death. Some mycotoxins also modify various types of adenosine triphosphatases (ATPase) which inturn disrupt certain important functions of the plasmamembrane.

4.18.1. Cellular respiration

Several mycotoxins are potent inhibitors of certain enzymes involved in the TCA cycle and the electron transport chain in mitochondria. The immediate cellular response to a disturbance of these metabolic processes is reduced oxygen consumption. Aflatoxins B_1, G_1 and M_1 inhibit electron transport between cytochromes b and c or c_1 (site II) in rat liver mitochondria (Doherty and Campbell, 1973). The *Fusarium* toxin, moniliformin, is a potent inhibitor of the mitochondrial pyruvate and ketoglutarate oxidation. In this way, moniliformin acts like fluoroacetate on the TCA cycle.

4.18.2 Uncoupling of oxidative phosphorylation

In addition to affecting cellular respiration, some mycotoxins also uncouple oxidative phosphorylation. Uncoupling of oxidative phosphorylation immediately depletes ATP in the cell. This loss of ATP may then cause mitochondrial swelling due to the failure of the cell to maintain sodium and potassium gradients. The uncoupling activity of mycotoxins may or may not accompany their ability to inhibit cellular electron transport.

Aflatoxin $B_1$and M_1 uncouples *in vitro* oxidative phosphorylation as well as inhibits electron transport. In contrast, luteoskyrin, emodin and secalonic acid D all uncouple oxidative phosphorylations more strongly than they inhibit the cellular respiration at site III.

The nephrotoxic mycotoxin, ochratoxin A, depletes cellular ATP by yet another mechanism. ATP is consumed in the uptake of ochratoxin A by mitochondria and more importantly, this toxin inhibits intramitochondrial phosphate transport, resulting in the impairment of oxidative phosphorylation.

4.18.3 Alteration of membrane permeability

Another way in which mycotoxins affect cellular energy production and utilization is the alteration of ATPase activity. These membrane-bound enzymes are responsible for the hydrolysis of ATP to generate the energy required for active transport of cations across the cell membrane. They are also involved in ATP synthesis and oxidative phosphorylation. These enzymes are especially vulnerable to mycotoxin attack, due to their association with the cell membranes. Binding of mycotoxins to ATPases may disrupt a variety of transport- or permeability-mediated physiological and biochemical functions of various types of cells.

The effects of mycotoxins on ATPases are enzyme and toxin specific, while the rat liver oligomycin-sensitive Mg^{2+} ATPase was inhibited by AFB_1 (Hayes and Hood, 1978) and the *Penicillium* toxins, patulin (Phillips and

Hayes, 1977) and rubratoxin B (Desaiah *et al.*, 1977), the oligomycin-insensitive Mg^{2+} ATPase is only inhibited by rubratoxin B and the Na+/K+ ATPase only by patulin. Similarly, secalonic acid D weakly enhances latent ATPase activity but suppresses 2,4-dinitrophenol-induced ATPase (Kawai *et al.*, 1983). The neurotoxic citreoviridin and its derivatives are also potent inhibitors of mitochondrial ATPase activity (Gause *et al.*, 1981).

Other types of membrane toxicity of mycotoxins have also been observed. Moniliformin causes disturbance of intracellular osmoregulation and hence severe intracellular oedema in the myocardium and other tissues, suggesting an involvement of suppression of ATP-derived transmembrane transport mechanisms (Thiel, 1978). Trichothecene mycotoxins may increase the permeability of blood vessel walls and intestinal mucosal epithelium and cause leakage of plasma contents into the intestinal lumen, which eventually leads to diarrhoea (Matsuoka and Kubota, 1981). The first sign of the ochratoxin A nephrotoxicity is also believed to be membrane associated. Thus, ochratoxin A may interfere specifically with the membrane transport system of p-hippuric acid, leading to the release of the membrane-bound alanine aminotranspeptidase followed by the release of leucine aminopeptide and -glutamyl transpeptidase (Endo, 1983).

4.19 INHIBITION OF MACROMOLECULAR SYNTHESIS

Inhibition of cellular biosynthesis of macromolecules is a major metabolic effect of mycotoxins at acute levels of exposure. This effect results in the cessation of replenishment of essential molecules, especially functional proteins, eventually leading to cell death. Tissue damage would then stimulate mast cells, resulting in inflammation. It also causes leakage of body fluids resulting in haemorrhage. If the exposed animals survive the acute effect, cell regeneration may promote expression of existing DNA lesions and development of cancer.

4.19.1 Inhibition of DNA synthesis

Inhibition of DNA synthesis is one of the first consequences of AFB_1 toxicity and possibly also by other genotoxic mycotoxins. In the case of AFB_1, inhibition of DNA synthesis in the liver, the target organ, occurs at toxin concentrations which are not inhibitory to RNA or protein synthesis indicating that inhibition of DNA synthesis is a primary biochemical effect. This effect is caused by covalent binding of AFB_1 to DNA and to proteins, resulting in modification of DNA template activity and inactivation of certain enzymes involved in DNA synthesis. Covalent binding of AFB_1 with membrane proteins may also reduce the transport of thymidine and other nucleotides required as building blocks in DNA synthesis.

4.19.2. Alteration in RNA metabolism

Alteration of RNA metabolism by mycotoxins will affect the role of RNA in the intricate transcription and translation processes of cellular protein synthesis. Rat liver nuclear RNA synthesis is rapidly and markedly inhibited by AFB1 (Lafarge and Frayssinet, 1970). Inhibition of RNA synthesis is primarily due to the reduction of DNA template activity, inhibition of nucleo-plasmic polymerase (polymerase II) and impairment of nucleotide transport. The nucleoplasmic polymerase II is primarily responsible for mRNA synthesis.

Aflatoxin B_1 also disrupts the mechanisms of post-transcriptional maturation or processing of nuclear RNA for the production of rRNA from nucleolar RNA precursors. It interferes with the cleavage of 45 sRNA precursors into the 18S and 28S rRNA components in rat liver possibly resulting in reduced production of 28S RNA and a consequent accumulation of 32S RNA fragments in the nucleolus. Similarly, tRNA processing is interrupted resulting in increased cytoplasmic levels of 5S precursor of tRNA.

4.19.3 Inhibition of protein synthesis

Mycotoxins inhibit cellular protein biosynthesis by modification of DNA template activity, inhibition of RNA synthesis and maturation, inactivation of biosynthetic enzymes and impairment of amino acid transport. The various mechanisms of inhibition of protein biosynthesis have been observed for AFB_1, ochratoxin A and trichothecene mycotoxins.

Ochratoxin A inhibits specifically and competitively the activity of phenylalanyl-tRNA synthetase, an enzyme involved in the initial step of protein synthesis. The trichothecene mycotoxins inhibit protein synthesis selectively in eukaryotic cells, through breakdown of the polyribosomal profile in the exposed cells.

Depending upon the sites of toxic action, the trichothecenes fall into two classes: the initiation inhibitors (I-type) which preferentially act on the initiation step of the ribosomal cycle, and elongation-termination inhibitors (ET-type) which preferentially act on the elongation and termination steps.

The I-type inhibitors, in the order of decreasing potency, include verrucarin A, fusarenon–X, nivalenol, T-2 toxin and diacetoxyscirpenol. The ET- Type inhibitors include deoxynivalenol, crotocin and trichodermin.

4.19.4 Alteration of immune responses

One consequence of the inhibition of protein synthesis caused by mycotoxins is the alteration of serum protein composition, resulting in the suppression of the production of nonspecific humoral substances important to native defence. Thus subacute doses of AFB_1 cause complement deficiency in guinea pigs and delay interferon production in turkeys. At higher doses,

AFB_1 lowers the levels of IgG and IgA in chicks resulting in decreased acquired immunity. Ochratoxin A is highly immunotoxic to BALB/c mice. A dose as low as 0.005 ug/kg is sufficient to suppress antibody formation (Haubeck *et al.* 1981).

Trichothecenes also inhibit antibody formation. When T-2 toxin or fusarenon-X is administered to mice, IgE and IgG_1 antibody formation *in vivo* is suppressed, as is the antibody formation of spleenic lymphocytes which is raised by T-dependent and independent mitogens.

4.19.5 Inhibition of gluconeogenesis

Reduction of hepatic glycogen and increase in serum glucose levels in the exposed animals is a common acute effect of mycotoxins such as AFB_1, Ochratoxin A and rubratoxin B. This effect is in part attributable to reduction of gluconeogenesis due to inhibition of the biosynthesis of the enzymes involved with these toxins. An outstanding example of this biochemical effect is the selective inhibition of the biosynthesis of a specific gluconeogenic enzyme, the renal cytosolic phosphoenolpyruvate carboxy-kinase (PEPCK), by ochratoxin A. Ochratoxin A greatly reduces the concentration of the mRNA coding for PEPCK in the rat kidney.

4.19.6 Effect on lipid metabolism

Alteration of lipid metabolism by mycotoxins generally occurs more readily than the effects on gluconeogenesis. The most pronounced effect is the accumulation of lipids in liver or fatty livers caused by mycotoxins such as, aflatoxins, ochratoxin A, luteoskyrin, rubratoxin B, T-2 toxin and citrinin. This effect is primarily a result of impaired lipid transport rather than of increased lipid biosynthesis. In fact, fatty acid and cholesterol biosynthesis is inhibited in most of the experimental mycotoxicoses.

4.20 EFFECT ON CENTRAL NERVOUS SYSTEM

Three classes of neurotoxic mycotoxins have been the subject of mode of action studies:

(a) Those causing paralysis and respiratory arrest, such as citreoviridin and asteltoxin;

(b) Those inducing trembling in animals, *e.g.* tremorgens such as penitrem A, verruculogen and fumitremorgin A.

(c) Those causing animals to vomit, e.g. vomitoxins such as certain trichothecenes.

The first classes of neurotoxins include potent inhibitors of mitochondrial energy generation and ATPase activity, as mentioned earlier. Their mode of action is related to nonspecific inactivation or killing of nerve cells due to disruption of energy supply by the toxins.

The tremorgens act on the central nervous system by altering functional states of neurotransmitters. Thus penitrem A stimulates the spontaneous release of endogenous glutamate, y-aminobutyric acid (GABA) and aspartate from sheep and rat cerebrocortical synaptosomes, while verruculogen increases the spontaneous release of glutamate and aspartate, but not that of GABA (Norris *et al.,* 1980). The primary site of action appears to be presynaptic and involves central synapses operated by amino acid transmitters. In mice, fumitremorgin A appears to affect the GABA generating function of the central nervous system.

The vomitoxins may act on the chemoreceptor trigger zone in the medulla oblongata of the central nervous system. They also change the level of biogenic amines. In chickens incubated with T-2 toxin, the brain dopamine concentration was significantly elevated, where as brain norepinephrine was decreased.

4.21. MODIFICATION OF CYTOSKELETON

A number of toxic fungal metabolites have been found to affect the functions of microtubules and microfilaments, the two cytoskeletons that regulate locomotion, cell division, cell membrane and other cellular functions. Some 40 fungal metabolites are known to possess the cytoskeleton-modifying activity and these are collectively called "cytochalasins". These toxins inhibit locomotion and cytoplasmic cleavage causing nuclear extrusion. Cytochalasins are the first chemicals discovered that consistently block cytoplasmic division, without interfering with nuclear division.

Cytochalasin B for example, inhibits gelation of actin filaments with high molecular weight actin-binding protein from a macrophage. Low levels of cytochalasin B inhibit both the rate of actin polymerization and the interaction of actin filaments in solution. At the cellular level, cytochalasins inhibit locomotion at a low concentration and cause ballooning of cells by inducing construction of actin fibres (stress fibres) at a high concentration. Cyclochlorotine induces deformation of the hepatocellular membranes of the isolated hepatocytes, which is characterized by many small balloons or blebs.

4.22. HORMONAL EFFECT

4.22.1 Activity of steroid hormones

Steroid hormones regulate cellular functions by specifically and non-covalently binding to cytoplasmic receptor proteins in target cells. The hormone-receptor complexes are activated and then transported to the nucleus, where they are bound to chromatin acceptor sites and are able to induce selective gene transcription and hence protein synthesis. The ability of AFB_1 to covalently bind chromatin components may lead to reduction of the nuclear acceptor sites for hormone-receptor complexes, and consequently decrease the activity of the hormones.

Another pronounced effect of AFB_1 on the activity of steroid hormones to induce protein synthesis is endoplasmic reticulum degranulation and polysomal disaggregation.

4.22.2. Estrogenic effect of zearalenone

Zearalenone, an estrogen, exhibits its physiological activity first by binding with cytosolic receptor protein and then by translocation of the estrogen-receptor complex into the nuclei, in a fashion similar to the action of steroid hormones. It possesses a high affinity for cytoplasmic estrogen receptors. An early event is the induction of specific proteins, accompanied by an increase in nuclear RNA polymerases I and II and protein synthetic activities.

In rats zearalenone is carried into the brain through the blood-brain barrier and binds with the estrogen receptors of both the hypothalamus and the hypophysis, suggesting that zearalenone affects the estrogen feed back system through the estrogen receptor of the rat brain. Reduction of zearalenone to zearalenol by the zearalenone reductase is apparently an activation of zearalenone, since the reduction product is about 10 times more estrogenic than the parent compound.

4.23. CARCINOGENIC EFFECT

It has been generally accepted that transformation of a normal cell to a cancerous cell involves two phases: the initiation and the promotion phases. In the initiation phase, biochemical lesions are produced in the components of cellular genetic materials, especially DNA by some genotoxic agents. These lesions become a "fixed" feature of the progeny of the initially altered cells through cell divisions. These altered cells are now potentially capable of being transformed into cancerous cells, but they must first undergo a promotion phase. In the promotion phase, "favourable" physiological conditions are provided by exogenous or endogenous stimuli, which accelerate the transformation of the altered cells into autonomous cancerous cells which will grow and develop independently of the systemic regulatory mechanisms. The biochemical events involved in the promotion phase are still largely unknown. Chemicals that only cause genetic damage which initiates the carcinogenic process are known as initiators (genotoxic). Chemicals that only promote the transformation of genetically modified cells to cancerous cells are known as promoters. Most chemical carcinogens are both initiators (genotoxic) and promoters, and are also known as complete carcinogens.

Well-known carcinogenic mycotoxins are aflatoxins B_1, G_1 and M_1, fumonisins, sterigmatocystin, versicolorin A, luteoskyrin and rugulosin. These mycotoxins are genotoxic and produce positive results in short-term genotoxicity assays such as the Ames mutagenicity test. Other similar genotoxic mycotoxins, such as emodin and fusarin C, are also likely to be carcinogenic. Trichothecene mycotoxins, ochratoxin A and zearalenone have questionable or no genotoxicity, but are definitely cancer promoters. Some biochemical events that have been observed in the initiation and promotion phases of mycotoxin-induced carcinogenesis will be described using specific examples.

4.23.1. Initiation

Aflatoxin B_1 has been extensively used as a model hepatocarcinogen to elucidate the mechanism of the initiation process of carcinogenesis in liver. Special attention has been focused on the covalent binding of the AFB_1 epoxide to the hepatocyte DNA and on the analysis of the aflatoxin-DNA adducts (Fig. 4.2).

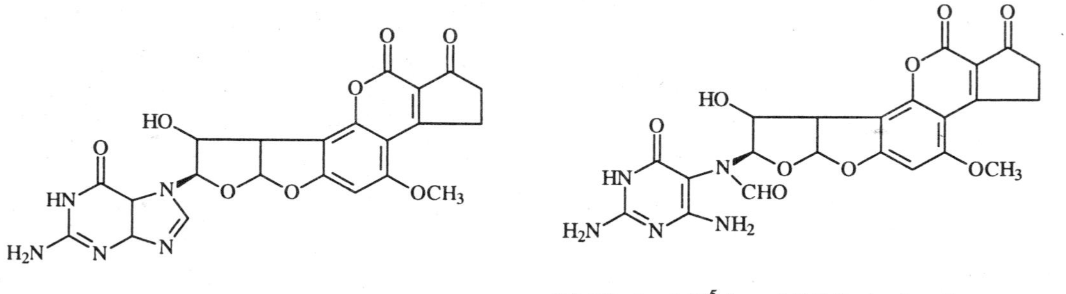

8,9-Dihydro-8-(N7-guanyl)-9-hydroxy-oflatoxin B_1 (AFB_1-Gua)

8,9-Dihydro-8-(N^5-formyl-2',5',6'-triamino-4'-oxo-N^5-pyrimidyl)-9-hydroxy-aflatoxin B_1 (AFB_1-triamino-Py)

Fig.4.2. Induction of cancer

4.23.2. Promotion

The mechanisms of promotion of carcinogenesis are largely unknown. However, the ability of a chemical to promote cancer development seems to be consistent with its ability to stimulate cell division, alter the immune system, or induce resistance to cytotoxicity. Nongenotoxic mycotoxin, such a trichothecenes and ochratoxin A, are cancer promoters due to their cytotoxity and immunotoxicity. They may induce development of cancer by selecting cells that are more resistant to cycotoxicity and by stimulating division of cells that may have been genetically altered. They may also promote carcinogenesis by damaging the immune system of an animal, thus undermining its natural defense mechanism against the development of abnormal cells.

CHAPTER - 5

EPIDEMIOLOGY OF MYCOTOXICOSIS RISK TO HUMAN HEALTH

Mycotoxin occurrence on commodities susceptible to mold infestation is largely unavoidable. Environmental factors such as temperature, humidity and extent of rainfall during the pre-harvest, harvest, and post-harvest periods influence mycotoxin production. Thus the incidence of mycotoxin contamination of a particular food crop can vary from region to region, country to country and year to year. Levels of contamination of a commodity with mycotoxin(s) vary with geographic location, agriculture and agronomic practices and the susceptibility of the plants to fungal invasion during different phases of growth (Wood and Trucksess, 1998). It has been estimated that approximately 25% of the world's food crops are affected by mycotoxins annually (Council for Agriculture Science and Technology, 1989). Hence, the potential for human consumption of mycotoxin-contaminated food leading to chronic or acute illness is not insignificant.

Mycotoxins may produce or be associated with various forms of toxicity in humans and susceptible animals, including acute toxicity, teratogenicity, mutagenicity, and or carcinogenicity (Wood and Trucksess, 1998). Classic epidemiological studies have been conducted for only a few mycotoxins; other less rigorous studies report on co-occurrence of a mycotoxin and a particular disease in humans.

Mycotoxins most commonly found in naturally contaminated foods include ergot, aflatoxins, trichothecenes, fumonisins, ochratoxin A, zearalenone and patulin. The relationships between these mycotoxins and human food-borne disease will be discussed in this chapter.

Mycotoxin formation in food stuffs is heavily influenced by environmental factors such as temperature and water activity which may account for pronounced annual variation in contamination by mycotoxins. Thus retrospective and prospective assessment of food stuff contamination by mycotoxins is difficult, if not impossible. Consequently the provision is inhibited of quantitative data on mycotoxin intake which are required for the calculation of risk association in retrospective studies of mycotoxin related diseases. This is the main reason for the paucity of epidemiological studies of mycotoxin related diseases. Primary hepatocellular carcinoma in man and the association with aflatoxin is the only topic so far subjected to analytical epidemiological studies and in none of the studies have all the risk factors been adequately considered.

Studies of Krogh (1976) on ochratoxin A associated nephropathy in swine have revealed the endemic occurrence of this disease with epidemic outbreaks related to climatic changes, such as excessive precipitation in previous harvest season. However, the most convincing evidence in the process of establishing causal association between factors and disease is usually obtained from analytical epidemiological studies from which the necessary quantitative expressions of association can be obtained (Lilienfed and Lilienfed, 1980). Within the mycotoxin research field analytical epidemi-ology has been applied on human disease associated with aflatoxins, where as no other mycotoxins have been dealt so far. The main reason for this paucity in epidemiological research data appears to be the difficulties encountered in attempts to quantify mycotoxin exposure in man and farm animals.

Health effects induced by mycotoxins are mainly caused by alimentary exposure to mycotoxin contaminated food stuffs, although the existence of other routes cannot be excluded, an example being air-borne exposure to mycotoxin containing spores and mycelial fragments. Usually fungi invade only small fraction of food stuffs which mainly contributes for pronounced variation in results from monitoring food stuffs for mycotoxin contamination. Mycotoxin formation in food stuffs is influenced by number of factors such as temperature and water activity, the latter reflecting in particular precipitation during crop harvest. As a result pronounced variation in mycotoxin contamination of food stuffs is evident even in the same crop from the same area when compared on animal basis. Thus it is difficult if not impossible retrospectively or prospectively to assess mycotoxin contamination of food stuff. Further, because of the variation from batch to batch even within the same crop year data on mycotoxin contamination of food stuffs cannot easily be translated into exposure data such as intake for man and animals. Indirect assessment of intake by measurements of tissue residues of mycotoxins has not been successful mainly because mycotoxins are retained only for short time in the human and animal body, before being metabolized. Thus ochratoxin A has a residual life (Rl_{50}) of 3-5d in swine tissue (Krogh, 1978).

During the years 1970-77 several epidemiological studies on the association of aflatoxins with Primary Hepatocellular Carcinoma (PHC) have been conducted in Africa and South East Asia in areas with high incidence rates of PHC and when the results are analysed a positive linear correlation was observed between the estimated daily intake of aflatoxins (3.5 – 22.2 mg/L body weight) and the annual incidence rate (1.2 – 13 cases, per 10^5 people), pointing to aflatoxin as a major disease factor in the causation of PHC.

However, there were limitations in this analysis such as that (1) population intake data not individual intake data were considered. (2) secondly the cross sectional study is not entirely adequate for elucidation of causal relationship of diseases such as PHC requiring considerable time for development and finally no other risk factors were considered known or suspects of causing PHC such as alcohol and hepatitis B virus inspite of the fact both of these factors are predominent in the areas studied.

5.1. ERGOT

Ergot is the common term for sclerotia of fungal species within the genus *Claviceps,* particularly *C. purpurea.* Florets of grasses and cereals are infected with the fungus. The florets are replaced with compact fungal structures or sclerotia, 2-20 mm long, strongly colored, and often purple-black. Consumption of these sclerotia by humans and animals results in toxicosis, called ergotism (IPCS, 1990).

Relationship to disease in humans

Generally, it is recommended that animals not be given feed containing more than 0.1% ergot (expressed on a weight basis). An estimate of human exposure to total ergolines of 5.1 μg/person in Switzerland has been performed, based on average concentrations of total ergolines found in wheat flour, rye flour, and by-products. Cleaning or milling processes remove the sclerotia. Various food processing techniques of flours containing ergolines have been shown to produce 50-100% reduction in total ergolines.

Epidemics of ergotism are being most numerous between the 9th and 18th centuries. Two types of disease have been noted: gangrenous and convulsive ergotism. The IPCS reported that the latest outbreaks of ergotism in Europe occurred in the United Kingdom and the USSR in 1926-28 (IPCS, 1990). Source did not agree on whether an outbreak with symptoms both bizarre and atypical in France in 1951 was due to ergot or another toxicant (Louria *et al.*, 1970).

The IPCS (1990) reported an ergometrine-related outbreak of ergotism in Ethopia in 1978. The episode occurred after 2 years of drought and involved the locally grown barley, the staple food, which became dominated by wild oats heavily contaminated with *C. purpurea* sclerotia. The amount of grain or ergot consumed by the affected individuals was unknown. The grain consisted of 0.75% ergot by weight and ergometrine was detected in the sclerotia by thin-layer chromatography. A total of 93 cases of ergotism was reported including 47 deaths. The male: female ratio was 2.5: 1, while more than 80% of affected persons were between 5 and 34 years of age. Examination of 44 cases revealed on going dry gangrene, feeble or absent peripheral pulses, swelling of limbs, desquamation of the skin, loss of one or more limbs, weakness, burning sensation, nausea vomiting and diarrhoea. A total of 50-60 infants and young children also died from starvation due to failure of the mothers to lactate.

The IPCS (1990) also reported several clavine-related outbreaks in India following ingestion of ergot from *C. fusiformis* in bajra or pearl millet. The most recent outbreak occurred in 1975 in the state of Rajasthan. The amount of grain consumed by affected individuals was not reported. Pearl millet from affected villages contained 15-174 g ergot/kg, resulting in contamination of the grain with 15-199 mg total ergolines/kg. Pearl millet from villages with no cases of intoxication contained 1-38 g ergot/kg with a total ergoline content of 15-26 mg/kg and it was suggested that the intake of 28 μg total ergolines/kg body weight would be nontoxic.

5.2. AFLATOXINS

Aflatoxins (AF) B_1, B_2, G_1 and G_2 are the mycotoxins that may be produced by three molds *A. flavus, A. parasiticus* and *A. nomius,* which contaminate plants and plant products. Aflatoxin M_1 (AFM_1) and M_2, the hydroxylated metabolites of AFB_1 and AFB_2 respectively, may be found in milk or milk products obtained from livestock. Of six aflatoxins, AFB_1 is the most frequent one present in samples. Most of the toxicological data relate to AFB_1. Dietary intake of aflatoxins arises mainly from contamination of corn and peanuts and their products (WHO, 1993).

Groopman Gallagher *et al.* (1994) wrote a review of the mechanisms of aflatoxin carcinogenesis. An extensive review of aflatoxin toxicology, carcinogenesis and the relationship of aflatoxin consumption to human liver cancer was preformed by the Joint FAO/WHO Expert Committee on Food Additives in 1997 (JECFA, 1998).

An excellent review on the cellular interactions and metabolism of aflatoxin has been published by McLean and Dutton (1995). Gorelick (1990) has written a review comparing metabolism of aflatoxins by different species. Guengerich *et al.* (1996) have discussed the involvement of cytochrome P-450, glutathione-s-transferase and epoxide hydrolase in the metabolism of aflatoxin B_1 (AFB_1) and the relevance to risk of human liver cancer. A wide variety of vertebrates, invertebrates, plants, bacteria and fungi are sensitive to aflatoxins, but the range of sensitivity is wide for reasons not yet fully understood (Cullen and Newberne, 1994). Two important factors in species and strain variation of sensitivity are (1) the proportion of AFB_1 that is metabolized to the 8,9-epoxide relative to other metabolites that are considerably less toxic and (2) the relative activity of phase II metabolism, which forms nontoxic conjugates and inhibits cytotoxicity. The 8,9-epoxide of AFB_1 is short-lived, but highly reactive, and is believed to be the principle mediator of cellular injury (McLean and Dutton, 1995).

The most important toxicological characteristics of the aflatoxins drawn from the voluminous literature may be summarized as follows:

1. Aflatoxin B_1 is the most toxic and most carcinogenic form. The acute toxicity correlates well to the susceptibility to hepatic cancer induced by AFB_1.
2. Rats are generally more sensitive than mice. The LD50 varies within species by strain, sex, age, route of administration, nutritional status of the animal and concurrent composition of the diet.
3. Males are generally more susceptible to acute and carcinogenic effects than females.
4. Large doses of AFB_1 lead to liver failure and death in animals. Smaller doses lead to cell death and regeneration. Single doses of AFB_1 do not cause liver cancer without extraordinary measures, but chronic exposure very effectively produces liver cancer.
5. Long-term administration of AFB_1 in animals induces metabolic enzymes that are involved in biotransformation of aflatoxin. The metabolic fate of AFB_1 in an animal never exposed may be different from that in an animal chronically exposed.
6. AFM_1, the metabolite that occurs in milk when dairy cattle are fed AFB_1-contaminated feed, is about an order of magnitude less toxic than AFB_1.

Relationship to disease in humans

Hendrickse (1991) has pointed out that kwashiorkor is limited to the tropics and has an increased incidence in the wet season, which can also be said of aflatoxin exposure. Studies in Kenya and Sudan have demonstrated high levels of urinary excretion of aflatoxin in affected children compared with marasmic or normal children (Hall and Wild, 1994). However, much of the information from these studies is anecdotal. Pockets of high-risk populations for liver cancer have been described in the Amazonian basin, among Eskimos, and in special populations like renal transplant patients. The incidence of liver cancer is consistently higher in men that in women with a sex ratio ranging from 2 to 3 in most countries. Within countries, further variation in incidence rates is observed across cancer registries. Men showing greater variation than women. World wide, the incidence of liver cancer in men and women shows a strong correlation. Migration from high-risk areas to lower-risk areas tends to reduce the risk to the levels of the host country, and this is observable within the first and second generations.

Yeh *et al.* (1989) examined the roles of the hepatitis B virus and AFB_1 in the development of Primary Hepatocellular Carcinoma (PHC) in a cohort of 7917 men 25-64 years old in southern Guangxi, China, where the incidence of PHC is among the highest in the world. After accumulating 30, 188 person-years of observation, Yeh *et al.* (1989) found 149 deaths, 76 (51%) of which were due to PHC. Ninety-one percent (69 of 76) were Hbsag + at enrollment into the study in contrast to 23% of all members of the cohort. Three of the four patients who died of liver cirrhosis were also Hbsag + at enrollment. There was no association between Hbsag positivity and other causes of death.

From the voluminous literature it can be concluded that aflatoxins are among the most potent mutagenic and carcinogenic substances known. Aflatoxins causes a cell to become tumorogenic by reacting rapidly with DNA to give DNA aducts. These aducts or their degradative products cause carcinogenic mutations. Glutathione-S-transferase catalyzed detoxification is a crucial factor in susceptibility to aflatoxins. Humans appear to be lacking significant GST-modulated protection against AFB_1. Further, biomarkers do not provide a reliable quantitative measure of long-term aflatoxin exposure in humans. Incidentally aflatoxin has not been shown to be an independent carcinogen in humans.

5.3. TRICHOTHECENES

The trichothecene mycotoxins are a large family of toxic metabolites produced by molds belonging to *Fusarium* and other genera including *Trichoderma, Trichothecium, Myrothecium* and *Stachybotrys*. Currently, 148 trichothecenes, having the same basic tetracyclic scirpenol ring system, have been described. The trichothecenes that have been shown to contaminate food or feed are deoxynivalenol (DON), nivalenol (NIV), diacetoxyscirpenol (DAS) and T-2 toxin; less commonly found derivatives include 3-acetoxy-DON, 15-acetoxy-DON, fusarenon-X and HT-2 toxin. The most common trichothecenes in food and animal feed is DON, often found with lesser amounts of NIV (IPCS, 1990). The most important toxicological characteristics of trichothecenes collected from the literature can be summarized as follows.

1. Acute systemic effects result from the more potent trichothecenes (e.g., T-2 and DAS) when administered orally parenterally or by inhalation to rodents, pigs and cattle.
2. Cytotoxic effects, as shown for example by T-2, include necrosis of the intestinal crypt epithelium and of lymphoids and hematopoietic tissues after oral, parenteral, or inhalation exposure. Trichothecenes

are toxic for actively dividing cells, by impairing protein synthesis or causing dysfunction of cellular membranes.

3. Haematological and coagulopathic abnormalities result from T-2 toxin and DAS exposure.
4. Suppression of cell-mediated and humoral immunity results from exposure to T-2, DON and DAS including decreased resistance to secondary infection by bacteria and viruses.
5. Immunostimulatory effects occur, which appear to involve impairment of normal regulatory mechanisms, such as superinducing IL-1 production by macrophages, stimulation of T helper cytokines, etc. Elevations in total serum IgA, but concurrent decrease in total IgM and IgG, occur.
6. Teratogenicity is exhibited (T-2 and DON in the mouse).
7. Weak clastogenic activity of T-2 is demonstrated in some assays, but there is general lack of mutagenic activity.
8. No evidence of carcinogenicity has been shown in long-term animal studies of those trichothecenes studied.

Relationship to disease in Humans: Historically, trichothecenes or *Fusarium* related disease outbreaks have been reported in various countries. A human disease known as Alimentary Toxic Aleukia (ATA) occurred in the USSR in 1931-47 and was reported to be related to the ingestion of over wintered grain invaded by toxic *Fusarium* species (particularly *Fusarium poae* and *F. sporotrichioides*) (IPCS, 1990). Pathological changes included necrotic lesions of the oral cavity, esophagus and stomach and a pronounced leukopenia; the primary lesion was bone marrow hypoplasia and aplasia. The disease was fatal in a high proportion of cases. The clinical symptoms reported in ATA, as well as the identified occurrence of *Fusarium* in food stuffs, suggest that it might have been associated with mycotoxins, identified years later in fungal culture of *Fusarium* species under laboratory conditions, possibly T-2 or Wortmannin.

From Korea and Japan during the period 1946-63, scabby grain toxicosis was reported in humans as well as farm animals. Common clinical symptoms were nausea, vomiting, diarrhoea, and abdominal pain, followed by recovery within a few days. *Fusarium* species, especially *F. graminearum*, were isolated from suspected cereals (IPCS, 1990).

Both ATA and scabby grain toxicosis have not been studied in epidemiological studies. Features of these human diseases were similar to trichothecene toxicosis as observed in experimental animals, notably symptoms caused by DON, NIV, DAS, and T-2 toxin. Recent outbreaks of trichothecene-related diseases have been reported in Japan (red mold disease), India (DON toxicosis), and China (fusariotoxicosis) (Rotter *et al.*, 1996).

Two outbreaks have been reported in some details by the IPCS (1990). In the Chinese outbreak of 1984-85, approximately 463 of 600 persons who consumed moldy corn and scabby wheat reported symptoms including nausea, vomiting, abdominal pain, diarrhea, dizziness, and headache with an onset of 5-30 min. Pigs and chickens fed with the molded cereals were also affected. DON was detected at a range of 0.34-0.93 mg/kg and zearalenone at a range of 0.004-0.59 mg/kg, while T-2 toxin and NIV were not found (IPCS, 1990).

In an Indian outbreak in 1987, illness was ascribed to consumption of bread made from flour that had molded in storage following unseasonal rains. *Fusarium* species were grown from the wheat. A total of 97 of 224 persons were affected with symptoms including abdominal pain, throat irritation, diarrhoea, blood in stools and vomiting, with a time of onset of 15 min to 1 hr. In 12 of 24 wheat flour samples, DON (0.35-8.4 mg/kg); Ac-DON (0.64-2.5 mg/kg); NIV (0.03-0.1 mg/kg); and T-2 toxin (0.55-08 mg/kg) were found; quantitative estimation of these toxins was obtained using HPLC and TLC but no rigorous confirmation of identity was undertaken (Bhat *et al.*, 1989; IPCS, 1990).

Analysis of 40 corn samples collected from agricultural stocks for human consumption in Haimen and Penlai, high- and low-risk areas from primary liver cancer (PLC) by Ueno *et al.* (1997) revealed that levels and rates of contamination of FBs and DON were significantly higher in Haimen than in Penlai. Henry and Bosh (2001) are of the opinion that trichothecenes, such as NIV, may promote AFB_1-initiated hepatocarcinogenicity as these authors have demonstrated in rats.

5.4. FUMONISINS

One of the most prevalent seed-borne fungus associated with corn (*Zea mays* L.) worldwide, *Fusarium moniliforme* Sheldon, produces fumonisins. Fumonisin B_1 has been demonstrated to have the following toxicological effects in animals, but has not been demonstrated to be mutagenic or genotoxic: LEM in horses, Pulmonary Edema Syndrome (PES) in pigs, Hepatotoxicity and liver cancer in rats, initiator and promotor of cancer in rat liver, cytotoxicity in mammalian cell cultures, phytotoxicity, sphingolipid biosynthesis inhibition and possible association with esophageal cancer in humans reported to be due to the toxic effect of fumonisins.

Relationship to Disease in Humans: Bhat *et al.* (1997) reported an outbreak in 1995 when unseasonal rains damaged the maize and sorghum crops harvested in a few villages of the Deccan plateau in India. Human consumption of those grains resulted in a food-borne disease outbreak characterized by abdominal pain, borborygmus and diarrhoea. A rapid epidemiological survey in the affected villages revealed persons in 27 of 50 villages surveyed were affected. Disease was seen only in households and subjects consuming the rain-damaged moldy sorghum or maize. The poorest of the poor segment of the population though aware of the mold damage of the grains, were forced to consume unleavened bread prepared from such grains owing to non availability of good grains at the household level and lack of purchasing power to buy grains from the market.

The mycological profile of samples from that region indicated that *Fusarium* sp., *Aspergillus* sp. and *Alternaria* sp. were the dominant mycoflora in sorghum, while *Fusarium* sp. and *Aspergillus* sp. were the dominant mycoflora in maize. Among the various mycotoxins, fumonisin B_1 (FB_1) was the most important mycotoxin detected. Levels of aflatoxin detected were extremely low. Levels of FB_1 in samples collected from unaffected households showed 0.07-0.36 mg/kg fumonisin B_1. A more accurate estimate of fumonisin exposure in affected individuals was not possible. The disease was self-limiting and disappeared when alternate food source became available. Diarrhoea was reproduced in day-old cockerels fed an incriminated maize sample containing 9 mg/kg FB_1.

Very high fumonisin concentrations have been reported from home grown maize in Transkei (118 mg/kg; Rheeder *et al.*, 1992) and China (150 mg/kg; Chu and Li, 1994). Lower levels have been reported from northern Italy (150-3760 mg/kg FB_1; Pascale *et al.*, 1995). There is indirect evidence for fumonisin effects on humans from epidemiological studies.

Rheeder *et al.* (1992) and Syndenham *et al.* (1990) have also established relationship between esophageal cancer and corn contaminated with mycotoxin(s) in the Transkei. In rural areas of the Transkei, maize porridge is the staple diet (up to 100% of calories). Adults also consume beer made from moldy maize deliberately selected by the housewife; such beers could contain a fumonisin concentration of 30 mg/L beer.

The highest mortality for oral, pharyngeal and esophageal cancer (Franceschi *et al.*, 1990) was attributed to contamination in corn meal of FB_1 at the concentration level of 150-3760 ng/ kg.

Seven of the cornmeal and corn grit samples from a supermarket in Charleston, South Carolina, were found to contain high amount of FB_1 which may be responsible for the highest incidence of esophageal cancer among blacks in the United States (Marasas, 1995). Abnormal climatic conditions occurred in many corn-producing areas during 1989-90 which promoted high levels of fumonisins and when these corn screenings were incorporated into commercial feeds, large numbers of horses died due to LEM and pigs due to PES (Marasas, 1995).

Corn is a staple food in parts of China including Linxian and Cixian Counties in Henan Province (Chu and Li, 1994) and mortality for esophageal cancer for males in the low-risk areas ranges from 26 to 36/100,000 and 76 to 161/100,000 in the high-risk areas. *F. moniliforme* has been reported to be in higher prevalence on corn in the high-risk areas (Chu and Li, 1994; Luo *et al.*, 1990). Ueno *et al.* (1997) reported that levels of fumonisins in a survey of three crop-years of corn samples (240 total samples) are significantly higher in Haimen (Jiangsu Country), a high-risk area for liver cancer, than in Penlai (Shandong Province), a low-risk area. Marasas (1995) feels that the contamination of corn with fumonisins represents a real threat to human and animal health, particularly in Africa where corn is the main staple diet of millions of people.

5.5. PATULIN

Patulin, a mycotoxin, produced by species of *Penicillium, Aspergillus* and *Byssochlamys* reported to occur in many moulded fruits, grains, cheese and other fruits.

The toxicological characteristics of patulin are LD_{50} in mice of 15-35 mg/kg, depending on mode of administration. Cytotoxic, antibacterial, antifungal and antiprotozoal, disturbs membrane permeability, disorganizes cytoplasmic micro-filaments, inhibits *in vitro* several enzymes including RNA and DNA polymerase, affects DNA directly and effects on transcription and translation. It lacks mutagenicity. It causes sister-chromatid exchange and chromatid breaks, isochromatid breaks and chromatid translocation. It also lacks carcinogenicity. But when administered subcutaneously, produced sarcomas probably due to local irritation effect. Foetoxicity and embryo-toxicity was observed in rodents.

Relationship to Disease in Humans

When patulin was evaluated at the 35th Joint FAO/WHO Expert Committee on Food Additives (JECFA) meeting in 1990, a maximum provisional tolerable weekly intake (PTWI) of 7 g/kg body weight/week was proposed.

Patulin has been tested as an antibiotic for treatment of the common cold in humans. Application was through the nasal route. The absence of reported cases of human disease associated with patulin inspite of ample opportunity for exposure. Though apples and apple products have been consumed in large quantities historically, but there are no reports in the literature to date of patulin intoxication of humans.

5.6. ZEARALENONE

The major toxicological characteristics of zearalenone are on reproduction, including reproductive organs and their function, leading to hyperestrogenism. Greater amounts of alpha-zearalenol, the more estrogenic metabolite are formed in humans and the pig compared to rodents. Biological half-life of these substances was longer in humans than in other species tested. Binding of zearalenone to estrogen receptors was approximately 20-fold lower than that seen with 17-beta-estradiol in several assays. Bioassays in mice and rats are considered to be "positive evidence of carcinogenicity" by the National Cancer Institute/National Toxicology Programme. IARC placed zearalenone in the category of "limited evidence of carcinogenicity". It did not induce SOS error-prone DNA repair in *E. coli*. Differential toxicity was shown in repair-deficient and proficient *Bacillus* strains. It did not induce mutation in *S. typhimurium* or mitotic crossing over in *S. cerevisiae*. But it induced sister chromatid exchange, chromosomal aberration and polyploidy in Chinese hamster cells.

Relationship to Disease in Humans

Several ecological studies have examined the relationship between exposure to *Fusarium* toxins and esophageal cancer. Most of the studies refer to mixture of many toxins from many species of fungi on corn. The studies by Marasas *et al.* (1988), Marasas (1995) and Syndenham *et al.* (1990) in the Transkei, South Africa with moldy corn and esophageal cancer failed to give clear evidence for carcinogenicity of zearalenone.

5.7. OCHRATOXINS

Ochratoxin A is a metabolic product of *Aspergillus ochraceus* and *Penicillium verrucosum*. The toxicological characteristics of ochratoxin observed are acute toxicity which ranges from 3.4 to 30 mg/kg (body weight) depending on species. In rats, females are more sensitive following oral administration of OA (Ochratoxin A). Kidney followed by liver, is primary target organ. Kidney tubular functions are impaired; capacity to concentrate urine is reduced. Immunological effects in rodents include reduced plasma fibrinogen, factors II, VII and X and reduced thrombocyte and megakaryocyte counts. In chickens lymphocytopenia developed and levels of serum immunoglobulins (IgA, IgC, and IgM) were reduced. In mice myelotoxicity developed. Embryotoxic effects in mice and rats treated with 3-5 mg/kg bw included increase prenatal mortality, decreased foetal weight, various foetal malformations. No OA activity in short-term tests *in vitro*, in bacteria, yeast, or cultured cells. DNA single-strand breaks observed in renal and hepatic tissues of mice and rats treated with single or chronic doses of OA. OA is carcinogenic in mice and rats, inducing kidney tumors, predominantly in male animals. Liver tumors were also observed in male and female mice. OA inhibits protein synthesis and tRNA synthetase in both microorganisms and mammalian cells may induce DNA single strand breaks by producing reactive oxygen species and enhancing lipid peroxidation in microsomes.

Relationship to disease in human

Acute gastrointestinal poisoning from food contaminated with high levels of OA has been described (IARC, 1993). In 1963 a new disease Balkan endemic nephropathy was recognized caused due to consumption of contaminated foods. This disease is characterized by bilateral, noninflammatory, chronic nephropathy in which the kidneys are extremely reduced in size, weight and show diffuse cortical fibrosis extending into the corticomedullary junction, hyalinized glomeruli and severely degenerated tubules. Characteristic progressive hyper creatininemia, hyperuremia and normo- or slightly hypochromoic anemia with generally nonelevated blood pressure are the clinical symptoms. Castegnaro *et al.* (1990) and Tatu *et al.* (1998) have summarized the reasons implicating ochratoxin A (or other mycotoxins) in the eteology of Balkan Edemic Nephropathy (BEN) and the associated urinary tract tumors (UTT) as follows:

1. BEN has a similar morphology and progressive course to the endemic porcine nephropathy observed in many countries and caused by ingestion of OA.
2. The kidneys and to a lesser extent liver, are the major target organs for toxicity and induction of neoplasms by OA in several mammalian species.
3. OA impairs the immune function in experimental animals as has been observed in patients with BEN and subjects living in the endemic areas.
4. OA when added to rat liver microsomes greatly enhances the rate of NADPH or ascorbate/Fe^{2+} dependent lipid peroxidation, as measured by malonalaldehyde formation; lipid peroxidation is similarly enhanced in kidney microsomes by OA. As Fe^{2+} enhances lipid peroxidation, higher levels of Fe^{2+}, manganese and chromium have been reported to occur in BEN-endemic villages. These heavy metals may act as cofactors by forming complexes with OA, leading to an increase in oxidative damage.
5. Environmental contamination by OA and citrinin of cereals and beans has been found to be more frequent in endemic than in control areas and also in affected than in non affected families.
6. Case-control studies have revealed more frequent and higher blood and urine contamination of OA in patients with BEN and/or UTT in

the Balkan area and also in Tunisia. A BEN-like nephropathy in Tunisia allegedly related to OA has recently been described (Maaroufi *et al.*, 1995a,b).

However, the implication of OA as sole risk factor for BEN and/or UTT has not been proven precisely. Castegnaro *et al.* (1991) feel that before implicating OA as a causal agent of BEN and/or UTT the following facts should be taken into consideration.

1. No direct epidemiological proof for a causal relationship between mycotoxins and BEN has yet been presented. Complex exposures to mycotoxins usually occur in conjunction with other poorly characterized environmental agents. Epidemiological studies at the individual level are required; both these and future laboratory studies should take into consideration the involvement of multiple risk factors.
2. Data collected on environmental contamination relate to periods after the time when the disease was likely to have begun.
3. It is not clear how difference in food storage or other dietary practices, as opposed to or in combination with individual susceptibility, can explain the unusual clustering of BEN and UTT. Villages afflicted in the past continue to be afflicted today, while nonendemic villages, sometimes located in close proximity to afflicted villages, have remained free of BEN.
4. BEN may be caused by long-term exposure to polycyclic aromatic hydro-carbons and other toxic organic compounds leaching into the well drinking water from low-rank coal underlying or proximal to the endemic settlements. These organic compounds have been observed in well water in endemic villages, but not in nonendemic villages (Tatu *et al.*, 1998).
5. A viral etiology for BEN is unlikely, but has not been ruled out. For example, there may be a relationship between Bolivian hemorrhagic fever, an arena viral hemorrhagic fever disease spread by rodents and BEN (Tatu *et al.*, 1998).
6. Phenotypic and genotypic difference in a number of activating and detoxifying enzymes have frequently been linked to individualized susceptibility to certain forms of cancer. The association between the efficiency of oxidative metabolism of debrisoquine and the risk for developing BEN and/or UTT, individuals suspected of having BEN, health controls from the endemic regions, and healthy controls from nonendemic regions. The extensive metabolizer phenotype was encountered in most of the BEN cases and subjects with BEN and UTT (Tatu *et al.*, 1998).

CHAPTER – 6

MANAGEMENT OF MYCOTOXIGENIC FUNGI

Mycotoxins are one of the important factors contributing for food loss. Based on the Global Environment Monitoring System and Food Contamination Monitoring and Assessment Programme (GEMS/FCMAP) data, mycotoxin contamination is widespread problem of the food supplies in most countries. Strict control of food and feed and appropriate public health measures are, therefore, of considerable importance in reducing the risk to human and animal health. As a result, many countries have enacted regulations to control level of mycotoxins in the nation food supply as well as in food moving in international trade. Eighteen countries all over the world have guidelines or regulations, which prescribe maximum acceptable limits for mycotoxins in foods and feeds. The limits prescribed for aflatoxins vary from 0 to 50 ppb in foods and from 0 to 1000 ppb in feeds in different parts of the world (Srivastava, 1987). In India the permitted level of AFB_1 in food commodities is 30 ppb but no such regulations are prescribed for the other mycotoxins. It is time to realize the problem and enact wisely by making regulations for other mycotoxins also to avoid health risks of man. In general all the agricultural commodities before export are subjected to stringent analysis for mycotoxins such as aflatoxins. Though health hazards from aflatoxin in developing countries such as India is more from maize than groundnut, the economic significance of aflatoxin contamination in terms of reduced export earnings and loss at different levels of the trade chain such as shelled groundnuts, edible oil and oil cake is more in groundnuts than in maize (Bhat, 1989b). Hence, the Governmental agencies should procure food grains confirming to prescribe specifications thereby minimizing the chances of contamination and thus ensure and supply of quality food grains to the consumers through a network of expert teams of save Grain Campaign.

There is an urgent need to form a National Guidelines for prevention and control of mycotoxins contamination specially aflatoxin in agricultural commodities, which should include. (1) harvesting of fully matured grains/cobs/ears and (2) drying to safe moisture level (less than 10%), (3) separating the healthy, sound grains/cobs/ears from damaged or rotted or infected, (4) drying the grains/cobs/ears on raised bamboo floors along with other appropriate drying methods as required for the merchants (Dry shelled maize on concrete drying floors). (5) using a water proof sheet (plastic not advisable) to cover drying produce for rain protection, even mechanical drying, if required. (6) storage at proper dry conditions, along with good ventilation, efficient sanitation practices to destroy or discard infested ears/cobs/etc. (7) and effective fumigation for control of insect problem.

Since last thirty years researches on mycotoxins are being carried out in different parts of India through various research projects being operated. Still we do not have a distinct picture of the impact of mycotoxins exposure and assessment of the risk. There is a need for co-ordination and demarcation of work on the basis of regional frontiers as well as infrastructure and expertise available in different laboratories.

The important factors such as physical (moisture, temperature, mechanical damage and length of storage etc.), chemical (level of CO_2 and chemical nature of substrate) and biological (plant varietal differences, availability of inoculums spore load, strain variation (Lectins) and microbial ecosystems) are reported to influence the mycotoxin contamination of foods, feeds and fodders. Therefore, ecology of moulds which produce toxins should be better known and measures should be taken to control the factors influencing mycotoxin production. Robens *et al.* (2002) have discussed various aspects and strategies to be developed for prevention of mycotoxins specially cantamination through cellular biological and environmental. Robens *et al.* (2002) elegantly discussed the molecular approaches for eliminating mycotoxins from foods and the role of fungal genomics to contain mycotoxin problem. Nichosson *et al.* (2004) discussed different strategies to be adopted in the management of mycotoxin problems of wheat and maize.

In most of the regions of India moist and humid weather, a characteristic feature of tropical climate exists which favor the growth and metabolic activity of moulds. The relative humidity remains fairly high during monsoon when the temperature shows considerable fluctuations. Further, unseasonal rains and flash floods are frequent which results in increased moisture level of the grains. Critical perusal of reports indicate that river basins of Ganga, Brahmapurtra, Godavari, Krishna as well as coastal areas of Andhra Pradesh, Tamil Nadu and Karnataka have high incidence of mycotoxins both on standing crop as well as in stored agricultural produce. Similarly almost every year coastal belt of A.P. is affected by floods which promote mould infestation damage both in stored and standing crops. Practically all food and feed commodities are prone to mould infestation. On the basis of reports of natural occurrence and levels of mycotoxins contamination from different parts of India, maize groundnut, rice, cotton seed and some millet can be categorized as high risk commodities. Insects and birds are reported to promote mould infestation due to the damage caused by infects to grains on standing crop (Sinha and Ranjan, 1989).

Toxic effects of mycotoxins on man and animals arise from the ingestion of food and feed stuff contaminated with mycotoxin. It is now realized that contamination of food stuffs is world wide phenomenon and the problem is more acute in tropical and sub-tropical countries like India. Further, people of developing countries are likely to be victims of first order due to increasing incidence of these mycotoxins over a wide range of foods and feeds. There is an urgent need to prevent infestation of food products by mycotoxin producing moulds by manipulation of the environment. Other control methods should be directed to either reducing the concentration of mycotoxin to safe level or to convert toxin to non-toxic degradative product without affecting the nutritive value of treated commodities. Park and Lee (1990) discussed merits and demerits of several methods in the management of aflatoxins in foods and feeds. Prawat-Tanboon (1991) suggested a number of method by which aflatoxins in maize can be controlled. Nicholsson *et al.* (2004) discussed different strategies to be adapted in the control of mycotoxins problems of wheat and maize. Goyal *et al.* (1991) discussed different methods of prevention and control of mycotoxins in India. Methods of mycotoxin control can, therefore, be classified into two main categories.

1. Prevention of mould contamination and growth
2. Detoxification of toxic products.

To design strategies for the reduction or elimination of mycotoxins, knowledge about their fungal sources are needed. The growth of fungi in crops and agricultural products is the main cause of toxin production. Many factors are involved in enhancing the formation of mycotoxins. They are: 1. plant susceptibility to fungal infestation, 2. suitability of substrate for fungal growth 3. temperature of climate, 4. moisture content and 5. physical damage of seeds due to insects and pests.

6.1. PREVENTION OF MOULD CONTAMINATION AND GROWTH

Toxin-producing fungi may invade seeds at pre-harvest period in the field, harvest time, at post-harvest handling and in storage.

According to the site where fungi infest grains, toxigenic fungi can be divided into three groups: (a) field fungi (b) storage fungi and (c) advanced deterioration fungi. The first category includes species of plant pathogenic fungi, namely, genus *Fusarium* (*F. moniliforme, F. roseum, F. tricinctum* and *F. nivale*). The storage fungi are principaly the genera *Aspergillus* and *Penicillium*. The advanced deterioration fungi normally do not infest healthy grains but easily attack damaged ones and require high moisture content. The examples of the third group are *A. clavatus, A. fumigatus, Chaetomium,*

Scopulariopsis, Rhizopus, Mucor and *Absidia*. The prevention of mycotoxins in our environment is a big task. This can be taken up at three steps.

6.1.1 Primary prevention

The step of prevention should be initially carried out before the fungal infestation and mycotoxin contamination. This level of prevention is the most important and effective plan for reducing fungal growth and mycotoxin production. Some of the practices recommended are to create unfavorable conditions for the growth of mycotoxin producing fungi.

1. Development of fungal resistant varieties of plants
2. Control of field infection by fungi of planting crops
3. Making schedule for suitable pre-harvest, harvest and post-harvest
4. Lowering moisture content of seeds at post-harvest and during storage
5. Store commodities at low temperature whenever possible
6. Use fungicides and food preservatives against fungal growth
7. Control of insect infestation in bulk stored grains with approved insecticides.

6.1.2 Secondary prevention

If the invasion of some fungi begins in commodities at early phase, the existing toxigenic fungi should be eliminated or its growth be stopped to prevent further deterioration and mycotoxin contamination. Some of measures suggested are

1. Stop growth of infesting fungi by redrying the products
2. Removal of contaminated seeds
3. Inactivation or detoxification of mycotoxins contaminated food grains
4. Protect stored products from any conditions which favour mycotoxigenic fungal growth.

6.1.3 Tertiary Prevention

Once the products are heavily infested by mycotoxigenic fungi, the primary and secondary preventions would not be then feasible. Only a few practices are recommended.

1. Complete destruction of the contaminated products.
2. Detoxification or destruction of mycotoxins to the minimal level.

It is foreseen that seeds rich in such antifungal enzymes likely resist the infestation and mycotoxin contamination. This enzyme for genotype evaluation is now being studied. If so, the seeds for breeding would be easily screened out and used as fungal resistant stock one.

Even then there are many technical problems in searching for the "super" plant against pathogenicity. The development of fungal-resistant plant varieties utilizing genetic resistance to mycotoxin contamination is still possible and encouraged.

(1) After the crops have been harvested, drying and proper storage and suitable transportation of the commodities are of prime importance. Several favorable factors contribute to the growth of fungi and mycotoxin production, namely high moisture content, humid climate, warm temperature (25-40°C), insect infestation and pest damage. Many means and measures to prevent a fungal contamination have been emphasized and practically done.

Food and Agriculture Organisation (FAO) also provides advisory services to member countries to assist them in assessing the importance of the mycotoxin problem and in formulating appropriate programmes and strategies for the prevention and control of contamination. FAO has organized two international conferences and the following recommendations were made

1. The development of feasible and practical methods to prevent post-harvest contamination of food crops and to improve post-harvest practices in particular at both local and large warehouse level so as to prevent fungal infestation and insect damage.
2. The extension and coordination of research into significant problems of mycotoxin toxicity, incidence, prevention and elimination.
3. The development of feasible, economical and safe means of decontamination of mycotoxin contaminated food products.
4. The strengthening of existing centers dealing with mycotoxin control and when necessary, the establishment of new facilities by the provision of equipment and training of laboratory personnel and extension staff
5. Development of application and harmonization of reliable, inexpensive and internationally accepted methods of sampling, sample preparation and analysis and standardization of legislation including the establishment of permissible levels of mycotoxins in food.

6.1.4. Methods of Prevention

Mould damage of stored grain may be prevented mainly by three kinds of treatments viz. (1) drying of grain, (2) controlled storage atmosphere and (3) chemical treatment.

(i) Drying of grain: It is an established fact that dry grain stores long, safe from insects and moulds because the requirements of moisture for their growth and development are not met. Exposure of aflatoxin contaminated groundnut oil to sunlight has given very promising results as it destroyed about 99 percent of the aflatoxins (Bilgrami and Sinha, 1984a). Other studies suggested that urea and sun light can detoxify aflatoxin in groundnut flour and cake (Shantha *et al.*, 1986). Other methods of grain drying include mechanical drying, in-bin drying, infrared, microwave or sonic and solar energy drying (Srivastava, 1986).

(ii) Controlled atmosphere storage: For most of the fungi especially aflatoxin producing fungi requires warm humid conditions (85°F is the optimum or 80-110°F is the range, RH of 85% or 62-99% range), while for fusarial toxins the optimum temperature is 45-75°F. So proper storage conditions are required for mycotoxins elaboration. Strategies to modify these conditions may help to fight against the mycotoxin problem. The practice of underground storage during some times may prove to be safe which results in grain cooling and depleting the oxygen content to the desired level whereby the microbes and insects cannot grow.

(iii) The use of natural products: constituents of numerous plants have been found to exhibit antifungal activity against storage fungi. Oregano, clove, cinnamon, garlic and thyme are some of such plants which are reported to inhibit different mycotoxins. Lack of unianimity regarding efficacy of these products may be due to assay technique, culture medium, plant source, time of harvesting, isolation of active ingredients.

(iv) Chemical treatment: Chemical control of fungal deterioration of stored grain conducted by Indian Grain Storage Institute showed that "Grain treat" (Mixture of propionic acid, acetic acid and benzoic acid) against species of *Aspergillus* and *Penicillium* failed to give effective protection. Phenol and 2,4-D inhibited the cyclopiazonic acid production by *Penicillium griseofulvum* (Krishna Reddy and Reddy, 1989). Similarly Laxma Reddy and Reddy (1992) have reported that p-nitrophenol and alpha-naphthol are effective inhibitors of satratoxin production by *Stachybotrys atra*. Clavero *et al.* (1993) could destroy aflatoxin with the help of hydrogen peroxide. Seed treatment with Bavistin and TMTD (Trimethyl thiuram disulphide) at 0.25% concentration gave 100 percent protection to wheat grain during one year storage against species of *Aspergillus* and *Penicillium* (Srivastava, 1986). Certain mild phenolics like ferulic acid and 0-Vanillin have also been reported to prevent aflatoxin production on rice, wheat, maize, groundnut and mustard seeds (Bilgrami and Sinha 1984). The application of oxalic acid (3%) on ground nut kernels was effective in inhibiting AFB_1 production. Bilgrami *et al.* (1981) have reported phenolic compounds to be useful in protecting the food grains from mycotoxin contamination. Phenol and 2,4-dichlorophenol inhibited the cyclopiazonic acid production by *P. griseofulvum* (Krishna Reddy and Reddy, 1992), while o-nitrophenol, p-nitrophenol, p-chloronitrtophenol and pentachlorophenol were effective in checking the growth of *P. griseofulvum. Stachybotrys atra* was sensitive to p-nitrophenol, B-naphthol, p-cresol and α-naphthol and failed to produce satratoxin (Laxma Reddy and Reddy, 1992). By decreasing level of oxygen and in its place CO_2 or N_2 are increased, creating anaerobic conditions discourages mould growth and mycotoxins contamination. Avantaggiato *et al.* (2005) reported that cholestryamine to be good absorbant material for detoxification of different *Fusarium* mycotoxins.

Some of the organic acids such as acetic acid (Buchanan and Ayers, 1979), propionic acid and butyric acid (Ghosh and Hagblom, 1985), malonic acid (Megalla and Hafez, 1982), benzoic acid (Uraiah and Honre, 1987), sorbic acid (Yousef and Marth, 1981), lactic acid (Reiss, 1979), citric acid and their sodium salts are reported to be effective in the control of mycotoxigenic fungal infestation and mycotoxin contamination. Benzoic acid derivatives such as onitrobenzoate, o-aminobenzoate, p-aminobenzoate, benzocain (ethyl aminobenzoate) ethyl benzoate, methyl benzoate and aspirin (O-acetoxy benzoic acid) are reported to be useful in control of mycotoxin problem (Davis and Diener, 1987).

Potassium sulfite and potassium fluoride (Davis and Diener, 1987), dichlorovos (Yao and Hsieh, 1967) and fumigant like ammonia and phosphine (Vandergraft *et al.*, 1975) are reported to be effective in the management of mycotoxin problem. Appletom and Tansey (1977) have reported allicin and related substances from garlic and onion extracts are effective in mycotoxin control. Chitosan or derivatives of chitin isolated from crustacean shells were also effective in checking mycotoxin contamination (Cuero *et al.*, 1988). Cinnamon extract containing trans-cinnamic acid, trans-cinnamaldehyde and ferulic acid (phydroxy-3 methyl cinnamic acid was found to be useful in controlling mycotoxin contamination (Bullerman *et al.*, 1977).

Certain short chain fatty acids have been shown to inhibit the growth of fungi Vandergraft *et al.* (1975) advocated the use of propionic acid to protect the high moisture corn without affecting its nutritive value. Dollear *et al.* (1968) suggested that the increase in acidity by feed ingredients can enhance the antifungal activity of propionic acid. Yousef and Marth (1981) have reported the efficacy of sorbic acid in the control of aflatoxin production by *A. parasiticus.* Similarly Garza *et al.* (1993) reported that potassium sulphate could inhibit the production of penicillic acid by *A. ochraceus* and *P. auruntiogriseum.* Dichlorvos was reported to inhibit the aflatoxin production by *A. flavus* without affecting its growth (Hsich, 1973). Zaki (1992) reported the inhibition of growth and aflatoxin production by *A. flavus* and *A parasiticus* by some natural antifungal agents. Hasan (1993) reported that mycotoxins like aflatoxins, DAS and zearalenone could be checked by some fungicides tried by them. Our investigations also revealed that propionic acid could check the production of cyclopiazonic acid, patulin, penitrem, roridin and satratoxin production by *P. griseofulvum*, *A. terreus, P. crustosum M. roridum* and *S. atra* respectively (Krishna Reddy *et al.,* 1997).

6.2 IMPROVED FARM MANAGEMENT

Recent studies have shown that fungal attack occurs in the field and, therefore, mycotoxin contamination on growing plant is of vital importance. This was confirmed by Bilgrami *et al.* (1991) on the standing crop of maize. Further, field contamination of mycotoxins is reported to be more under stress conditions. Studies of Bilgrami *et al.* (1991) have revealed that agronomic practices such as planting date, plant densities and irrigation levels of crops are reported to influence mycotoxin contamination in general and aflatoxin in particular. Therefore, good farm management practices such as using fungus free seeds for planting, controlling insects and plant diseases, improved irrigation practices should be employed. In addition, harvesting the crop at maturity and proper moisture, operation of harvesting equipment to avoid damage to seeds should also be employed. Good storage facilities and proper environmental conditions can be created to prevent mould growth and mycotoxin contamination. Strict control of food and feed and appropriate public health measures are, therefore, of considerable importance in reducing the risks to human and animal health. Mun Kvold (2003) excellently discussed different aspects of cultural and genetic approaches in managing mycotoxins in maize.

6.3 SEPARATION OF INFESTED GRAINS

Physical separation of infested grains is an efficient and feasible method of minimizing mycotoxin contamination. This is effected either by manual operation or with the help of an electronic sorter. Fungal infection of seeds or grain usually imparts characteristic colour or other physical properties. Physical screening and subsequent removal of damaged kernels by air blowing, washing with water or use of specific gravity methods have shown to effective by some mycotoxins like fumonisins and aflatoxins (Trenholm *et al.,* 1992).

6.4 VOLATILE COMPOUNDS

Volatile compounds which move fast leave no residues and highly active at low concentrations are being advocated in the protection of food grains at post-harvest stage (Rytych and Zyska, 1977; Girisham and Reddy, 1987). Vapours of cyclohexane, ethyl acetate inhibited the patulin production by *A. terreus,* while terreic acid production was inhibited by vapours of n-hutanol

(Girisham and Reddy, 1988). *Penicillium griseofulvum* was inhibited for the production of cyclopiazonic acid by vapours of acetic anhydride, n-butanol, formic acid and formaldehyde (Krishna Reddy and Reddy, 1989). *P. crustosum* was sensitive to vapours, of aniline, acetone, ethyl acetate and isoamyl alcohol and failed to produce peniterm (Surekha and Reddy, 1992) *B. satratoxin* production by *S. atra* was totally inhibited by vapours of acetic acid, formaldehyde and propionic acid, (Laxma Reddy and Reddy, 1992).

6.5 FOOD PRESERVATIVES

Use of food preservative against mycotoxigenic fungi has also been advocated. Shashi and Kapoor (1980) reported the efficiency of boric acid in protecting maize seeds. Similarly Hackler *et al.* (1982), Lee *et al.* (1986) and Krishna Reddy and Reddy (1990) have suggested the use of food preservatives in protection of different food grains from mycotoxin contamination. The use of acids protects the moist wheat grains from microbial colonization (Singh *et al.,* 1987). Ghosh and Haggblom (1985) have reported the control of aflatoxin and growth of *Aspergillus flavus*. Some food additives could inhibit the production of ochratoxin A by *A. ochraceus* (Kim *et al.,* 1992). Crystal violet was effective against the growth of *A. terreus* and patulin production (Girisham and Reddy, 1986). Sodium metabisulphite and propionic acid could suppress production of aflatoxin, cyclopiazonic acid, patulin, terreic acid, penitrem B, roridin and satratoxin by respective fungi (Krishna Reddy and Reddy, 1990; Girisham and Reddy, 1988; Surekha and Reddy, 1992; Laxma Reddy and Reddy, 1992a). The production of T_2 toxin by *Fusarium tricinctum and F. sporotrichioides* was prevented by propionic acid (Muller and Thaller, 1982).

6.6 ANTIOXIDANTS

Antioxidants are commonly added to human food and animal feed to protect unsaturated lipids and other materials from spoilage by oxidative degradation. Nicotinic acid, methylxanthine, kaemperol, flavaniods, polyphenols, parabenzoquinone, vitamin A,C,E,B, carotene, butylated hydroxy anisole (BHA) and propyl gallate (PG) are important antioxidants. Thompson (1993) reported that phenolic antioxidants can effectively check the growth of mycotoxigenic fungi. Antioxidant rich cabbage seed extract and garlic extract gave a profound protective effect on against fumonisins. This they attributed to inhibiton of microsomal lipid peroxidation, enhancement of antioxidant driven detoxification systems and or scavenging free radicals (Abdel-Wahab *et al.*, 2004).

BHA was effective in checking the growth and mycotoxin production by many mycotoxigenic fungi while BHT was ineffective against species of

Fusarium and respective mycotoxin production (Giridhar and Reddy, 1995), PG was also ineffective against fusarial and ochratixin A production. Carvacol could inhibit the fusarial toxins production but it had no effect on the species of *Aspergillus and Penicillium.* Parsone *et al.* (2007) reported the antiaflatoxigenic property of food grade antioxidants in peanut grains.

6.7 POLYAMINE BIOSYNTHETIC INHIBITORS

Polyamines which regulate the growth and development of living organisms are key metabolites for the very existence of an organism. These polyamines are reported to be synthesized, atleast in two biosynthetic pathways. Fungi are reported to have only one pathway (Bitonti and Mc Cann, 1982), while plants have two pathways (Florey *et al.,* 1989). Taking the advantage of this difference in polyamine biosynthesis, Rajam *et al.* (1985) and Rajam and Galston (1985) have suggested to exploit these for protecting the plants from fungal attack. Metcalf *et al.* (1978), Rajam (1993), Machatschka *et al.* (1990) and Singhania *et al.* (1991) have successfully checked the intensity of fungal diseases studied by them with the help of polyamine biosynthetic inhibitors.

BCHA could effectively check the elaboration of aflatoxins, ochratoxin A., patulin, terreic acid, citrinin, penitrem B and cyclopiazonic acid by respective fungi (Giridhar *et al.,* 1997; Krishna Reddy *et al.,* 1997). MGBG was differential in its activity. It was more effective against *A flavus* and *A ochraceus,* while *A terreus, P. citrinum, P.crustosum* and *P. griseofulvum* were comparatively less sensitive (Giridhar and Reddy, 1997). However, all these fungi were totally inhibited at higher concentrations. DFMO was highly effective against *A flavus, A.ochraceus,* while *A. terreus* and *P. citrinum* required higher dosage of DFMO.

6.8 ANTAGONISTS

Many microorganisms such as bacteria, actinomycetes including fungi inhibit the growth of mycotoxigenic fungi or mycotoxin production. Still some organisms are reported to degrade the mycotoxin already produced. *Absidia repens* can convert toxic aflatoxin to non-toxic aflatoxicol with in 3 – 4 days (Detroy and Hesseltine, 1969). Similarly *Lactobacillus acidophilus, L. bulgaricus* and *L. plantarum* could prevent the mould growth which in turn may pave the way for inhibition of mycotoxin production (Karunaratne *et al.,* 1990). Further, moulds producing mycotoxins may also be responsible for degrading the toxin which depends on multitude factors (Doyle *et al.,* 1982). The aflatoxin degrading activity of *A parasiticus* was attributed to the production of peroxidase (Doyle and Marth, 1979). Similarly, species of *Fusarium* deacetylate or acetylate trichothecene mycotoxins converting them to non-toxic. *Citrobacterium* was able to assimilate T-2 toxin as a carbon and energy source (Ueno, 1983). *Morganella morganii*, an intestinal organism, can degrade aflatoxin B_1 (Itoh and Aibara, 1992). Many fungi, especially yeasts and bacteria have been shown to be effective in controlling post-harvest diseases of fresh agricultural produce. However, the biological control of fungi present on dry agricultural produce has been studied only to a limited extent. The possible mechanisms by which fungal growth and mycotoxin production inhibition in a competitive environment include competition for nutrients and space. Induction of deference mechanisms, hyperparasitic interactions and antibiosis. *A. parasiticus, A. niger, C. brachyspora, F. solani and M. roridum* inhibited the patulin production by *A. terreus* (Girisham and Reddy, 1986). Chary *et al.* (1986) reported the inhibition of citrinin synthesis by *Penicillium citrinum, A. niger and T. viride. P. griseofulvum and A. niger* completely prevented the production of penetrim B by *P. aurantiogriseum* (Reddy and Surekha, 1994).

Several microorganisms such as *Trichoderma viride* (Choudhary, 1992; Aziz, 1997), *Lactobacillus* sp. (Gauramma and Bullerman, 1995), a toxigenic strain of *Aspergillus flavus* (Choudhary and Sinha, 1995) and *Flavobacterium aurantiacum* (Line and Brackett, 1995) are reported to inhibit aflatoxin production by *A. flavus*. Similarly *Streptococcus salvarius* (Skringar *et al.*, 1996) could eliminate ochratoxin A from yoghurt sample. Yates *et al.* (1999) reported that *Trichoderma viride* inhibited growth and fumonisin production by *F. moniliforme. Saccharomyces cervisiae* inhibited the fumonisin production by *F. moniliforme*. Etcheverry *et al.* (1998) reported that *Aspergillus parasiticus* and *Bacillus subtilis* could suppress alternariol and alternariol mono methylether and tenuazoic acid by *A. alternata*. Castoria (2002) observed that *Aureobasidium pullulans* could metabolise patulin produced by *Penicillium expansum*. Rasic *et al.* (1991) could reduce aflatoxin with the help of *Lactobacillus delbrukii*. Species of *Aspergillus, Mucor, Helminthosporium, Rhizopus, Absidia, Trichoderma* and *Dactylum* are reported to transform aflatoxin B_1 to aflatoxicol (Machatschka *et al.*, 1990). Peeri *et al.* (2004) reported the degradation of ochratoxin A by astraxanthin producing yeasts. Species of *Aspergillus niger* group present on the grape fruit surface inhibited ochratoxin A production (Valvano Fabio *et al.*, 2001). Yabe *et al.* (2008) observed that *Achromobacter xylosoxidans* reduced aflatoxin production by *A. parasiticus* by producing inhibitory substance called cyclo.

Though the feasibility of biological control in mycotoxin management is not yet clear, but a major benefit which might emerge from studies on antagonistic fungi would be discovery of compounds inhibitory to storage fungi. Shanta (1991) discussed the influence of different microorganisms on the production of aflatoxins by *A. flavus* and *A. parasiticus*. Dorner (2008) suggested the control of mycotoxins problem by employing non-toxigenic strains against toxigenic strains.

Integrated approach to contain mycotoxins in food and feeds is desirable. It is now realized that this approach should include efforts both to prevent contamination and detoxify contaminated grain, especially where food resources are limited. Analytical quality assurance in agricultural export produce is required (Sashidhar *et al.*, 1992a). Biological detoxification methods such as fermentation can eliminate some classes of mycotoxins. Cell free extracts of *Phoma* sp. were able to degrade AFB_1 upto 99% (50 μg/100 ml of culture media). Various methods of management of mycotoxins suggested in India are precised in Table 6.1.

Table 6.1. Mycotoxins management practices in India

Item	Area of research	Reference
Cooking method	Detoxification of AFB_1	Fasiha Rehna and Basappa (1990). Basappa and Shantha (1996), Mitra and Joseph (2000).
Aqueous extract of tea and coffee	Impact on AFB_1 toxicity	Shantha and Rati (1996).
Modified buchner funnel filtration	Mycotoxin decontamination process for oils	Kavita Waghrey and Uma Reddy (1996)
Impact of light	AFB_1 stability in grains	Shrivastava *et al.* (1987)
Influence of cropping pattern	AFB_1 contamination	Bilgrami *et al.* (1992)
Biological control	AFB_1	Shantha and Rati (1990); Choudhary (1992)
	Citrinin	Giridhar and Reddy (1997)
Plant extracts	AFB_1	Bilgrami *et al.* (1979; 1980b); Ghewande and Nagaraj (1987); Kumar and Prasad, (1992); Sinha *et al.* (1993); Prasad *et al.* (1994).
	Penitrem	Surekha and Reddy (1992)
	Satrotoxin	Laxma Reddy and Reddy (1992b)
	Citrinin	Giridhar and Reddy (1996)
Food preservatives	AFB_1 other mycotoxins	Chourasia (1993); Giridhar and Reddy (1996a).
Cloves	AFB_1 inhibition	Sinha *et al.* (1993).
Storage structure on toxic fungi	Grains and oil seeds	Ranjan *et al.* (1992)
	Black pepper	Giridhar and Reddy (2002)
	Sorghum	Hareeshvardhan Rao *et al.* (1996)

AFB_1 production	Effect of vitamins and alcohol	
AFB_1 production	Effect of vitamins and alcohol	Basappa *et al.* (1967)
	Ferulic acid	Bilgrami *et al.* (1981d)
	O-vanillin	Bilgrami *et al.* (1982).
	Polyamine biosynthesis inhibitors	Giridhar *et al.* (1999)
Polyamine biosynthesis inhibitors		Giridhar *et al.* (1997)
		Krishna Reddy and Reddy (1997)
Growth regulators influence	Citrinin production	Giridhar and Reddy (1997c)
		Giridhar and Reddy (1997f)
Vitamins and chemicals mutagens		Giridhar and Reddy (1997).
Influence of bacteria		
Polyamine biosynthesis inhibitors	Terreic acid and patulin	Giridhar and Reddy (1998)
CPA	Food preservatives	Reddy and Reddy (1990)
	Black pepper seeds, phenolic antioxidants	Laxma Reddy and Reddy (1992b)
		Giridhar and Reddy (1996c)
		Giridhar and Reddy (2001c)
Interaction between various fungi	Mycotoxins production	Choudhary and Sinha (1993)
Detoxification	AFB_1	Shantha and Sreenivasamurthy (1981); Sharma *et al.* (1988, 1989).
Elicitation of phytoalexins	Influence of AFB_1, CPA and satratoxin on plants	Giridhar *et al.* (2001)
Volatile compounds and environmental factors	Patulin and terreic acid	Girisham and Reddy (1989 a,b)
H_2O_2 and irradiation	AFB_1 inactivation	Sreenivasamurthy *et al.* (1967); Patel *et al.* (1989).
Spice oils	Mycotoxin production inhibition	Patkar *et al.* (1994)
Pressure parboiling	AFB_1, mycotoxins	Rajendran *et al.* (1987)
Sodium bicarbonate	AFB_1 inhibition	Singh and Chand (1993)

6.9 GENETIC ENGINEERING APPROACH

In recent times, efforts are being made to develop mould resistant varieties which will be mould free not only in the field on standing crop but also during storage and restrict development of moulds and mycotoxin elaboration. Out of 15 commonly cultivated maize varieties screened for aflatoxin production, 4-varieties viz., Him-123, Ganga-2, Ganga 5 and EH-

2420 proved to be resistant to aflatoxin production (Bilgrami *et al.*, 1980). Similarly Krishna Reddy and Reddy (1992) during screening of maize cultivars for prevention of cyclopiazonic acid expressed its feasibility. Girisham and Reddy (1987) have reported the resistance of some of the cultivars of bajra to patulin contamination and *Aspergillus terreus* infestation. However, evaluation and selection for resistance to mycotoxins at post-harvest stage may be more difficult than it has been for plant resistance to field diseases.

Lillehoj *et al.* (1975) reported the varying resistance of corn to mold growth and aflatoxin production and its possible exploitation in the management of aflatoxin problem.

6.10 RAPID SCREENING TECHNIQUES

Mycotoxins in contaminated seeds are generally found only in small number of kernels. In order to ensure that quality food systems, a rapid screening test for mycotoxins is desirable. Manual, mechanical and electronic devices are suggested to minimize aflatoxin problem in processed peanuts (Heathcote and Hibbert, 1978). Microscopic observations (Dickens and Welty, 1968) and BGY fluorescence are known to help in the detection of *A flavus* growth and aflatoxin production. Seggregation of healthy corn on the basis of buoyancy reported to reduce the aflatoxin contamination to a minimum level (Huff and Hagler, 1982). Mills *et al.* (1989), Figueira *et al.* (1990) and Kamimura *et al.* (1990) have suggested ELISA technique for detection of deoxynivalenol, aflatoxin and ochratoxin A respectively.

6.11. CONTROL OF ENVIRONMENTAL CONDITIONS

Environment of storage place is of paramount importance in keeping the food commodities free from mould infestation and mycotoxin contamination. Among various factors, moisture, temperature, presence of contaminating microorganisms and the nature and physiological state of the commodity are important in minimizing the mycotoxin contamination (Davies and Diener, 1977).

6.11.1 Moisture

Moisture is a critical factor for keeping quality of stored product. Drying of seed and other agricultural commodities to the safe moisture levels is a critical factor (9% kernels and 13.5% for corn) in the control of mycotoxin. The important parameter when considering moisture and fungal growth is not water content, but water activity. Water activity is a measure of the fraction of the water content of a material which is "free". Since in any natural material water will partition into "bound" and "free" states. "Bound" water is attached chemically or physically to the material, and is, therefore, not available to support fungal growth. The remainder is free water, (a_w) which is immediately available for fungal growth. The growth limit for the most xerophilic fungal types is around 0.7. The relationship between water content and water activity for a given material can be determined by constructing "moisture sorption isotherms" which are graphs of moisture content plotted against water activity. Once the relationship between water content and water activity is known, safe water activity levels can be translated into safe moisture content. This is important in the commodity supply situation because the concept of moisture content is far more readily understood, and can be measured without the rather specialized equipment needed to assess water activity accurately. It is important to note that water activity is a function of temperature so moisture sorption isotherms are always constructed at a specified temperature. Infact, for a given water content, water activity increases with increasing temperature. The practical implication for commodity storage is that a temperature rise may lead to an increase in water activity to unsafe levels, and allow the inhibition of fungal growth. Hence the importance of temperature control in the storage situation.

Safe water content vary from one commodity to another. For maize it is around 14% and for wheat around 15% and for groundnuts around 7%. This is a reflection of the chemical makeup of the commodity and the relative proportion of "free" water. Generally moisture and temperature are interrelated and are likely to regulate mould growth and mycotoxin contamination. Generally storage moulds are known to be xerophils and can grow at 13% and above. The moisture of 13% is equivalent to 70% RH and a_w 0.7. The available water (a_w) is more important in permitting the growth of moulds. An increase in moisture content results in the ten fold increase of mould growth. During storage even though major portion of seed lot is dry, certain pockets may have seeds with high moisture content which persists for long time in view of lack of aeration. Thus, those pockets become foci for mycotoxin elaboration and spread to other regions. Aeration of stored products is important to prevent mould growth, mycotoxin elaboration and biodeterioration. Maintenance of the container or warehouse at low temperature and humidity is another way of protecting food grains from mould infestation and mycotoxin contamination.

6.11.2 Temperature

Temperature is again an important factor for mould growth and mycotoxin production. *A. flavus* produces aflatoxin over a wide range of temperature ranging from 7.5°C to 35°C. If commodities are stored at 4°C, one can prevent *A. flavus* growth and aflatoxin production. Air tight storage is not desirable which may provide ideal moisture encouraging mould growth.

6.11.3 Modified Gaseous Atmosphere

If normal composition of gases of the atmosphere is changed the moulds get discouraged. Generally oxygen level is reduced and in its place CO_2 and N_2 are increased resulting in anaerobic condition. This atmosphere is called modified atmosphere packing (MAp) and Controlled Atmosphere Storage (CAS). CAS is generally used for bulk storage for a period of 6 to 9 months and by this method it was possible to control the growth of *A. flavus* and aflatoxin elaboration in peanuts (Diener and Davies, 1972). Further, it can check mould growth indirectly by controlling insect infestation in CAS. MAP is the enclosure of food products in gas barrier materials which results in reduced microbial growth.

6.12 DETOXIFICATION OF TOXIC PRODUCTS

Detoxification process involves in the destruction, degradation or inactivation of mycotoxins by physical, chemical or biological methods. Dowd *et al.* (1992) reported the detoxificaiton of mycotoxins by some insects. However, the treatment has its own limitations, since the treated products should be health safe from the chemicals used and their essential nutritive value should not be deteriorated.

Though He *et al.* (1992) reported the microbial transformation of deoxynivalenol, it has limitations. The process must be economical and it must remove all traces of toxin and at the same time the nutritional quality of the commodity must not be impaired.

Although certain treatments have been found to reduce the levels of specific mycotoxins, no single method has been developed that is equally effective against the wide variety of mycotoxins which may occur together in various commodities. The ideal decontamination procedure should include.

1. Completely inactivate, destroy or remove the toxin or reduce its concentration to acceptable level.
2. Not to produce or leave toxic residues in the food / feed
3. Not to alter the acceptability or the technological properties of the produce.
4. Preserve nutritive value of the food/feed
5. Destroy fungal spores and mycelia so as to prevent revival and toxin production.
6. Be integrated, if possible, into the regular food-processing and preparation steps.
7. Be cost effective
8. Be easy to use

9. Not to destroy or damage equipment or pose a health hazard to workers.
10. Be approved by regulatory agencies

The topic of decontamination of different toxins has been reviewed from time to time by several workers including Basappa and Shanta (1996), Scott (1998), Sinha (1999), Karlovsky (1999) and Danicke (2002). The following methods are suggested to be applied for effective decontamination of some mycotoxins.

Organic solvents (chloroform, acetone, hexane and methanol) have been used to extract aflatoxins from agricultural products, but mainly in vegetable oil refining process (Vorster, 1985).

Heating and cooking under pressure can destroy nearly 70% of aflatoxin in rice compared to under atmospheric pressure only 50% destroyed. Dry and oil roastings can reduce about 50-70% of aflatoxin B (Fenell, 1966). Baking reduced aw levels by upto two thirds, but never removed entirely (Abbas *et al.*, 1985).

Ionizing radiations such as gamma-rays can stop the growth of food spoilage organisms, including bacteria, molds and yeasts. It also inactivates pathogenic organisms including parasitic worms and insect pests. It has been reported that gamma irradiation (5-10 M-rad) caused reduction of aflatoxin (Osmmer and Fortcage, 1969). They destroy the toxin and its mutagenicity. Therefore, combination of gamma irradiation and ammoniation should, therefore, be attempted for more aflatoxin decontamination.

Chemical treatment has been used as the most effective means for the removal of mycotoxins from contaminated commodities. The method should be sure that the detoxification system is capable of converting the toxin to a nontoxic derivative(s) without deleterious change in the raw product. Mutagenicity of the treated products should be assessed. Some of these chemicals include acetic acid (Wilson *et al.* 1981). Ammonia gas or ammonium salts, 3-5% (Brekke *et al.*, 1977), calcium hydroxide (Codifier *et al.*, 1976), formaldehyde (Mann *et al.*, 1970), hydrogen peroxide (Sreenivasa Murthy *et al.*, 1967), methylamine (Park *et al.*, 1989), ozone gas (Dwarakanath *et al.*, 1968), phosphoric acid and phosphine gas (Mann *et al.*, 1970) are highly toxic sodium bicarbonate (Mashaly *et al.*, 1981), sodium bisulfite (Mashaly *et al.*, 1981) and sodium hvpochlorite (Yang, 1972) which have successfully employed in the decontamination of mycotoxins of different food commodities studied by them.

Certain conditions such as moisture content, heat, ultraviolet or gamma irradiation, sunlight and pressure at different treatment-periods have been simultaneously combined with the chemicals for the enhancement of detoxification. Inactivation methods can be achieved by mixing, packing, fumigation and immersion with the chemical used. It can be achieved by (a) Physical methods (b) Chemical methods and (c) Biological methods.

6.12.1 Physical Methods

It can be achieved by one of the following methods.

(i) Solvent extraction (ii) Heat inactivation (iii) Irradiation and (iv) Adsorption.

(i) Solvent Extraction: The aflatoxins from agricultural commodities can he extracted without leaving any traces of aflatoxin by 95% ethyl alcohol, 90% aqueous acetone, 80% isopropyl alcohol, hexanol-ethanol and hexane-methanol (Anonymous, 1980). However, it is not only costly and special treatment for removal of solvent, loss of nutrients and problem in disposal of extract also pose some difficulties. Moss and Smith (1985) feel that materials treated in this way can only be used for animal feed. Ammoniation process for detoxification of aflatoxins in oil seed meals has been widely accepted on a commercial scale (Basappa and Shanta, 1996).

(ii) Heat inactivation: Heating at 25°C may degrade aflatioxins. Increase of moisture and pressure may improve the efficiency of aflatoxin removal (Mann *et al.,* 1976). Cooking at atmospheric pressure can destroy about 50 percent of mycotoxins. Dry roasting and oil roasting of groundnut reduces aflatoxins to a significant degree. Cooking rice under 15 lbs. pressure for 5 minutes gave maximum destruction of aflatoxins (72%) as compared to ordinary cooking or cooking with excess water. Detoxification by various cooking methods is highly effective in detoxification of AFB_1 (Fasiha Rehana and Basappa, 1990). Light has also been employed successfully to destroy aflatoxin in crude groundnut oil. Studies have shown that visible light is more effective than either ultra-violet or infra-red light (Pruthi, 1978). Similarly oven drying and autoclaveing detoxified AFB_1 by more than 65% (Bilgrami *et al.*, 1984b). Detoxification was also possible by bulb and tube light treatments to the extent of 51-45 in black cumin seeds and almond respectively (Bilgrami *et al.*, 1984a). It was observed that on cooking, 17-56% reduction in AFB_1 concentration in maize based roti (NIN, 1992). However, the above said drying treatment efficiency varied with mycotoxins. Some of the major cooked products such as roti and gutka (poridge) prepared from contaminated maize and sorghum flour respectively showed about 29% reduction in extractable fumonisins B_1 in maize and no reduction in sorghum grains.

(iii) Irradiation: Irradiation with UV in the control of mycotoxin is of doubtful utility. However, gamma irradiation on large scale may be considered, but recently Casadio and Niola (1993) reported the destabilization of patulin in fruit juices with microwaves. Patel *et al.* (1989) reported the reduction of aflatoxins from high moistured food by treating with hydrogen peroxide combined with gamma irradiation.

(iv) Adsorption: Masimango *et al.* (1979) have suggested the use of bentonite for removal of aflatoxins from the solution. Doyle *et al.* (1982) reported the removal of aflatoxin M_1 from milk. Alumina: silica, alumino silicates adsorb mycotoxins from solution. Hydrated sodium and calcium, aluminosilicates have high affinity to aflatoxin B_1. Philips *et al.* (1988) and Carson and Smith (1983) have reported the adsorption of T-2 toxin by intestine wall by addition of bentonite to feed.

6.12.2 Chemical methods

A wide range of chemicals have been tried as reagents for degradation of mycotoxins. Moss and Smith (1985) have reported the formation of hemiacetals from Aflatoxins B_2, G_2 by some acids through incorporation of water. Inorganic or organic bases are inexpensive and can degrade aflatoxins more effectively (Goldblatt and Dollear, 1977). Camaron and Colvin (1989) evolved one step procedure for the selective removal of the 12,13-epoxide ring of diacetoxyscirpenol and deoxynivalenol. Athapol-Noomhorm and Cardona (1991) suggested chemical treatment to prevent mycotoxin contamination. Hydrated Na–Ca alumino-silicate can degrade aflatoxin and diacetoxy-scirpenol (Kubena *et al.,* 1993). Aldehydes such as formaldehyde also reported to be effective in reducing aflatoxin level in groundnut. Of various bases ammonia gas seems to be the most effective and economically feasible for reduction of aflatoxin content in variety of food stuffs. Ammonia is now being used on commercial scale in Senegal, France and USA for removal of aflatoxins from food commodities. However, a disadvantage of this process is that it lowers the protein decarboxylation of the β-keto acid results in the formation of aflatoxin B_1 at high temperature and pressure. Though Mann *et al.* (1971) suggested the use of methylamine and ethylamine in the management of mycotoxin contamination, they are not gaining acceptance due to formation of toxic products. Treatment of ground nut with hydrogen peroxide resulted in the removal of aflatoxin upto 97% (Srinivasa Murthy *et al.,* 1967). Sen *et al.* (1988) reported the total removal of aflatoxins from spilled food by treating with sodium hypochlorite. Partial reduction of aflatoxins during fermentation of Yogurt (Megalla and Hafez, 1982).

(i) Oxidizing agents: Dwarakanth *et al.* (1968) reported that storage of cotton seed and peanut meal in ozonized air resulted in complete destruction of aflatoxin B_1 and G_1 in one hour but had no effect on B_2. Further, this process decreased the amino acid lysine and protein efficiency ratio of the meal (Dollear *et al.,* 1968). Hydrogen peroxide is another oxidising agent which has a potential to destroy aflatoxin (Clavero *et al.,* 1993). Doyle and Marth (1979) have shown that lactoperoxidase in the presence of hydrogen peroxide and chloride ions can degrade aflatoxin B_1 and G_1. Treating the corn with 0.5% H_2O_2 resulted in the decrease of aflatoxin to a significant level. Doyle *et al.* (1982) reduced the aflatoxin in milk by treating with 0.04 g/10 ml.

(ii) Bisulfite: It is commonly used as a food preservative in beverages, fruits and vegetables for several purposes such as inhibition of browning as an antioxidant, reducing agent and to inhibit the microbial growth. Bisulfite has also been used to degrade aflatoxin B_1 and G_1 (Roborts and Mc Weeny, 1972). Hackler *et al.* (1982) have reported the aflatoxin destruction in corn with sodium bisulfite. Its degradation is more effective than ammonia (Moerc *et al.,* 1980). Control of mycotoxin contamination likely to involve a combination of methods and a search of all possible avenues are needed. The approach should be an integrated one involving genetics, biochemistry, ecology and plant breeding. Though factors affecting mycotoxin production are known, biochemical basis of these effects is unknown. The genes responsible for inhibition of mycotoxins in some of the resistant varieties have been identified but plant breeders do not have markers for resistance. Mycotoxin problem can he managed by engineering a biological control agent and identifying the compounds that block mycotoxin formation and compounds in the resistant variety that inhibit the fungus.

6.12.3. Biological methods

Plant Products

Misra *et al.* (1987) and Anasari and Srivastava (1991) reported the inhibition of aflatoxin production by *A. flavus* by essential oil, eugenol of *Cinnamornum tamala* and Eucalyptus oil respectively. Sakai *et al.* (1988) recorded the inhibition of aflatoxin in herbal drugs screened. Garlic extracts (Garcia and Garcia, 1988) and *Ocimum basalicum* (Dube *et al.,* 1989) are reported to inhibit aflatoxin production by *A. flavus* and *A. parasiticus*. Essential oils of *Cuminum cyminum* (Dube, 1991), *Cardamum* (Badei, 1992) inhibited aflatoxin production by *A. flavus*. Sinha (1990) reported the inhibition of aflatoxin production by *A. flavus* in the presence of tannic acid, caffeic acid and phloroglucinol and some plant extracts. Saxena and Mathela (1996) recorded inhibition of aflatoxin production by *A. flavus* with the help of iridial-β-monoenol acetate isolated from essential oil of *Nepala leucophylla*. Maryamma *et al.* (1999) recorded amelioration of aflatoxicosis of ducks with extracts of *Tinospora cordifolia*.

Other herbs such as thyme, staranise seeds (Hilokoto *et al.,* 1978), black and white pepper (Madhyastha and Bhat, 1984), *Plumbago indica* are reported to be effective in the control of mycotoxin contamination. Flavnoids from *Vingilia orobiodes* and chlorophorin from *Chlorophora excela,* a drimane type sesquiterpinoids from tree, *Warburgial salutaris* have potential and offer safe strategies for reducing toxin levels (Odhav *et al.*, 2001). Campbell *et al.* (2001) reported that the constituents of walnut seed coat specially from the cultivar Tulare are potent inhibitors of aflatoxin biosynthesis is capable of rendering aflatoxigenic *A. flavus* virtially atoxigenic.

6.13. QUALITY CONTROL AND REGULATORY MEASURES

Careful control of mycotoxins should be started and administered by the Government of each country through ministries and organizations such as the Ministry of Health, the Ministry of Agriculture, Food and Drug Administration, National Environment Committee Board and Consumer Protection Committee Board. The control programme may be set up by a special administrative committee and the legislative body who regulates the national policy of food safety and the maximum tolerance limits for mycotoxins. Farmers, middlemen, food and feed factories and exporters should be well educated about mycotoxins and encouraged to prevent and control the growth of mycoflora and their health-hazardous mycotoxins in the food commodities as much as possible.

International cooperation for the mycotoxin regulation in trading products or commodities is also needed. The countries should establish quality control limits for certain commodities intended for export or import. The producer countries would be stimulated to be aware of mycotoxin contamination in their exported susceptible commodities. For example, the European Economic Community (EEC) has already established certain maximum tolerance limits of aflatoxins for animal feeds, i.e. not more than 20 ppb for complete feed for figs, poultry and feed supplements for dairy animal; 50 ppb per produce to be processed into mixed feed and complete feed for cattle, sheep and goats. Tolerance levels for different mycotoxins fixed by different countries are precised in Table 6.2.

Table 6.1. Tolerance levels to different mycotoxins in different countries

COUNTRIES	PRODUCTS	MYCOTOXINS	(µg/kg)
Argentina	Baby food	B_1	0
	Groudnut, maize and by products	B_1	0
		$B_1B_2G_1G_2$	20
	Liquid milk, powdered milk	M_1	0.05
	Milk products	M_1	0.5
Australia	All foods	$B_1B_2G_1G_2$	5
	Peanut butter, nuts and the nut portion of products	$B_1B_2G_1G_2$	15
Austria	All foods	B_1	1
		$B_1B_2G_1G_2$	5

	Milling and shelled products and derived products	B_1	2
		$B_2G_1G_2$	5
	Children foods (in prepared food)	$B_1B_2G_1G_2M_1$	0.02
	Wheat, rye	Ochratoxin A	5
		Deoxynivalenol	500
		Zeralenone	60
	Durum wheat	Ochratoxin A	5
		Deoxynivalenol	750
		Zeralenone	60
	Fruit juice	Patulin	50
	Milk (products)	M_1	0.05
	Whey powder	M_1	0.4
	Whey, liquid whey products	M_1	0.025
	Cheese	M_1	0.25
	Butter	M_1	0.02
	Pasteurized fresh milk for infants/children; children foods	M_1	0.01
	Powdered milk (products), condensed milk, milk concentrates	M_1	0.04
Bahamas	All foods, all grains	$B_1B_2G_1G_2$	20
Barbados	All foods	$B_1B_2G_1G_2$	20
	Milk	M_1	0.05
Belgium	Peanuts	B_1	5
	Milk	M_1	0.05
Belize	Maize, groundnut	$B_1B_2G_1G_2$	20
Bosnia & H	Wheat, maize, rice, cereals	B_1G_1	1
	Beans	B_1G_1	5
Brazil	All food stuffs	B_1	15
		$B_1B_2G_1G_2$	30
	Imported foodstuffs	B_1	5
		$B_1B_2G_1G_2$	10
	Industrially prepared foodstuffs for children from 0-2 years and for school meals	$B_1B_2G_1G_2$	3
	Maize	Zearalenone	200
	Maize and groudnut	B_1G_1	30
	Milk (products)	M_1	0.5
	Imported milk (products)	M_1	0.1

Bulgaria	Peanut (products), kernel (products), cocoa beans, cocoa butter, cocoa powder	$B_1B_2G_1G_2$	5
	Cereals (products)	$B_1B_2G_1G_2$	2.5
	Liquid milk products	M_1	0.5
	Powdered milk	M_1	0.1
	Powdered milk for dietetics and infant feeding	M_1	0
	Cheese and similar products	M_1	0.5
Canada	Nut (products)	$B_1B_2G_1G_2$	15
	Uncleaned soft wheat	Deoxynivalenol	2000
China	Rice and edible oils	B_1	10
	Maize, peanut (products), maize and peanut oil	B_1	20
	Wheat, barley, oats, beans, sorghum, other grains, fermented food stuffs	B_1	5
	Cow milk, milk products (calculated on the basis of milk)	M_1	0.5
Colombia	Foods	$B_1B_2G_1G_2$	20
	Cereals (sorghum/millet)	$B_1B_2G_1G_2$	30
	Oil seeds	$B_1B_2G_1G_2$	10
Costa Rica	Maize	$B_1B_2G_1G_2$	35
Cuba	Foods	$B_1B_2G_1G_2$	5
	Cereals, groudnuts	$B_1B_2G_1G_2$	5
Cyprus	Cercals, pulses, dried fruit, nuts, sesame and food produced exclusively from these, caraway seeds, poppy seed, seeds used in bakery products and confectionery	$B_1B_2G_1G_2$	10
	Milk, dairy products	All mycotoxins	0.5
Czech Republic	All foods	B_1 $B_1B_2G_1G_2$	5 10
	Infant/children foods	B_1 $B_1B_2G_1G_2$	1 2
	All foods	Patulin Ochratoxin A	50 20
	Children foods	Patulin Ochratoxin A	30 5
	Infant foods	Patulin Ochratoxin A	20 1

	All foods	M_1	5
	Milk	M_1	0.5
	Infant/children foods	M_1	1
	Infant foods on milk basis	M_1	0.1
		B_1	0.1
		$B_2G_1G_2$	0.2
Denmark	Peanut (products)	B_1	2
		$B_1B_2G_1G_2$	4
	Brazil nuts	B_1	2
		$B_1B_2G_1G_2$	4
	Dried figs	B_1	2
		$B_1B_2G_1G_2$	4
	Pig kidneys	Ochratoxin A	25
	Pig kidneys	Ochratoxin A	10
	Cereal (products)	Ochratoxin A	5
Dominican Republic	Maize (products), groudnut, soya, tomato (products)	B_1G_1	0
	Imported maize	$B_1B_2G_1G_2$	20
Egypt	Peanut (products), oil seed (products), cereal (products)	$B_1B_2G_1G_2$	10
		B_1	5
	Maize	$B_1B_2G_1G_2$	20
		B_1	10
	Starch and its derivatives	$B_1B_2G_1G_2$	0
		B_1	0
	Milk and dairy products	$G_1G_2M_1M_2$	0
		M_1	0
Finland	All foods	$B_1B_2G_1G_2$	5
		Patulin	50
France	All foods	B_1	10
	Peanuts, pistache nuts, almonds, oil seeds, children foods	B_1	1
	Wheat meal	B_1	3
	Wheat bran	B_1	10
	Vegetable oils, cereals, wheat meal (complete)	B_1	5
	Apple juice	Patulin	50
	Cereals, vegetable oils	Zearalenone	200

	Cereals	Ochratoxin A	5
	Milk, milk powder (calculated on reconstituted product)	M_1	0.05
	Milk, milk powder for infants under 3 years	M_1	0.03
Germany	All foods	B_1	5
		$B_1B_2G_1G_2$	4
	Enzyme (preparation) intended for the production of foodstuffs	$B_1B_2G_1G_2$	0.5
	Foods for infants and young	$B_1B_2G_1G_2$	0.05
	Milk	M_1	0.05
	Foods for infants and young children	M_1	0.01
Greece	Peanuts, hazelnuts, walnuts, cashewnuts, pistaches, almonds, pumkin seeds, sun-flower seeds, pine seeds, apricot seeds	B_1 $B_1B_2G_1G_2$	5 10
	Maize, dried figs, dried apricots, dried prunes, dates and raisins	B_1 $B_1B_2G_1G_2$	5 10
	Raw coffee beans	Ochratoxin A	20
	Apple juice, apple products	Patulin	50
Guatemala	Maize, kidney beans, rice, sorghum	$B_1B_2G_1G_2$	20
	Groudnuts, groudnut butter	$B_1B_2G_1G_2$	20
Honduras	All foods	$B_2G_1G_2$	1
	Maize (grounded or whole grain)	B_1	1
	Baby food	$B_1B_2G_1G_2$ M_1	0.01
	Milk (products)	M_1	0.05
	Cheeses	M_1	0.25
Hungary	All foods	B_1	5
	Groundnut kernels	B_1	30
	Preserved foods	All mycotoxins	0
	Groundnuts	$B_1B_2G_1G_2$	5
India	All foods	B_1	30
Indoneisa	Penuts, maize, herb, seeds	(being prepared	
Ireland	All foods	B_1	5
		$B_1B_2G_1G_2$	30
Israel	Nut (products), peanut (products), maize, flour (products), fig (products)	B_1 $B_1B_2G_1G_2$	5 15
	Milk, milk powder (calculated on the basis of milk)	M_1	0.05

Italy	All foods	$B_1B_2G_1G_2$	10
		B_1	5
	Dried figs	$B_1B_2G_1G_2$	10
		B_1	5
	Spices	$B_1B_2G_1G_2$	40
		B_1	20
Jamaica	Food, grains	$B_1B_2G_1G_2$	20
Japan	All foods	B_1	10
Jordan	Almonds, cereals, maize, peanuts, pistachio nuts, pine nuts, rice	B_1	15
		$B_1B_2G_1G_2$	30
Kenya	Peanut (products), vegetable oils	$B_1B_2G_1G_2$	20
Luxembourg	Peanut (products)	B_1	5
Macedonia	Wheat, maize, rice, cereals	B_1G_1	1
	Beans	B_1G_1	5
Malawi	Peanuts (export)	B_1	5
Malaysia	All foods	$B_1B_2G_1G_2$	35
Mauritius	All foods	B_1	5
		$B_1B_2G_1G_2$	10
	Groudnuts	B_1	15
		$B_1B_2G_1G_2M_1M_2$	
Mexico	Flours	All aflatoxins	20
Morocco	Codex texts followed		
Netherlands (The)	All foods and food ingredients except groudnuts used for the preparation of peanut oil	B_1	5
	Cereal (products, pulse (products), legume (products)	All mycotoxins	0
	Milk (products), milk power calcd reconstituted product	M_1	0.05
	Cheese	M_1	0.2
	Butter	M_1	0.02
	Infant foods on milk basis	M_1	0.05
New Zealand	All foods	$B_1B_2G_1G_2$	5
	Peanut butter, shelled nuts, nut portion of products containing nuts	$B_1B_2G_1G_2$	15
Nigeria	All foods	B_1	20
	Infant foods	B_1	
	Fluid milk	M_1	

Norway	All foodstuffs	$B_1B_2G_1G_2$	5
	Brazil nuts, bulk wheat	$B_1B_2G_1G_2$	5
	Apple juice (concentrated)	Patulin	50
Peru	All foodstuffs	$B_1B_2G_1G_2$	10
Philippines	Nut (products)	$B_1B_2G_1G_2$	20
Poland	All foods	B_1	0
Portugal	All foodstuffs	B_1	20
	Peanuts	B_1	25
	Infant foods	B_1	5
Romania	All foods	B_1	0
		Patulin	30
		Ochrtoxin A	5
		Zearalenone	30
	Milk, dairy products	M_1	0
Russia	Animal fats	B_1	0
		M_1	0.5
	Bottles/canned/potted fruits and berries	Patulin	50
	Bottles/potted/canned vegetables	B_1	5
		Patulin	50
	Casein	B_1	0
		M_1	5
	Cereals (wheat of hard and strong types), flour, wheat bran	B_1	5
		Zearalenone	1000
		T2-Toxin	100
		Deoxynivalenol	1000
	Fruits, berries and vegetables (bottled/ canned/potted juices and puree), cacao, cacao powder, chocolate, coffee, eggs, dehydrated egg, meat and poultry (fresh/chilled Frozen/tinned/potted/bottled). Sausage and culinary products from meat and poultry, subproducts of farming animals and poultry, sweets	B_1	5
	Leguminous, protein isolators and concentrators, vegetable oil	B_1	5
		Zearalenone	1000
	Nuts (kernel)	B_1	?
		Zearalenone	1000
	Milk, sour dairy products, concentrated milk, cheese and cottage cheese products, cow butter	B_1	0
		M_1	0.5
Salvador	Foods	$B_1B_2G_1G_2$	20

Serbia	Wheat, maize, rice, cereals	B_1G_1	1
	Beans	B_1G_1	5
Singapore	All foods	B_1	0
		$B_1B_2G_1G_2$	0
South Africa	All foods	B_1	5
		$B_1B_2G_1G_2$	10
Spain	All foods	B_1	5
		$B_1B_2G_1G_2$	10
Srilanka	Foods	All aflatoxins	30
	Foods intended for children up to 3 years of age	All aflatoxins	1
	Milk (products)	All aflatoxins	1
Surinam	Maize	$B_1B_2G_1G_2$	30
	Groundnut (products), legumes	B_1	5
Sweden	All foods	$B_1B_2G_1G_2$	5
	Berries, fruits, juices	Patulin	50
	Liquid milk products	M_1	0.05
Switzerland	All foods (except maize/cereals/herbs)	B_1	1
		$B_2G_1G_2$	5
	Maize, cereals (granular or ground)	B_1	2
		$B_2G_1G_2$	5
	Herbs	B_1	5
		$B_2G_1G_2$	5
	Baby/infant foods	$B_1B_2G_1G_2$	0.01
	Cereals products	Ochratoxin A	2
	Maize (products)	Fumonisins B_1B_2	1000
	Fruit juice	Patulin	50
	Milk (products)	M_1	0.05
	Whey (products)	M_1	0.025
	Cheese	M_1	0.25
	Butter, baby/infant food	M_1	0.02
Taiwan	Cereals	$B_1B_2G_1G_2$	50
Thailand	All foods	$B_1B_2G_1G_2$	20
Trinidad et Tobago	Foods	$B_1B_2G_1G_2$	10
United Kingdom	Nut (products), dried fig (products)	$B_1B_2G_1G_2$	20

Uruguay	Foods and spices	$B_1B_2G_1G_2$	20
	Texturized soy protein products; flour, starch, concentrate, isolate	$B_1B_2G_1G_2$	30
	Peanuts, dried fruit (products)	$B_1B_2G_1G_2$	30
	Cocoa beans	$B_1B_2G_1G_2$	10
	Infant foods, produced industrially	$B_1B_2G_1G_2$	3
	Rice, barley, beans, coffee, maize	Ochratoxin A	50
	Maize, barley	Zearalenone	200
	Fruit juice	Patulin	50
	Milk (products)	M_1	0.5
USA	All foods	$B_1B_2G_1G_2$	20
	Finished wheat products	Deoxynivalenol	1000
	Whole milk, low fat milk, skim milk	M_1	0.5
Venezuela	Rice flour	$B_1B_2G_1G_2$	5
Zimbabwe	Foods	B_1	5
		G_1	4
	Groundnuts, maize, sorghum	B_1	5
		G_1	4

6.14. HACCP IN MANAGEMENT OF MYCOTOXINS

The application of HACCP principles to Mycotoxin control is relatively new approach, which are based on some very sound reasoning of why we should wish to prevent mycotoxin formation in the first place

1. Once formed, mycotoxins are relatively difficult to remove using available decontamination methods
2. There are significant problems associated with obtaining representative samples for mycotoxin analysis from large bulk stores'
3. Precise mycotoxin analysis is costly and often too slow to be of real use in many commodity chains
4. Remember that ultimately, a properly conducted HACCP scheme largely takes away the need for extensive quality control type testing.

6.14.1.Pre-harvest and harvest (field) situation

For some commodities and mycotoxin types, contamination is essentially a field event. Important examples of pre-harvest mycotoxins are the trichothecenes such as deoxynivalenol (DON) and T-2 toxin produced by species of

Fusarium on growing cereal crops. Understanding mycotoxin contamination at this stage is, therefore, the realm of the agronomist, mycologist and plant pathologist, and experts in these disciplines will need to be part of the HACCP team

HACCP systems are sometimes perceived as being over-complicated and "difficult", and in the present case, where we are considering a holistic approach, this must be acknowledged as being a real danger. If we are to avoid a system that is too cumbersome to be of any real use. We need some other systems to be in place before we begin. At the field stage the most important quality system is Good Agricultural Practice (GAP) which is a loose collection of codes of practice designed to ensure safe primary food production. GAP features in the Codex Code of General Principles on Food Hygiene (1997). Some aspects of GAP, together with related codes of practice, have been published in the UK by OEFRA and the Health and Safety Commission. Most crop production in the UK is within some form of crop proficiency scheme, and these schemes will normally set down detailed GAP type principles to be followed in growing the crops. Examples of such schemes are the Assured Combinable Crops Scheme (England and Wales) and Scottish Quality Cereals (Scotland). The practical implication of the use of GAP within a HACCP scheme is that fewer codex code principles (CCP's) are likely to appear at the field stage. The following are examples of controls that will be within the remit of GAP, and, therefore, not necessarily CCP's for the purposes of developing a HACCP plan:

1. The use of fungus or pest resistant cultivars.
2. Appropriate land preparation; crop rotation, trash removal, correct ploughing and fertilizer application
3. The control of insect and other pest damage (since this can encourage fungal infection)
4. The control of fungal infection.
5. The prevention of drought and weed stress,
6. Harvesting at appropriate moisture content and stage of maturity.
7. Appropriate handling and storage immediately after harvest

There are sound economic reasons for producing healthy crops if only for the concomitant increased yields and value and it follows that crops free from fungal infection are also free from mycotoxin contamination. Here again we return to the point that mycotoxin control often occurs as a result of the attention to control of more obvious fungal disease. However, with regard to mycotoxin contamination, there are some complications that fall outside the scope of GAP. For example improper use of fungicides can in some instances lead to increased mycotoxin contamination. This may happen if the application of fungicide not sufficient to quickly eliminate the infection, and the stressed fungus responds by producing increased quantities of mycotoxins.

In some instances it will be appropriate to identify CCP's at the field stage. This will normally be in situations where control steps are put in place that is specifically designed to prevent mycotoxin development. A number of examples of HACCP plans where CCP's have been identified at the field stage are given below

(i) Aflatoxin contamination in pistachio nut cultivation in S.W. Asia

The seggregation and removal of visibly damaged nuts at harvest was considered a CCP, since damaged nuts are most susceptible to fungal attack and aflatoxin production. The step was, therefore, designed specifically to control mycotoxin contamination.

(ii) Aflatoxin contamination in peanut cultivation for peanut butter manufacture in Southern Africa

Wind rowing and sun drying of harvested peanut pods were both considered CCP's since they were designed to bring the crop to a precise moisture level where aflatoxin development was not possible. Wind rowing and sun drying is a two stage process where the pods are first dried in heaps and then in thin layers on a flat surface. This combination allows the most rapid drying without the use of drying equipment

It is important to note that when dealing with products originating in developing countries, certain control measures such as the use of expensive chemical treatments may not be possible for socio-economic reasons. This further emphasizes the need for a thorough understanding of the constraints operating within a process before an effective HACCP plan can be established

Post-harvest stage is the commodity supply chain in that operations carried immediately following harvest and leading up to the first processing step. This will typically involve storage, drying (if necessary) and transportation steps. In fact the post-harvest stage can become very complex as the commodity passes through the hands of a number of separate intermediaries from farm to primary processor.

The post-harvest situation will normally be characterized by mycotoxins produced by the storage fungi such as *Aspergillus* and *Penicillium* species that are able to grow in relatively dry conditions.

6.14.2 Codes of Practice

We have already seen that, in the pre-harvest stage, HACCP schemes can be simplified in part by invoking Good Agricultural Practice (GAP) and the related Good Hygenic Practice (GHP) are also relevant to the commodity immediately after harvest and are concerned with the initial on-farm treatment of crop material. Further, there is a need to follow good storage production (GSP). Storage structure, sanitation, prevention of moisture ingress (e.g. use of moisture barriers on floors), storage conditions such as temperature and aeration and prevention of invertebrate infestation and prevention of access by rodents and birds are some of the good storage production practices.

6.14.3. Post-harvest control

Control at the post-harvest stage is dominated by single parameter, moisture level. The importance of drying and moisture control during storage is generally well understood by the industry, in terms of the importance of prevention of fungal contamination. As in the case of pre-harvest controls, the concern has not been with the occurrence of mycotoxins but the far more obvious implications of allowing the commodity to become visibly mouldy namely the negative effects on acceptability and price.

The objective of drying and subsequent moisture control is to maintain the commodity at a safe level where fungal growth and mycotoxin production are not possible. It should be borne in mind that some growers especially in developing countries like India may not have access to technologically advanced drying equipment. In many regions out door sun drying is the most commonly used method, and this may fail in unsuitable weather conditions.

In the storage environment the relevant fungi are usually those species capable of growing in relatively dry conditions. Fungi that are able to grow in these conditions are termed xerophilic or xerotolerant. One complicating factor is that once growth is initiated the metabolic water produced by the growing fungus can perpetuate and amplify the process and even allow the development of other more moisture dependent fungal species.

Prevention of pest damage is relevant to mycotoxin control because pests, feeding on stored material, can cause local heating and moisture generation due to their metabolic activity. These warmer, damper areas provide ideal conditions for the initiation and development of local fungal growth and mycotoxin production. Localised effects such as this lead to the development of heterogeneous conditions in bulk storages.

6.14.4. Post-harvest CCP's

The designating of CCP's at the post-harvest stage will normally follow from examination of the system in question and the results of any additional surveillance studies and controlled experiments. Stages critical to mycotoxin control can then be identified and the appropriate control measures are implemented.

Production of yellow maize for animal feed formulation in South East Asia. This example was the result of a project carried out in the mid 1980's when exports of this commodity were threatened due to persistent aflatoxin levels above the regulatory limits (eg. 20mg kg-1 in the EU).

Yellow maize is produced in South East Asia as wet and dry season crops. Dry season maize is found to be more susceptable to aflatoxin contamination at preharvest, due to the increased risk of drought damage, while the larger wet season crop is more likely to be subject to contamination at post-harvest, due to problems with sun drying. The harvested crop was typically stored initially on farm, and then passed through a local (primary) then regional (secondary) trader before passing to large feed mills or export silos. All of these represent storage sites. "Shelling" of the maize ie removal of the kernels from the cob, normally occurred after a period of on-farm storage of harvested cobs.

The main concern is overwhelmingly with drying and moisture control. It can also be seen that mycotoxin analysis itself does not figure importantly in the supply chain until the latter stages of bulk storage. This is very typical for many commodity types in developing countries. This probably arises from a lack of finance and relevant skills. Even in the western world the time required to carry out analyses, limits their usefulness as a monitoring tool, although specific analysis can be used to advantage in verifying a HACCP plan.

Because most of these control measures are not very effective. Rigorous programmes for preventing human and animal exposure to contaminated foods and feeds have been established and effective methods for monitoring the toxin level in the foods have been developed. Although risk assessment has been carried out for some mycotoxins, more epidemiological data for human exposure are needed for establishing toxicological parameters and safe close in humans. Mycotoxin problems may co-exist with use for long time and there are many mycotoxicoses that need to be further studied. It is unrealistic to expect completely to eliminate the problem only through multiple approaches and good farm management; we can minimize the problem and enhance human and animal health.

6.15. THE INTEGRATED APPROACH

Mycotoxin management may be achieved by the synergy of combination of different control methods each at a lower level than if it were used alone. Integrated approach includes application of modified atmosphere and gamma irradiation, fungistatics, natural products and bacteriocins. Although this approach could become important because each component is employed at only a relative low level, it still involves known disadvantages. Biotechnological approaches such as replacing aflatoxigenic strains with non-aflatoxigenic ones in the field, incorporation of antifungal genes into crop plants, breeding resistant varieties and prevention of aflatoxin biosynthesis are some of the strategies which may provide long lasting solutions to mycotoxin problems.

International organizations such as FAO, WHO and UNEP in the UN system are engaged in providing essential information on various aspects of prevention and control of mycotoxin problems to all the countries. Guidelines for international trade include: (a) procedure of sampling and analysis, (b) surveillance and food control inspection systems, (c) use of contaminated produce in feeding of different animals, (d) protocols for detoxification and the quality control of the products. Conferences, symposiums, trainings and workshops on current information of mycotoxins should be promoted. Low--cost technology for assessment, prevention and control of environmental mycotoxins could be then transferred from developed countries to developing ones.

Recent attempts led to the isolation of genes involved in biosynthesis of mycotoxins. Information gained through enzymology and molecular biology of these mycotoxigenic fungi can be used in several ways in the management of these obnoxious fungi in one of the following ways:

1. To engineer a biological control agent and
2. To identify the compounds which block mycotoxin formation and to identify the compound those inhibit the mycotoxigenic fungus.

Another control strategy is to identify compounds which specifically inhibit the transcription pathway of genes or inhibit the pathway of enzymes themselves. This would enable plant breeders to select the plant that contain inhibitory compounds or plant that contain inducing compounds.

In India, the problem of mycotoxin is being realised only recently and intensive surveys at geographical and climatic region needs to be undertaken. A strategy to control mycotoxins of that region has to be taken up accordingly. There is an urgent need for intensive and extensive survey on the natural incidence. Intricacies involved in the elaboration of these mycotoxins needs to be taken up at the molecular level. An integrated approach in the management of this problem needs special attention.

Chapter-7

CARCINOGENIC MYCOTOXINS

7.1. AFLATOXINS

Aflatoxins are a group of structurally related secondary fungal metabolites that are carcinogenic, hepatoxic, teratogenic and immunosuppressive. The era of mycotoxin research has developed as a result of epoch discovery of mysterious disease in poultry and fish in diverse localities during 1960 which took heavy toll of turkey poults in UK as a result of unknown etiological agent which is named as turkey X disease. Disease of similar symptoms also occurred in farm animals. A common factor in the out-breaks was the incorporation of Brazilian groundnut meal. The toxic factor was identified to be aflatoxins (Sargeant *et al.*, 1961). Aflatoxins were subsequently found to be widespread representing a toxic and carcinogenic threat to variety of farm animals and potentially a threat to humans consuming contaminated commodities. Aflatoxins in substantial doses produce an acute toxicity and chronic exposure to the level can result in cancer (Marquardt, 1996).

7.1.1. Physico-chemical characteristics

Aflatoxins contain a coumarin nucleus fused to bifuran and either a pentanone (AFB_1 an AFB_2) or six membered lactone (AFG_1 and AFG_2). AFB_1 and AFG_1 are more toxic than either to AFB_2 or AFG_2. A similar pattern holds for its carcinogenicity potency and $AFB_1 > AFG_1 > AFB_2 > AFG_2$ (Fig. 7.1.1). These are produced by *A. flavus* and *A. parasiticus* only. Hydroxylated derivatives of aflatoxin called AFM_1 and AFM_2 were reported by Wilson *et al.* (1968) in milk of cows fed toxic rations (Fig. 7.1.2). These derivatives were subsequently isolated from lactating rats, rat liver and sheep urine, liver and kidneys as well as mouldy peanuts and corn (Shotwell *et al.*, 1976). AFM_1 a hydroxylated derivative of AFG_1 was isolated from *A. flavus* culture. Other derivatives produced by *A. flavus*, the 2 hydroxy aflatoxins are known as AFB_2 and $AFG_{2}a$ (Dutton and Health cote, 1968). Pohland *et al.* (1970) have reported AFB_2 hemiacetal and aflatoxin. Hydroxy dihydroaflatoxin B has been reported by Ciegler and Peterson (1968). AFB_3 or parasiticol (Fig. 7.1.1) is a natural metabolite of *A. flavus* and *A. parasiticus* and it is as toxic as AFB_1 to ducklings, while for chick embryos it is much less toxic.

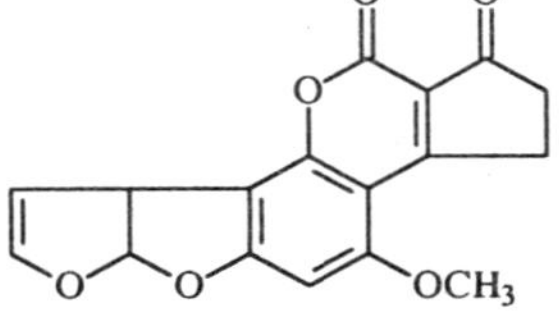

Aflatoxin B_1

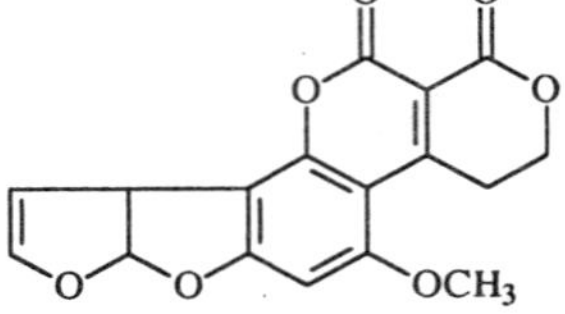

Aflatoxin G_1

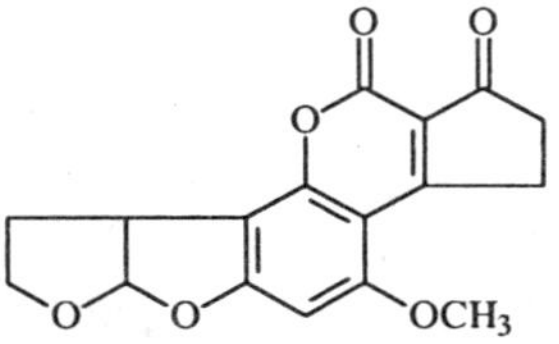

Aflatoxin B_2

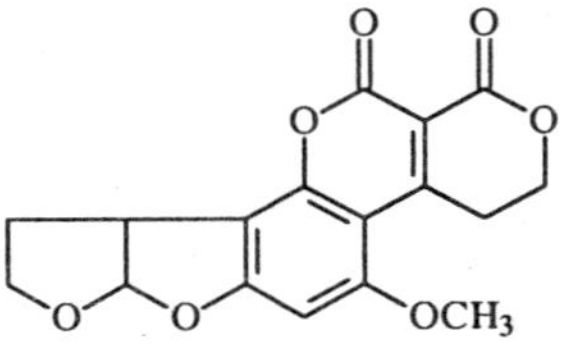

Aflatoxin G_2

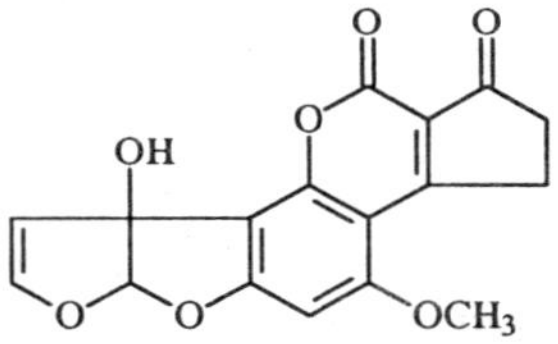

Aflatoxin M_1

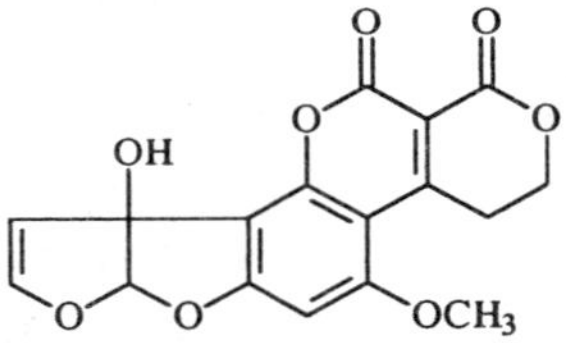

Aflatoxin M_2

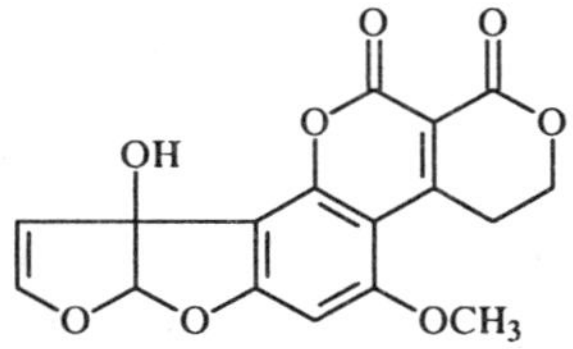

Aflatoxin GM_1

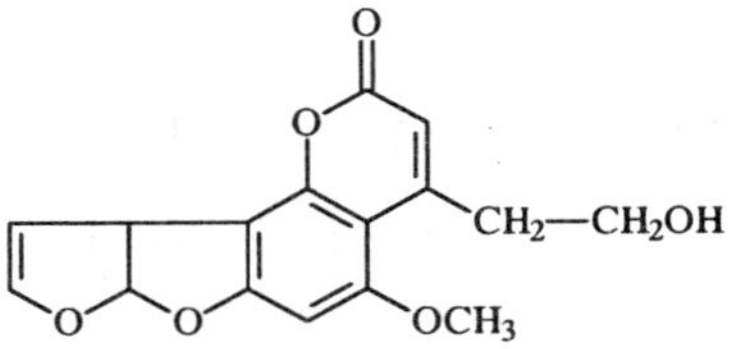

Aflatoxin B_3
(Parasiticol)

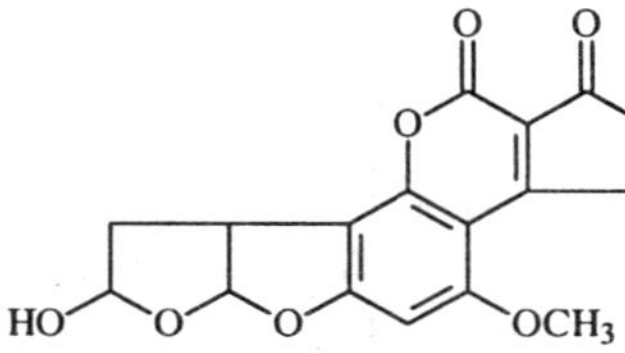

Aflatoxin $B_{2}a$

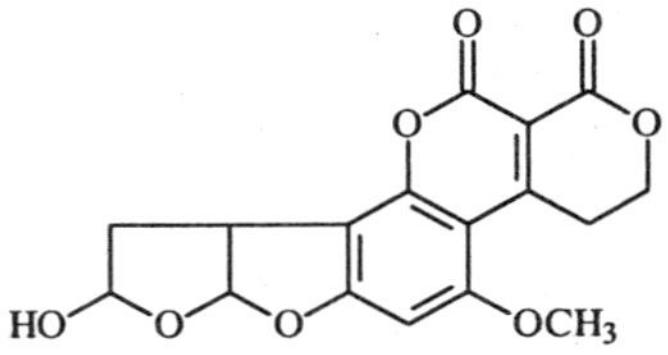

Aflatoxin $G_{2}a$

Figure 7.1.1: Structure of aflatoxins and their hydroxylated derivatives

Reduction of pentanone of AFB_1 by microorganisms yielded aflatoxicol (Fig. 7.1.1) which is also demonstrated to be produced by animals and it is highly toxic and carcinogenic (Schoenhard *et al.*, 1976). Animals metabolise AFB_1 by o-demethylation to produce AFP_1 (Fig. 7.1.2) which is less toxic than AFB_1. *In vitro* metabolism by liver microsomes from monkey and man produced a major metabolite of AFB_1 called AFQ_1 that contained hydroxyl group at C_3 position (Fig. 7.1.3). Detoxification of AFB_1 by using ammonia produced two major derivatives of AFB_1 they are called AFD_1 and AFD_2 which are resulted due to disruption of six membered lactone ring (Fig. 7.1.4 and 5). All the above structures described represent most of the major metabolites and derivatives of the four original aflatoxins. Aflatoxin is stable once it is formed in grain and is not degraded during normal milling and storage (Brown, 1996).

Some of the important physico-chemical properties of aflatoxins are precised in Table 7.1.1.

Table 7.1.1. Physical and chemical properties of some aflatoxins

Aflatoxin	Molecular Formula	Molecular weight	Melting point	UV absorption maxima (ε) in methanol	
				265 nm	360-362 nm
B_1	$C_{17}H_{12}O_6$	312	268-269	12400	21800
B_2	$C_{17}H_{14}O_6$	314	286-289	12100	24000
G_1	$C_{17}H_{12}O_7$	328	244-246	9600	17700
G_2	$C_{17}H_{14}O_7$	330	237-240	8200	17100
M_1	$C_{17}H_{12}O_7$	328	299	14150	21250 (357)
M_2	$C_{17}H_{14}O_7$	330	293	12100 (264)	22900 (357)
Aflatoxicol	$C_{17}H_{14}O_6$	314	230-234	10800 (261)	14100 (325)

Aflatoxins are crystalline substances freely soluble in moderately polar solvents such as chloroform, methanol and dimethylsulphoxide and they dissolve in water to the extent of 10-20 mg/L. They fluoresce under UV irradiation although aflatoxin B_1 and aflatoxin G_1 need derivatisation to enhance the fluorescence to a similar level to that of aflatoxin B_2 and G_2. Crystalline aflatoxins are extremely stable in the absence of light particularly UV radiation even at temperature excess of 100°C. Solutions prepared in chloroform or benzene are stable for years if kept cold and in the dark. The lactone ring makes aflatoxins susceptible to alkaline hydrolysis and processes involving ammonia or hydrochlorite have been tried but toxicity of breakdown products are coming in the way of their use in protection of foods and feeds by this method. Oxidising reagents react and make the aflatoxins loose their fluorescence property. Aflatoxins are resistant and stable in foods under different processes operations (Smith *et al.*, 1994).

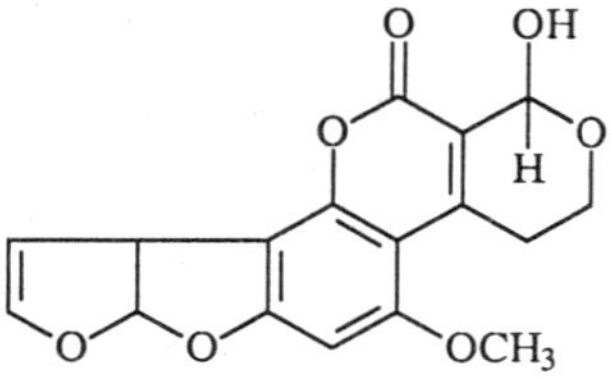

1 Aflatoxicol (AFL or AFR_0)

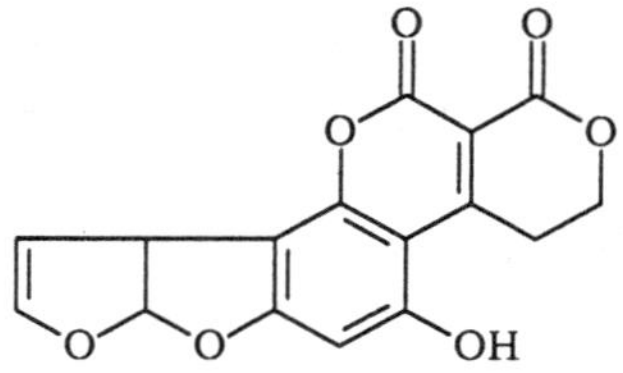

2 Aflatoxin P_1 (AFP_1)

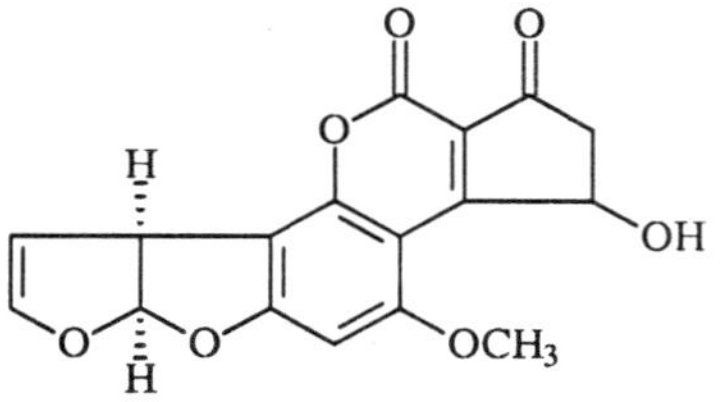

3 Aflatoxin G_1

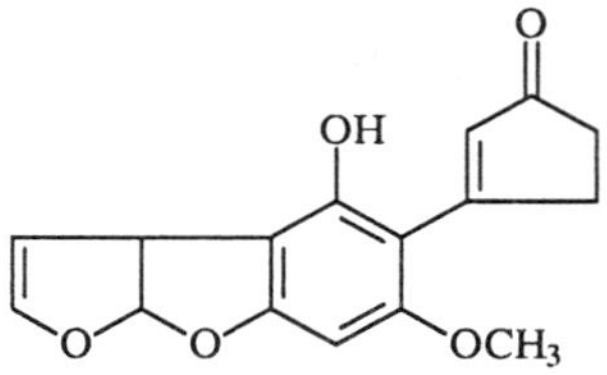

4 Compound D_1

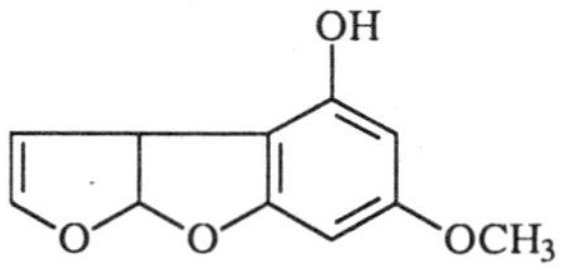

5 Compound D_2

Figure 7.1.2. Metabolized forms of aflatoxin B_1 (1-3) and degradation products (4 and 5)

7.1.2. Organisms

A. flavus and *A. parasiticus* are the organisms which produce aflatoxins. Moss (1996) added a third species, *A. nanius* as producer of aflatoxins. Reports of production of aflatoxin by other fungi have not been confirmed. Wilson *et al.* (1968) have suggested several ways in which false positive results could have been obtained by inadvertent contamination of substrates with prior fungal growth or glassware contaminated with aflatoxins. Another possibility was that other fluorescent metabolites had R_f values on TLC plates were similar to aflatoxins. Further, majority of strains of *A. parasiticus* are aflatoxigenic, where as only 35% of strains of *A. flavus* produce aflatoxins. Indeed some strains of *A. flavus* employed for koji production and more widely used. *A. oryzae* is probably a domesticated form of this species. Stressing host plants by insect damage, drought, poor nutrition or delayed harvest increases aflatoxins production (Brown, 1996). Goto *et al.* (1996) reported the production of aflatoxin by *Aspergillus tamorii*. After development of confirmation tests it was possible to distinguish between aflatoxins and other metabolites that appear similar on TLC plates.

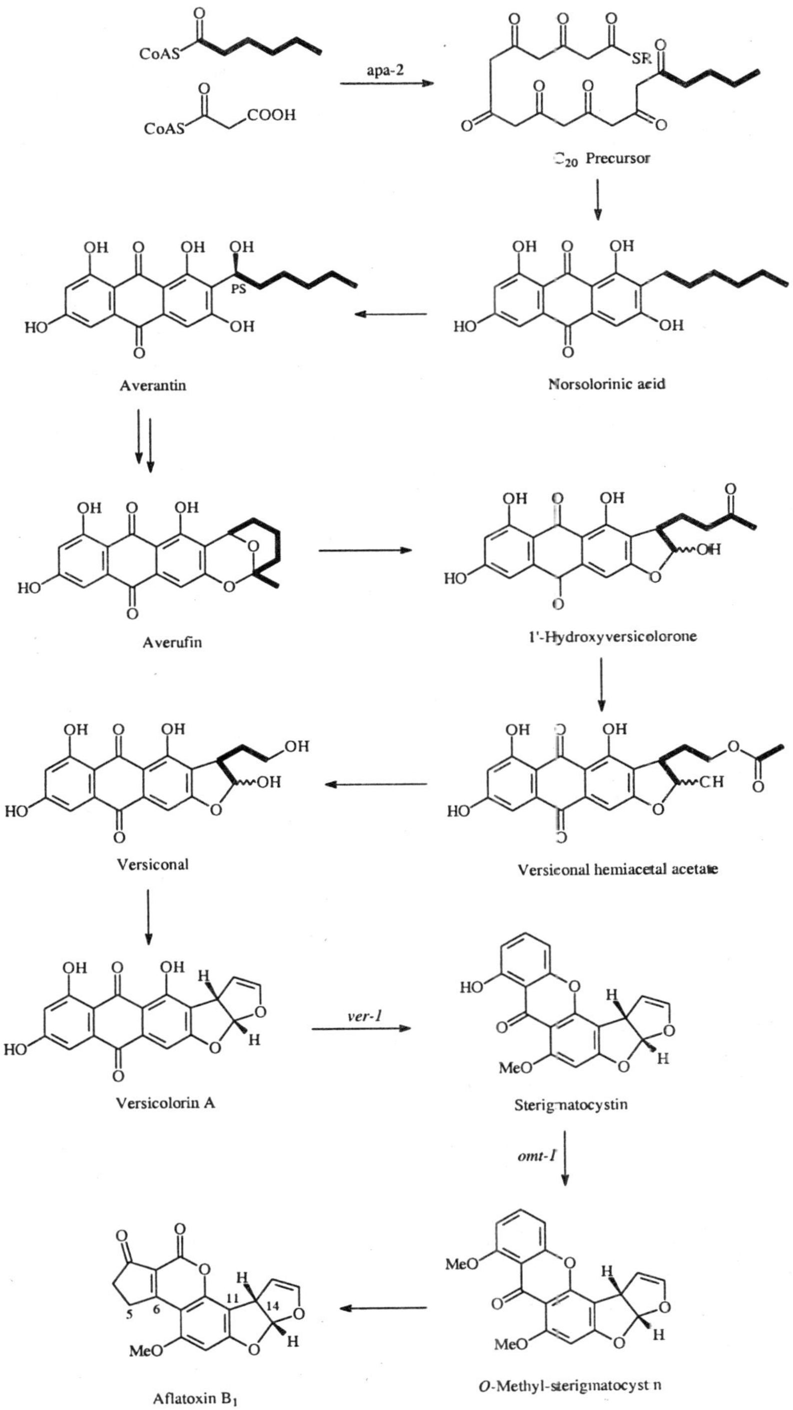

Figure 7.1.3: Biosynthetic pathway of aflatoxin B_1 (Steyn *et al.*, 1982)

7.1.3. Biosynthesis and Metabolism

Extensive studies have been conducted on biosynthesis of aflatoxins using radiolabelled isotopes (Biollaz *et al.*, 1970), stable isotope labels (Samson *et al.*, 1982) and mutants and enzyme inhibitors (Steyn *et al.*, 1979).

With the help of ^{14}C lebelling Buchi *et al.* (1974) established the polyacetyl origin of AFB_1 and revealed an anamalous pathway as the toxin contained two unusual carbon-carbon linkages viz C(5) – C(6) and C(11) – C(14) in which both carbon atoms are derived from the carboxy and methyl-carbon atoms of acetate respectively.

Steyn *et al.* (1982) with the help of ^{13}C NMR spectroscopy and labeling experiments (^{13}C, ^{18}O, ^{2}H) postulated the dekaketide origin of aflatoxins. The sequence : Norsolorinic acid – averufin – versiconal acetate – versicolorin A – Sterigmatocystin – Aflatoxin B_1 – Aflatoxin G_1 (Fig. 7.1.3) are also involved a common polyketide folding pattern based on advanced labeling experiment of these metabolites (Simpson *et al.*, 1983).

Townsend *et al.* (1992) postulated the aflatoxin biosynthetic pathway proceeds through five phase (1) An assemblege stage in which polykelide synthase is initiated by hexanonyl – CoA to form norsolorinic acid (2) A redox stage in which the side chain of norsolornic acid is transformed to form the internal ketal of averufin (3) The dihydrobisuran synthesis stage involving versiconal acetate as an intermediate and leading to the creation of versicolorin B (4) The conversion of the vericolorin A (anthraquinone) to sterigmatocystin (oxanthone) (5) The final stage in which sequential methylations lead to the conversion of the xanthone sterigmatocystin into the coumarin (Aflatoxin B_1).

The other probable biosynthetic pathways for aflatoxins and related compounds are depicted in Fig. 7.1.4.

7.1.4. Isolation and detection

Numerous methods exist for qualitative and quantitative analysis of aflatoxins. Generally they employ extraction with an organic solvent for removal of or separation from interfering compounds and observation of the characteristic fluorescence at 254 nm.

A number of glass minicolumns containing one or more layers of absorbent through which an aliquot of a solvent extract was passed. The fluorescing bands are compared to those on a column on which an aflatoxin standard has been added. The procedures vary depending on the commodities being tested.

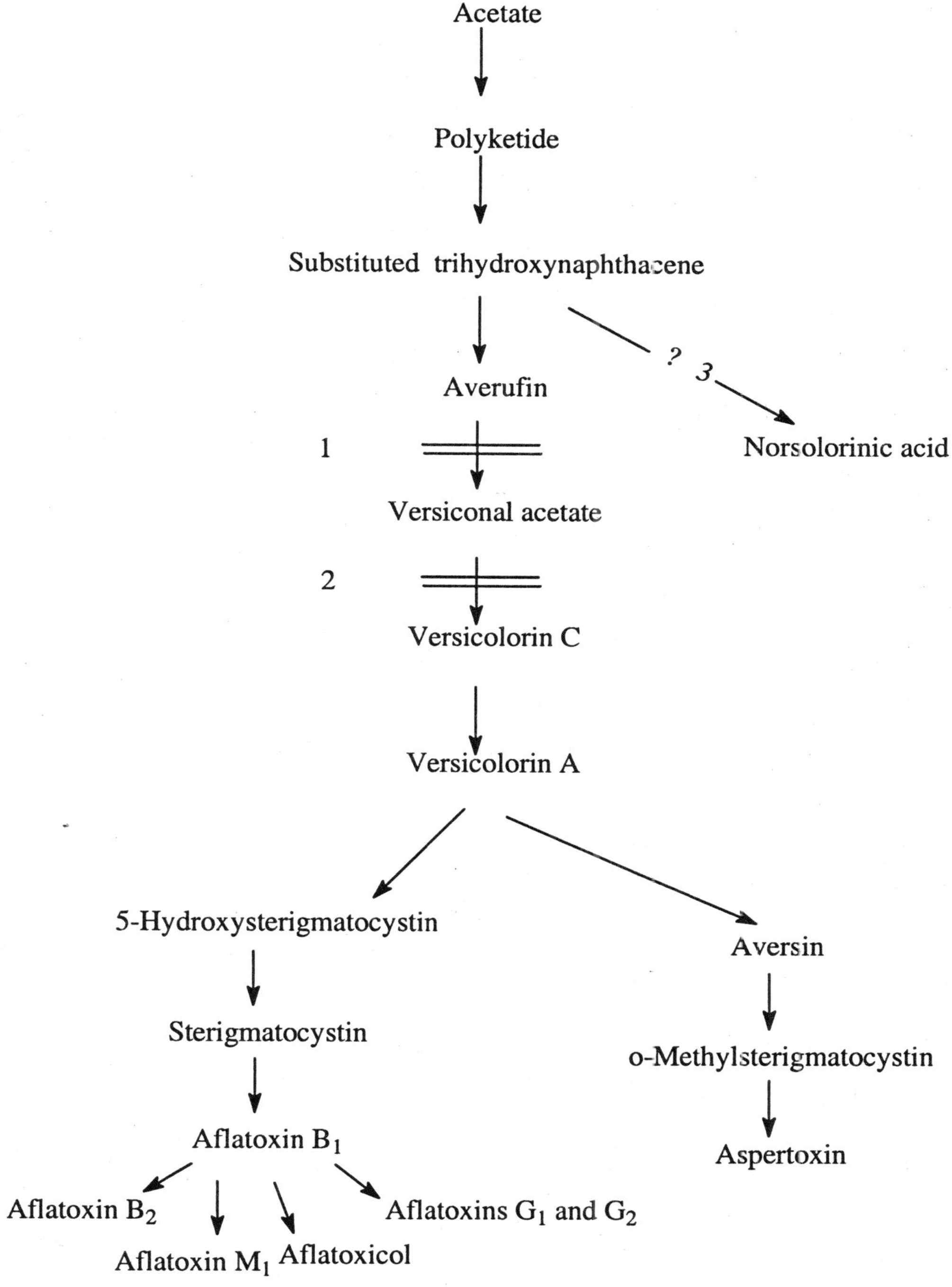

Fig. 7.1.4: Probable biosynthetic pathways for aflatoxin and related compounds. 1. Averufin mutant; 1, the result of inhibition by dichlorvos; 3. nitrosoguanidine – induced mutation (Townsend *et al.,* 1992)

Table 7.1.2. Methods for the Determination of Aflatoxins in Various Foods

Food	Extraction	Cleanup	Quantitation	Detection (μg/kg)	Ref.
Beef liver (B_1,M_1)	Homogenize sample with saturated NaCl solution (2ml/g); shake mixture 10 min with 80:20 methanol/water (12.5 ml/g)	C_{18} reverse-phase SPE cartridge (×2)	HPCL Column: Spherisorb silica (5 μm, 25 cm × 4.6 mm i.d.) Mobile phase: 95:1:4 Water-saturated chloroform/s-propanol/ tetrahydrofuran Detection: Fluorescence (silica gel packed flow cell)	0.1	Qian and Yang (1984)
Black olives (B_1)	Shake ground olivers 5 min with 60:40 methanol/water (6 ml/g)	Liquid-liquid partition	TLC Plate: Silica gel G Solvent: (a) 3:1 Benzene/hexane (b) 60:60:15 Toluene/ ethyl acetate / formic acid Detection: Visualization under long wave UV light	5-7	Tutor *et al.* (1984)
Brazil nuts, pistachio nuts	Blend 3 min with 71:29 methanol/water (3.5 ml/g)	Liquid-liquid partition	TLC Plate:Silica gel sheet Solvent: (a) Diethyl ether; cut sheet above aflatoxins (b) 80:12:0.2 Chloroform/acetone/water in the opposite direction Detection: Fluorodensitometry (ex: 365 nm; em: >400 nm)	0.5-100	Steinert *et al.* (1992)
Cocoa beans	Freeze beans with liquid nitrogen; grind into powder; defat 50 g powedered beans with hexane in Soxhlet apparatus; dry overnight and add 25 ml silver nitrate solution (25 g/100 ml water) and 250 ml chloroform; shake 30 min at low speed; filter chloroform layer and collect first 75 ml; add 7.5 ml silver	Silica gel column chromatography	TLC Plate: Silica gel 60 Solvent: 90:10 Chloroform/acetone Detection: Visualization under long-wave UV light		Scott (1995)

	nitrate solution and shake 25 min; collect 50 ml chloroform layer				
Coconut, copra copra meal	Blend 50 g for 3 min at high speed with 4 g NaCl, 100 ml water, and 250 ml chloroform	Silica gel column chromatography	TLC Plate: Silica gel 60 Solvent: 90:10 Chloroform/acetone Detection: Visualization under long-wave UV light		Scott (1995)
Corn	Shake 50 g ground sample 30 min with 25 ml water, 25 g diatomaceous earth, and 250 ml chloroform	Silica SPE cartridge	HPCL Column: Reverse-phase C_{18} (5 µm, 10 cm x 4.6 mn i.d.) Mobile phase: 80:10:6 Water/ acetonitrile/ tetrahydrofuran Detection: Fluorescence (ex: 365 nm; em: >418 nm) Derivatization: Precolumn with TFA	1.0	Hutchins *et al.*, (1989)
Corn	Shake 50 g ground sample 30 min with 25 ml water, 25 g diatomaceous earth, and 250 ml chloroform	Silica gel column chromatography	HPCL Column: µ Bondapak C_{18} (30 cm x 3.9 mm i.e.) Mobile phase: 1:1 Methanol/0.005 g/ml solution of β-cyclodextr in water Detection: Fluorescence (ex: 365 nm; em: 418 nm) Fluorescence enhancement: Post-column addition of β-cyclodextrin solution	1.0	Francis *et al.* (1988)
Corn, peanuts	Shake 30 min on mechanical shaker with 85:15 methanol/water (4 ml/g); filter	Liquid-liquid partition and silica gel column chromatography	TLC Plate: Silica gel 60 Solvent: 90:10 Chloroform/acetone Detection: Fluoro-densitometry or visualization under long-wave UV light.	1.5-50	Scott (1995)
Corn, peanuts	Blend 1 min at high speed with 80:20 methanol/water (2 ml/g)	Precipitation and liquid - liquid partition	Minicolumn Packing: Layers of Florisil, silica gel, alumina Elution solvent: 9:1 Chloroform/	10	Scott (1995)

			acetone Detection: Blue fluorescent band at top of Florisil layer under long-wave UV light		
Corn, peanut butter	Blend 3 min at high speed with 80:20 methanol/0.1 N HCl (5 ml/g)	Liquid-liquid partition and silica gel column chromatography	HPCL Column: Reverse- phase C_{18} (5 µm, 15 cm x 4.6 mm i.d.) Mobile phase: 70:17:17 Water/ acetonitrile/ methanol Detection: Fluorescence (ex: 360 nm; em: 440 nm) Derivatization: Precolumn with TFA	0.3	Scott (1995)
Corn, peanut butter	Shake 50 g ground sample 30 min with 25 ml water, 25 g diatomaceous earth, and 250 ml chloroform	Silica gel column chromatography	HPCL Column: Reverse-phase C_{18} (3 mm, 7.5 cm x 4.6 mm i.d.) Mobile phase: 39:9:7 0.10 M KH_2PO_4/ acetonitrile/ methanol Detection: Fluorescence (ex: 365 nm; em: 440 nm) Derivatization: Postcolumn with iodine		Thiel *et al.* (1986)
Corn, peanuts, peanut butter	Blend 2 min at high speed with 70-30 methanol/water (5 ml/g) plus NaCl (0.04 g/ml)	Immunoaffinity column (Aflatest, Vicam, Watertown, MA)	(1) Fluorimetric measurement of brominated eluate and/or (2) HPCL Column: Reverse-phase C_{18} (5 µm, 25 cm 4.6 mm i.d.) Mobile phase: 60:20:20 Water/ acetonitrile/ menthanol Detection: Fluorescence (ex: 360 nm; em: >420 nm) Derivatization: Postcolumn with iodine	10.0	Scott (1995)
Corn, peanuts, peanut butter (B_1, B_2, G_1)	Corn and peanuts: Blend 3 min at high speed with 80:20 methanol/ water and balances(ml/g) Peanut butter: Blend 3 min at high speed with 55:45 methanol/water (5 ml/g) and hexane (2 ml/g)	Filtration	ELISA cup: Color development indicates aflatoxins <20 ng/g; no color indicates aflatoxins>20ng/g (International Diagnostic Systems Corp., St. Joseph, MI)		Scott (1995)

Corn, peanuts, peanut butter, others	Shake 30 min with 85:15 methanol/water (4 ml/g)	Liquid-liquid partition and silica gel column chromato-graphy	Two-dimensional TLC Plate: Aluminum silica gel foil Solvent: (a) Diethyl ether; (b) 90:10 chloroform/acetone Detection: Fluorescent densitometry (ex: 365 nm; em: 430 nm)	0.5	Maferus and Zakaria (1992)
Corn, roasted peanuts (B_1)	Shake with 55:45 methanol/water (5ml/g)	Filtration	ELISA (Agri-Screen; Neogen Corp., Lansing, MI)		Scott (1995)
Eggs (B_1)	Blend with saturated NaCl solution and acetone (amounts vary depending on % water present in egg product)	Precipitation, liquid-liquid partition, and silica gel column chromatography	Two-dimensional TLC Plate: Silica gel GHR Solvent: (a) 96:3:1 Ethyl ether/methanol/water; (b) 9:1 chloroform/acetone Detection: Visualization or densit-ometry under long-wave UV light		Scott (1995)
Figs	Blend 3 min with 71:29 methanol/ water (3.5 ml/g)	Liquid-liquid partition	TLC Plate: Silica gel sheet Solvent: (a) Diethyl ether; cut sheet above aflaoxtins; (b) 80:12:0:2 chloroform/acetone/water in the opposite direction Detection: Fluoroden-sitometry (ex:365 nm; em: >400 nm)	0.2	Steiner *et al.*, (1988)
Liver (B_1, M_1)	Treat 10 g ground meat with 10 ml of 20% citric acid solution and 20 g diatomaceous earth; shake with 200 ml methylene chloride	Silica gel column chromatography	Two-dimensional TLC Plate: Silica gel 60 Solvent: (a) 95:4:1 Ethyl ether/methanol/ water; (b) 87:10:3 chloroform/ acetone/ isopropanol Detection:Visualization or densit-ometry under long-wave UV light	$\leq$0.1	Scott (1995)
Medicinal herbs	Sonicate ground sample 30 min with 70:30 methanol/water (4 ml/g)	Immunoaffinity column chromatography (Aflaprep, Coring System, Darmstadt, Germany)	HPCL Column: Reverse-phase C_{18} (4 µm, 10 cm x 8 mm i.d. Radial-PAK cartridge)	0.05	Reif and Metzger (1995)

			Mobile phase: 58:42 Water/methanol plus 119 mg KBr/L and 100 µl HNO_3/L Detection: Fluorescence (ex:360 nm; em: 460nm) Derivatization: Postcolumn with electrochemically generated bromine		
Nutmeg, cinnamon	Blend 5 min with 55:45 methanol/ 1% NaCl solution	Liquid-liquid partition	Two-dimensional TLC Plate: Silica gel 60 Solvent: (a) 96:3:1 Ethyl ether/ methanol/ water; (b) 9:1 chloroform/ acetone Detection: Visualization or densitometry under long-wave UV light		Tuinstra *et al.* (1990)
Peanuts	Blend 3 min with 85:15 acetone/ water (10 ml/g)	Phenyl-bonded column chromatography	HPLC Column: Reverse-phase C_{18} (25 cm x 4.6 mm i.d.) Mobile phase: 60:30:10 Water/ acetonitrile/ menthol Detection: Fluorescence (ex: 363 nm; em: 418 nm) Derivatization: Precolumn with TFA		Roch *et al.* (1992)
Peanuts	Blend 2 min with 80:20 methanol/ water (2 ml/g)	Liquid-liquid partition	HPLC Column: Reverse-phase phenyl (4 µm, 10 cm x 8 cm i.d. Radial-PAK cartridge) Mobile phase: 80:20 Water/ tetrahydrofuran Detection: Fluorescence (ex: 365 nm; em: 418 nm) Derivatization: Postcolumn with iodine	0.1	Dorner and Cole (1988)
Peanuts	Blend 2 min with 80:20 methanol/ water (2 ml/g)	Liquid-liquid partition	HPLC Column: Reverse- phase C_{18} (4 µm, 15 cm x 3.9 cm i.d.) Mobile phase: 60:40 Water/ methanol plus 100 µl HNO_3 and 60 mg KBr/L	0.1	Dorner and Cole (1993)

			Detection: Fluorescence (ex: 365 nm; em: 440 nm) Derivatization: Postcolumn with electrochemically generated bromine		
Peanuts (B_1)	Blend 2 min with 60:40 acetonitrile/ water (2 ml/g)	Immunoaffinity column (EASI-EXTRACT, Oxoid Ltd., Basingstoke, UK)	TLC Plate: Aluminum-backed silica gel Solvent: 90:10 Chloroform/acetone Detectiion: Visualization under long-wave UV light		Carvzal *et al.* (1990)
Peanuts, peanut products	Blend 1 min at high speed with 55:45 methanol/water (5 ml/g), hexane (2 ml/g), and 4 g NaCl	Liquid-liquid partition of aqueous methanol extract and chloroform (1:1); concentrate chloroform phase and dissolve in 98:2 benzene/ acetonitrile for TLC	TLC Plate: Silica gel 60 Solvent: 90:10 Chloroform/acetone Detection: Visualization under long-wave UV light		Scott (1995)
Peanuts, peanut products	Shake 50 g ground sample 30 min with 25 ml water, 25 g diatomaceous earth, and 250 ml chloroform	Silica gel column chromatography	TLC Plate: Silica gel GHR Solvent: 90:10 Chloroform/acetone Detection: Visualization under long-wave UV light		Scott (1995)
Peanut butter	Blend 2-3 min with 50:50 acetonitrile/ water (5 ml/g)	Filtration	ELISA (Biokits Total Aflatoxin Kit, Cortecs Diagnostics Ltd., Deeside Industrial Part, Clwyd CH5 2NT, UK)		Scott (1995)
Peanut butter	Blend 2-3 min with 55:45 acetonitrile/ water (5 ml/g) and hexane (2 ml/g)	Liquid-liquid partition	HPLC Column: Reverse-phase C_{18} (5 μm, 20 cm x 4.6 mm i.d.) Mobile phase: 60:24:16 Water/ methanol/ acetonitrile Detection: Fluorescence (ex: 365 nm; em: 440 nm) Derivatization: Postcolumn splitflow iodine from a solid-phase iodine column	1.0	Jansen and Jansen (1987)

Peanut butter	Shake 30 min with 60:40 acetonitrile/ water (3 ml/g)	Immunoaffinity column (Biocode, York, UK)	HPLC Column: Reverse-phase C_{18} (5 µm, 20 cm x 4.6 mm i.d.) Mobile phase: 58:30:12 Water/ methanol/ acetonitrile Detection: Fluorescence (ex: 364 nm; em: 434 nm) Derivatization: Postcolumn with iodine		Patey *et al.* (1991)
Peanut butter	Blend 3 min with water (1.5 ml/g); add 200 ml acetone to 100 g slurry and blend 3 min	Phenyl SPE cartridge	Biridectional HPTLC Plate: Aluminum-backed silica gel 60 Solvent: (a) Diethyl ether; (b) 6:3:1 cjp; prpfpr/ xylene/acetone Detection: Fluorodensitometry		Dett *et al.*, (1990)
Peanut butter (B_1)	Blend 3 min with 55:45 methanol/ water (5 ml/g)	Filtration	ELISA kit (Quantitox B_1, May and Baker Diagnostics, Glasgow, UK)		Mortimer *et al.* (1988)
Peanuts, corn	Blend 3 min with 75:25 methanol wter (3 ml/g) and NaCl (0.06 g/ml)	Automated immunoa-ffinity columns as part of HPLC system	HPLC Column: Reverse-phase C_{18} 30 cm x 3.0 mm i.d.) Mobile phase: 56:26:18 Water/ methanol/ acetonitrile with 1 mM KBr and 1 mM HNO^3 Detection: Fluorescence (ex: 360 nm; em: 440 nm) Derivatization: Postcolumn with electrochemically generated bromine	0.03 -0.1	Urano *et al.* (1993)
Peanuts, peanut butter, corn, tree nuts	Shake 50 ground sample 30 min with 25 ml water, 25 g diatomaceous earth, and 250 ml chloroform	Silica gel column chromatograpy	HPLC Column: Reverse-phase C_{18} (10 µm, 25 cm x 3.2 mm i.d.) Mobile phase: 75:15:10 Water/ acetonitrile/ methanol Detection: Fluorescence (ex: 365 nm; em: 450 nm) Derivatization: Precolumn with TFA		Beebe (1978)

Peanuts, peanut butter, figs	Peanuts: Blend 3-5 min with 60-40 acetonitrile/ water; dilute with water and blend for 2 min Figs: Slurry with water (5:3) and blend 80 g with 120 ml of 68.75:31.25 acetonitrile/ water	Immunoaffinity column chromatography run by a robotics system	HPLC controlled by the robotics system Column: Reverse-phase C_{18} (5 µm 25 cm x 4.6 mm i.d.) Mobile phase: 58:30:12 Water/ methanol/ acetonitrile Detection: Fluorescence (ex: 364 nm; em: 434 nm) Derivatization: Postcolumn with iodine		Shepherd and Gilbert (1984)
Peanuts, tree nuts, peanut butter	Blend 3 min with 80:20 methano l/0.1/N HCl (5 ml/g)	Liquid-liquid partition and silica gel column chromatography	HPLC Column: Reverse-phase C_{18} (10 µm, 25 cm x 4.6 mm i.d.) Mobile phase: 70:17:17 Water/ acetonitrile/ methanol Detection: Fluorescence (ex: 365 nm; em: >418 nm) Derivatization: Precolumn with TFA	0.3	Tartar *et al.* (1984)
Pistachio nuts (B_1)	(a) Shake with chloroform (10 ml/g), water (1 ml/g), and an equal weight of diatomaceous earth (b) Blend with 55:45 methanol/water (10 ml/g), hexane (4 ml/g) and ca 4g NaCl	(a) Silica gel column chromatogoraphy (b) Liquid-liquid partition of aqueous methanol extract and chloroform (1:1); concentrate chloroform phase and dissolve in 98:2 benzene/acetonitrile for TLC	TLC Plate: Silica gel 60 Solent: 90:10 Chloroform/acetone Detection: Visualization under long-wave UV light		Scott (1995)
Spices	Blend 2 min with 80:20 methanol/ water (4 ml/g)	Immunoaffinity column (Biocode Ltd., York, UK)	HPLC Column: Reverse-phase C_{18} (5 mm, 25 cm x 4.6 mm i.d.) Mobile phase: 60:22.2:17.8 Water/ acetonitrile/methanol Detection: Fluorescence (ex: 362 nm; em: 418 nm) Derivatization: Postcolumn with pyridinium bromide perbromide	1.0	Garmer *et al.* (1993)

Various	Blend 2 min with 90:10 acetonitrile/ water (2 ml/g)	Multifunctional column (Romer Labs., Inc., Union, MO)	HPLC Column: Reverse-phase C_{18} (5 µm, 10 cm x 4.6 mm i.d.) Mobile phase: 80:20 Water/ Acetonitrile Detection: Fluorescence Derivatization: Precolumn with TFA	<0.5	Wilson and Romer (1991)
Various	Blend 2 min with 70:30 methanol/ water (2 ml/g)	Immuno- affinity column chromatoGraphy run by a robotics system	HPLC controlled by the robotics system Column: Reverse-phase C_{18} (5 µm, 25 cm x 4 mm i.d.) Mobile phase: 50:50 Water/methanol Detection: Fluorescence (ex: 360 nm; em: 450 nm) Derivatization: Postcolumn with iodine	0.03-0.1	Niedwe-tzki *et al.* (1994)
Vario-us	Blend 3 min with 85:15 acetone/ water (5 ml/g) or shake 45 min on mechanical shaker at fast speed	Precipitation and liquid- liquid partition	Minicolumn Packing: Layers of Florisil, silica gel, alumina Elution solvent: 9:1 Chloroform/ acetone Detection: Blue fluorescent band at top of Florisil layer under long-wave UV light	5-15	Scott (1995)
Vegetable oils	Mix oil with equal volume of hexane for extraction and cleanup by column chromatography	Silica gel column chromatography	Two-dimensional TLC Plate: Silica gel 60 Solvent: (a) 9:1 Chloroform/ acetone; (b) 96:3:1 ethyl ether/ methanol/ water Detection: Visualization under long-wave UV light HPLC Column: Silica (5 µm, 25 cmx4.6 mm i.d.) Mobile phase: Mix 80:24 chloroform/ cyclohexane with 32:22:0.7 acetonitrile/ 2-propanol/ water *Detection*: UV (360 nm lamp)		Miller *et al.* (1985)

Wines	Shake 200 ml wine with 200 ml methanol and 200 ml dichloro-methane for 1 min in separatory funnel	Silica gel-alumina column chromatography	HPLCColumn: Reverse-phase C_{18} (10 µm 25 cm x 3.2 mm i.d.) Mobile phase: 15:3:2 Water/ acetonitrile/ methanol Detection: Fluoreoscence (ex: 363 nm; em: >400 nm) Derivatization: Precolumn with TFA	0.02 µg/L	Takahashi (1977)

Table 7.1.3 Methods for the Determination of Aflatoxin M_1 in Dairy Products

Food	Extraction	Cleanup	Quantitation	Detection (µg/kg)	Ref.
Cheese	Blend 5 g with 1 ml saturated NaCl solution, 5 g diatomaceous earth, and 100 ml chloroform	Silica gel column chromatography	TLC Plate: Silica gel GHR Solvent: 87:10:3 Chloroform/ acetone/ isopropanol Detection: Visualization or densitometry under long-wave UV light		Scott (1995)
Milk	Fluid milk: Shake 50 ml with 10 ml saturated NaCl solution, and 120 ml chloroform for 60 sec Powedered milk: Reconstitute 5 g with 50 mL water and proceed as with fluid milk	Silica gel column chromatography	TLC Plate: Silica gel GHR Solvent: 87:10:3 Chloroform/ acetone/ isopropanol Detection: Visualization or densitometry under long-wave UV light		
Cheese	Blend 10 g with ml dichloromethane and 5-10 g Hyflosupercel for 2 min at high speed; defat with hexane	Immunoaffinity column (Aflaprep M; Rhone-Poulenc Diagnostics, Lyon, France)	HPLC Column: Reverse-phase C_{18} (25 cm x 4.6 mm i.d.) Mobile phase: 25:75 Acetonitrile/ water Detection: Fluorescence (ex: 360 nm; em: 435 nm)	0.02	Dragacci *et al.* (1995)
Cheese	Blend 20 g at high speed for 2 min with 75 ml chloroform, 1 ml saturated NaCl	Evaporate, defat, immunoaffinity	HPLC Column: Reverse-phase C_{18}	0.005	Sharman *et al.* (1989)

	solution, and 5 g Celite 545	column (Aflatoxin, M1 EASI-EXTRACT, Oxoid, Basingstoke, UK)	(5 μm, 25 cm x 4.9 mm i.d.) Mobile phase: 60:10:30 Water/ acetonitrile/ methanol Detection: Fluorescence (ex: 364 nm; em: 434 nm)		
Cheese	Blend 25 g 1 min with 100 ml chloroform and 5 ml saturated NaCl solution	Silica gel column chromatography followed by C_{18} SPE cartridge	HPLC Column: Reverse-phase C_{18} (10 μm, 15 cm x 4 mm i.d.) Mobile phase: 35:65 Methanol/ 0.01 M KH_2PO_4 Detection: Fluorescence (ex: 365 nm; em: 410 nm) Derivatization: Precolumn with TFA	0.01	Hisata *et al.* (1984)
Cheese	Blend 2 min with 3:1 acetone/ water (8 ml/g)	C_{18} and silica cartridge	1. Two-Dimensional TLC Plate: Silica gel 60 Solvent: (a) 95:4:1 Ethyl ether/ methanol/water (b) 70:30 Chloroform/acetone Detection: Fluorodensitometry (ex: 366 nm; em: >430 nm) 2. HPLC - Column: Reverse-phase C_{18} (10 μm, 25 cm x 4.6 mm i.d.) Mobile phase: 60:10:30 Water/ acetonitrile Detection: Fluorescence Derivatization: Precolumn with TFA for confirmation	0.01	Bijl *et al.* (1987)
Cheese, yogurt	Cheese: Blend with 30% methanol/ PBS Yogurt: Blend with 15% methanol/ PBS	Filter and dilute	Bead ELISA Incubate diluted extract with antibody-coated polystyrene beads. Measure optical density photometrically.	0.01-0.05	Aman (1994)

Milk	Rotobics-operated C18 SPE cartridge used to extract toxin from milk	Rotobics-operated C_{18} SPE cartridge	HPLC Column: Reverse-phase C_{18} (5 µm, 25 cm x 4.9 mm i.d.) Mobile phase: 60:10:30 1% Acetic acid/acetonitrile/ methanol Detection: Fluorescence (ex: 355 nm; em: 433 nm)	<1.0	Gifford *et al.* (1990)
Milk		Automated dialysis and concentration on a Perisorb RP_{18} cartridge	HPLC Column: Reverse-phase C_{18} (3 µm, 10 cm x 4.6 mm i.d.) Mobile phase: 26:74 Acetonitrile/ water Detection: Fluorescence (ex: 345 nm; em: 435 nm)	0.02	Tunistra *et al.* (1990)
Milk[b]		Warm milk to 35-37°C; filter; pass through immunoaffinity column (Rhone Poulenc Diagnostics Ltd., Glasgow, Scotland); elute with acetonitrile	HPLC Column: Reverse-phase C_{18} (5 µm, 25 cm x 4.6 mm i.d.) Mobile phase: 75:25 Water/ acetonitrile Detection: Fluorescence (ex: 360 nm; em: 435 nm)		Tuinstra *et al.* (1990)
Milk	Mix 50 ml milk gently with 10 ml saturated NaCl solution and 120 ml chloroform	Liquid-liquid partition	TLC Plate: Silica gel G-25HR Solvent 95:4:1 Ethyl ether/methanol/ water Detection: Visualization under long-wave UV light	0.05	Serralheiro *et al.* (1985)
Milk	Mix 40 ml with 1 g NaCl and centrifuge; filter skim portion	Pass 25 ml skim milk through immuno affinity column (Vicam, Watertown, MA)	Fluorimetric measurement of brominated eluate	0.05	Hansen (1990)
Milk, milk	Shake 50 ml milk, after basification, with 70 ml 50:20 methanol/acetone;	Silica gel column chromatography	TLC	0.02-0.1	Dominguez *et al.* (1987)

products	centrifuge; reextract with chloroform		Plate: Silica gel 60 Solvent: 85:10:5 Chloroform/acetone/ isopropanol Detection: visualization under long-wave UV light		
Parmesan cheese	Shake 20 g with 10 g Celite and 100 ml chloroform	Liquid-liquid partition	HPLC Column: Reverse-phase C_{18} (5 µm, 25 cm x 4.6 mm i.d.) Mobile phase: 80:20 Water/acetonitrile Detection: Fluorescence (ex: 360 nm; em: 420 nm)	0.025	Barbieri *et al.* (1994)
Powdered milk	Reconstitute 10 g in 100 ml water; defat by centrifugation	Pass 20 ml defatted milk through immunoaffinity column (EASI-EXTRACT, Oxoid, Basingstoke, UK); column can be reused after washing with DMSO-phosphate buffer	HPLC Column: Reverse-phase C_{18} (5 µm, 25 cm x 4.9 mm i.d.) Mobile phase: 60:30:10 Water/ acetonitrile/methanol Detection: Fluorescence (ex: 360 nm; em: 435 nm)	0.05	Toyoda *et al.* (1994)
Fluid milk	Dilute 20 ml milk with 20 ml hot water; run sample through C_{18} SPE cartridge and wash with 95:5 water/acetonitrile (10 mL); elute with ether	Silica gel column chromatography	HPLC Column: Reverse-phase C_{18} (5 µm, 25 cm x 4 mm i.d.) Mobile phase: 80:12:8 Water/acetonitrile/ methanol Detection: Fluorescence (ex: 360 nm; em: 440 nm) Derivatization: Precolumn with TFA	0.07	Scott (1995)

Official quantitative methods for aflatoxins are established for various commodities by collaborative testing supervised by the AOAC and their official referees. The AOAC procedures employ a representative sampling technique, solvent extraction (one phase or two phase systems), use of precipitating agents (Removal of proteins and pigments), cleanup, column chromatography and TLC separation of the aflatoxins with measurement by flouro-densitometry or visual methods.

Adequate sampling is the key to accurate measurements of aflatoxins in agricultural commodities. Often aflatoxin contamination is localized with in a lot with individual kernels containing high levels of toxin and others none at all. Evaluation of true level of contamination of the load would necessitate extraction of entire. Since this is impossible, selection of large representative samples is achieved by probe or stream splitters. Samples then are ground and blended for uniform toxin distribution prior to 50 g aliquots being removed for analysis. Use of high pressure liquid chromatography (HPLC) is an attractive alternative to TLC flourodensitometry and visual analytical procedures because of the high resolution potential, rapidity of separations and potentially improved quantitative accuracy and precision (Table 7.1.2 & 7.1.3).

7.1.5. Natural incidence

Besides large number of workers, Vrabcheva (2000), Bassa *et al.* (2001), Reddy *et al.* (2001) and Cast (2003) have reported aflatoxins in corn, peanuts and peanut products, cotton seeds, pipers, rice, pistachios, tree nuts, pumpkin seeds, sunflower and other oil seeds, copra, spices and dried fruits from different parts of the world. Widstrom (1996) recorded high level of infestation of *Aspergillus flavus* in peanuts, corn and cotton seed, while they are in the field and proliferate during their storage due to ideal conditions prevailing in the store house. On the other hand, Jelineck *et al.* (1989) are of the opinion that soybeans, pulses, cassava, grain sorghum, millets are comparatively resistant and moderately susceptible to *A. flavus* and aflatoxin contamination. Rustom (1997) reported high level contamination of aflatoxin B_1 in some lots of rice, figs, spices and pecans. Incidences of field contamination of crops by Aflatoxin B_1 have been well documented for the southern states of USA (Hagler, 1990).

7.1.6. Biological activity

Aflatoxin B_1 is an acute toxin as well as being carcinogenic and immunosuppressive but in contrast to T_2 toxin, the precise nature of the response to the consumption of aflatoxin is dependent on species, sex and age. Zuezada *et al.* (2000) found that one week old boilers were more susceptible to AfB_1 than

4-week old. A diet containing as low as 4 µg/ml aflatoxin caused hepatoma in 25% of rainbow trat with in 12 months. There have been a number of isolated cases of deaths in humans attributed to alfatoxins in food. A tragic demonstration of the acute toxicity of aflatoxins in humans was reported in India in 1974 when a large outbreak of poisoning occurred involving 1000 people of whom nearly 100 died (Moss, 1996). LD_{50} for AFB_1 in humans is approximately 5 mg/kg body weight thus lying between that for the dog and the rat. Patients with primary liver cancer in Czechoslovakia during 1987 were found to contain AFB_1 in liver of 27 patients out of 34 screened. Although aflatoxin exposure occurs more in India than in S. Africa, the incidence of liver cancer is less than that occurs in S. Africa which may be attributed to genetic difference. However, more data from different parts of the world is needed before reaching any such conclusion indicating AFB_1 in liver diseases.

After absorption from the small intestine, AFB_1 readily binds to plasma albumin which serves as the major transporter of AFB_1 in blood (Yatin and Sachan, 2001). AFB_1 is oxidized by microsomal mixed function oxidase (cytochrome P_{450}) to several water-soluble metabolites. It is generally believed that the formation of AFB_1 expoxide an active metabolite but its subsequent covalent binding to DNA, RNA and protein plays a critical role in both acute and chronic toxicity (Eaton and Groopman, 1994). Animals which fail to produce the epoxide are relatively resistant to both but those animals which produce epoxide but do not effectively metabolise it further may be at the highest risk to the caranogenic activity of AFB_1 but relatively resistant to the acute toxicity. Those animals that not only produce the epoxide but also effectively remove it with a hydrolase enzyme thus producing a very reactive hydroxy actinal are most sensitive to the acute toxicity (Moss, 1996). The extreme sensitivity of turkeys to aflatoxins may be attributed to the combined activity of cytochrome P_{450} and deficient detoxification as a result of the absence of AFB_1 – specific glutathione S-transferase activity – the liver (Klein *et al.*, 2000). The glutathione S-transferase isozyme Mla-la appears to be the most active in confuguting 8,9-epoxide in humans (Johnson *et al.*, 1997).

The epoxide of AFB_1 is known to react with guanine residues in DNA and can cause subsequent depurination, while the hydroxyacetate derivative reacts with protein through such residues as lysine. The parent molecule may thus be seen as a very effective delivery system having the right properties for absorption from gut and transmission to the liver and other organs of the body. Thus the response of the animal depends on how it is metabolized subsequently (Moss, 1996). An alternative biotransformation of AFB, to aflatoxicol is mediated by a reductase and the product can be reconverted to AFB_1 by a dehydrogenase (Fig. 7.1.5). Thus aflatoxicol may serve as reservoir of AFB_1 in some animals and play critical role in aflatoxicosis (Lee *et al.*,

2001). On the other hand, feels that aflatoxin is rapidly excreted in the bile and urine and does not accumulate or persist in body tissues.

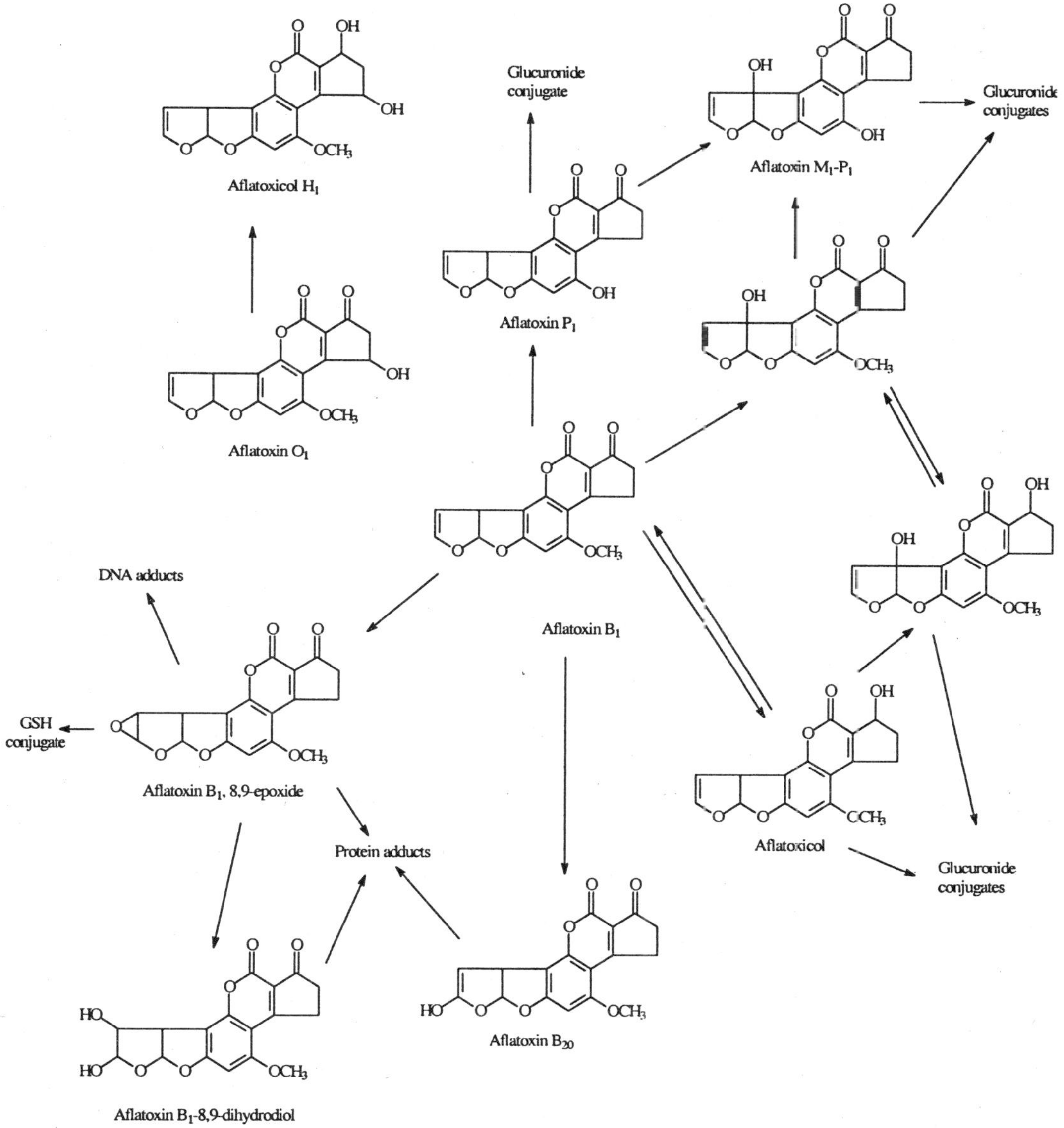

Fig. 7.1.5. Biotransformation pathways for aflatoxin B_1

Acute aflatoxin toxicity has been demonstrated in a wide range of mammals, fish and birds, rabbits, dogs, primates, ducks, turkeys and trout are highly susceptible which varies with age, sex and nutritional status. Young and males are more susceptible than old and females. The liver is the principal target organ although site of hepatic effect varies with the species.

Lungs, myocardium and kidneys are also affected but it accumulates in the brain. Teratogenic effects were also observed in hamstered resulting high proportion of malformed, dead or resorbed foetuses following single intraperitonial doses of 4 mg/kg aflatoxin B on 8th day of ingestation. Teratogenic effects in chicks and mice were also reported but not in pigs when given feed containing 0.45 mg/kg aflatoxin B_1. Human aflatoxicosis due to consumption of most frequently maize, rice or cassava or corn products such as parata or peanut meal (Oyelami *et al.,* 1996). During a chinese festival in Malaysia in which approximately 40 persons were affected and 13 children died (Chao *et al.,* 1991) after eating noodles highly contaminated with aflatoxin and boric acid.

Aflatoxins have been implicated in sub-acute and chronic effects in humans. These effects include primary liver cancer, chronic hepatitis, jaundice, hepatomegaly and cirrhosis through repeated ingestion of aflatoxin. It is also considered that aflatoxins may play a role in number of diseases including Reye's syndrome (Krishna Machari *et al.,* 1975) and Kwashiokor (Coulter *et al.,* 1986). Aflatoxins can also affect the immune system (Pier, 1991). The carcinogenicity and mutagenicity of aflatoxins are considered to arise as the result of the formation of a reactive epoxide at the 8,9 position of the terminal furan ring and its subsequent covalent binding to nucleic acid.

AFB_1 is known as a potent hepatotoxin and heaptocarcinogen and the liver is considered to be the primary target for aflatoxin (Towner *et al.,* 2000). Aflatoxin B_1 metabolism in liver is depicted in Fig. 7.1.6. The toxic metabolites of aflatoxins bind to nucleic acids and nucleoproteins essential to cellular viability and result in an excessive build up of hepatic lipids with enlargement of the liver, proliferation of bile duct epithelium (Adav and Godinwar, 1997), necrosis and hepatocellular carcinoma (Bammler *et al.,* 2000). The resistance of mice to AFB_1 hepatocarcinogenicity appears to be due solely to the constitutive expression of a specific form of glutathione-S-transferase, m-GSTA 3-3, with unusually high catalytic activity towards AFB-2,3/AFB-8, epoxide (Beuther *et al.,* 1992). A different aspect of toxicology of aflatoxins is reviewed by Moss (1998).

With excessive exposure to aflatoxins, that accumulates as clear velenoles in the cytoplasm of hepatocytes in a dose and time dependent fashion and results in grossly evident large, friable and yellow livers. With continued exposure intra-hepatic bilary epithelial hyperplasia occurs as an attempt to regenerate the hepatic parenchyma when the parenchymal cells themselves have lost their capacity. Such hepatobiliary hyperplasia resulted in a significant increase of alanine aminotransferase, gamma-glutamyltransferase and total bilurubin. Zaky *et al.* (1998) recorded an increase in liver microsomal protein, electron transport components and drug metabolizing enzymes due to administration of AFB_1. Verma and Raval (1997) observed reduced glycogen content of liver and skeletal muscles of young rabbit after consumption of 15 mg AFB/kg feed for 60 days which they attributed to enhanced glycogenolysis. Extensive cellular necrosis in liver and kidney was resulted due to impairing protein synthesis and secretion (Verma *et al.*, 1998). In fish also the liver is the primary target of aflatoxins (Hendricks, 1991). Effects of aflatoxins other than liver are secondary. Factors such as route of administration, dose and frequency of dose and species strain, age and sex of animals appears to affect the degree of extra-hepatic involvement (Coulumbe, 1994). Several surveys have shown that AFB_1 is virtually always present in grain dusts at levels often exceeding those found in contaminated food. According to Coulumbe (1994) AFB_1 inhaled may be a significant health hazard but further studies are necessary to establish the risk associated with inhaled AFB_1.

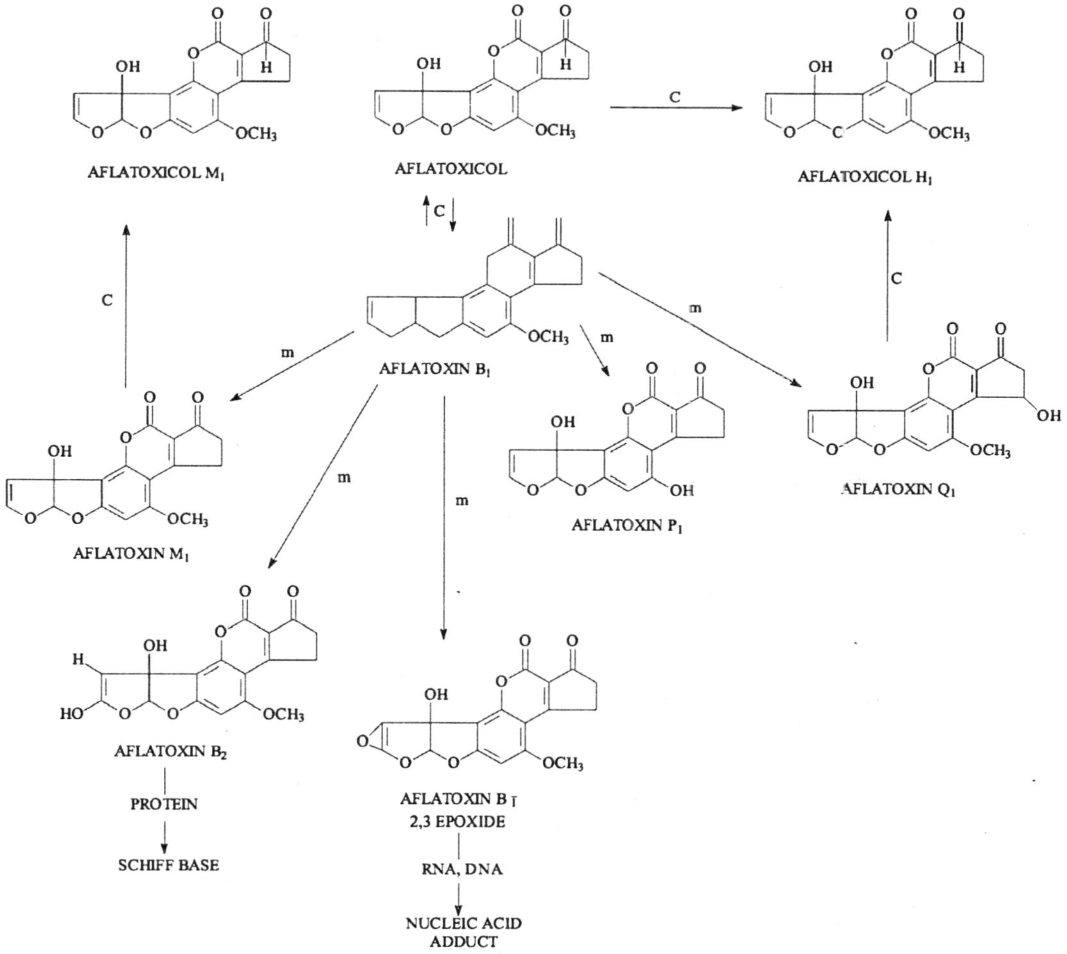

Fig. 7.1.6. Aflatoxin B_1 metabolism in the liver

Epidemiological studies have shown a strong positive correlation between AFB_1 levels in diet and primary hepatocellular carcinoma (Wild and Hall, 2000). AFB_1 has been found to be a potent carcinogen in many animal species (Smela *et al.*, 2001), while role of AFB_1 in carcinogenesis in humans is complicated by hepatitis B virus infections (Wild *et al.*, 1992). Epidemiological studies have shown a strong positive correlation between AFB_1 levels in diet and primary hepatocellular carcinoma incidence in some parts of the world (Eaton and Groopman, 1994; Wild and Hall, 2000).

Epstein *et al.* (1969) demonstrated that oral AFB_1 is a renal carcinogen. Verma and Kolhe (1998) recorded increased level of creatinine in the serum and urine of aflatoxin fed rabbits is due to increased transformation of phosphocreatine in muscle which might be due to lower utilization of phosphocreatine in muscle contraction. Increased concentrations of creatinine in serum indicate severe kidney damage. Increased phosphatase activity in kidneys of rabbits fed with aflatoxins may be due to reabsorption of nitrogenous materials other than urea and uric acid from lumen (Verma and Raval, 1997).

Carry over of aflatoxin from layer feed to eggs has been demonstrated where 500-600 mg/kg AFB_1 in the diet of layers resulted in a concetration of 100 mg/kg (Oliveira *et al.,* 2000) and 300 mg/egg AFB_1 (Sudhakar, 1992) respectively. Aflatoxins have been reported to cause a variety of disturbances in chick embryos from early embryonic death to organ proliferations (Kemper and Lucpke, 1986) resulting in serious economic loss and decreased productivity in the hatchery (Celik *et al.,* 2000). Trace amounts of aflatoxins found in eggs may cause functional defects in the immune system of the developing embryo with subsequent immunodeficiency (Neldon-Ortiz and Qureshi, 1992). Singh *et al.* (1994) also recorded decrease in WBC count in blood of chick receiving aflatoxin.

Verma *et al.* (1998) reported a rise in intracellular concentration in the organs of aflatoxins-fed rabbits. The rise in cytosolic calcium may affect the cytoskeleton and thereby damage cellular integrity either directly or through calcium activated proteases. Verma and Raval (1991) recorded an increase in the rate of haemolysis under the influence of AFB_1 which they attributed to lipid peroxidation of plasmamembrane, permeability alterations and cell lysis.

Consumption of aflatoxins contaminated feed lowers disease resistance in domestic animals (Coulumbe, 1994). Immune system is a highly sensitive indicator of aflatoxicosis in poultry (Giambrone *et al.,* 1985b). Immuno suppression by aflatoxins may be due to depressed protein synthesis via inhibition of RNA polymerase which results in suppression of specific immunoglobulin synthesis (Giambrone *et al.,* 1985b). Non-specific factors such as complement, interferon and non-specific serum protein concentrations also decreased as a result of liver damage (Stewart *et al.*, 1985). Long term AFB_1 treatment of weanling rats caused in addition to a depletion of cell populations of the thymus reduced bone marrow and red and white blood cell counts of the thymus, a depression of macrophage numbers with reduced phagocyte capacity (Raisuddin *et al.*, 1990). Total lymphocyte and T lymphocyte counts were significantly lowered in chicks fed 0.3 and 1 mg AFB_1, 1 kg feed (Ghosh *et al.*, 1991) and 2.5 mg total aflatoxin/kg feed (Celik *et al.*, 2000). Maternal transfer of aflatoxins residues to eggs reduced the post hatch immunological potential of chicks and their resistance to disease challenges (Dureshi *et al.*, 1998). According to Nicolas-Bolnet *et al.* (1995) the exposure of embryos to aflatoxin metabolites via maternal transfer may affect the differentiation and maturation processes of immature progentor cells which is crucial for the establishment of various haematopoiet lineage cells such as lymphocytes and macrophages. Sahoo and Mukherjee (2001) reported decrease in total protein, globulin levels and bacterial agglutination titres in a fish treated with aflatoxins for 90 days.

Aflatoxins in broiler chicks decreased the plasma, oxycarotenoid concentration and caused pale skins under field conditions (Osborne *et al.*, 1982). Aflatoxins probably decrease the ability to form micelles in the amount and character required for optimal absorption of lutein from intestinal contents (Schaeffer *et al.*, 1988). Aflatoxin intoxicated adult has have decreased egg production and the hatchability of those eggs is reduced. However, insemination of hens with semen from affected males failed to show consistent results (Brown, 1996).

Aflatoxins are reported to induce a variety of physiological and biochemical disorders in different plant groups including higher plant forms, algae, fungi and bacteria. This aspect has been very well discussed by Bilgrami and Sinha (1992). Aflatoxins inhibited seed germination in *Lepidium sativum* (Schoental and White, 1965), chlorophyll deficiency (Reiss, 1971), Aflaroot disease syndrome in ground nut (Mehan and Chohan, 1974), chlorophyll synthesis inhibition in seedlings of *Abelmoschus esculentum* (Klang, 1978). Chohan (1983) feels that aflatoxins inhibit chlorophyll synthesis by blocking growth hormone induced synthesis of RNA, DNA and protein in the leaf. Aflatoxins at its subtoxic concentration inhibited chromatic bound DNA dependent RNA polymerase activity. Bilgrami *et al.* (1986) recorded mitoinhibitory activity of aflatoxins at lower concentrations, while at higher concentration it promoted mitotic division in *Allium cepa*. Crisan (1973) feels that aflatoxin B_1 acts as antimetabolite that binds to DNA and interferes with RNA and protein synthesis.

Aflatoxins are also reported to inhibit spore germination and seedling development in *Onclea sensibilis* (Cahill *et al.*, 1978), growth of *Chlorella pyrenoidosa* (Cahil *et al.*, 1978), inhibition of growth of *A. awamon*, *Penicil-*

lium chrysogenum (Lillihoj *et al.*, 1967). The principal manifestations of AFB_1 induced toxicity are lipid peroxidation and oxidative DNA damage which could be mitigated by antioxidant (Verma and Nair, 1999).

7.1.7. Factors affecting AFB_1 toxicity

Numerous factors can influence the magnitude of response of invertebrates to exposure to AFB_1. One of the most striking modulating effects on AFB_1 toxicity is associated with dietary components that the animals consume prior to or following exposure to AFB_1 (Cullen and Newberne, 1994). Protein and vitamin content of diet, hormonal status and treatment with pharmacologically active compounds modify the response of experimental animals to acute AFB_1 toxicity. Valdivia *et al.* (2001) demonstrated that the broiler treated with AFB_1 plus N-acetylcystine were partially protected against the deleterious effects on performance of liver and kidney damage and biochemical alterations induced by AFB_1 in broiler chickens. High protein diet protected monkeys from aflatoxin liver injury (Cullen and Noewberne, 1994). Raisuddin *et al.* (1994) observed that protein A provided protection to rats from AFB_1 – induced immunotoxicity. Zaky *et al.* (1998) recorded that protein A counter acted the toxic effect of AFB_1 on haematocrit globulin and bilerubin in broilers. According to Riasuddin *et al.* (1994), the protective effect of vitamin A on AFB_1 toxicity is caused by inhibition of the activity of liver cytochrome P_{450} mixed function oxidase system that activated AFB_1 to form the active metabolite that binds to cell DNA and RNA. Total aflatoxin-DNA aduct levels in the choline deficient animals were increased significantly during the multiple-dose schedule. Carnitive combination with coenzyme D_{10} offered significant protection against oxygen radicals induced by aflatoxins in bacteria (Atrosin, 1998). AFB_1 binds to serum albumin and hepatic DNA in a dose dependent manner (Sabbioni *et al.,* 1990). AFB_1 has been shown to have highest affinity for liver microsomes followed by the cytosol, mitochondria and nuclei (Ewaskiewioks *et al.,* 1991). Yatin and Sachan (2001) found that carnitive increased the retention of AFB_1 by plasma protein and lowered the binding of AFB_1 to DNA which they attributed to less free AFB_1 available for uptake and metabolism by liver cells.

7.1.8. Management

Reen *et al.* (1997) demonstrated the participation of hepatic microsomal enzymes such as P_{450} in the metabolism of AFB_1 (Fig.7.1.6). Similarly adminstration of Phenobarbital intermittently in the drinking water of aflatoxin-fed chickens caused substantial increase in cytochrome P_{450} content and consequently enhancing the metabolism and elimination of AFB_1 (Dalvi and McGown, 1984).

Hao and Brackett (1988) have reported the degradation of AFB_1 in food products by *Flavobacterium aurantiacum*. Line *et al.* (1994) using ^{14}C-labelled AFB_1 demonstrated that the removal of AFB_1 was through enzymatic degradation rather than by binding to the bacterial cell wall. Smiley and Dranghan (2000) attributed the degradation of AFB_1 to a protein with enzyme characteristics. Huff (1980) reported the removal of low density grains with sucrose gradient reduced the aflatoxin content by 90%.

Philips *et al.* (2008) reported significant reduction in aflatoxins in feeds without affecting availability of vitamins and other micronutrients by us of novasil clay. The details of management of aflatoxins are also discussed in chapter management of mycotoxins.

Control of aflatoxins need a integrated approach in view of their incidence in the field, in stored commodities or as a result of passage through the food chain to contaminate foods that had not been mouldy (Lopez-Garcia *et al.*, 1999). Such programmes include control of formation in the field by preventing insect damage, alleviating drought stress and ensuring that potential inoculum does not build up in plant residues after harvest. Following harvest it is essential that crops be dried to reduce their water activity and store in cool and dry conditions. In case of relatively large particulate commodities such as ground nuts and pistachio nuts it may be possible to reduce the average aflatoxin contamination of this commodity from 12 μg/kg for product going to the market for human consumption (Schatzki and Pan, 1996).

7.2. FUMONISINS

Fumonisins are novel toxic metabolites of *Fusarium verticelloides* and related species of *Fusarium*. Though toxicity of *F. moniliforme* to horses is reported as back as 1904 (Marasas *et al.,* 1976) and other experimental animals such as monkeys, pigs, sheep and rats (Kirek *et al.,* 1981), it is Wilson *et al.* (1985), who implicated naturally infested corn samples by *F. moniliforme* in the field outbreak of Leucoencephalomalacia (LEM) and for hepatocarcinogenicity in rats. Marasas *et al.* (1979) and Marasas *et al.* (1984) recorded *F. moniliforme,* an ubiquitous endophyte of corn as most prevalent fungus associated with human and animal dietary corn. Li *et al.* (1980) and Yang (1980) have also implicated *F. moniliforme* infested corn in the incidence of human esophageal cancer in China. Gelderblom *et al.* (1984) could isolate and identify the cancer promoting activity of *F. moniliforme* (fusarin C), a mutagenic metabolite. Thiel *et al.* (1982) and Gelderblom *et al.* (1984) could detect moniliformin, zearalenone, deoxynivalenol and fusarinin in corn naturally contaminated with *F. moniliforme* from high esophageal cancer risk area of Tanskei, South Africa. However, Gelderblom *et al.* (1986) were of the opinion that fusarinin elaborated by *F. moniliforme* is not involved in the hepatocarcinogenicity. It is, Gelderblom *et al.* (1988) who have isolated two compounds which have cancer promoting activity and identified them to be fumonisin B_1 and B_2. Bezuidenhout *et al.* (1988) have elucidated the structure of fumonisins. Presently, approximately 20 fumonisins analogs have been isolated and characterized (MacKentzie *et al.,* 1998). Marasas *et al.* (1988) could demonstrate the leukoencephalomalacea in horse by intravenous injection of fumonisin B_1 isolated from *F. moniliforme* (MRC 826). Association of these toxins with porcine pulmonary edema in swine (Harrison *et al.,* 1990) and cancer promoting activity in rats have been recorded (Rheeder *et al.,* 1992). Gelderblom *et al.* (1992) have further created special interest in the study of this group of fusarial toxins. Bullerman *et al.* (2001), Riley *et al.* (2001) and Jackson *et al.* (2004) have excellently reviewed different aspects of fumonisins in particular and fusarial toxins in general. According to Abbas *et al.* (2000) aflatoxins are a chronic problem in the south eastern USA, while the fumonisins are a chronic problem worldwide and therefore, may represent a greater challenge to control. Marasas *et al.* (1984) aptly state that if the new era of research on food-borne diseases caused by mycotoxins was introduced by the aflatoxins in 1960, the twentieth century closed with worldwide attention focused on the fumonisins. The next chapter to be written on food-borne diseases caused by *Fusarium* mycotoxins may be called "Fumonisins – Food-borne. *Fusarium* toxins for the third millinium" have excellently reviewed different aspects of fuminosins.

7.2.1. Physico-chemical characteristics

Fumonisins ($C_{34}H_{59}NO_{15}$) are diesters of propane-1,2,3-tricarboxylic acid (tricarboxylic acid) and 2-amino-12,16-dimethylpolyhydroxydicosanes, in which the C_{14} and C_{15} hydroxyl groups are esterfied with the terminal carboxyl group of tricarboxylic acid (Fig. 7.2.1). There are as many as six structurally related fumonisin analogues. Seo *et al.* (1999) reported N-acetyl derivatives of the type C fumonisins by *F. oxysporum*. Akiyama *et al.* (2000) have reported a novel FB_1 analogue. Branham and Plattner (1993) have reported fumonisin C_1 (FG) which lacks the amino and terminal methyl group (CH_3) of FB_1. Wang *et al.* (1998) reported that *F. moniliforme* which showed dimorphism produce FB_1.

Of the six known fumonisins, FA_1 and FA_2 the acetyl derivatives of FB_1 and FB_2 respectively are produced in the lowest yields in cultures of *F. moniliforme* and have the lowest toxicities. These two structural analogues and FB_4 do not occur under natural conditions. Several other fumonisins-related derivatives have been found in corn or are created as a result of chemical and/or thermal processing of food (Fig. 7.2.1). More details can be obtained from Blackwell *et al.* (1994), Sydenham *et al.* (1996) and Apsiman (2001).

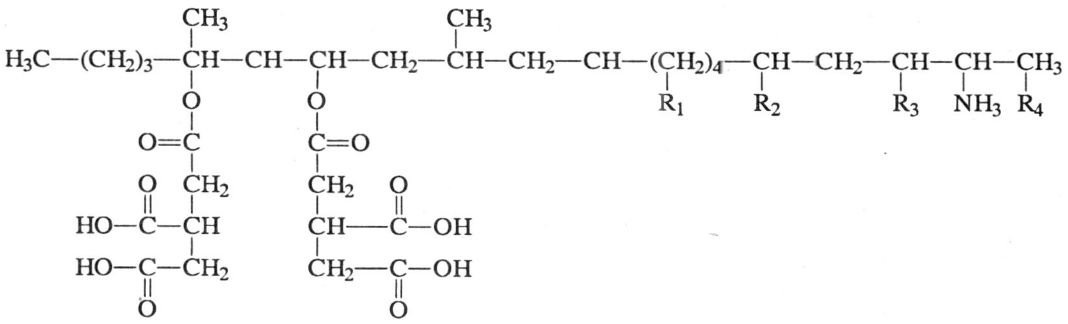

```
                 CH3                    CH3
                  |                      |
H3C—(CH2)3—CH—CH—CH—CH2—CH—CH2—CH—(CH2)4—CH—CH2—CH—CH—CH3
                  |      |                       |        |         |    |    |
                  O      O                       R1       R2        R3  NH3  R4
                  |      |
             O=C        C=O
                  |      |
           O  CH2        CH2   O
           ||  |         |     ||
     HO—C—CH           CH——C—OH
                  |      |
     HO—C—CH2          CH2—C—OH
           ||                  ||
           O                   O
```

Fumonisins	Group						Formula	M.W.
	R_1	R_2	R_3	R_4	R_5	R_6		
FA_1	TCA	TCA	OH	OH	$NHCOCH_3$	CH_3	$C_{36}H_{61}NO_{16}$	763.87
FA_2	TCA	TCA	H	OH	$NHCOCH_3$	CH_3	$C_{36}H_{61}NO_{15}$	747.87
FA_3	TCA	TCA	OH	H	$NHCOCH_3$	CH_3	$C_{36}H_{61}NO_{15}$	747.87
FAK_1	=O	TCA	OH	OH	$NHCOCH_3$	CH_3	$C_{36}H_{53}NO_{11}$	603.74
FB_1	TCA	TCA	OH	OH	NH_2	CH_3	$C_{34}H_{59}NO_{15}$	721.83
FB_2	TCA	TCA	H	OH	NH_2	CH_3	$C_{34}H_{59}NO_{14}$	705.83
FB_3	TCA	TCA	OH	H	NH_2	CH_3	$C_{34}H_{59}NO_{15}$	705.83
FB_4	TCA	TCA	H	H	NH_2	CH_3	$C_{33}H_{57}NO_{15}$	721.83
FC_1	TCA	TCA	OH	OH	NH_2	H	$C_{39}H_{52}NO_{16}+$	707.80
FP_1	TCA	TCA	OH	OH	NH_2	CH_3	$C_{39}H_{52}NO_{15}+$	800.91
FP_2	TCA	TCA	H	OH	NH_2	CH_3	$C_{39}H_{52}NO_{15}+$	784.91
FP_3	TCA	TCA	OH	H	NH_2	CH_3	$C_{28}H_{53}NO_{10}$	784.91
PH_{1a}	TCA	OH	OH	OH	NH_2	CH_3	$C_{28}H_{53}NO_{10}$	563.72
PH_{1b}	OH	TCA	OH	OH	NH_2	CH_3	$C_{22}H_{47}NO_5$	563.72
$AP_1(HFB_1)$	OH	OH	OH	OH	NH_2	CH_3	$C_{36}H_{61}NO_{16}$	405.61
$NCMFB_1$	TCA	TCA	OH	OH	$NH(C_2H_3O_2)$	CH_3	$C_{36}H_{61}NO_{16}$	779.87
$NDFB_1$	TCA	TCA	OH	OH	$NH(C_6H_{11}O_5)$	CH_3	$C_{36}H_{61}NO_{16}$	883.97

Fig. 7.2.1: Chemical structure of different fumonisins produced by *F. moniliforme*.

7.2.2. Organisms

Fusarium moniliforme which is known to be a producer of fumonisins was reported on corn earlier as *Oospora verticelloides* in 1881 in Italy. Subsequently Sheldon (1904) incriminated this as a cause of mold-corn toxicosis of animal in USA and subsequently described as *F. moniliforme*. It is Gelderblom *et al.* (1988) who isolated fumonisins from cultures of *F. moniliforme* and conclusively proved it to be responsible for encephalomala ceionephropathy. Subsequently Ross *et al.* (1990) reported that *F. proliferatum* also produces this toxin. Thiel *et al.* (1992) and Nelson *et al.* (1992) reported that fumonisins are produced even by *F. nygamei* and *F. anthophylum* respectively. Klittich *et al.* (1997) reported that *F. napiforme* also produces fumonisins. Sydenham *et al.* (1997) added another species of *Fusarium*, *F. globosum* to fumonisins producers. Production of fumonisins by *F. polyphiadium* and *F. oxysporum* (Abbas *et al.*, 1995) and *Alternaria alternata* (Chen *et al.*, 1992) require confirmation. Large number of isolates of representative species of both liseola section and other sections of *Fusarium* are to be screened before reaching any decisive conclusions. Production of these toxins by *Alternaria alternata* if proved to be correct, gives ample scope of production of these toxins by other group of fungi.

7.2.3. Biosynthesis

The biosynthetic pathway of fumonisins is completely understood. Cladas *et al.* (1998) have tried to trace biosynthetic pathway of fumonisins through radiolabelled amino acids, water and molecular oxygen. They concluded that O_2 of tricarboxylic acid moieties of fumonisins was derived from water, while lipid back bone of hydroxyl fumonisins. Using ^{14}C isotope concluded that biosynthesis of fumonisins involves addition of methionine, glutamate, serine or alanine to the hydrocarbon backbone supporting the hypothesis that as opposed to being a modified lipid (Blackwell *et al.*, 1994). Twenty carbon chain backbones of fumonisins resemble those for fatty acids and linear poly-ketides (Proctor *et al.*, 2000). Fumonisins are the inhibitors of *de novo* sping-olipid biosynthesis and alters the sphingolipid metabolism (Voss *et al.*, 1994). Detailed analyses of *Gibberella fuzikuroi* have uncovered 15 genes that are associated with fumonisins production and these genes are clustered toget-her. Gene 2 and 3 are associated with interconversion of different fumonisins.

7.2.4. Isolation and Detection

The development of analytical methods for the quantification of naturally occurring levels of fumonisins was difficult because they do not possess any chromatophores and, therefore, do not absorb UV or visible light nor do they produce fluorescence. Alberts *et al.* (1990) described quantification of fumonisins by HPLC analysis of the malyl derivatives. But with this method of fumonisins detection in culture filtrate at level of 17.9 g/kg can only be detected. Shephard *et al.* (1990) developed HPLC technique which involves fluorescence detection of O-phtholdialdehyde (OPA) derivatives proved to be a suitable method and has been evaluated in international collaborative studies. Cawood *et al.* (1991) described the method of isolation and quantific-ation of FB_3 and FB_4 and FA_1 and FA_2 in addition to known FB_1 and FB_2. Scott and Lawrence (1992) by employing 4-flouro-7-nitrobenzofurazan (4-flouro-7-nitrobenzo-2-oxo-1,3-diazole) and naphthalene-2,3-dicarboxyald-ehyde potassium cyanide determined the amount of FB_1 in corn. Conseque-ntly, the methods of Sherphard *et al.* (1996) as modified by Sydenham *et al.* (1996) and Thiel *et al.* (1996) were recommended for the quantitative analysis of FB_1 and FB_2 in corn based foods and feeds. Gelderbalm *et al.* (1998) have made an attempt to isolate and purify fumonisins B_1 and B_2 from culture of *F. moniliforme*. Extraction of the fumonisins was achieved by methanol: Water (3:1) followed by a solvent partitioning step using chloroform. Subsequent purification of the aqueous phase was subjected to amberlite αAD-2, silica gel and reverse phase (C_{12}) columns yielding FB_1 and FB_2. Quantitation of fumonisins was also accomplished by liquid secondary ion mass spectrometer (Koof Macher *et al.*, 1991), post hydrolysis GC (Sydenham *et al.*, 1990), GC-MS (Plattener *et al.*, 1990), fast bombardment (Koof Macher *et al.*, 1991), HPLC (Thiel *et al.*, 1993) and TLC (Shelby *et al.*, 1994).

Stack and Eppley (1992) described a liquid chromatographic method after derivatizing with O-phtholdialdehyde and 2-mercaptoethanol. Holcomb *et al.* (1993) described method of fumonisins B_1 detection in rodent feed through acetonitrile: water (80:50) and clean up with tandem C_{18} sep-pak vac and strong anion-exchange (SAX) columns and eluting with acetic acid (1%) in methanol and quantified in a gradient elution HPLC using per column derivatization with FMOC and fluorescence detection. Margos and Richard (1994) employed ELISA and liquid chromatography for detection of fumonisins in milk. ELISA technique though simple and useful for routine analysis but was of low sensitivity. Several immunochemical methods for the determination of FB_1 have also been developed including ELISA with polyclonal antibodies (Azcona Olivera *et al.*, 1992a and Usleber *et al.*, 1994), monoclonal antibodies (Azcona Olivera *et al.*, 1992) and anti-idiotype antibodies (Chu *et al.*, 1995). Schneider *et al.* (1995) have developed a competitive direct enzyme immunoassay (EIA) and an enzyme linked immuncfilteration assay (ELISA) for detection of FB_1. Immunoaffinity columns which use monoclonal antibodies are commercially available and have detection limits at ppm level. Liquid chromatography of fumonisins following extraction with methanol-acetone and strong anion exchange followed by derivatization with naphthalene-2,3-dicarboxyaldehyde was sensitive with detection limits of 5 μg/ml. Sydenham *et al.* (1996) while comparing the sensitivity of HPLC and monoclonal antibody-based (MAB) competitive direct enzyme linked immunosorbent assay (CD ELISA) for detection of fumonisins in corn concluded that HPLC is more sensitive and reproducible than ELISA probably due to the lipid based matrix effect. Xie *et al.* (1997) detected two naturally occurring structural isomers of partially hydrolyzed fumonisins B_1 in corn by on-line capillary liquid chromatography and fast atom bambardment mass spectrometry. Karuna and Sashidharan (1999) have developed ion exchange chromatography method for detection of fumonisins. Poling and Platner (1996) developed a rapid purification of fumonisins B_1 and B_2 by loading onto a log NH phase extraction cartridge and eluting fumonisins in acetic acid (5%) in methanol. The elutants were diluted by 1.5 times with water and loaded into a log to 18 cartridges. Some of the common methods of extraction, purification and detection of fumonisins are précised in Table 7.2. Isolation and characterization of fumonisins was reviewed by Merdith (2000). Kulisek and Hazebroek (2000) have recommended the use of phosphate buffered saline (PSB) as an extraction solution for fumonisins in maize in conjunction with ELISA as an alternative to methanol-water or acetonitrile-water. Ho *et al.* (2000) have developed a novel immunoassay for fumonisins using fumonisins liposome conjugates filled with flourophore (Sulfortodamine). Groland *et al.* (2001) while comparing different methods of extraction and clean up procedures for detection of fumonisins, felt that the use of acetonitrile and water (1:1) as an extraction solvent and immunoaffinity column for clean up proved to be better for recovery of fumonisins and chromatographic resolution when compared to methanol: water (3:1) and strong anion exchange (SAX) respectively.

Table 7.2.1. Analytical methods for detection of fumonisins

Method	Reference	Solvent System
Liquid chromatography	Sydenham *et al.* (1992)	Methanol : Water (3 : 1) 0.5% acetic acid in methanol Methanol, Sodium hydrogen phosphate
HPLC	Sydenham *et al.* (1992)	Acetonitrile : Water (1:1) Acetonitrile : Water (70:30) 40% acetonitrile + 60% 0.05 M KH_2PO_4
HPLC	Sydennam *et al.* (1992)	Methanol : Water (3:1) O-phthaldialdehyde
HPLC	Shephard *et al.* (1990)	Methanol : Water (3:1) Acetonitrile : Water : Acetic acid (47:53:0.5)
HPLC	Holcomb *et al.* (1993)	Precolumn derivatization with FMOC and flouroscence detection
HPLC	Thiel *et al.* (1993)	--
TLC and competitive immunoassay	Shelby *et al.* (1994)	--
TLC	Sydenham (1989)	Methanol : Water (3: 1)
TLC	Cawood (1991)	-- Spray : 0.5% p-anisoldehyde
Enzyme immunoassay	Usleber *et al.* (1994)	--
LC-ELISA OPLC	Stack and Epply (1992)	
Reverse phase	Poling and Platner (1996)	Solid phase extraction column
Chromatography	Cawood *et al.* (1991)	--
Ion exchange Chromatography	Karuna and Sashidharn (1999)	Purified Dowex – 1 Methanol : Water (3:1)
TLC-LSD LSIMS (Liquid second ion mass spectrometry)	Koof Macher *et al.* (1991)	
Post hydrolysis GC	Sydenham *et al.* (1990)	--
GC-MS	Plattner *et al.* (1990).	--

Thiel *et al.* (1996) reported that limits of detection to be 100 ppm for TLC 50 ppb for LC-MS and 200 ppb for ELISA. Maragos (1997) described a method for measuring fumonisin level in corn and equine serum by capillary zone electrophoresis. Mullett *et al.* (1998) developed a surface plasmon resonance (SPR) immunosensor to quantify FB_1. Xie *et al.* (1997) could detect isomers of partially hydrolyzed fumonisin B_1 by on line capillary liquid chromatography fast atom bombardment mass spectroscopy. Shetty and Bhat (1998) developed a sensistive method for detection of fumonisins in human urine. Wilkes *et al.* (2001) have developed a elctrospray mass spectrometry for detection of fumonisins. Maragos *et al.* (2001) have developed fluorescence polarization method for detection of fumonisins in maize. Details of some common methods of determination of fumonisins are depicted in Table 7.2.2.

Table 7.2.2 Methods for the Determination of Fumonisins in Corn, Corn-Based Foods and Milk

Food	Extraction	Cleanup	Quantitation	Detection (μg/kg)	Ref.
Corn	Shake ground sample 60 min with 50:50 acetonitrile/water	Strong anion-exchange SPE column (important to control flow rate)	Gradient HPLC Column: Reverse phase C_{18} (5 μm, 25 cm x 4.6 mm i.d.) Mobile phase: (a) 99:1 Acetonitrile/ acetic acid (b) 99:1 water/acetic acid Gradient: 60% (a) + 40% (b) for 8 min; 80% (a) + 20% for 16 min Detection: Fluorescence (ex: 420 nm; em: 500 nm) Derivatization: Precolumn with NDA		Bennelt & Richard (1994)
Corn	Blend ground sample 2 min with 80:20 methanol / water (2 ml/g) and NaCl (0.05 g/ml)	Immuno-affinity column chromatogr-aphy (Fumonitest, Vicam, Water-town, MA)	HPLC Column: Reverse-phase C_{18} (5 μm, 25 cm x 4.6 mm i.d.) Mobile phase: 60:40:1 Acetonitrile/water/acetic acid Detection: Fluorescence (ex: 420 nm; em: 470 nm) Derivatization: Precolumn with NDA-NaCN	4-10	Ware *et al.* (1994)
Corn	Shake ground sample 15 min with 3:1 methanol/ water (5 ml/g)	Strong anion-exchange column chromato-graphy	Gradient HPLC Column: TSKgel ODS-80Ts (15 cm x 4.6 mm i.d.) Mobile phase: (a) 50:50 0.05 M NaH_2PO_4/methanol; (b)75:25 Acetonitrile/water Detection: Fluorescence (ex: 450	10	Akiyama *et al.* (1995)

			nm; em: 590 nm) Derivatization: Precolumn with DBD-F		
Corn	Blend ground sample 5 min with 3:1 methanol/ water (2 ml/g)	Strong anion-exchange cartridge	HPLC Column: Reverse-phase C_{18} (25 cm x 4.6 mm i.d.) Mobile phase: 80:20 Methanol/0.1 $MNaH_2PO_4$ Detection: Fluorescence (ex: 335 nm; em: 440 nm) Derivatization: Precolumn with OPA/mercaptoethanol	50	Thiel *et al.* (1993)
Corn	Shake ground sample 30 min with 50:50 acetonitrile/ water (2 ml/g)	C_{18} SPE cartridge	TLC Plate: Reverse-phase C_{18} (Whatman KC18) Solvent: 60:40 Methanol/1% aqueous KCl Detection: Visualization under long-wave UV light Derivatization: Spray with 0.1 M sodium borate buffer, fluorescamine, and 40:60:0.01 M boric acid/acetoni-trile	100	Rottinaham *et al.* (1992)
Corn	Blend ground sample with ethyl acetate; filter and extract residue 3 times with 3:1 methanol/water	Liquid-liquid partition	1. TLC Plate: Silica gel (Whatman K6F) Solvent: 60:30:10 Chloroform/ methanol/ acetic acid Detection: Blue-violet spot at R_f 0.24 Derivatization: p-Anisaldehyde spray and heat 2. GC-MS of hydrolyzed fumonisin Column: DB1 fused-silica capillary (0.25 μm, 15m) Temp. profile: 120°C; 10°C/min to 270°C Derivatization: Tri-Sil TBT	4	Plattner *et al.* (1990)
Corn[a,b]	Blend ground sample 2 min with 3:1 methanol/ wter (2 ml/g)	Strong anion-exchange cartridge	HPLC Column: Reverse-phase C_8 (5 mm, 12.5 cm x 4 mm i.d.) Mobile phase: 68:32 Methanol/0.1 M NaH_2PO_4 Detection: Fluorescence (ex: 335 nm; em: 440 nm) Derivatization: Precolumn with OPA/mercaptoethanol	50	Sydenham *et al.* (1992)

Corn (B_1)	Shake ground sample with 50:50 acetonitrile/ water (5 ml/g)	Strong anion-exchange cartridge	GC-MS of hydrolyzed fumonisin Column: DB5 fused-silica capillary (0.25 µm, 30m) Temp. profile: 80°C for 1 min; 20°C/min to 200°C; 5°C/ min to 270°C Detection: Electron capture negative chemical ionization	10	Plattner *et al.* (1990)
Corn (B_1)	Blend ground sample 5 min with 3:1 methanol/ water (2 ml/g)	Strong anion-exchange cartridge	HPLC Column: Reverse-phase C_{18} (25 cm x 4.6 mm i.d.) Mobile phase: 45:55 Acetonitrile/pH 7 KH_2PO_4 + 1 mM Na_2EDTA Detection: Fluorescence (ex: 345 nm; em: 435 nm) Derivatization: Precolumn with OPA/*tert*-butyl thiol		Holocomb *et al.* (1994)
Corn, cereal products	Shake ground sample 30 min with 50:50 acetonitrile/water (5 ml/g)	C_{18} SPE column chromatography	HPLC Column: Reverse-phase C_{18} (3µm, 3 cm x 4.6 mm i.d.) Mobile phase: 40:60 Acetonitrile/ 0.05 M KH_2PO_4 (pH 3.3) Detection: Fluorescence (ex: 335 nm; em: 440nm) Derivatization: Precolumn with OPA/ mercaptoethanol		Rice and Ross (1999)
Corn, cornbased foods	Shake (15-30) min or blend (5 min) ground sample with 3:1 methanol/water (2 ml/g)	Strong anion-exchange column chromatography	Gradient HPLC (2 methods) 1. Column: Reverse-phase C_{18} (5 µm, 25 cm x 4.6 nm i.d.) Mobile phase: (a) 50:50 Methanol/ 0.05 M NaH_2PO_4 (pH 4.4); (b) 80:20 acetonitrile/water Gradient: 100% (a) for 5 min; 50% (a) 50% (b) for 15 min Detection: Fluorescence (ex: 460 nm; em: 500 nm) Derivatization: Precolumn with NBD-F 2.Column: Reverse-phase C_{18} (5 µm, 25 cm x 4.6 nm i.d.) Mobile phase: (a) 55:45:1 Acetonitrile/water/acetic acid; (b) 80:20 acetonitrle/ water Gradient: 100% (a) for 10 min; 20%, 80% (b) for 7 min Detection: Fluorescence (ex: 420 nm; em: 500 nm)	100	Scott and Lawrence (1994)

			Derivatization: Precolumn with NDA-KCN		
Milk	Pass 5 ml milk through immunoaffinity column (Fumonitest, Vicam, Watertown, MA)		Gradient HPLC Column: Reverse phase C_{18} (5 µm, 25 cm x 4.6 nm id.) Mobile phase: (a) 55:45 Methanol/0.05 M NaH_2PO_4 (pH 3.5); (b) 80:20 acetonitrile/water Detection: Fluorescence (ex: 335 nml em: 440 nm) Derivatization: Precolumn with OPA/ mercaptoethanol	3-7	Scott *et al.* (1994)
Milk	Shake 5 ml milk 10 min with 15 ml 50:50 methanol/acetone	Strong anion-exchange cartridge	Gradient HPLC Column: Reverse phase C_{18} (5 µm, 22 cm x 4.6 nm id.) Mobile phase: (a) 75:24:1 Methanol/water/acetic acid; (b) 75:24:1 Acetonitrile/ water/ acetic acid Detection: fluorescence (ex: 250 nm; em: 470 nm) Derivatization: Precolumn with NDA	5	Maragos and Richard (1994)
Sweet corn (B1)	Blend ground sample 2 min with 80:20 methanol/water (2 ml/g) and NaCl (0.05 g/ml)	Immuno-affinity column chromato-graphy (Fumoni-test, Vicam, Watertown, MA)	HPLC Column: Bondapak C18 (30 cm x 3.9 nm id.) Mobile phase: 50:50:1 Water/ acetonitrile/acetic acid Detection: fluorescence (ex: 335 nm; em: 440 nm) Derivatization: Precolumn with OPA/ mercaptoethanol	4	Trucker *et al.* (1995)

7.2.5. Natural incidence

Ever since characterization of fumonisins of *F. moniliforme*, its natural incidence has gained paramount importance in view of its carcinogenicity and causal agent of leukoencephalomalaesia. *F. moniliforme* which is ubiquitous and reported to contaminate wide variety of agricultural commodities and elaborate fumonisins from different parts of the world. Incidence and toxigenic potentialities of species of *Fusarium* are précised in Table 7.2.3. Out of different commodies, corn is most susceptible and supports the elaboration of fumonisins B_1 and B_2. It is also reported that fumonisins contaminate animal feeds and fodders. Groves *et al.* (1999) recorded fumonisins B_1, B_2, B_3 in 19, 25 and 6% of the corn samples respectively collected from Shandong. Martins *et al.* (2001) reported fumonisins B_1 and B_2 in black tea and medicinal plants from Portugal. Oho *et al.* (2002) feel that the seed moisture of freshly harvested maize at crop harvest stage is critical for infestation and elaboration of fumonisins by *F. verticelloides*. Abbas *et al.* (1999) reported fumonisins in rice sheath rot caused by *F. proliferatum* both *in vitro* and *in vivo*. Fumonisins have been detected in virtually all types of corn based foods produced or sold in most areas of the world including Argentina (Hennigen *et al.*, 2000), Brazil (Machiwrki and Soares, 2000), China (Groves *et al.*, 1999), Denmark (Petersen and Thorup, 2001), Germany (Seefeldar *et al.*, 2001), Gautemala (Meredith *et al.*, 1999), Hungary (Fazekas *et al.*, 1996), India (Jindal *et al.*, 1999), Iran (Shephard *et al.*, 2000), Italy (Cirillo *et al.*, 2003), Japan (Ueno *et al.*, 1993) Kenya (Kedera *et al.*, 1999), Korea (Sohn *et al.*, 1999), Nepal (Ueno *et al.*, 1993), The Netherlands (De Nijs *et al.*, 1998), Poland (Doko *et al.*, 1995), South Africa (Schlechter *et al.*, 1998), Spain (Sanchis *et al.*, 1994), Switzerland (Pitter *et al.*, 1992), UK (Scudamore and Patel, 2000) and USA (Castelo *et al.*, 1998). Countries with relatively cold climates such as Canada and eastern and northern Europe have reported lower levels of fumonisin contamination of corn, while higher levels of toxin have been reported in corn grown in countries with warmer climates such as Italy, Portugal and Zambia (Doko *et al.*, 1995).

Table 7.2.3. Natural Incidence of species of *Fusarium* and fumonisins in different agricultural commodities

Food	Species	Mycotoxins	Country	Reference
Corn	*F.moniliforme*	FB_1	Argentina	Solovey *et al.* (1999)
	F.proliferatum			Magnoli *et al.* (1999)
	F.nygamai			Torres *et al.* (2001)
	F.moniliforme	FB_1	Brazil	Hirooka *et al.* (1996)
	F.moniliforme	FB_1	China	Groves *et al.* (1999)
		FB_2		
		FB_3		
	F.moniliforme	FB_1 and FB_2	Croatia	Segvic and Pepelijnjak
	F.moniliforme	FB_1	Egypt	(2003)
		FB_2		
	Fusrium sp.	Fumonisins	Europe	Bruno Doko *et al.* (1995)
	F. verticilloides	FB_1	France	Melonk and Akem (1998)
	F. proliferatum			
	Fusarium sp.	FB_1	India	Bhatt *et al.* (1997)
	F. moniliforme	Fumonisins	Iran	Shephard *et al.* (2002)
	F. proliferatum	FB_1 and FB_2	Italy	Pascale *et al.* (1995)
	F. moniliforme	Fumonisins		
	F. moniliforme	Fumonisins	Korea	El-Mezzami *et al.* (1999)
	F. moniliforme	FB_1, FB_2, FB_3	South Africa	Sydenham *et al.* (1992)
				Odhav *et al.* (2001)

	Alternaria alternata	Fumonisins	Southern USA	Abbas *et al.* (2000)
	F. verticilloides	FB_1	Spain	Castella *et al.* (1999)
	F. moniliforme	Fumonisins	Venezula	El-Mezzami *et al.* (1999)
	F. proliferatum	Fumonisins	USA	Dejardins *et al.* (2000)
	F. graminarium	B_1 B_2		
	F. moniliforme	FB_1		
	F. anthophilum		USA	Abbas *et al.* (2000)
	F. dlamini			
	F. napiforme			
	F. nygamai			
	F. moniliforme	FB_1 and FB_2	Uruguayan	Maya & Silva (1997)
	F. moniliforme	FB_1	Western Kenya	Kedera *et al.* (1999)
	Fusarium spp.	FB_1, FB_2 and FB_2	W. Africa Zambia	Albert *et al.* (1990)
Cereal Grains	*Fusarium spp.*	Fumonisins	Europe	Visconti and Doko (1994)
	Fusarium spp.	Fumonisins	Italy	Campbell *et al.* (2003)
	Fusarium spp.	Fumonisins	Canada	
Rice	*Fusarium spp.*	Fumonisins	USA	Abbas *et al.* (2000)
	Fusarium spp.			Desjardins *et al.* (2000)
	F.moniliforme			
	F. proliferatum			
Wheat	*Fusarium spp.*	Fumonisins	USA	Dejardins *et al.* (2000) Castella *et al.* (1999)
Poultry feed	*F. nygamai* *F. proliferatum*	FB_1, FB_2, FB_3	Argentina	Magnoli *et al.* (1999)
Sorghum millet	*F. moniliforme*	Fumonisins	Africa	Nelson *et al.* (1992)
	F. subglutinans		Asia	Visconti and Doko (1994)
	F. napiforme			
	F. nygamai			
Agricultural Commodities	*F. moniliforme* *F. proliferatum*	FB_2	Japan	Akiyama *et al.* (2000)
Swine corn	*F. moniliforme*	Fumonisins	Spain	Castella *et al.* (1999)
Black Tea Medicinal Plants	*F. moniliforme*	FB_1 and FB_2	Portugal	Martins *et al.* (2001)

The range of a_w and temperature limits for growth (3-37°C) was much wider than that for fumonisin production (10-37°C) depending on isolates (Marin *et al.*, 1999). Similarly minimum a_w requirements for growth was 0.9 and a_w 0.93 for fumonisin production (Marin *et al.*, 1999). Marin *et al.* (1999) reported negligible amount of fumonisins in barley and wheat under similar conditions suggesting that the substrate composition affects fumonisin biosynthesis which may be the reason for low incidence in these cereals.

Chulze *et al.* (1996) recorded *F. moniliforme* and *F. proliferatum* at different stages of ear development of corn. Fumonisins production was maximum during late stages of development of ear of corn. Corn ear damage caused by European corn borer increased the incidence of *F. moniliforme* and fumonisins production (Low *et al.,* 1991). Shelby *et al.* (1994) recorded more amount of fumonisins in corn hybrids grown under stress conditions than grown under optimal conditions. Rice *et al.* (1999) recorded fumonisins in feccal samples of cattle, sheep and rats fed with diets mixed with *F. moniliforme*. Castella *et al.* (1999a) could record FB_1 and FB_2 in maize from Spain. Yamashita *et al.* (1995) have reported the incidence of fumonisins in corn and different cereals from Philippines and Thialand respectively. Chatterjee and Mukherjee (1994) and Chourasia and Shelby (1996) reported fumonisins in maize ears visibly infected with *F. moniliforme* collected from a local field in India. Hennigen *et al.* (2000) recorded high level contamination of corn flour with fumonisins but could not find any difference in fumonisins level between brand and unbranded corn flour samples. Silva *et al.* (2007) reported fumonisins in maize and maize base foods intended for human consumption. Data published on the incidence of fumonisins in corn in different countries are presented in Table.7.2.4.

Table 7.2.4. Natural incidence of fumonisins in corn and corn products

Commodity	Country	Publication Year	No. of samples	Incidence	Range (μg/kg)
Corn feeds	UK	1998	67	97%	25-27000
Corn feeds	UK	1997	50	80%	50-4550
Corn foods	UK	1997	266	29%	10-2124
Corn foods	USA	1997	121	20%	42-350
Corn foods	Taiwan	1997	153	34%	73-2395
Corn kernels	China	1997	240	56%	50-34868
Corn kernels	India	1997	35	74%	100-4740
Corn kernels	Uruguay	1997	22	50%	50-5787
Corn kernels	Argentina	1996	50	100%	225-37000
Corn kernels	Brazil	1996	48	98%	600-37650
Corn kernels	Costa Rica	1996	64	89%	550-6320
Corn kernels	Germany	1996	317	27%	6-7132
Corn kernels	Zambia	1995	20	100%	20-1710

7.2.6. Biological Activity

Fumonisin B is primarily a hepatotoxin and carcinogen in rats (Shier, 2000). The involvement of food-borne *Fusarium* mycotoxins in hemorrhage, feed-refusal, emetic, estrogenic, neurotoxic and pulmonary diseases in animals and the association of food-borne *Fusarium* mycotoxin with ATA, SG and EC in humans have been reviewed by several workers. It is much more difficult to establish relation between human diseases and food-borne *Fusarium* toxins.

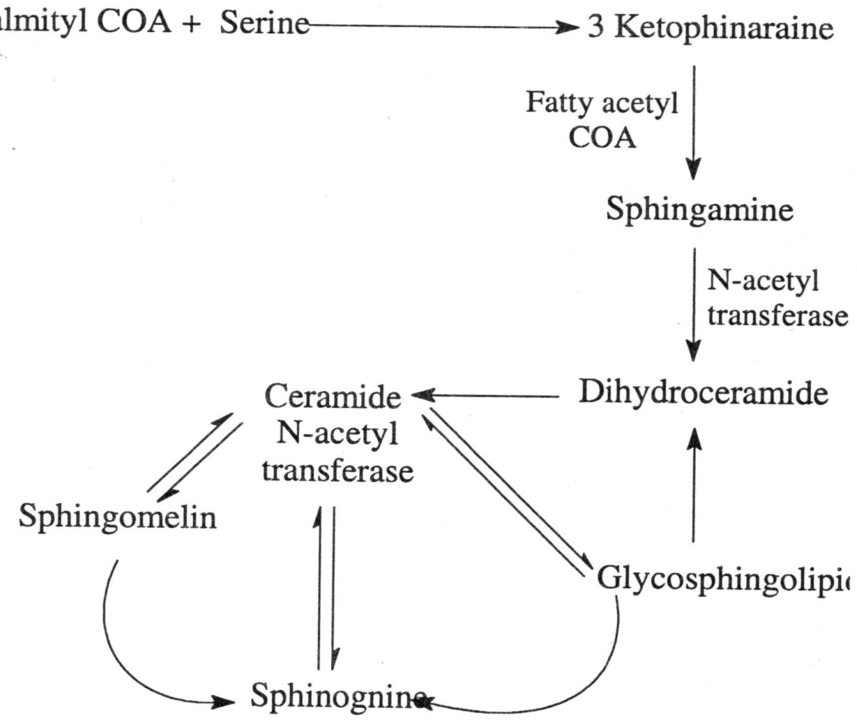

Fig. 7.2.2: Mechanism of inhibition of sphingolipid synthesis by fumonisin B_1

Outbreak of encephalomalacia due to the consumption of a mouldy corn has been known for several decades. Neurotoxic and hepatotoxic symptoms may occur singly or in combination in horses. Feed refusal, lameness, atoxia, oral and facial paralysis and recumbency begins with in a few days after initial consumption of moldy corn and rapidly followed by seizers and morbidity. Histopathology of liver shows centrilobular necrosis, fibrosis and bile duct proliferation with increased mitosis, acute inflammation and fatty degeneration in fumonisins fed horses. Marasas *et al.* (1988) and Kellerman *et al.* (1990) have experimentally induced leukoencaphalomalacia in horses both by intravenous injection and oral feeding of fumonisins. Haschek *et al.* (1992) experimentally induced pulmonary edema by administering higher doses of fumonisins, while at lower doses slowly progressive hepatic disease was most predominent. Gelderblom *et al.* (1992) induced liver lesions in rats with fumosins B_1 and B_2. Wang *et al.* (1991) and Wang *et al.* (1992) discussed the mechanism of toxicity of fumonisins. According to them fumonisins are the inhibitors of *de novo* sphingolipid biosynthesis and alters the lipid metabolism as depicted in Fig. 7.2.2.

Bhat *et al.* (1997) reported association of fumonisins contaminated maize and sorghum with a food-borne disease in humans. Shetty *et al.* (1997b) reported the outbreak of poultry disease due to ingestion of feed contaminated with fumonisins. Water damaged kernels developed high concentrations of fumonisins which on consumption by horses resulted in an outbreak of leucoencephalomalacia in horses (Horne *et al.*, 1991). Fumonisins have been reported to effect on the expression of cytokines and chemokines in human DC and also reported to be immunotoxic to humans (Stockmann-Jlivala *et al.*, 2008).

Wang *et al.* (1991) feel that fumonisins inhibit the conversion of sphingamine to N-acyl-(c)-sphinganines. This is further confirmed by the observations that fumonisins inhibited the activity of sphingosine N-acyltransferase ceramide sulphase in rat liver microsomes and reduced the conversion of sphingosine to ceramide by intact hepatocytes. The disruption of the sphingolipid biosynthetic pathway leads to increased levels of sphingolipid precursors and decreased levels of complex sphingolipids. As a result, tissue levels of sphinganine become elevated initiating a cascade of cellular alternations that may result in toxicity and carcinogenicity (Desai *et al.*, 2003). The mechanism by which this cascade results in carcinogenicity is not known. The increase in intracellular sphinganine coupled with the concomitant decrease in ceramides may induce apoptosis and the cell proliferation that follows may lead to increased cancer incidence (Kodell *et al.*, 2001).

Rats fed with *F. moniliforme* infested feed when examined revealed development of hepatic nodules, cholantiofibrosis or cholangiocarcinomas with in 6 months. They also showed development of liver tumors and primary hepatocellular carcinoma and chronic interstitial nephritis in kidneys. No lesions were observed in the esophagus, heart or fore-stomach of fumonisins treated rats. Fumonisins exert strong effect on point mutation events with primary action on tumor promotion and selection of initiated cells on human cells caused antiproliferative effect after increased apoptotic cell death as opposed to decreased cell proliferation (Wolf, 1994). Li *et al.* (1980) have recorded a close correlation between the consumption of fermented corn and unfermented corn pancakes and the incidence of elevated stomach cancer mortality in Rurallinque country in Shandong province in China. Buchi *et al.* (1996) could induce leukoencephalomalacia and haemorage in the brain of rabbits with fumonisins. Bane *et al.* (1992) have tried to correlate fumonisins contamination of feed and mysterious swine disease. Voss *et al.* (1990) could find a correlation between hepatotoxicity and FB_1 and FB_2 content of aqueous extract and methanolic extracts of *F. moniliforme*. Weiblug (1993) have studied the effect of FB_1 and FB_2 on broiler chick. Colvin and Harrison (1992), Osweiler *et al.* (1992) and Colvin *et al.* (1993) have recorded porcicine pulmonary edema (PPE) and hydrothoracea in swine. Some of the mycotoxicoses of animal and man attributed to the involvement of fumonisins and *Fusarium* infestation are precised in Table 7.2.5. Rosiler *et al.* (1998) have reported the outbreak of leukoencephalomalacia in Mexico due to the consumption of fumonisins contaminated corn. On the other hand, You *et al.* (1988) feel that stomach cancer increased by 30% among those consumed corn pancakes atleast daily. Groves *et al.* (1999) feel that fumonisins contamination does not increase the risk of gastric cancer. Chatterjee and Mukherjee (1994) recorded significant reduction in the viability and phagocytic potential of macrophages for chicken peritonial exudate cells. Javed *et al.* (1992) recorded lesions in the liver, kidneys, heart and lungs and subsequent death of broiler chicks. Gelderblom *et al.* (1996) feel that FB_1 causes lipid peroxidation. Gelderblom *et al.* (1997) feel that fumonisins disturb membrane components, the fatty acid storage pool and the accumulation of long-chain fatty acids within the cell which could eventually lead to the disintegration of membrane structure and cell death apart from inhibition of sphingolipid biosynthesis. FB_1 has also been found to effect the synthesis of cellular lipids by altering the incorporation of palmitic acid (Gelderbolm *et al.,* 1996). Increased levels of polyunsaturated fatty acids are associated with lipid peroxidation in FB_1 exposed primary hepatocytes suggests their indirect role in carcinogenicity. Gelderblom *et al.* (1994) feel that prolonged exposure to fumonisin B_1 is essential for initiation of cancer rather than short exposure. IARC (1993) feel that fumonisins are carcinogenic. The current data suggest that they may have more impact on the health of farm animals than on humans. The co-occurence of fumonisins with carcinogenic mycotoxins such as AFB_1 or nitrosamine may play an important role in carcinogenesis in humans.

Marassas (1995) felt that the contamination of corn with fuminosins presents a real threat to human and animal health in Africa. Henry and Bosch (2001) have elegantly discussed the relationship between fumonisins incidence and mycotoxicoses epidemiology. Fumonisins are reported to act synergistically with aflatoxins (Carlson *et al.*, 2001).

Fumonisins are non-mutagenic according to the *Salmonella* test and non-genotoxic according to the DNA-repair assay with *Escherichia coli* (Gelderblom *et al.,* 1996) and do not induce unscheduled DNA synthesis in primary hepatocytes (Norred *et al.,* 1990). Fm B_1 is reported to alter cardio-vascular function (Gumprecht *et al.*, 2001). FB_1 mimics genotoxic carcinogens in both cancer initiation and promotion (Gelderblom *et al.,* 1994). Studies of Thiel *et al.* (1992), Ueno *et al.* (1997), Wang *et al.* (1998) and Shephard *et al.* (2002) could demonstrate the etiological role of exposure to fumonisins to ingestion of moldy corn to human liver and esophageal cancer. Sun *et al.* (2007) could also link between high contamination rates of FB_1 in home grown corn collected from South Africa and China and high risk areas for esophageal and liver cancer. Several studies indicate etiological role of exposure to fumonisins in human liver and esophageal cancer (Sydenham *et al.*, 1991; Thiel *et al.*, 1992; Ueno *et al.*, 1997; Wang *et al.*, 1998; Shephard *et al.*, 2000, 2002). FB_1 induces Y-glutamyl transpeptidase (GGT) and the placental form of glutathione-S-transferase (STP).

Table 7.2.5. Mycotoxicosis of animal and man due to fumonisins

Animal	Country	Symptoms	Mycotoxin	Reference
Rat (world wide)	South Africa	Equine leukoencephalom alacia (LEM) Porcine pulmonary edema (PPE)	FB1, FB2, FB3	Sydenham *et al.* (1992)
Rat	-	Foetotoxic	FB1	Lebepe –Mazur *et al.* (1995)
Rat	USA	Heaptocarcinogenic esophageal cancer, hepatocelluar carcinoma	FB1 moniliformin	Gelderblom *et al.* (1991)
Rat	USA	Liver tumors	FB1	Gelderblom *et al.* (1991)
Rat		Skin tumor(Carcinogen) Kidney or liver apoptosis Foetotoxic	FB1	Lebepe –Mazur *et al.* (1995)
Rat	USA	Hepatotoxicity, Nephrotoxicity	FB2, FB2	Cawood *et al.* (1994)
Rat	South Africa	Carcinogenicity, Neurotoxicity	FB1,FB1	Benzuidenhout *et al.* (1988) Jackson *et al.* (1996)
Rat	Hungary	Reproductive failure	FB1	Rati (1991)
Horse		Leukoencephalomalacia	FB1	Marasas *et al.* (1988)
Horse	Brazil	Leukoencephalomalacia	Fumonisins	Mallamann *et al.* (1999)
Scoine Horse	-	Pulmonary edema Encephalomalacia	Fumonisins	Benzuidenhout *et al.* (1988)
Pig	-	Hepatic lesion syndrome and immunodepression	Fumonisins	Ballarini (2002)

Pig	-	Cardiovascular changes Hepatotoxic and nephrotoxic	FB1, FB1	Mathur *et al.* (2001)
Horse and donkeys	Mexico,USA, Egypt, South Africa	Brain tissue necrosis Leukoencephalomalacia (Liauefacttive necrosis of the white matter of brain)	Fumonisins	Gelderblom (1988)
Pig	Spain	Pulmonary edema syndrome lung disease	Fumonisins	Harrison *et al.* (1990) Marasas *et al.* (1988)
	Italy	DNA damge in fibroblasts (hear condition)	FB1	Galvano *et al.* (2002)
Humans	South Africa China	Cancer of Oesophagus	FB1, FB2	Sydenham *et al.* (1990); Yang (1980)
Broiler chicks	-	Lesions in the liver, kidney hear and lungs	FB1	Javed *et al.* (1993)
Soyabe an and Tomato	-	Phytotoxic	FB1, FB2	Abbas *et al.* (1992)
Corn and Tomato seedling	-	Phytotoxic (reduction of root and shoot length)	FB1	Lamprecht *et al.* (1994)

Several studies indicate their etiological role of exposure to fumonisins in human liver and esophageal cancer (Sydenham *et al*., 1992; Thiel *et al*., 1993; Ueno *et al*., 1997; Wang *et al*., 1998).

It is established that FB_1 induces completely different toxic effects in different animal species. Mallir *et al.* (2001) reported that fumonisins B_1 to be heptotoxic and nephrotoxic in milk fed calves. Galvano *et al.* (2002) reported that fumonisin B_1 caused DNA damage in human fibroblasts. Moodley *et al.* (2001) reported fumonisins B_1 as an etiologic agent of preeclampsia in pregnant women of S. Africa. Hepatoxicity and hepatocarcinogenicity in rats suppress both growth and foetal bone development (Lebepe-Mazur *et al.,* 1995). Prolonged exposure to FB_1 affected kidney in rats (Gelderblom *et al.,* 1992). Lemmer *et al.* (1999) have investigated histopathology and gene expression changes in rat liver fed by FB_1. Fumonisins induced apoptosis in human erythroleukamia cell lines (Minervini *et al.,* 2004). Fumonisins caused early apoptosis in rats (Moon *et al.,* 2000). Seefelder *et al.* (2003) reported that fumonisin B_1 could induce apoptosis in cultured cells of human proximal tube cells. Bohni (2000) reviewed the role of fusarial toxins in animal health. Animals can be commonly treated against the chronic affects of mycotoxins by strengthening their immune system in conjunction with the beneficial microorganisms or enzymes. Vitamin and carotenoid supplemen-tation, short term administration of proteins or amino acids and addition of antioxidant trace elements such as selenium also promotes immune system.

FB_1 is an inhibitor of sphinganine N-acyltransferase and the increase in the sphingamine/sphingosine ratio in the urine or serum and has been proposed as biomarker to evaluate exposure to fumonisins (Qiu and Liu, 2001). Males are more sensitive than females to FB_1 in the disruption of sphingolipid metabolism (Qiu and Liu, 2001). Howard *et al.* (2001) feel that FB_1 inhibits ceramide synthesis and caused carcinogenesis in 2 year rodent. Foreman *et al.* (2001) observed some subtle neurological signs in horses suffering from fumonisins toxicosis. Constable *et al.* (2000) observed consistent cardiovascular changes in fumonisins treated horses with sphingosine mediated calcium-channel blackade of heart and vasculature. Delongchamp and Young (2001) feel that fumonisins increase age specific tumour incidence. Fumonisins are probably first example of apparently non-genotoxic (non-DNA reactive) agent producing tumours through a mode of action involving apototic necrosis, atrophy and consequent regeneration (Cohen *et al.,* 2001).

Alteration in the n-6 fatty acid profile and decrease in the free cholesterol membrane associated levels which resulted in higher phosphatidyl-choline and cholesterol ratio suggesting a more rigid membrane structure FB_1 fatty acid storage pool and the accumulation of long-chain fatty acids with in the cell which could lead to the disintegration of membrane structures and eventually result in cell death (Gelderblom *et al.,* 1997).

FB_1 and FB_2 are phytotoxic and cause damage to weed and crop cultivars of soybean and tomato. The primary centre of toxicity is plasmalemma or tonoplast. Abbas *et al.* (1992) recorded light dependent cytoplasmic degeneration and chloroplast disruption. FB_1 is more toxic than FB_2 and FB_3 to corn or tomato seedlings and cause reduction in shoot and root length (Lamprecht *et al.,* 1994). Abbas *et al.* (2000) reported the necrosis, growth inhibition and death of *Datura strematium* seedlings by fumonisin B_1. Zomo and Vurro (1999) recommended fumonisins as herbicide to kill *Striga hormanthica*. Fms are also toxic to simson weed, black nightshade, duckweed and tomatoes (Abbas and Riley, 1996). Kritzinger *et al.* (2006) reported that fumonisins inhibited the seed germination, root and shoot elongation FB_1 treated embryonic tissues showed compaction in the protoplasm and separation of plasmalemma from the cell. Similarly Kroschel and Elzein (2000) have reported the herbicidal effects of fumonisins produced by *Fusarium nygamaon* and *Striga heormanthica.*

7.2.7. Management of Fumonisins

Galvano *et al.* (2001) discussed elegantly the strategies to be adapted to contain fumonisins contamination of food grains. Fumonisins are generally concentrated on the outer pericarp layer of corn kernels (Syndenham *et al.*, 1993b). Syndenham *et al.* (1993) could reduce the fumonisins level very significantly by sorting out fine particals of less than 3 mm size. Wet milling, the processing of converting corn into starch is an effective method for reducing fumonisins level. Broggi *et al.* (2002) reported that dry milling of fumonisin contaminated corn tended to concentrate the fumonisins in the bran and germ fractions and produced grits relatively free of contamination. Fumonisin contaminated corn kernels have a low density and upto 86% of the toxin was removed in the bouyant fraction when naturally contaminated corn was washed with an aqueous saturated sodium chloride solution (Shetty and Bhat, 1998).

Syndenham *et al.* (1993) reduced the fumonisins level in corn by treating with 0.1 m calcium hydroxide for 24 hrs at 25°C. Norred *et al.* (1991) could reduce the fumonisins level in corn by ammoniation. Norred *et al.*, (1991) reported fumonisin decontamination with ammoniation, while Scott and Lawrence (1994) could reduce fumonisins by alkaline hydrolysis. Detoxification of fumonisins in maize by ammonia could reduce to 30% (Chourasia, 2001). Sampapundo *et al.* (2007) reported the effective management of infestation of *Fusarium proliferatum* and fumonisins production in corn, by treating with 1% ammonium bicarbonate. *Streptococcus* and *Enterococcus* could bind FB_2 to the extent of 62% and could help to decrease its toxicity in animals (Niderkor *et al.*, 2007). Bhat (1997) recorded variation in fumonisins B_1 level in hybrids and hybroid of maize. Thompson (1996) reported the inhibition of growth of *F. moniliforme* in the presence of butylated hydroxyanisole or carvacrol. Juglal *et al.* (2002) reported that clove oil (engenol) followed cinnamonthymol to be inhibitory to the production of fumonisins by *F. verticelloides*. Yamgish *et al.* (2001) could produce mutant of *F. proliferatum* which produced little or no fuminosins and can be used as a competitor for fumonisins producing strains and thus minimise fumonisin contamination. Bacon *et al.* (2001) suggested that the exploitation of *Trichoderma* sp. and *Bacillus subtilis* in the control of fumonisin B_1 production by *F. verticelloides*. Castelo *et al.* (2001) recorded reduction of fumonisin B_1 in extruded and baked corn based foods with sugars to the extent of 92.1%. Reynoso *et al.* (2002) reported that antioxidants like PP and BHA were effective in arresting the growth of *F. verticelloides* and *F. proliferatum* at three a_w 0.99, 0.98 and 0.95. Trihydroxy butyrophenone and butylated hydroxytoluene (BHT) were less effective either in single or in combination. Use of combined mixture of antioxidants (BHA and PP) could be good strategy to minimize the entry of fumonisins into animal feed and human food.

Charlubruns (2003) suggested crop management is the main strategy to control fumonisins by growing of adapted hybrids, proper plant nutrition, irrigation and insect control or use of transgenic hybrids help in reducing contamination of fumonsin B_1. Shim-Wan *et al.* (1999) feel that strategies that target the regulatory elements of nitrogen metabolism may be effective of reducing the risk in fumonisins contamination in food. Beekrun *et al.* (2003) feel that naturally occurring phenolics of plants such as chlorophorin, iroka, maakiarin, benzoic acid, caffeic acid, ferulic acid and vanillic acid are effective inhibitors of fumonisins production by *F. verticelloides*. Bacon *et al.* (2001) developed a biological control of contamination of corn in the field using the endophytic bacterium, *Bacillus subtilis.* King and Scott (1981), Shelby (1994) and Visconti (1996) have advocated cultivation of resistant varieties of maize in relation to geographical location in order to minimize fumonisin contamination.

Duvick (2001) feels that gentic modification of corn through plant breeding or by transgenic means offers several approaches for contributing ear mould and myctoxin formation such as development of corn hybrids that are resistant to environmental streses such as heat, drought and insect damage; resistant to fungal infections and disease prevent mycotoxin from being formed and detoxify mycotoxins. Ono *et al.* (2002) feel that the decreased seed moisture of freshly harvested maize pre-drying of storage grains can control fungal growth and fumonisin contamination.

Cahagneir *et al.* (1995) and Marin *et al.* (1999) reported that 300 fold increase of fumonisins with 10% of increase of a_w. Shetty *et al.* (1999) reported the effectiveness of sodium chloride (30%) solution for physical decontamination of fumonisins contaminated maize. Bullerman *et al.* (2001) are of the opinion that post harvest control of fumonisins in corn based foods and feed may be achieved through proper storage of corn before processing such as physical treatment that cause, redistribution, removal or destruction of fumonisins, chemical treatments to remove that destroy the toxins and biological treatment to remove or destroy fumonisins. Microwave popping of popcorn from reject streams of the cleaning operation that cotained fumonisins resulted in significant reduction of the mould toxin (Dovidio *et al.*, 2007)

Spoilage of corn by *F. proliferatum* and *F. verticelloides* and fumonisins contamination could be checked with the help of modified atmosphere containing 10% CO_2 in initial head space (Samapundo *et al.*, 2002). Musser and Plattner (1997) reported decreased fumonisins production by *Fusarium verticillioides* with the increased concentration of CO_2 in sealed bags. Scott and Lawrence (1994) succeeded in obtaining about 80% reduction by heating at higher temperature. However, caution is required as it has been reported that a break down product (hydrolysed fumonisins) is just as toxic as the parent compound (Hogkins and Murghy, 1993).

Wet milling, the processing of converted corn into starch is an effective method of reducing fumonisins level. USFDA have fixed the levels for total fumonisins in human food and animal feed to protect the human from fumonisin toxicity (Table 7.2.5).

Table 7.2.6. US Food and Drug Administration guidance levels for total fumonisins in human food and animal feed

Food / Feed	Total fumonisin level (FB_1+ FB_2+ FB_3)
Human food	
Degermed dry-milled corn products (e.g. flaking grits, corn meal, corn flour with fat content < 2.5%, dry weight basis)	2μg/g
Whole or partially degermed dry-milled corn products (e.g. flaking grits, corn meal, corn flour with fat content ≥ 2.5%, dry weight basis)	4μg/g
Dry-milled corn bran	4μg/g
Clean corn intended for masa production	4μg/g
Cleaned corn intended for popcorn	3μg/g
Animal feed	
Equids and rabbits	5μg/g (no more than 20% of diet)
Swine and catfish	20μg/g (no more than 50% of diet)
Breeding ruminants, breeding poultry and breeding milk	30μg/g (no more than 50% of diet)
Ruminants > three months old being raised for slaughter and mink being raised for pelt production.	60μg/g (no more than 50% of diet)
Poultry begin raised for slaughter	100μg/g (no more than 50% of diet)
All other species or classes of livestock and pet animals	10μg/g (no more than 20% of diet)

The incidences of fumonisins producing fungi are more widespread under varied climatic conditions than visualized. Though large number of reports is available on the contamination of maize at different stages from production to consumption, the field incidence and meteorological and agronomic practices will favour infestation of species of *Fusarium* and fumonisins contamination. Location of fumonisins in pericap needs to be examined through intensive and extensive survey both at pre- and post-harvest stage. Transgene mediated having commercial potential corn hybrids had to be developed. Inspite of accurate and sensitive method of detection of fumonisins are available, a need is felt for developing a simple and sensitive method by which fumonisins can be detected easily and accessable to common seed research laboratories. Pharmacokinetic studies need to be strengthened. Only preliminary reports are available for reducing the level of fumonisins. A simple and more effective method of management of these mycotoxins is the need of the hour.

Fungal growth and mycotoxins production result from a complex interaction of many factors and understanding each factor is essential to understanding to overall process and to predict and prevent mycotoxin production. Therefore, a clear understanding the nature of the asymptomatic infection of corn by *F. verticelloides*, where and how fumonisins are produced by the fungus in the corn plant.

Similarly plant molecular biotechnology techniques to breed mycotoxin resistant cereals and nut producing cultivars should be evolved. More work is needed to determine when low and where fumonisins are produced by the fungus in the corn plant. There are no available epidemiological data to link between fumonisins exposure and esophageal cancer, liver cancer and neutral tube defects and there is an urgent need of a effective biomarker for this toxin and need to be identified. More work is needed to identify and determine toxicological effects of products resulting from physical decomposition or chemical modification of the fumonisins

Efforts should be focused on the molecular genetics of fumonisins producing fungi, mechanism of their action, difference in the metabolism and pharmacokinetics, immunobased and physicochemical techniques for their quantification and analysis of risk involved in the exposure of man and domestic animals to fumonisins and associated regulation for the control of fumonisins contamination need to be formulated. More work is needed to verify the link between fumonisin consumption and the incidence of cancer and other chronic diseases.

CHAPTER - 8

NEPHROTOXIC MYCOTOXINS

8.1. OCHRATOXIN A

The discovery of the hepatotoxic aflatoxins in the early 1960, intensified research on mycotoxins and their role in human and animal health. Ochratoxins are first major group of toxins discovered after aflatoxins. Vander Merwe *et al.* (1965) were the first to characterize Ochratoxin A and B. It gained prominent importance owing to its frequent occurrence in nature. Its established role in Danish porcine nephropathy, in poultry mycotoxicoses, its implicated role in Balkan endemic nephropathy and urinary system tumors in North Africa received increased attention.

Ochratoxins are a group of dihydro seven isocoumarin derivatives linked with an amide bond to the amino group of L-phenylalanine (Fig. 8.1). Ochratoxin A is a 7-carboxy-5-chloro-8-hydroxy 3, 4-dihydro-3 R-methylisoc-oumarin, which is linked through the carboxyl group to L-phenylalanine.

Ochratoxin A is the predominant and most toxic member of the series and majority of the literature has been concerned with this mycotoxin. Ochratoxin A is produced in large amounts and it is frequently encountered as a natural contaminant in foods and feeds. Ochratoxin A and its methyl and/ethyl esters (Ochratoxin C) were approximately equitoxic and were considerably more potent than Ochratoxin B. Ochratoxin B is essentially non-toxic. The presence of chlorine atom in Ochratoxin A has been considered to be a primary determinant of toxicity of this compound. Different aspect of this toxin has been reviewed by Aish *et al.* (2004).

8.1.1. Physico-chemical characteristics

Ochratoxins are a group of structurally related compounds of which Ochratoxin A is the most important and most commonly occurring. They consist of a polypeptide derived dihydro-iso-coumarin mostly linked through the 12-carboxy group to phenylalanine. It is a colourless crystalline compound exhibiting blue fluorescence under UV light. The sodium salt is soluble in water. As acid it is moderately soluble in polar organic solvents such as chloroform, methanol and acetonitrile and dissolves in dilute aqueous sodium bicarbonate. On acid hydrolysis it yields phenylalanine and an optically active lactone acid, Ochratoxin α. Reaction in methanol and hydrochloric acid yields the methyl ester.

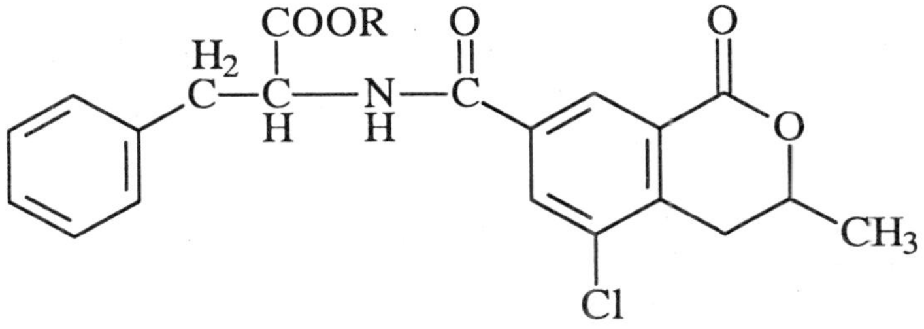

	R
Ochratoxin A	H
Ochratoxin A methyl ester	CH_3
Ochratoxin A ethyl ester	CH_3

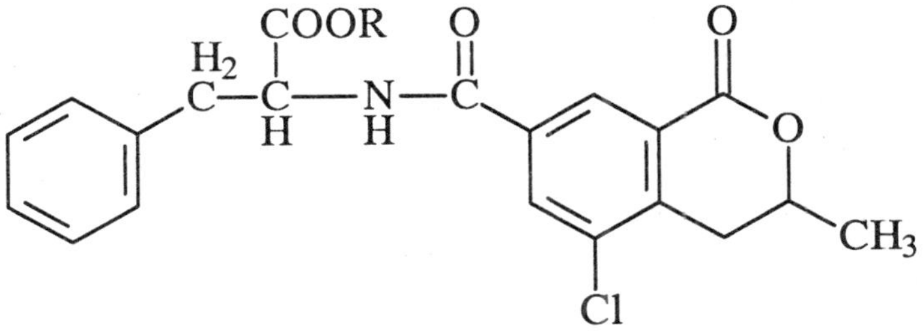

	R
Ochratoxin B	H
Ochratoxin B methyl ester	CH_3
Ochratoxin B ethyl ester	CH_3

Fig. 8.1.1 Structure of Ochratoxins A and their derivatives

Ochratoxin A is a moderately stable molecule and will survive in most food processing operations. Scott (1996) reviewed different aspects of ochratoxin A fate during malting and brewing (Baxter, 1996), bread making, processing in cereals (Alldrick, 1996) and processing coffee, meat and a few processing products (Garcies, 1996; Scudamore *et al.*, 1998).

8.1.2. Biosynthesis of Ochratoxin A

The Ochratoxins are a family of mycotoxins composed of polyketides and the (modified) amino acid, phenylalanine which are linked via peptide bond. A possible scheme leading to the biosynthesis of OTA was published by Huff and Hamilton (1979). According to them three distinct steps occur in OTA biosynthesis. The first part is polypeptide synthesis of Ochratoxin α via mellein involving a polypeptide synthase. This step is followed by acyl activation. Mellein is methylated and oxidized to 7-carboxy-mellein (OTB). Chlorination by a chloroperoxidase leads to OTα. This component is then transformed into mixed anhydrides an activation reaction using ATP. The second precursor, phenylalanine, is synthesized via the sikkimic acid pathway followed by ethylester activation so that it can participate in the subsequent acryl displacement reaction. In the third step linkage of the activated precursors via a synthetase takes place-generating OTC, an ethyl ester of OTA. Desterification by an esterase or transesterification is the last step in postulated biosynthetic pathway (Fig. 8.1.2).

8.1.3. Organisms

Scott (1965) first isolated Ochratoxin A from *Aspergillus ochraceus* (k-804) isolated from sorghum grains. At present a total of 10 species of *Aspergillus (A. sulphureus, A. sclerotiorum, A. citricus, A. fonsecacus, A. alliaceus, A. melleus, A. ochraceus, A. ostinus* and *A. pelrakii)* are reported to elaborate this toxin. Similarly after the report of production of this toxin by *Penicillium viridicatum* by Van Walback *et al.* (1969), *P. palitans, P. commune, P. purpurscens, P. verrucosum, P. variable* and *P. cyclopium* are also reported to produce Ochratoxins. However, *A. ochraceus* and *P. viridicatun* are leading producers under natural conditions. According to Carlton and Krogh (1979), *A.ochraceus* and members of *A. ochraceus* group are probably the most significant producers of Ochratoxins in the tropical and sub-tropical regions, where as *P. viridicatum* appears to be predominant in colder areas of the world. *Petromyces allicens* and *A. citricus* have also been found to produce ochratoxin A (Abarca *et al.,* 1994; Ono *et al.*, 1995). The discovery of the capability of *A. niger* for Ochratoxin A production led to the suggestion that all the *A. niger* strains of group should be screened for Ochratoxin A production before its use in industrial fermentations (Schuster *et al.*, 2002).

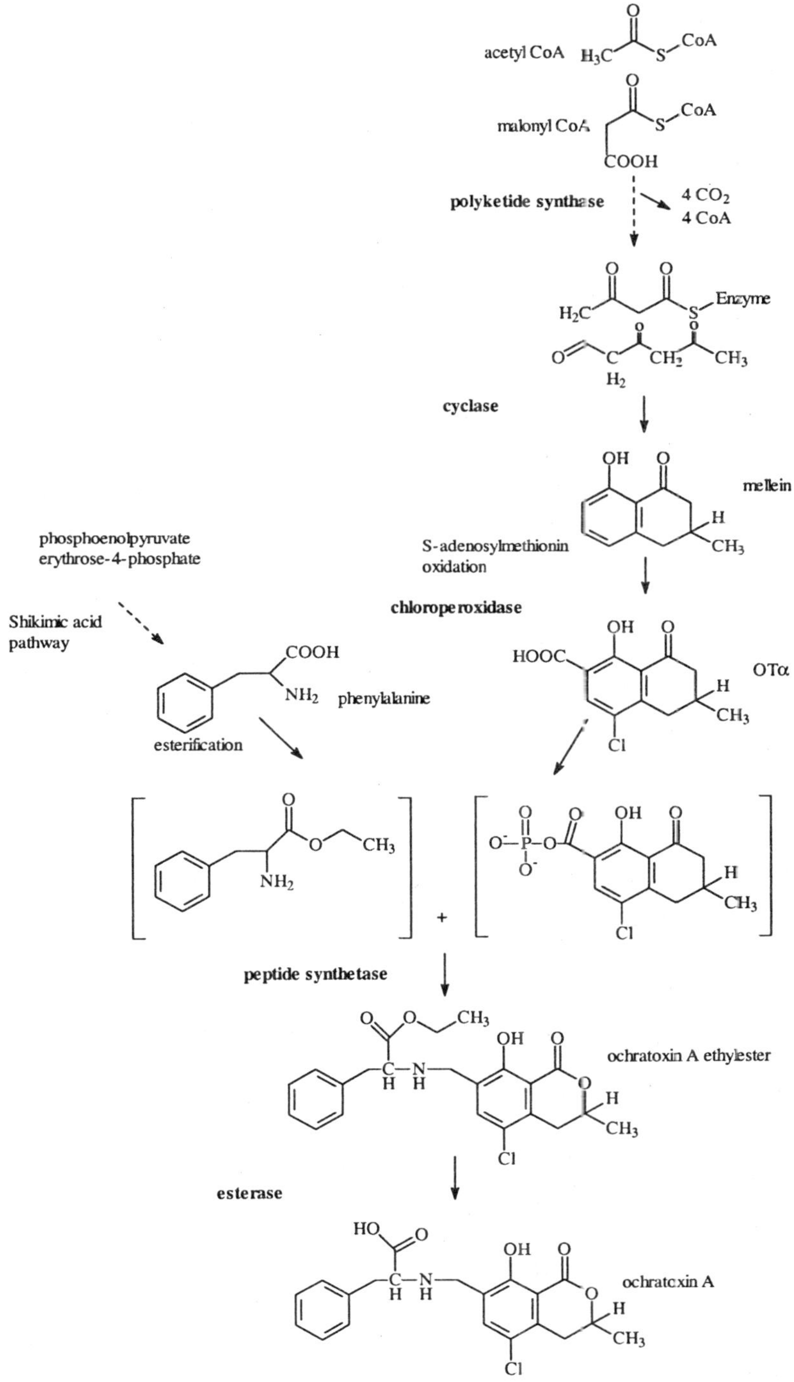

Figure 8.1.2. Schematic representation of the hypothetical OTA biosynthetic pathway as proposed by Huff and Hamilton (1979).

8.1.4. Isolation and Detection

The techniques involved in purification and detection of Ochratoxins were excellently reviewed by Nesheim (1976). The presence of the carboxyl group in the Ochratoxin molecules was utilized in most of the isolation procedures. The following methods are employed in isolation and detection of Ochratoxin A.

Traditional analysis by TLC (Nesheim, 1973).

1. Modified TLC method for detection in animal tissues (Krogh *et al.*, 1974).
2. Elling *et al.* (1975) first used modified method to detect Ochratoxin A in chicken muscle.
3. Fluorescence measurement in buffer (pH 7.5) after enzymatic conversion of Ochratoxin A to Ochratoxin (Hult and Gatenbeck, 1976).
4. Liquid-liquid extraction and final clean up preparative HPLC (Peterson and Ciegler, 1980).
5. HPLC and a different clean up procedure which involves enzymatic degradation of proteins in the sample to get protein bound Ochratoxin A free into solution (Hunt *et al.,* 1979).
6. Two dimensional TLC methods with special emphasis on the analysis of pig kidneys with lower concentration limit of 0.5 ng (Paulsch *et al.*, 1982).
7. Immunological methods such as RIA or ELISA for much smaller samples (Chu, 1984).

Three extraction techniques are used for solid samples.

(a) Soxhlet extraction.

(b) Blending with a solvent in a warning blender for 1-3 min.

(c) Shaking for 10-30 min. with the solvent.

The following solvents are in use for extraction of Ochratoxin A.

1. Chloroform: Methanol (1:1).
2. Methanol: water: hexane (55:45:40).
3. Methanol: water: hexane (55:40:40).
4. Chloroform: methanol: hexane (8:2:1).
5. Chloroform: water (10:1).
6. Chloroform: methanol (7:3).
7. Acetonitrile: water: hexane (9: 1:5).
8. Acetone: water (85:15).

(1) TLC determination of Ochratoxin A in food stuffs
(Schematic representation of procedure)

Extract with 25 ml, 0.1 mol H_3PO_4 + 250 ml Chloroform

↓

Filter through Whatman No.1 filter paper and Collect 50 ml filtrate

↓

Partition 50 ml of filtrate on sodium bicarbonate-celite column
(retain remainder for confirmation of identity)

↓

Wash column with 70 ml hexane

↓

Elute Ochratoxin A with 100ml acetic acid: benzene (5:95)

↓

Evaporate elute to near dryness

↓

Transfer elute to 8 ml vial with acetic acid: benzene (1:99)

↓

Evaporate extract to dryness

↓

Dissolve extract in 0.5 ml acetic acid: benzene (1:99)

↓

Separate by TLC

↓

mobile phase, benzene: acetic acid: methanol (90:5:5)

↓

Quantify by visual or densitometer
under long-wave UV

Fig. 8.1.3. Detection and quantification of Ochratoxin A by TLC

(2) Fermented product (wheat)

↓

Grind in Warning blender with equal weight of water

↓

Extract twice with methanol

↓

Methanolsolution

↓

Add NaCl (3M) extract twice with methylene chloride.

↓

Methylenechloride solution

↓

Extract with 0.2m $NaHCO_3$

↓

Hydrogen Carbonate solution

↓

(acidify to pH 2.5 with conc. HCl)

↓

Extract with methylene chloride

↓

Methylene chloride solution

↓

Filter through $(Na)_2S0_4$

↓

Evaporate to dryness at room temperature

↓

Ethylacetate solution, filter and evaporate to dryness

↓

Ethyl acetate solution (redissolved) and inject into HPLC

Fig. 8.1.4. Detection of Ochratoxin A by HPLC

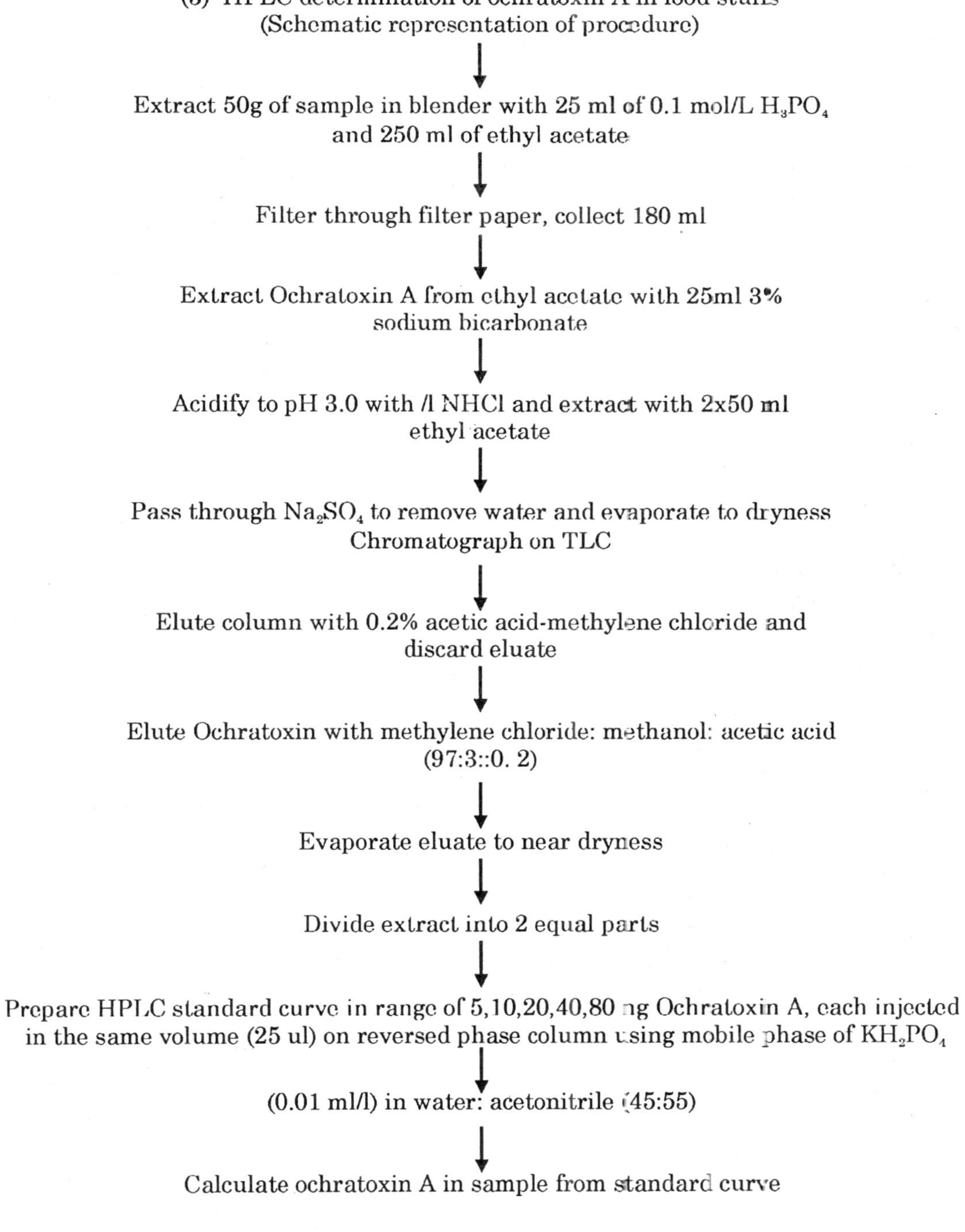

Fig 8.1.5. Determination of ochratoxin A by HPLC

Vander Merwe *et al.* (1965) extracted Ochratoxin A and B from the crude extract (in chloroform) with aqueous $NaHCO_3$. The Ochratoxins are obtained from the aqueous layer by acidification and finally extracted into chloroform. Ochratoxin containing mixtures can also be separated on formamide impregnated cellulose powder and elution with mixtures of hexane and benzene. Scott *et al.* (1971) reported that Ochratoxin A was eluted from silica gel, oxalic acid (95 : 5) by elution with chloroform: acetone (99 : 1 or 98 : 2) and applied methodology to the large scale preparation of Ochratoxin A. Peterson and Ciegler (1980) reported the purification of Ochratoxin A using liquid-liquid extraction and the final clean up with preparative HPLC.

Nakajima *et al.* (1997) could estimate ochratoxin A in green coffee beans with the help of HPLC linked with immuno affinity chromatography. Array biosensors, fluorescence, polarization, immunoassay, screen printed electrode immunoassay, surface plasm on resonance biosensors and Quartz crystal microbalance biosensors are some of the rapid assay methods of Ochratoxin (Visconti and Girolamo, 2005).

Some of the rapid field assays without special laboratory equipments include lateral flow devices, membrane-based flow through portable immuno-assays and test cards (Dietrich *et al.,* 1995, Van Peteghem, 1999; Danks *et al.,* 2003; Zheng *et al.,* 2005 and Saha *et al.,* 2006). Lobean *et al.* (2007) developed a quick clean up tandem assay for detection of Ochratoxin A in cocoa powder. Different analytical techniques in practice for detection of ochratoxin A in varous foods are précised in Table 8.1.1 and Table 8.1.2, respectively.

Table 8.1.1. Analytical methods for the detection of Ochratoxin A and Ochratoxin α

Method	Matrices	Extraction/ clean-up	LOD or LOQ[1] (ppb)	Reco-very/%	Reference
HPLC[2]-fluorescence	Wheat, maize, red pepper, cheese, wine	Chloroform	8	80	Aboul-Enein *et al.* (2002)
HPLC-fluorescence	Barley	Immunoaffinity column	0.2	93	Entwisle *et. al.* (2000)
HPLC-fluorescence	Baby food	Immunoaffinity column	0.1	108	Burdaspal *et al.* (2001)
HPLC-fluorescence	Ham	Immunoaffinity	0.7	83	Chiavaro *et al.* (2002)
HPLC-fluorescence	Liver pate	Acetonitrile	0.6	86	Jimenez *et. al.* (2001)
ELISA	Spices	Methanol	0.035	>90	Thirumala-Devi *et al.* (2001)

HPLC-fluorescence	Animal feed	Immunoaffinity column	10	95	Dalcero *et al.* (2002)
HPLC-fluorescence	Coffee	Immunoaffinity column	0.2	85	Entwisle *et al.* (2001)
Enzyme immunoassay	Coffee	Methanol	4	73-87	Sibanda *et al.* (2002)
Thin layer chromatography	Green coffee	Immunoaffinity column	0.5	98-104	Santos and Vargas (2002)
HPLC-fluorescence	Wine, beer, fruit juice	Toluene	0.01	80-90	Filali *et al.* (2001)
HPLC-fluorescence	Wine	Immunoaffinity column	0.01	98	Shepard *et al.* (2003)
HPLC-fluorescence	Wine	Immunoaffinity column	20-45	90	Castellari *et al.* (2000)
GC[4]-photodiode array	Wine, beer	Immunoaffinity column	0.1	83-94	Soleas *et al.* (2001)
HPLC-fluorescence	Serum	Dichloromethane	100	90	Curtui *et al.* (2001)
LC-MS/MS	Plasma	Solid-phase extraction	0.5	Not given	Lau *et al.* (2000)
HPLC-fluorescence	Human milk	Ethyl acetate	100	80-87	Miraglia *et. al.* (1995)

Table 8.1.2. Methods for the Determination of Ochrotoxin A in Various Foods

Food	Extraction	Cleanup	Quantitation	Detection (μg/kg)	Reference
Barely	Shake 50 g ground barley 30 min with 25 ml 0.1 M phosphoric acid and 250 ml chloroform	Aqueous sodium bicarbonate, diatomaceous earth column	TLC Plate: Silica gel GHR Solvent: 18:1:1 Benzene/methanol/ acetic acid Detection: Visualization or densitometry under long-wave (ochratoxin A) or short-wave (ochratoxin B) UV light	12	Scott (1995)
Corn Barley	Blend 50 g for 3 min with 25 ml 0.1 M Phosphoric acid and 250 ml chloroform;	Liquid-liquid partition and C_{18} column chromatography	HPLC Column: Reverse-phase C_{18}		Scott (1995)

	add 10 g diatomaceous earth near end of blending		(5 µm, 25 cm x 4.6 mm i.d.) Mobile phase: 99:99:2 Water/ acetonitrile/ acetic acid Detection: Fluorescence (ex: 333 nm; em: 460 nm)		
Green coffee	Shake ground sample (wet with water) 30 min with chloroform (5 ml/g)	Basic diatomaceous earth column chromatography	TLC Plate: Silica gel GHR Solvent: 5:4:1 Toluene/ ethyl acetate/ formic acid Detection: Visualization of greenish-blue fluorescent spot under long-wave UV light Confirmaion: Spray with alcoholic $NaHCO_3$ or alcoholic $AlCl_3$; spot becomes blue and more intensely fluorescent		Scott (1995)
Black olives	Shake ground olives 5 min with 60:40 methanol/water (6 m/g)	Liquid –liquid partition	TLC Plate: Silica gel G Solvent: (a) Benzene/ hexane; (b) 60:30:15 toluene/ethyl acetate/ formic acid Detection: Visualization under long-wave UV light before and after treatment with NH_3 vapor	20	Tutour *et al.* (1984)
Cer-eals	Blend sample 3-5 min with 50:50 methanol/ phosphate-buffered saline (4 ml/g)	Immunoaffinity column chromatography (Easi-extract, Biocode, York, UK); method can be automated with a commercially available automated sample prep system	HPLC Column: Reverse-phase C_{18} (5 µm, 25 cm x 4.6 mm i.d.) Mobile phase: 99:99:2 Water/ acetonitrile/ acetic acid Detection: Fluoresce-nce (ex: 333 nm; em: 477 nm)	0.2	Sharman *et al.* (1992)

Meat	Blend ground sample 3 min with 50:25:25 chloroform/ 2 M sodium chloride/0.5 M phosphoric acid (10 ml/g)	Centrifugation and silica gel column chromatography	HPLC Column: Reverse-phase C_{18} (5 μm, 25 cm x 4.6 mm i.d.) Mobile phase: 248: 248:5 Water/ aceton-itrile/acetic acid Detection: Fluoresc-ence (ex: 330 nm; em: 470 nm)	0.5	Scott *et al.* (1991)
Coffee	Sonicate finely ground coffee beans or instant coffee 15 min with 1% aqueous sodium bicarbonate (16 ml/g)	Immunoaffinity column chromatography	HPLC Column: Reverse-phase C_{18} (25 cm x 4.6 mm i.d.) Mobile phase: 64:33:3 Acetonitrile/ water/0.2 M Phosphate buffer Detection: Fluoresc-ence (ex: 365 nm; em: 450 nm)	0.025-0.5	Nakajima *et al.* (1990)
Barley, wheat	Shake ground sample 30 min with 12:1 chloroform/0.1 M H_3PO_4 (6 ml chloroform/g)	Reverse-phase TLC Plate: Whatman KC-18 (200 μm, 5 x 10 cm) Solvent: (a) Hexane; (b) 70:30 methanol/ water Detection: Visuali-zation under long-wave UV light; elute blue fluorescent spot at R_f 0.76 for quantitative analysis	1. Spectrofluorometry (ex: 340; em: 249) 2. HPLC Column: Reverse-phase C_{18} (5 μm, 25 cm x 4.6 mm i.d.) Mobile phase: 70:30 Methanol/water Detection: Fluoresce-nce (ex: 333 nm; em: 418 nm)	100-200	Frohlich *et al.* (1988)

8.1.5. Factors influencing ochratoxin A production

Several factors determine the production of Ochratoxins. They are strain of fungus, substratum and temperature and water activity. Although most OTA producers can grow in the range of 4-37°C and at an a_w as low as 0.78, optimal conditions for toxin production are narrower with temperature of 24 – 25°C and a_w values less than 0.97.

Ochratoxin production by either *A. ochraceus* or *P. viridicatum* is mainly influenced by two factors such as water activity (a_w) and temperature.

Northolt (1979) determined the influence of temperature and water activity on Ochratoxin A production by A. *ochraceus.* A minimum temperature at which Ochratoxin production was noticed is 4°C for *P. viridicatum* and 12°C for A. *ochraceus* and optimum temperature recorded for its production was 28°C. *P. viridicatum* is reported to grow over a temperature range of 4°C - 30°C and optimum was 20°C. Similarly *A. ochraceus* became dominant in corn stored for several weeks at 25°C and 18.5 per cent moisture (Lopez and Christenson, 1967).

Raulin- Thom and YES medium were used for production of Ochratoxin by *A.ochraceus* and *P.viridicatum* respectively. Addition of phenylalanine (1%) greatly stimulated Ochratoxin production. Similarly combined addition of trace elements Fe, Zn, Bo, Mn and Mo increased the synthesis of Ochratoxin A (Ferreria, 1968). Minimum water activity (aw) required for production of Ochratoxin A was 0.83-0.87 (Northolt, 1979). The solid substrate fermentation technique appears to be most suitable for the large-scale production of Ochratoxin A.

8.1.6. Natural incidence

Ochratoxin A has been detected in several plant products, particularly cereals and occasionally groundnut and beans. Shotwell *et al.* (1969) first reported its natural occurrence in corn. Subsequently Scott *et al.* (1970) reported it on Canadian wheat. OTA has been widely detected in cereals (Trucksess *et al.,* 1999), green coffee (Leomi *et. al.,* 2000), bread, grape juice (Zimmerli and Dick, 1995), wine (Miraghia and Brera, 2002 and Blera *et al.,* 2004), rice and dried fruits (Zinedine *et al.,* 2007; Gonzalaz-Osnaya *et al.,* 2000) and coffee bean (Vander stegen *et al.,* 1997; Nakajima *et al.,* 1997). It is now a known natural contaminant of various food stuffs and feeds (Table 8.1.3 and 8.1.4) from different parts of the world (Pittet, 1998). OTA has been found in European and North American cereal crops (Trucksers *et al.,* 1999). A high mean concentration and incidence of contamination in breast milk samples from sierradeone provides indirect evidence of OTA contamination of food stuffs of African origin (Jonsyn *et al.,* 1995). Further to this, very high level of OTA was detected in wheat, barley, mixed cereals, dried vegetables and olives collected in Tunisia from families with one or more patients with nephropathy (Maaroufi *et al.,* 1995a).

Barley appears to be most common commodity of contamination of ochratoxins producing organisms in temperate countries but in tropical and sub-tropical regions a much wider range of food stuffs may be contaminated including coffee, cocoa, wine, fruits and spices (Pittet, 1998). Reiss (1981) reported the occurrence of *A. ochraceus* in cake and bread. Scott *et al.* (1972) reported the presence of Ochratoxin A in dried beans destined for human consumption and in samples of mouldy peanuts. Krogh (1972) reported

Ochratoxin A in 19 samples of barley and oats that were destined to fed swine. Ochratoxin A residues were also detected in kidney, liver and adipose tissues of swine that feed with Ochratoxin contaminated barley and oat grains. Ochratoxin residues (67 μg/kg) have been detected in the muscle, kidney and fat of pigs and in poultry muscle. A suggestion has been made that endemic Balkan nephropathy, a human disease of unknown etiology, reported from Bulgaria, Rumania and Yugoslavia may be associated with elevated Ochratoxin. The above findings indicate that human exposure to Ochratoxin A may occur not only through consumption of mouldy plant materials but also through consumption of animal products. The presence of Ochratoxin A residues in animal products is of concern because it binds tightly to serum albumin and has long half life in animal tissues and fluids (body).

OTA has been widely detected in cereals (Trucksess *et al.* 1999) green coffee (Leomi *et al.*, 2000), grape juice (Zimmeli and Dick, 1995), wine (Miraglia and Brera, 2002 and Blera *et al.*, 2004), rice and dried fruits (Zinedine *et al.*, 2007) and coffee beans (Panneruselvum *et al.*, 2000; Nakajima *et al.*, 1997). OTA has also been reported to contaminate spices such as turmeric, coriander, ginger (Thirumala Devi *et al.*, 2001) and chillies (Thirumala Devi *et al.*, 2000) from the Indian subcontinent.

Table 8.1.3. OTA in different foodstuffs

Foodstuff	Country of origin	Range of concentrations detected (mean) μg/kg or μl/l	LOQ (LOD)[1] μg/kg or μl/l	Number of OTA-contaminated samples	Reference
Cereals					
Wheat	Germany 1997	0.6-0.7 (0.7)	0.4	2/14	Birzele *et al.* (2000)
Wheat	Germany 1998	0.6-0.8 (0.6)	0.4	7/29	Birzele *et al.* (2000)
Wheat	Hungary	0.12-0.5 (0.29)	(0.1)	3/36	Fazekas *et al.* (2002)
Wheat flour	Hungary	0.11-0.15 (0.13)	(0.1)	2/16	Fazekas *et al.* (2002)
Maize course meat	Hungary	0.4(0.4)	(0.1)	1/6	Fazekas *et al.* (2002)
Oats	Norway	(0.065, maximum 0.47)	0.01	3/22	JECFA (2000)
Beer					
Home brewed beer	South Africa	3-2340	(1)	10/32	Odhav and

					Naicker (2002)
Beer	Europe	(0.025, maximum 0.47)	0.004	40/40	JECFA (2000)
Coffee					
Pure roasted coffee and Coffee products	Purchased in Hungary	0.17-1.3 (0.57)	(0.1)	33/50	Fazekas *et al.* (2002)
Coffee products	Europe	(0.9)	0.2-1	299/633	JECFA (2000)
Cocoa and chocolate					
Chocolate	UK	(0.16)	0.02	30/40	JECFA (2000)
Cocoa powder	UK	(1.67)	0.2	20/20	JECFA (2000)
Conventional & organic bread					Gonzalez-Osnaya (2007)
Wine					
Red wine	Morocco	0.04-3.24 (median 0.912)	0.01	20/20	Filali *et al.* (2001)
Red wine	UK	(0.074, maximum 0.46)	0.02	28/50	JECFA (2000)
White/Rose wine	Morocco	0.028-0.18 (median 0.117)	0.01	10/10	Filali *et al.* (2001)
White wine	Italy	(0.57, maximum 8.86)	0.01	7/21	JECFA (2000)
Dried wine fruit					
Raisins	UK	(2.79, maximum 20)	0.2	17/20	JEFCA. (2000)
Currants	UK	(4.97, maximum 40.8)	0.2	96/100	JEFCA. (2000)
Meat and offal					
Pig kidney	Romania	(0.54)	0.01	41/52	Curtui *et al.* (2001)
Pig kidney	Denmark	0-15 (0.5)	0.09	284/300	Jorgensen and Petersen (2002)
Pig meat	Romania	(0.15)	0.01	9/52	Curtui *et al.* (2001)
Pig meat	Denmark	0-2.9 (0.12)	0.06	228/300	Jorgensen and Petersen (2002)
Pig liver	Romania	(0.16)	0.01	39/52	Curtui *et al.* (2001)

Herbs, spices and sweets					
Chilli powder	UK	(Maximum 50.4)	0.1	?/4	JECFA (2000)
Curry powder	UK	(Maximum 21.3)	0.1	?/10	JECFA (2000)
Liquorice	Purchased in Germany	Sweet 0.4-3.0 Root 0.3-216.5	(0.3)	Sweet 18/19 Root 9/19	Bresch *et al.* (2000)

Table 8.1.4. Natural incidence of Ochratoxin A in feeds

Commodity	Country	No. of samples analysed	Contaminated samples (in percentage)	Ochratoxin A level (range in pg/kg)
Barley, Wheat, Oats, rye and maize	Poland	150	05.3	50-200
Mixed feeds	Poland	203	49	10-50
Maize feeds	Yugoslavia	191	25.7	45-5125
Barley, oats	Sweden	84	8.3	16-409
Wheat, hay wheat, oats	Canada	95	7.4	30-6000
Barley, Rye	Canada	32	56.3	30-27000
Barley, oats	Denmark	33	57.6	28-27500

8.1.7. Biological activity

Ochratoxin A at 166 to 216 μg/kg affected the chick by causing diarrhea and hemorrhage in the liver, kidneys and proventricules. Sublethal doses of Ochratoxin A produced dose related growth depression. Birds dying after administration of the toxin had visceral gout and acute nephrosis. Hepatic lesions occurred less often than renal lesions and induced mild, diffused vacualisation of hepatosis and foci of necrosis. Surviving and dead birds showed suppression of haeluoises in the bone marrow and depletion of lymphoid elements from the spleen and bursa fabriciuss. The birds also suffered enteritis. Laying hens fed diets at 2 or 4-ppm levels of OTA for six months showed 18 and 64 per cent mortality respectively. Growth depression was also noted in male broiler chicks fed diets containing Ochratoxin A. A dose dependent increase in relative weight of the carpo, pancreas, liver and gizzard and significant regression of the bursa Fabricius was recorded. Experimental feeding of graded levels of Ochratoxin A in diets has a detrimental effect on growing chicks (Prior *et al.*, 1980; and Kubena *et al.*, 1985). A drastic decrease in total proteins, albumins, cholesterol and aspartate amino transferase activity was reported by Kubena *et al.* (1988) in growing broiler chicks fed with Ochratoxin A. OTA is not directly genotoxic but may act to stimulate oxidative damage of DNA.

One of the manifestations of cellular damage in OTA toxicity is due to enhancement of lipid peroxidation. OTA toxin impairs cellular Ca^{++} homeostasis AMP internal epithelial cells. OA interferes with hormonal Ca^{+2} signaling and leads to altered cell proliferation.

Toxicity of OTA has been attributed to inhibition of mitochondrial succinate dehydrogenase and cytochrome oxidase (Meisner and Chan, 1974), modifies mitochondrial membrane transport and inner membrane ATPase (Meisner and Chan, 1974), inhibits enzymes involved in phenylalanine metabolism (Creppy *et al.,* 1980).

Ochratoxin A attacks the kidney and liver in animals causing nephropathy enteritis and immunosuppressive symptoms (Smith and Ross, 1991). A synergistic toxic activity of Ochratoxin A with aflatoxins was recorded by Raju and Devagowda (2000).

Ochratoxin A interferes with several processes in the cell such as protein synthesis, enhancement of lipid peroxidation, impairment of Ca^{++} and AMP homeostasis energy metabolism and several enzyme activities.

Human exposure to Ochratoxin A has been clearly demonstrated by its detection in blood (Bauer *et al.,* 1986; Breitholtz *et al.,* 1991) and breast milk (Miraglia *et al.,* 1993). Risk assessment has been carried out by Kuiper-Goodman *et al.* (1993) and Kuiper-Goodman (1996). The presence of Ochratoxin A in foodstuffs is clearly undesirable. Thuvander *et al.* (1996) feel that low level exposure to OTA alters the relative and absolute number of lymphocytes subpopulations in lymphoid organs but does not affect the immune functions. Maarunfi *et al.* (1995a) could find a correlation between food and blood contamination by OTA and chronic interstitial nephropathy in Tunisia.

Ochratoxin A has high affinity with some enzymes and proteins and binding to the renal proteins is highly specific to a new organic anion transporter or more likely a cytosolic binding component of unknown functions present in human, rats, mice and possibly pigs (Heussnel *et al.,* 2002)

Ochratoxin A induced impaired phagocytic ability of heterophs (Chang and Hamilton, 1980) and thymus regression in broiler chicken (Chang *et al.,* 1981). Bodnarchuk and Kaspruk (1984) reported the presence of haemorrhages in various tissues and catarrhal gastritis and enteritis. Feeding Ochratoxin A from 0 to 21 day of age was associated with proximal tubular intranuclear membrane bound inclusions elongated tortuous and ring shaped mitochondria, enlarged mitochondrial matrical granules with hyaline centers and an increase in size and membrane peroxisomes and secondary lyposomes (Brown *et al.,* 1986).

Recently Burns and Dwivedi (1986) made a review on Ochratoxicosis in poultry. In liver mild degenerative changes, necrosis and periportal fibrosis with inflammatory reaction was reported (Dwivedi and Burns, 1984;

Subramaniam and Govinda Swamy, 1985). The haemotological investigations showed that there is significant decrease in PCV, Hb, total RBC count, total leucocytic count and differential counts in toxin (0. 7511lg/kg body weight) fed birds as against controls. There was further significant drop in all these values with increased dosage of toxin (Vikram Reddy, 1990).

The following are the principal changes in chicks (Dwivedi and Burns, 1984).

1. Necrosis of the renal tubular epithelial periportal liver cells and enteritis.
2. Growth depression, visceral gout, renal and hepatic lesions and enteritis.
3. Delayed sexual maturity
4. Reduced egg production and feed efficiency.
5. Enlargement of kidney, liver and lymphoid organs.

Zebra fish and brine shrimp larvae were sensitive to Ochratoxin A (Abedi and Scott, 1969). Most of the bacteria were insensitive (Arai and Otomo, 1969), while actinomycetes were inhibited at the concentration of 100-500 µg/ml (Arai and Otomo, 1969). Ochratoxin exhibited cytotoxicity to Hela cells. It also induced mitotic abnormalities in monkey epithelial cells in tissue culture (Broce *et al.,* 1970). Ochratoxin A at 100 µg/30g of duckling caused changes in liver characterised by swelling and structural damage of mitochondria. Morphological changes in the endoplasmic reticulum and fatty infiltration of hepatocytes were reported (Theron *et al.,* 1966).

Ochratoxin A at 100 µg developed focal areas of liver cell necrosis and hyaline degeneration in rats, proliferation of smooth ER with formation of masses of SM with in the liver cell cytoplasm referred to as finger prints. Ochratoxin A at LD 50 values of 20-22 mg/kg induced necrosis of the renal tubules and periportal cells of the liver. Glycogen accumulation was observed in liver cells. On oral feeding for 90 days at 0.2 ppm concentration developed distal convoluted tubules of kidney cells.

Speijers and Van Egmond (1993) could find a correlation between Ochratoxin A present in food and feed and human nephropathy (Table 5). Dirheimer (1998) considers OTA as possible human carcinogen that through accumulation of OTA remains prolonged in tissues that OTA affects the immune system is cyto-and geno-and neurotoxic and that OTA passes unchanged through the food chain although many effects are the result of prolonged exposure to or accumulation of OTA. Acute toxicity also plays an important role with susceptibility varying from species to species. Stormer (1992) discussed elegantly different toxicological aspects of ochratoxin A.

Table 8.1.5. Natural occurrence of Ochratoxin-A in human blood samples

Country	Year	Incidence of positive samples	Mean concentration (ng/ml)	Concentration range of positive samples (ng/ml)
Bulgaria	1984-90	9/125 (7)[b]		1.0 – 10.0
Canada	1994	144/144 (100)	0.88	0.29 – 2.37
Croatia	1997			
Zagreb		29/50 (58)	0.26	0.20 – 1.28
Rijeka		18/50 (36)	0.17	0.20 – 0.82
Osijek		50/50 (100)	0.68	0.20 – 1.65
Split		27.49 (55)	0.25	0.20 – 1.39
Varazdin		24/50 (48)	0.59	0.20 – 15.9
Czechoslovakia	1990	35/143 (24)	0.14	0.10 – 12.6
Czech Republic	1994	734/809 (91)	0.23	0.10 – 13.7
	1995	404/413 (98)	0.24	0.10 – 1.9
Denmark	1986-88	78/144 (54)	1.8	0.10 –13.2
France	1991-92			
Alsace		97/500 (19)		0.10 – 12
Aquitaine		385/2055 (19)		0.10 – 160
Rhone-Alpes		75/515 (15)		0.10 – 4
Federal Republic of Germany	1977	84/164 (51)	0.4	0.1 – 4
	1985	89/141 (63)	0.3	0.1 – 2
	1988	142/208 (68)	0.75	0.1 – 8
Hungary	1995	291/355 (82)		0.2 – 10
	1997	213/277 (77)		0.1 – 1.4
Italy	1992	65/65 (100)	0.5	0.1 – 2
Japan, Tokyo	1992-96	156/184 (85)	0.068[c]	0.004 – 0.278
Poland	1983-84	25/397 (6)	0.2	1-13
	1984-85	52/668 (8)	0.3	1-40
Sierra Leone	1996	12/36 (33)[d]		1.5 – 18
Sweden	1989			
Visby		29/99 (29)	0.26	0.3 – 7
Uppsala		3/99 (3)	0.02	0.3 – 0.8
Ostersund		6/99 (6)	0.03	0.3 – 0.8
Switzerland	1992 – 93			
North of the Alps	251/252 (100)			0.06 – 2.14
South of the Alps	116/116 (100)			0.11 – 0.75
Tunisia	1993-95	73/140 (52)		0.1 – 8.8

Human exposure to ochratoxin A has been clearly demonstrated by its detection in blood (Bauer *et al.*, 1986; Breitholtz *et al.*, 1991) (Table 8.1.5). Risk assessment have been carried out by Kuiper-Goodman *et al.* (1993) and Kuiper-Goodman (1996).

8.1.8. Pharmacokinetics and metabolism

Ochratoxin A is absorbed by the intestinal tract of all animals and mostly in the region of proximal jejunum (Wyllie and Morehouse, 1978). OTA readily binds to serum proteins and transported in the blood predominantly in bound form. Ochratoxin A was detectable in blood and liver, excretes in urine and faces over 72 hrs period (Purchase and Nel, 1967). Maximum concentration of Ochratoxin A was found in serum of rat and chicken after 30 min. of ingestion. Two other compounds, Ochratoxin A and 4-hydroxy Ochratoxin A formed due to loss of phenylalanine from Ochratoxin A were also present in the urine. These hydrolytic activities in the rat were predominantly localized in small intestine preparation with significantly lesser amounts found in large intestine and liver. Purchase and Theron (1968) have reported degenerative changes in the rat liver mitochondria including dilation at 100mg/kg of Ochratoxin A. Intramitochondrial ATP pool was rapidly depleted by 80% within 2 min of treatment with 18 nmol Ochratoxin μg protein. Higher toxin levels caused total ATP depletion.

Ochratoxin A was strongly nephrotoxic to dogs (Szezech *et al.,* 1972). Ochratoxin A was detected in decreasing order in liver, kidneys, muscle and placenta of pregnant pigs fed for 8 days. Ochratoxin A and B were also present in the serum but only Ochratoxin A was excreted in urine and faeces. Rutqvist *et al.* (1978) have observed Ochratoxin A in 25% of the kidneys of pigs in Sweden with a concentration up to 109 μg/g. Krogh (1977) reported 35 percent kidneys with Ochratoxin A in Denmark with concentration up to 68 μg/g. Rutqvist *et al.* (1978) have observed Ochratoxin A in the liver and higher concentration in the blood and serum. Such high concentrations of Ochratoxin A were also reported from. Germany (Bauer *et al.,* 1984) and Poland (Golinski *et al.,* 1985). Principle pathological changes found in pigs due to Ochratoxin A are anorexia, depression, diarrhea, fever, polydipsia, polyurea, dehydration and prostration. LD_{50} values reported for different animals are precised in Table 8.1.6.

In ruminants rumen flora has good capability to hydrolyse Ochratoxin A into non-toxic Ochratoxin metabolite preventing/the absorption of the toxin. However, a cow fed for 7 weeks with a feed containing 1125g/ kg Ochratoxin A was detected in milk and urine. OTA was found to induce nitric oxide production in cells which causes DNA damage and tumor formation (Aish *et al.*, 2004).

Table 8.1.6. LD_{50} values of Ochratoxin A for some animals

Animal	Route	LD_{50}
Duckling	Oral	150 μg/kg
Rat	Oral	20-22 μg/kg
Trout	Intra peritoneal	4.64 μg/kg
Chick	oral	166 μg/kg

8.1.9. Management

Ascorbic acid supplementation (300 mg/kg) of laying hen diets which contained 3mg Ochratoxin A/kg partially ameliorated the toxicity effects including the negative effects on egg mass production and rigidity of egg shells. Munimbazi *et al.* (1997) reported the inhibition of Ochratoxin A production by a fungicide, prodione. Malaveille *et al.* (1994) reported that a water soluble form of vitamin E completely quenched the genotoxic effects of Ochratoxin A in *E. coli.* Piperine and natural alkaloids of cloves (*Piper longman*) inhibited Ochratoxin A production by *Aspergilli* screened by Lee *et al.* (2007). Similarly curcumin and sesamin could inhibit Ochratoxin A production by *Aspergillus auricomus* and *A. sclerotiorum.* Basilico and Basilico (1999) reported the inhibitory effect of some spice essential oils on growth and Ochratoxin A production by *A. ochraceous.*

Frank (1999) has demonstrated the importance of the drying process in the production of Ochrotoxin A free coffee beans and has attempted to apply the Hazard Analysis Critical Control Point (HACCP) approach to the study of coffee production.Visconti *et al.* (2008) discussed critical points which need to be taken care to reduce ochratoxin A contamination in grapes.

An adsorption mechanism has been suggested for Ochratoxin A removal by lactic acid bacteria, (Bejaovi *et al.,* 2005) and yeasts (Bejaovi *et al.,* 2005). *A. niger, A. carbonarius* and biseriate strains of *Aspergillus* inhibited the Ochratoxin A. On the other hand, *Eurotium amstelodomi* stimulated the OTA production. *Phaffia rhodozyma* could degrade Ochratoxin A up to 90%, which has been attributed to enzyme carboxy peptidase (Peteri *et al.,* 2004).

Several bacteria (Wegst and Lingens, 1983; West Lake *et al.,* 1987; Hwang and Drangham, 1994), yeasts (Bohm *et al.,* 2000; Molnar *et al.,* 2004) and filamentous ascomycetes (Varga *et al.,* 2000, 2002; Abrunhora *et al.,* 2000) are reported to detoxify Ochratoxin A. (Peteri *et al.* 2004) reported the degradation and adsorption of OTA by astraxanthin producing yeast. Lopeg Garcia *et al.* (2008) designed and implemented integrated management practice for the control of ochratoxin A in coffee production chain.

8.2. CITRININ

Citrinin, a toxic secondary metabolite was first isolated from cultures of *Penicillium citrinum* by Hetherington and Raistrick (1931). Later it was isolated from the leaves of Australian *Crotolaria crispata* plant (Ewart, 1933) and characterised as an antibacterial antibiotic (Raistrick and Smith, 1941). Subsequently other species belonging to *Aspergillus* and *Penicillium* are reported to produce this toxin (Betina *et al.,* 1964; Hesseltine, 1969; Krogh *et al.,* 1970; Scott *et al.,* 1970; Pravindra Chary and Reddy, 1986). Other structurally related compounds were also isolated from mutants of *P. citrinum* but their biological effects are unknown (Turner, 1971). A similar compound ascohitin was isolated from cultures *Aschochyta pisi* which proved to be a phytotoxic metabolite. An isochroman -3,7- dimethyl-8-hydroxy-methoxy isochroman macotoxin was isolated from *P. stickii* (Cox *et al.,* 1979)

8.2.1. Physico-Chemical characteristics

IUPAC name of citrinin is (3R, 4S)-4, 6-Dihyrlro-8hydroxy-3,45-trimethyl-6-oxo-6-3H-2 benzopyran 7-carboxylic acid (Fig. 8.1.1). Its other name is Antimycin, S-52, Citrinin, p-quinone methide. It forms yellow needles when crystallized from absolute alcohol or benzene-cyclohexane. Its mol.wt. is 250 Daltons, Melting point is 175°C and thermostable. It is soluble in hot ethanol, ethyl acetate, benzene, acetone, chloroform and insoluble in water. Structure of citrinin ($C_{13}H_{14}O_5$) was elucidated by Hetherington and Raistrick (1931) and confirmed by Brown *et al.* (1948).

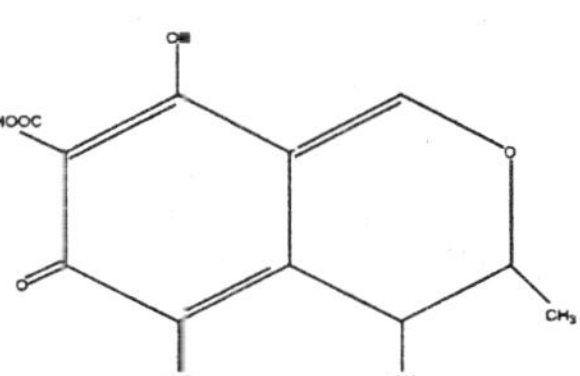

Fig. 8.1.1. Structure of citrinin

8.2.2. Organisms

Citrinin is produced by several species belonging to *Penicillium* and *Aspergillus*. It is reported to be produced by 20 species of *Penicillia*, 3 species of *Aspergilli*. *P. viridicatum* is the main producer of citrinin in cereals in temperate countries. *P. sartoryi, P. citrinum, P. lividum, P. islandicum, P. aurantio-violaceum, P. fellytanum, P. implicatum, P. chrazaszszi, P. virldic-atum, P. steckii, P. jensemi, P. carybphillum* (Jabbar and Rahim, 1962), *P. citreoviridae, P. expansum, P. velutinum* (Hald and Krogh, 1973), *P. canescens, P. purpuresecens, P. claviforme, P. roqueforti* and *P. janthinellum* (Betina and Binovska, 1979) are the other species of *penicillium* which produce citrinin. Among the species of *Aspergillus, A. candidus, A. terreus* and *A. flavipes* are citrinin producing species. Citrinin has also been isolated from culture filtrates of *Pythium ultimum* (Endo and Kuroda, 1976). Citrinin is also produced by *Manascus rubber* (Hajjaj *et al.*, 2000).

8.2.3. Biosynthesis

Citrinin is synthesized through the acetate-malonate pathway (Yamazaki, 1978). Five molecules of acetic acid combine head-to-tail to form polyacetic acid. According to Smith and Moss (1985) and Barber *et al.* (1987) citrinin is synthesized through pentaketide pathway. C-methyl group is incorporated into a pentaketide during the biosynthesis of citrinin.

8.2.4. Isolation and detection

Taira and Yamatodani (1947) for the first time detected citrinin by using ferric chloride in "ooh-han-mai" (Yellowed rice) problem during post war in Japan. Later flourometric method was developed by Kawashiro *et al.* (1955) for detecting citrinin in rice.Scott *et al.* (1970) for the first time developed TLC method for detecting citrinin in fungal extracts. Davis *et al.* (1975) developed the method of extraction of citrinin from culture filtrates of *P. citrinum*. A multitoxin TLC method was described by Roberts and Patterson (1975) for the detection of citrinin in animal feed stuffs. Martin *et al.* (1978) recommended chloroform-methanol-hexane (64 + 1 + 35) as developing solvent for detection of citrinin.

Chalam and Stahr (1979) described TLC detection of citrinin as detailed below. Citrinin feed is extracted with methanol + water and made it alkaline with 10% Na_2CO_3. Extracted with chloroform aqeuous layer acidified with 2N HCl and extracted with chloroform. Developed in chloroform + acetone + ethanol + water (60 + 40 + 10 + 1) and quantified with florodensitometric method. The detection limit by this method was 0.5 μg/gr.

Wu *et al.* (1974) described a flourodensitometric method for detection and quantitation of citrinin. Cultures of *P. viridicatum* grown in potato dextrose broth were extracted with chloroform. The crude extract was filtered and the chloroform layer was seperated. The extraction was repeated 3 times with 500 ml of chloroform. The chloroform extracts were combined and concentr-ated. The concentrated extracts were diluted to 5ml with chloroform. Citrinin was separated from the extracts by TLC on adsorbosil using toluene: ethyl acetate: formic acid (6:3:1) as developing solvent. Gimeno (1979) reported the TLC method for detection of citrinin along with 16 other mycotoxins from mixed feeds, food products and feed stuffs. Minimum detectable concentration of citrinin is 600-750 μg/kg. He subsequently improved solvent systems. Gimeno (1980) recommended the use of Benzene: Methanol : acetic acid (180: 10 : 10), Ethyl acetate : acetone : water (80 : 80 : 32), Ethyl acetate : Methanol

: water : 90% formic acid (190 : 8 : 2 : 2), Toluene : Ethyl acetate : Chloroform : 90% formic acid (70 : 50 : 50 : 20) and Toluene : Ethyl acetate : 90% formic acid (150 : 50 : 2). An elegant and simple procedure for the isolation and quantification of citrinin was developed by Damodaran *et al.* (1973) which is described below:

Culture filtrate was acidified to pH 1.5 and citrinin was extracted repeatedly with CH_3Cl and pool extracts and evaporate chloroform and redissolved in known volume of chloroform or 0.1 M hydrogen carbonate buffer (pH 10). The extract was spotted on TLC plates and developed in Toluene: Ethyl acetate: formic acid (5:4:1). Greenish yellow flouresence under UV light (R_f value = O.8) was taken as positive, Marked, scraped flouresent portion into tube containing 3ml of 0.1M carbonate-hydrogen carbonate buffer (pH 10) and allowed the suspension to stand for 1 hour with intermittent shaking. Centrifuged and took clear supernatant of known volume and add 1ml of 15% Na_2CO_3 and 1 ml of Folin reagent. Incubate the reaction mixture for 15 min. at 37°C. Read the intensity of blue colour at 37°C and calculate the citrinin from standard graph.

Martin *et al.* (1978), Scudamore and Atkins (1985) and Lepom (1986) have described HPLC method for detection and quantification of citrinin. They reported that 0.25N phosphoric acid and acetonitrile or methanol were acceptable as an elution system for citrinin extraction from corn. Phillips *et al.* (1980) modified the above method to quantitate the citrinin in urine, bile and plasma. Reversed phase chromatography is recommended for the seperation of the lipophilic compound, citrinin. Hald and Krogh (1973) also developed a method for detecting citrinin in contaminated plant material. A screening method for the detection of citrinin and four other mycotoxins was described by Wilson *et al.* (1976). Dorner (1996) gave two most common methods of detection of citrinin (Table 8.2.1).

Citrinin in culture broth of *Penicillium janthinellum* was determined by a microbiological diffusion plate method using a highly sensitive strain of *Bacillus subtilis* (Betina *et al.,* 1973). This method is most convenient for monitoring the course of citrinin production. Vail and Homann (1990) recommended the use of rapid and sensitive method for detection of citrinin by HPLC.

8.2.5. Natural incidence

Citrinin has been found as a natural contaminant of rice, wheat, corn, rye, barley and oats in Canada. Takahashi *et al.* (1972), Krogh *et al.* (1973); Hald *et al.* (1983), Reddy *et al.* (1986), Mislivec *et al.* (1975) reported that all isolates of *P. citrinum* screened and three out of 15 isolates of *P. viridicatum* from dried beans produced citrinin. Buckle (1983) reported the occurrence of citrinin along with other mycotoxins in cereals and animal feed stuffs from England. Pravindra chary and Reddy (1986) detected citrinin in 10 samples of paraboiled rice out of 12 samples screened.

Table 8.2.1. Methods for determinatioin of Citrinin.

Toxin	Food	Extraction	Cleanup	Quantification	Detection (μg/kg)	Reference
Citrinin	Cereals	Mix 40 g ground sample 5 min with 150 ml chloroform and 20 ml 0.1 M phosphoric acid	Partition column chromatography	HPLC Column: Silica buffered with 0.2 M citric acid (5 μm, 30 cm x 4.6 mm i.d.) Mobile phase: 60:40 Hexane/ chloroform Detection: Fluorescence (ex: 360 nm; em: 500 nm)	0.1	Zimmererti *et al.* (1995)
Citrinin	Corn and barley	Shake 60 g ground sample 30 min with 180 ml acetonitrile and 20 ml 10% glycolic acid solution in water	Liquid-liquid partition	TLC Plate: Silica gel G-25HR impregnated with glycolic acid Solvent: (a) 70:90:40:2 Ethyl ether/ hexane/ethyl acetate/90% formic acid (b) 70:50:50:20 Toluene/ethyl acetate/chloroform/90% formic acid (c) 50:100:50 Ethyl ether/hexane/ethyl acetate Develop a different plate with each solvent Detection: Visualization of fluorescent spotunder 254 nm and 366 mm UV light before and after spraying with aluminum chloride or treating with NH_3 vapor	30-40	Gimeno and Quintailla (1983)

It has been detected in naturally contaminated peanuts and grains in tropical regions (Scott *et al.,* 1972; Krogh *et al.,* 1973; Subramanyam and Rao, 1974). Smith and Moss (1985) reported that *P. citrinum*, *P. expansum* and *P. viridictum* are most important toxigenic species of *Penicillium* commonly found in cereals and other foods. Citrinin has been identified as a contaminant of yellow peanut kernels from damaged pods (Subramanyam and Rao, 1974). Kumari and Nusrath (1987) reported the occurence of citrinin in coconut products from Andhra Pradesh. El-Banna *et al.* (1987)

reported production of citrinin by different species of *Penicillium*. Mills *et al.* (1989) reported citrinin and ochratoxin A production by *penicillium* species from stored durum Wheat. Pande *et al.* (1990) reported the natural occurence of mycotoxins including citrinin in some cereals like wheat, rice and maize at Nainital of U.P. Mills *et al.* (1990) reported the contamination of cereals by *Penicillium citrinum* in Western Canada. Abdelhamid *et al.* (1990) reported the occurence of citrinin along with other mycotoxins in various Egyptian feeds.

8.2.6. Factors influencing citrinin production

Citrinin is produced in liquid media under static or submerged conditions and on cereals or fruits. *P. janthinellum* produced maximum amount of citrinin in a medium containing sucrose, K_2HPO_4 and corn steep liquor when compared with other four different media (Betina *et al.,* 1973).

Damodaran *et al.* (1973) used Timonln's medium and Rlichard's medium for production of citrinin. Pravindra Chary and Reddy (1986) have reported Richard's medium and SMKY medium to be the best for production of citrinin by *P. citrinum*. Davis *et al.* (1975) reported maximum production of citrinin (upto 1.75g/lit) by *P. citrinum* as stationary cultures. Ciegle *et al.* (1977) recommended YES broth for the growth and citrinin production by *P. expansum* at 25°C in 14 days. *P. citrinum* reported to produce citrinin over pH range 3.0 to 7.0 (Pravindra Chary and Reddy, 1986) with a maximum at pH 3.0.

Lin *et al.* (1980) reported that *P. viridicatum* produced maximum citrinin at 25-30°C. Pravindra Chary and Reddy (1986) reported that temperature of 30-35°C was favourable for citrinin production by *P. citrinum*. Montani *et al.* (1988) have also reported that *P. citrinum* produced maximum citrinin at 30°C and rapid decrease in citrinin was observed at 37°C. *P. viridicatum* produced citrinin at a_w 0.80 and above (Patterson and Damoglan, 1986). Pravindra Chary and Reddy (1986) reported that sorbitol followed by sucrose were good carbon sources, while potassium nitrate and glycine were preferred nitrogen sources for citrinin production by *P. citrinum*.

8.2.7. Biological activity

Citrinin has antibacterial, antifungal (Melonk and Akem, 1987) and antiprotozoal properties (Betina, 1984). LD50 values in mice and in rabbits were 35 and 19 mg/kg respectively (Spector, 1957). Citrinin caused teratogenic eftects in chicken embryos, with malformations primarily in the extremities showing the head and neck twisted to the left instead of the right as normally occurs (Ciegler *et al.,* 1977).

The LD_{50} of citrinin has been reported for mice, rats, rabbits and guinea pigs. The subcutaneous LD_{50} varies among these species from 35-67 mg/kg body weight. Intraperitoneally the LD_{50} is 30 mg/kg in mice and about 50 mg/kg in rabbits (New borne, 1987). Citrinin acts synergistically with ochratoxin A (Vrabcheva *et al.*, 2000).

Citrinin has been suggested as a causative factor in renal disease and death among livestock, poultry and perhaps humans. It is commonly found in field samples along with ochratoxin A. These two mycotoxins have been suggested as the cause of procine nephropathy in Denmark. Feeding of grain contaminated with *P. citrinum* caused nephrotoxicity in swine. Acutely lethal doses of citrinin administered to rabbits, guinea pigs, rats and swine caused swelling of the kidneys and acute tubular necrosis. The nephropathy produced in swine by citrinin has structural and functional similarities to the total human renal disorder. Endemic Balkan Nephropathy prevalent in Danube river basin in Yugoslavia, Bulgaria and Rumania (Betina, 1984). The kidney is unquestionably a primary site of citrinin (Phillips *et al.,* 1979) and in addition it impairs liver metabolism (Ramdoss and Shanmugasundaram, 1973).

Stark (1980) reported citrinin as mutagenic mycotoxin. Besides renal damage citrinin causes acetylcholine or pilocarpine like responses including vasodilation and contraction of bronchi (Newborne, 1987). Endo and Kuroda (1976) have shown inhibitory effect of citrinin on cholesterol and triglyceride biosynthesis in rat liver.

Out of several enzymes acetoacetyl CoA, thialase and 3-hydroxy-3-methylglutaryl CoA reductase were specifically inhibited. *In vitro* inhibition of bovine liver glutamate dehydrogenase by citrinin was reported by Ramadoss and Mukherjee (1972). Creppy *et al.* (1990) reported the inhibition of DNA synthesis by citrinin. Arai and Hiblno (1983) reported citrinin to cause renal tumors in rats. In chicken embryos an additive effect on embryo toxicity (Vesela *et al.,* 1983). In beagle dogs given combined doses of ochratoxin A and citrinin (0.1-0.2 mg/kg) synergism was observed in terms of severity of clinical disease and mortality (Kitchen *et al.,* 1977a).

In neonatal rats exposed peritonially to a single dose of ochratoxin A at 1mg/kg body weight and of citrinin at 25 mg/kg body wt. or to both doses with in 24 hours of birth, a synergistic effect of the two mycotoxins was observed on cytochrome-P450, NADPH-dependent dehydrogenase and NADP-cytochrome reductase (Siraj *et al.*, 1981).

Citrinin caused permeability changes in fungal and plant membranes. It caused leakage of riboflavin from growing mycelium of *Eremothecium ashbyi* (Betina and Barthova, 1968) and of anthocyanins from *root* slices of *Beta vulgaris.* (Betina and Rockova, 1972). It was also found to inhibit leucine uptake and its incorporation into protein by cucumber cotyledon discs (white and Truelove, 1972). Smith and Moss (1985) reported the variable levels of mutagenicity. Complete information on the chemical and physical data and biological effects *of* citrinin is available (Anonymous, 1986).

Gupta and Pandey (1986) studied the effect of citrinin on some hormones and enzymes and substrates of EMP in mice. Cortisol, tricodotyrosine and tyrosine levels of serum increased but ATPase, hexokinase, LADH, LAPA decre-ased in blood, liver, kidney and brain. Insulin decreased glucose in blood, kidney and brain increased, while glycogen content of liver decreased.

Hanika *et al.* (1986) recorded and reported ultrastructural alteratlons in rabit. The proximal tubular alterations were brush border, disruption, cytoplasmic retention and swelling of interdigitating process. At higher dose mitochondria condensed and distorted medullary and straight cortical distal tubules had marked distinction of the intercellular spaces and disorganisation. Leukocyte infiltration was associated with damaged tubules indicating a contribution of inflammation to the development of lesions.

Smith and Moss (1985) concluded that citrinin is a potent toxin inducing enlargement and tubular necrosis of the kidney Beasely *et al.* (1986) reported that blood urea, nitrogen, creatine in blood levels were lower in rats receiving high levels of citrinin. Low levels of citrinin induced diuresis and water consumption.

Hantow and Widerman (1985) reported that citrinin acts directly on kidneys to alter several tubular transport processes and that the effect of citrinin is irreversible in single comb white leghorn pullers. Citrinin exhibited low cytotoxicity to cultured renal cells. Madin-Darby bovine kidney (MDBK) were more sensitive than primary foetal bovine kidney (PFBK) cells. Cells got elongated, flattened, swelled and rounded (Yoneyama *et al.,* 1986). Carlton and Tuite (1986) studied comparative pathological changes in ochratoxicosis and citrinin toxicosis.

Kirby *et al.* (1987) studied citrinin toxicity in young chicks and observed the increased water intake and urine excretion. Glatin and Widerman (1987) studied the effect of citrinin in chicks and reported that citrinin caused acute dose related increase in urine flow, free water clearance, and fractional Na excretion and dose related decrease in urine osmolality. Citrinin had no direct effect on grandular filtration rates, fractions of K excretion or fractional inorganic phosphate excretion. Citrinin showed the inhibition of growth of *Sclerotina minor* and other pathogens (Melonk and Akem, 1987).

Citrinin affects cellular metabolism leading to cytotoxicity in renal cells (Yoneyama and Sharma, 1987). Haraguchi *et al.* (1987) have suggested that citrinin affects the mitochondrial electron transport system, co enzymes Q during its fungicidal action. Haraguchi *et al.* (1987) explained the mechanism of antifungal action of citrinin. Reddy *et al.* (1988) suggested that citrinin mildly stimulates the immune system but does not have constant immunotoxic effects.

Glahn *et al.* (1988) studied the effect of ochratoxin A alone and in combination with citrinin on kidney function of single comb white leghorn pullet. Citrinin infusion caused increased urine flow rate/GFR/glomeruli filtration rate/free water clearance (FWC), Na excretion and decreased osmolality. It is concluded that citrinin and ochratoxin A do not have additive diuretic effects during first 48 hours of toxin exposure. Brakhage *et al.* (1988) reported the induction of base substitution mutations by citrinin. Chandrashekar *et al.* (1990) reported the effect of citrinin on behaviour of cockroach.

Abdelhamid *et al.* (1990) recorded citrinin residues upto 10 μg in fresh muscle and egg yolk and 6 μg in egg albumin of Mamonra chicken receiving 100 μg/kg. Griffiths and Done (1991) reported citrinin as a possible cause of the praritis, pyrexia, haemorrhagic syndrome in cattle.

8.2.8. Management

The main entry of citrinin into the human and animal food chain is, from agricultural products. From a practical stand point, the best means of restricting citrinin contamination is by prevention, excluding or reducing toxigenic mould growth in the raw and processed materials. (a) Reducing fungal infestation in growing crops, (b) Rapid drying and correct storage of harvested crops, (c) Removal of contaminated material by mechanical methods and (d) Detoxification by physical, chemical or biological techniques are some of other methods to contain citrinin contamination of agricultural products. Pravindra Chary and Reddy (1986) recommended bavistin and brassicol in the control of citrinin production by *P. citrinum*.

Rytych and Zyska (1977) recommended the use of large number of volatile compounds to check the mould growth. Normann (1968) feels that volatile compounds applied will sublimate and form a protective atmosphere for combating the growth of microorganisms. Chary and Reddy (1985) suggested the use of formic acid, acetic anhydride, propionic acid, isoamyl alcohol, aniline and secondary butanol in controlling the growth and citrinin production by *P. citrinum*. About 99.5% citrinin degradation by using sodium hypochlorite was reported (Castegnaro *et al.,* 1991). Degradation of citrinin using ammonia is recommended in the solid substrates, aqueous solution in volatile organic solvents, solutions inorganic solvents misible with water (Castegnaro *et al.,* 1991). Destruction of citrinin was affected by treatment with ammonia at 100°C in a tightly closed metal container such as a bomb calorimeter.

In view of systemic activity and non-phytotoxic nature use of plant extracts have been advocated as chemotherapeutants. Chary and Reddy

(1986) reported that flower extracts of *Punica granatum* and leaf extracts of *Lawsonia inermis*, *Sapindus emarginatus*, *Eucalyptus* sp., *Casurina equesetifolia*, *Allium cepa*, *Artabotrys uncinatus*, *Ricinus communis* and *Hibiscus cannabinus* could effectively check the production of citrinin by *P. citrinum*. Further, Chary and Reddy (1986) reported that latex of *Vallaris solanacea*, *Tabernaemontana divaricata* and *Ficus bengalensis* checked the citrinin production by *P. citrinum*.

Camonite, cinnamon and potassium benzoate prevented the formation of citrinin in maize and reduced fungal population (Abdelhamid *et al.*, 1985). Kirby *et al.* (1987) reported that heating of corn at 76°C for 20 min caused loss in toxicity of citrinin but heating to 105°C for 16 min. destroyed the diuretic effect of citrinin. Some of the food preservatives such as benzoic acid, potassium metabisulphate were also effective inhibitors of citrinin elaboration (Pravindra Chary and Reddy, 1986b). *A. niger* and *T. viride* are reported to completely suppress the citrinin production by *P. citrinum* (Pravindra Chary and Reddy 1986) and may be exploited in the biological control of citrinin contamination.

CHAPTER - 9

CYCLOPIAZONIC ACID

9.1. INTRODUCTION

Holzapfel (1968) was first to isolate a toxic tetramic acid named cyclopiazonic acid (CPA) from cultures of *Penicillium cyclopium* infecting groundnut seeds. Holzapfel *et al.* (1970) subsequently reported two new relatively non-toxic indole derivates namely, cyclopiazonic acid- imine and bisdehydrocyclopiazonic acid of CPA. Betina (1984) gave detailed account of incidence and biological activity of this mycotoxin. Ostry and Polster (1989) have excellently reviewed the properties of CPA and its producers in foods.

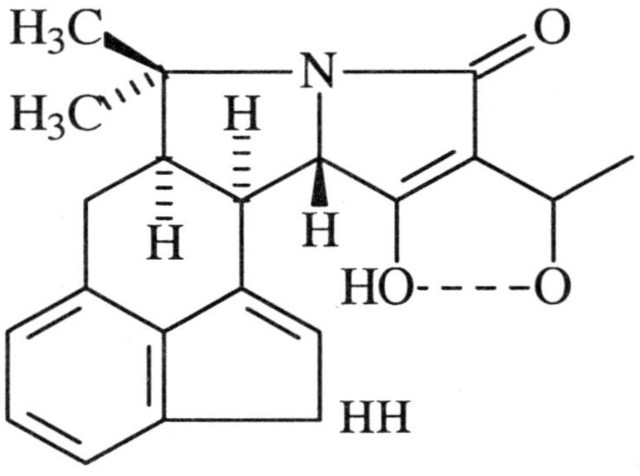

Fig. 9.1. Structure of cylcopiozonic acid

9.2. PHYSICO-CHEMICAL CHARACTERISTICS

Cyclopiazonic acid ($C_2OH_2ON_2O_3$) is an optically active, colourless, crystalline metabolite and is characterised by indole and tetramic acid moities (Fig. 9.1). Crystallization of pure cyclopiazonic acid (CPA) from methanol solution gives needle like crystals. It is an optically active colourless crystalline compound that is soluble in chloroform, dichloromethane, methanol, acetonitrile and sodium bicarbonate. Melting point is 245-246°C. UV spectrum shows max. CHOH 225, 253: 275 (sh) 284. 292 (sh) nm, lR analysis reveals major absorptions at max. CHCl 3478, 3200-2600, 1708 and 1618 cm

1. High mass spectral analysis shows a molecular ion at mlz 336.1463 and shows fragment ions at mlz 196.1123, 182.0808, 181.075, 155.07027 and 154.0651. It reacts with 0.1 N sulphuric acid in methanol gives a blue violet Ehlrich colour reaction and an orange red reaction with ferric chloride.

9.3. ORGANISMS

Intensive search for moulds producing CPA subsequent to its isolation by Holzapfel (1968), revealed that species of both *Penicillium* and *Aspergillus* are potential producers of this mycotoxin. Among species of *Penicillium*, *P. patulum, P. viridicatum, P. cyclopium, P. curstosum* and *P. puberulum* (Leistner and Pitt, 1977), *P. camembert;* (Le Bars, 1979), *P. griseofulvum* (Hermansen *et al.,* 1984) and *P. verruculosum* var. *cyclopium* (Malik *et al.,* 1986), while among species of *Aspergillus*, A. *versicolor* (Ohonomo *et al.,* 1973), A. *flavus* (Luk *et al.,* 1977) and A. *oryzae* (Orth, 1977) are reported to produce CPA. Hermansen *et al.* (1984) screened 62 isolates of *Penicillium* and *Aspergillus* and have reported *Penicillium camemberti* group II, *P. griseofulvum* and *Aspergillus flavus* to produce CPA. Trucksers *et al.* (1987) have isolated CPA from *A. flavus, P. camemberti* which inturn were isolated from dried beans, corn meal, macaroni and pecans.

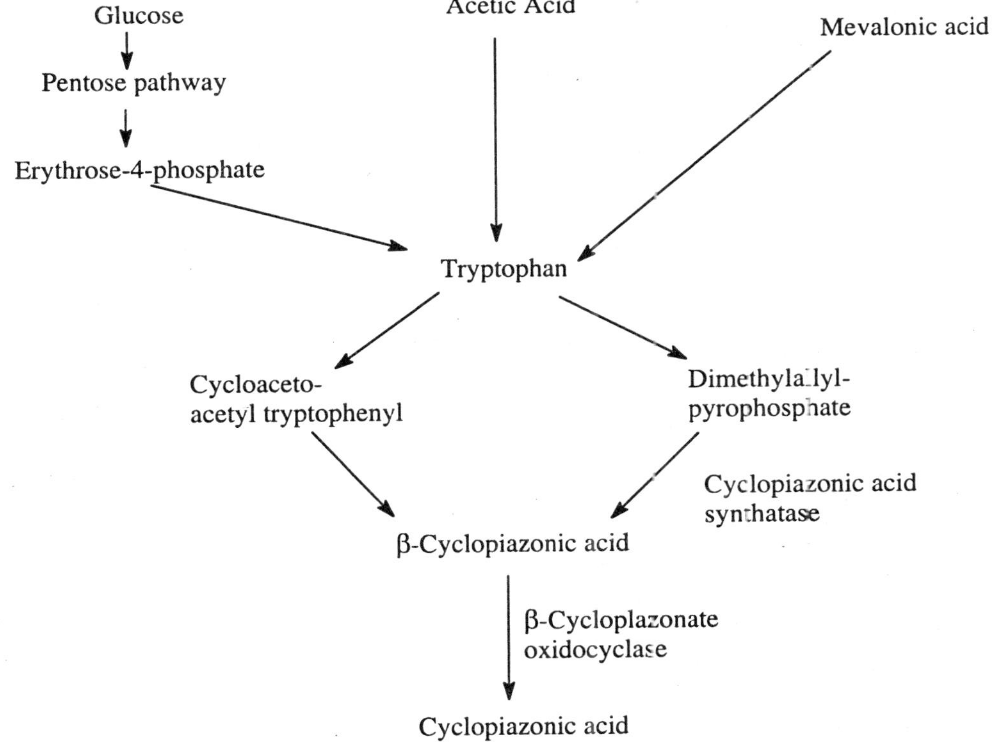

Fig. 9.2. Biosynthesis of Cyclopiazonic acid

9.4. BIOSYNTHESIS

The carbon skeleton of CPA probably is derived either from tryptophan, a C_5 unit formed from mevalonic acid and two molecules of acetic acid or from tryptophan and two C_5 units formed from Mevalonic acid. Mc Grath *et al.* (1976) proved following double labelling experiments that all the carbon atoms of the tryptophan were incorporated directly into cyclopiazonic acid. α-CPA is derived either from acetate or from mevalonic acid by chemical degradation.

β-CPA may be a precursor for the CPA synthesis (Holzapfel and Wilkins, 1971). β-CPA was already present in the mycelium and it falls rapidly when formed. This was proved using ^{14}C acetate labelled β-CPA. The enzyme responsible for the conversion of β-CPA into α-CPA was subsequently isolated by Schabort *et al.* (1971). Fig. 9.2 explains the process of CPA biosynthesis.

9.5. ISOLATION AND DETECTION

Analytical methodology for the determination of cyclopiazonic acid in agricultural products has received considerable attention in the past few years. The following methods have been suggested for extraction and estimation of CPA in contaminated commodities :

1. Thin layer chromatographic methods (Gallagher and Wilson, 1978; Le Bars, 1979; Gorst-Allman and Steyn, 1979; Lansden and Davidson, 1983; Lansden, 1984)
2. Liquid chromatographic method (Lansden, 1984)
3. Densitometric quantitation (Ohmono *et al.,* 1973)
4. 2-dimensional development on specially pretreated TLC plates (Popken and Dose, 1983)
5. Colorimetric method (Rathinavelu and Shanmugasundaram, 1984; Rao and Husain, 1987).
6. Purification of CPA by Liquid Chromatography (Peterson *et al.*, 1989).
7. Improved spectrophotometric determination (Chang-Yen and Bidasee, 1990).

Solvents used for extraction of cyclopiazonic acid include butanol, chloroform, chloroform-methanol and dichloromethane. Most of the TLC methodology relies on the reaction of CPA with Ehrlich reagent (2, 4 dimethyl amino benzaldehyde) in acidic conditions.

Cyclopiazonic acid, like other tetramic acids, readily forms a stable complex with iron, copper and other metal cations. Thus, the ability to form metal complex provides presumptive evidence for the presence of a tetramic acid moiety and may be useful for purification purposes.

Table 9.1 Methods for the Determination of Cyclopiazonic Acid

Food	Extraction	Cleanup	Quantitation	Detection (μg/kg)	Ref.
Corn	Shake 50 g ground sample 30 min with chloroform/methanol (4 ml/g)	Liquid-liquid partition and silica gel colum chromatography	Spectrophotometry Detection: Measure absorbance at 580 nm Derivatization: Mix final extract in acetone with ρ-dimethylamino-benzaldehyde (4 g in 100 ml 5.5 M HCl)	80	Chang-gen and Bidasse (1990)
Corn, peanuts	Blend ground sample 3 min with 70:30 methanol/2% $NaHCO_3$	Liquid-liquid partition and silica SPE cartridge	Gradient HPLC Column: Reverse-phase C_{18} (5 μm, 15 m x 3.9 mm i.d.) Mobile phase: (a) 85:15 Methanol/ water (b) 85:15 Methanol/water with 4 mM $ZnSO_4.7H_2O$ Gradient: 100% (a) to 100% (b) in 10 min; 100% (b) for 10 min Detection: UV (279 nm)	50-100	Urano *et al.* (1992)
Corn, peanuts	Defat 50 g ground sample; dry; shake 30 min with 250 ml 80:20 chloroform/methanol with 0.5 ml phosphoric acid	Liquid-liquid partition	TLC Plate: Silica gel 60 F-254 HPTLC Solvent: 50:15:10 Ethyl acetate/2-propanol/ ammonium hydroxide Detection: Reflection densitometry at 540 nm Derivatization: Spray with 1% ρ-dimethylam-inobenzaldehyde 75:25 ethanol/HCl	125	Lansden (1986)
Meat	Blend ground sample 3 min with 80:20 chloro-form/methanol plus 0.5 ml 6 N HCL (4 ml/g)	Liquid-liquid partition and silica gel mini-column chro-	HPLC Column: Reverse-phase C_{18} (25 cm x 4.6 mm i.d.)		Norred *et al.* (1987)

		matography	Mobile phase: 40:30: 20:10 Acetonitrile/isopropyl alcohol/water solution of 10% ammonium acetate, 0.25% 4-dodec-ylidiethylenetriamine, and 0.01 M zinc acetate Detection: UV (284 nm)		
Milk and eggs	Shake defatted samples 30 min with 80:20:0.25 chloroform/ methanol/5 N HCL	Liquid-liquid partition	TLC Plate: Silica gel 60 F-254 HPTLC Solvent: 85:15:10 Ethyl acetate/methanol/ammo nium hydroxide Detection: Bluish-purple spot in visible light at R_f 0.36 Derivatization: Spray with 1% ρ-dimethylam-inobenzaldehyde in ethanol followed by 50% ethanolic sulfuric acid	5	Dorner *et al.* (1994)

The cyclopiazonic acid was detected directly under UV light or after spraying with concentrated sulphuric acid, Ehrlich reagent or 1% ethanoloic iron chloride and heating at 110°C for 10 min. Cyclopiazonic acid turns violet-red on prolonged standing on the silica gel plates impregnated with oxalic acid shows a dark spot under UV light and a red-brown spot when sprayed with either sulphuric acid or iron chloride solution. The best spray for the detection of cyclopiazonic acid and related indoles is Ehrlich's reagent. Different methods adapted for determination of CPA in different foods are precised in Table 9.1.

9.6. NATURAL INCIDENCE

The economic significance of CPA as a potentially important food contaminant was enhanced considerably when it was found to occur naturally in corn samples from South Eastern USA. It was also found that a large number of isolates of *A.flavus* isolated from various agricultural commodities (Luk *et al.,* 1977; Gallagher *et al.,* 1978; Kirksey and Cole, 1973) could produce this toxin. Still *et al.* (1978) reported the production of CPA by the isolate of *P. camemberti* which was isolated from meat. Le Bars (1979) reported that all 20 strains of *P. camemberti* freshly isolated from 20 commercial cheese brands produced cyclopiazonic acid.

Gallagher *et al* (1978) reported that 28 of 54 isolates of *Aspergillus flavus* produced CPA when grown on different agricultural commodities. A preliminary screening of some aflatoxins samples for the first time revealed the dual occurrence of CPA in agricultural commodities.

Lansden and Davidson (1983) have reported natural contamination of peanuts with CPA. The various foods and feeds from which CPA producing fungi have been isolated include cheese, corn, guinea pigs ration, turkey ration mixed feeds, walnuts and pea nuts. The association of *P. aurantio-griseum* with the Kasan cheese at the stage of consumption produced CPA (Aran and Eke, 1987). Rao and Husain (1987) have reported the natural contamination of CPA in kodo millet causing "Kodua" poisoning in man. Norred *et al.* (1988) reported the occurrence of CPA in meat meal for chickens.

9.7. FACTORS INFLUENCING CPA PRODUCTION

The best yield of CPA was obtained by growing CPA producing strains on Czapek's medium for 7 days at 25°C under shake cultures (150 rpm) condition. Large scale production was carried out by using maize meal (Corn meal) inoculated with most toxic isolate of *P. cyclopium* (CSIR 1082). The strain of *A. versicolor* has the ability to produce indole derivatives in submerged and surface cultures at 30°C for 30 days. Yield obtained was 63 g/lit. Our studies on *P. griseofulvum* showed that 16 days incubation on glucose-asparagine medium is most suitable for production of CPA (Reddy and Reddy, 1987a).

Luk *et al.* (1977) have demonstrated ten fold increase in CPA production by A. *flavus* strain if grown for 10 days at 30°C on rotary shaker as compared to the static rice technique. The optimum temperature for CPA production in malt extract agar was 25°C (Still *et al.,* 1978). Freshly isolated 20 strains of *P. camemberti* from 20 commercial cheese brands (Le Bars, 1979) produced CPA in two different media at 4°C and 25°C. Reddy and Reddy (1987a) have demonstrated the production of copious amounts of CPA at 25°C and 7.5 pH by *P. griseofulvum.*

CPA was found in every strain of fungus grown under every culture condition, but the amount of toxin was found to vary depending on strain, medium, incubation period, temperature and time (Betina, 1984). Malik *et al.* (1986) reported maximum CPA production in a medium containing 2% yeast extract and 2.5% sucrose. They further reported that curdled milk stimulated the production of CPA by *P. verruculosum* var. *cyclopium. P. griseofulvum* produced maximum amount of CPA when glycerol and L-aspargine was served as carbon and nitrogen source respectively (Reddy and Reddy, 1988).

2,4-D is reported to stimulate the CPA production by *P. griseofulvum* (Reddy and Reddy, 1988).

9.8. BIOLOGICAL ACTIVITY

Though CPA is most frequently associated with several agricultural commodities, studies on this toxin are limited. Most of the studies were limited to toxicity on laboratory animals. This has been reported to produce liver necrosis and carcinogenic effects in male rats Frank *et al.,* 1977 and Orth 1977). Riley and Gorger (1992) feel that CPA is a very specific inhibitor of key enzymes in animals cells. It is a lipophilic and highly charged molecule that can cause alterations in electrical properties of animal cell membrane, has antioxidant property and very good iron chelator. CPA can co-occur with aflatoxins (Takashi *et al.*, 1992) and may enhance the over all toxic effect when this happen (Cole, 1986).

Purchase (1971) studied the toxicity of CPA on rats. Intraperitoneal injection produced hyperaestheria and convulsions followed by death in about 30 minutes. The LD_{50} in males were 2.3 mg/kg. Oral administration resulted in death within 1 to 5 days. The LD_{50} oral was 36 mg/kg in males and 63 mg/kg in females. Voss *et al.* (1990) reported subchronic oral toxicity of CPA in male Spragne-Dawley rats. The CPA is a specific inhibitor of calcium dependent ATPase in the Sarcoplasmic reticulum that results in altered cellular Ca levels (Duran *et al.*, 2007).

The lesions produced in rats included degenerative changes and necrosis in the liver, spleen, exocrine and muscles, pancreas, kidney, salivary glands, myocardium and skeletal muscles. The unusual lesions seen in various ducts was characterised by enlargement of nucleus with margination of chromatin. At higher dosage levels, CPA induced focal necrosis in most of the organs, whereas at lower dosage levels ducts or organs were affected. Testosteron secretion has also been found to be slightly affected by CPA (Fink-Gremmels, 1990).

Dorner *et al.* (1983) have reported that the broiler chickens receiving feed with 100 ppm of CPA for 7 weeks had high mortality, decreased weight gain and poor feed conversion when compared to birds receiving 10 and 50 ppm. Their post-mortem examinations revealed that the chickens which were given higher dose of CPA, had proventricular lesions characterised by mucosal erosion and hyperemia. The birds also produced numerous yellow foci in their liver and spleen. Histological lesions along with necrosis and haemorrhage, hyperplasia of the mucosa of proventriculus and hepatocellular vacuolisation, skeleton muscle degeneration, increase of creatin kinase activity was observed by Culler *et al.* (1988). Porter *et al.* (1988) have reported the neurochemical effects of CPA in chickens. Kumar and Balachandran (2005) recorded a synergistic effect on weight loss of chick receiving aflatoxin and cyclopiazonic acid. Giridhar *et al.*, 2000 reported the acute toxicity of cyclopiazonic acid as it caused significant changes in haematology and biochemical composition in vital organs of chick.

There is a dearth of human exposure data. However, 'Kodua' posiouing in India resulting from the ingestion of contaminated millet seeds has been linked to this toxin. It has similar pharmacological properties to the antipsycholic drugs chloropromazine reserpine in mice and rabiits.

Cole *et al.* (1988) explained the CPA toxicity in the lactating ewe and laying hens. Miller *et al.* (1990) studied *in vitro* effects of CPA on Turkey papillary muscles. Moririy *et al.* (1984) and Van Rensberg (1984) have reported CPA to be mutagenic. Nishic *et al.* (1987) have also reported the toxicity of CPA on pregnancy of CF mice.

CPA which induced variety of chromosomal aberrations like disturbed spindle plate, extreme fragmentation, and bridge formation at metaphase and bridges at anaphase CPA was illustrated to be mitotic inhibitor by Reddy *et al.* (1989). Enlargement of nucleus with extreme vacuolisation was also a common mitotic aberration following CPA treatment.

Cole (1986) feels that some of the clinical sysptoms in turkey X disease may be due to CPA which is a frequent product of *Aspergillus flavus*. Hence, he feels that in view of wide spread contaminant of large number of agricultural commodities, studies on the chronic toxicity of CPA to vertebrate animals are urgently needed. Meanwhile, an additional knowledge on cyclopiazonic acid toxicity and natural occurrence in our country is urgently needed.

9.9. MANAGEMENT

Le Bars (1979) while studying the production of CPA by all the 20 strains of *P. camembert* isolated CPA from cheese, advocated the choice of strains and the observance of correct storage temperature compatable with manufacturing technology as a means of prevention of infection by CPA producing organisms. Reports of isolation of economically important fungi capable of producing CPA from a variety of commodities from different parts of the worldare available. In conjunction with the reports of the presence of this mycotoxin in cheese and corn, place CPA in the category of a potentially serious mycotoxin (Betina, 1984). Duran *et al.* (2007) recorded that a mutant of *A. flavus* (VeA) showed decreased production of CPA.

Hence, there is a need to prevent the contamination of CPA at different levels such as at production, storage and consumption by following common preventive strategies:

1. Agronomic practices.
2. Development of post harvest practices.
3. Use of chemicals to minimize fungal contamination. Education and extension activities.
4. Enactment and implementation of legislature concerning mycotoxins. Economic strategy for CPA prevention.
5. Continuous surveillance for CPA.
6. Development and popularisation of analytical methods for detection of CPA

CHAPTER – 10

TREMORGENIC MYCOTOXINS

10.1. INTRODUCTION

The implication of fungal tremorgens in animal diseases is becoming more obvious with the emergence of new information. Since the first report of a tremorgenic toxin produced by *Aspergillus flavus* by Wilson and Wilson (1964), this group of fungal toxins have been implicated in the etiology of some mysterious animal diseases. Most of the tremorgens have been implicated in neural disease syndromes, such as rye grass staggers, bermuda grass staggers and marsh staggers in Newzealand, U.S.A. and Britain. Fungi capable of producing tremorgenic metabolites represent species from taxonomically diverse and unrelated groups. Atleast 20 tremorgenic mycotoxins are now recognised including the ergot tremorgens isolated from *Claviceps paspali* and others from species of *Penicillium* and *Aspergillus* commonly found in soil. Since tremorgens are more consistently present on natural forage, ruminants generally animals, most likely to ingest sufficient quantities to cause neurological disorders. Moreau (1990) has excellently reviewed various aspects of these mycotoxins.

10.2. PHYSICO-CHEMICAL CHARACTERISTICS

An early review grouped the tremorgens on the basis of the number of nitrogen atoms per molecule i.e. one nitrogen molecule (the penitrem type), three (the verruculogen-fumitrogen type) and four (the tryptoquivaline type). This classification reflects their biosynthetic origins. The tremorgens contain the basic biologically active moiety. This moiety is derived from geranylgeranol and tryptophan. Difference among various tremorgens is due to various added isoprenylations and subsequent modifications of these isoprenylations (Acklin *et al.,* 1977; Tanabe, 1977). Recently the tremorgenic mycotoxins are divided into four different chemical groups. The largest group consists of aflatrem, paxilline, paspalitrems, lolitrems, penitrems and janthitrems. The other 3 chemical groups include the verruculogen, fumitremorgen group (represented by verruculogen), the tryptoquivaline group (tryptoquivaline) and the territrem group (basic territrem structure). The later represents only group that is not derived trom tryptophan and, therefore, does not contain nitrogen. The physical and chemical properties of tremorgenic mycotoxins and their detection are precised in Table 10.1 and 10.2. respectively.

Table 10.1. Physico-chemical characteristics of tremorgenic toxins

Name of the toxin	Physical and Chemical properties	Reference
Penitrems	The chlorine atom present in penitrems A, C and F is replaced by hydrogen atom in B, D and E. Penitrems A, B, E and F contain the 23rd, 24-epo oxide has common structural feature whereas penitrem C and D contain 23, 24-double bond. Except penitrem d all are amorphous forms.	Vesonder *et al.* (1980)
Penitrem A	Penitrem A MI. Wt. 633.280, micro-crystals from ethylether, m.p. 237-239°C, soluble in ethylether, chloroform, acetone and methanol. Unstable in acid, Penitrem B, mol. Wt. 583.329, Fine needles, m.p. 185-195°C.	Steyn and Vleggar (1985)
Janthitrems	Janthitrem A (Mol. Wt. 601), Janthitrem B (Mol. Wt. 585), Highly fluorescent when irradiated with long wave uv light.	Mantle (1977)
Aflatrem	Colorless needles, m.p. 222-224°C, UV maxima in ethanol are at 231, 282 and 292 nm. IR : 3365, 2920, 2840, 1195 and 1120 cm	Wilson (1971)
Paxiline	Mol. Wt. 435.240, Colorless prism, m.p. at 252°C, at UV maxima (methanol) 230 and 281 nm. IR absorption at 1650 cm.	--
Paspaline	Mol. Wt. 421.298, Colorless crystalline compound, m.p. 264°C, UV maxima 228-231, 282 and 291 nm. IR (KBr) 2950, 1387, 1330 cm	Steyn and V leggar (1985)
Paspalicine	Mol. Wt. 417.230, Colorless crystalline compound, m.p. 230°C, UV maxima, 231, 250 and 275 nm. IR (Chloroform 3450, 2930, 1670, 1440 cm), Paspaline is a biosynthetic precursor of paspalicine.	Steyn and V leggar (1985)
Paspalinine	Mol. Wt. 433.225, UV maxima 232, 250 and 274 nm	Steyn and Uleggar (1985)
Paspalitrem A	Mol. Wt. 501.287 UV maxima (methanol), 234 and 278 nm.	Cole and Cox (1981)
Fumitremorgins	Fumitremorgin-A Mol. Wt. 579.294, m.p. 206-209°C soluble in chloroform and ethylacetate. Unstable to prolonged exposure to light. Fumitremorgin-B	

	Mol. Wt. 479.242, colorless needless, m.p. 211-212°C, UV and IR spectra of both the toxins are similar.	
Territrems	Do not contain nitrogen Territrem A – m.p. 288-290°C Territrem A – m.p. 200-203°C Territrem B – m.p. 222-224°C	Ling *et al.* (1984)
Verrucosidin	Lacks indole moiety, colourless plates, m.p. 90-91°C.	Burka *et al.* (1984)
Verruculotoxin	m.p. 152°C	MacMillan *et al.* (1976)
Verruculogen	Mol. Wt. 511.231, colourless plates, m.p. 233-235°C, soluble in benzene, ethylacetate and acetone, UV max (ethanol 0 226, 277 and 295 nm.	--

10.3. ORGANISMS

(a) *Aflatrem* is produced by *Aspergillus flavus.*

(b) *Paspaline, Paspalinine, Paspalicine* and *Paspalitrem A* are produced by *Claviceps paspali.*

(c) Paxilline is produced by *Penicillium paxilli.*

(d) Fumitremorgin A and B are isolated from strains of *Aspergillus fumigatus, A. caespitosus* and *Penicillium lonosum.*

(e) *Verruculogen* is produced by *Penicillium verruculosum, P. paxilli, P. paraherguei, P. piscarium* and *P. estinogenum, P. simplicissimum.*

(f) Tryptoquivaline and Tryptoquivalone are produced by *Aspergillus clavatus.*

(g) Penitrems A-F are produced by *Penicillium crustosum, P. aurantio-griseum, P. palitans, P.lanoso-coerulum, P. commune, P. cyclopium, P. granulatum, P. clavigerum, P. canes cens* and *P. janthinellium.*

(h) *Janthitrems* are produced by *P. janthinellium.*

(i) *Lolitrems* are produced by *Acremonium sp. Lolium perenne.*

10.4. ISOLATION OF AND QUANTIFICATION

10.4.1. Penitrems

Biosynthesis: The indole part of the tryptophan contributes to the aromatic part of the penitrem. NMR spectrum of acetate derived penitrem A reveals that six mevalonate units are involved in the formation of non-indole part of the compound (de Jesus *et al.,* 1983, a,b,c,). Steyan and Vleggaar (1985) proposed a biosynthetic pathway for penitrem A (Fig. 10.1.2).

Table 10.2. Chemical detection of tremorgenic toxin

Name of the toxin	TLC			Chemical detection									
	Absorbent	Solvent system	UV	$CeSO_4$ 1% in 6N H_2SO_4	2,4-DNP	Fe Cl_3 3% in ethanol	2% $FeCl_3$	Ehrlich Reagent	50% H_2SO_4 in ethanol	3% phosphomolybedic acid in ethanol	Van Urk Reagent	20% ethanolic $Alcl_3$	RF
Aflatrem	Silica gel G-HR	Chloroform-acetone (93:7)	Dark Area	--	--	--	--	--	Blue spot	Orchid to violet	--	--	0.65
Fumitremorgin A	Silica gel G-HR	Chloroform-acetone (93:7)	Light purple	Yellow brown	Light orange	Orange	--	--	Gray blue (v) mustard (UV)	--	--	--	0.3
B	Silica-gel 60-HR	Diethyl ether	--	--	--	--	--	--	--	--	--	Blue fluroscent	0.67
Janthitrem A, B, C	Silica-gel 60-F	Toluene-ethylacetate acetone (3:2:1)	Purple blue	--	--	--	--	Gray green	--	--	--	--	A-0.61 B-0.54 C-0.74
Paspaline	Silica-gel G-HR	Toluene-ethylacetate acetone (5:4:1)	--	--	---	--	--	--	Dull fluroscent	--	Pale green	--	0.76
Paspalicine	Silica-gel G-HR	Toluene ethylacetate formic acid (5:4:1)	--	--	--	--	--	--	Dark spot (UV)	--	--	--	0.85
Paspa-linine	Silica-gel G-HR	Chloroform Acetone (93:7)	--	--	--	--	--	--	Blue spot	--	--	--	0.57
Paspa-litrem A	Silica-gel G-HR	Chloroform Acetone (93:7)	--	--	--	--	--	--	Blue spot	--	--	--	0.60

B	Silica-gel G-HR	Chloroform Acetone (93:7)	--	--	--	--	--	--	Blue green	--	--	--	0.20
Penitrem A	Silica-gel G	Chloroform Methanol (95:5)	--	--	Purple	Grey	Green	--	Blue spot	--	--	--	0.40
B	Silica-gel G	Chloroform acetone (93:7)	--	--	Purple	Grey	Green	--	Blue spot	--	--	--	0.60
Verru-culogen A	Silica gel G-HR	Chloroform acetone (93:7)	Light purple	Yellow Brown	Light orange	Orange	--	--	Gray blue (V) Mustard (UV)	--	--	--	0.30

Production: Penitrems can be produced both on corn and liquid medium. These are intracellular toxins. Malaiyandi *et al.* (1976) developed a large scale production method using Czapek Dox medium. Penitrem production was better with yeast extract-sucrose medium (Wagener *et al.*, 1980). Mass *et al.* (1981) used Czapek medium enriched with yeast extract for the production of the toxin. Richards medium with D-xylose as supplement carbon source enriched the production of penitrem B. Similarly potassium nitrate was the best nitrogen source for the maximum production of penitrem B (Surekha and Reddy, 1992). Penitrem B production was maximum at 25°C and pH 5.5 (Surekha and Reddy, 1991).

Isolation and detection

Ten day old stationary cultures were filtered and the mycelium macerated with acetone. The aqueous acetone solution was concentrated 'and partitioned between dichloromethane and water. The residue from the organic layer was partitioned between 90% methanol and hexane. The crude extract thus obtained was percolated through silica gel column with benzene: acetone (4: 1). The mixture of penitrems was purified by chromatography on silica gel using benzene-acetone (85:15) to yield six penitrems in a order of decending Rf values (Betina, 1984). Hou *et al.* (1971) suggested a method for the quantitative estimation of penitrems. Different chemical determination of tremorogenic mycotoxin.

10.4.2. Janthitrems

Biosynthesis: Although no data has been published on the biogenesis of janthitrems, the diterpenoid part of the molecules must be biosynthetically very closely related to penitrems (Steyn and Vleggaar, 1985). Wilkins *et al.* (1992) elucidated the structure of Janthitrem B produced by *Penicillium janthinellum.*

Production: The strains of *P. janthinellum* can be cultivated in different media (Gallagher *et al.*, 1980b)

(1) Potato broth containing dehydrated mashed potato, skim milk powder and sucrose for 16-18d at 25°C.

(2) Potato -dextrose- agar- tryptophan mixture for 12-15 days at 25°C.

(3) de Jesus *et al.* (1984) used modified Czapek medium for 5 days at 25°C and for 15 days at 18°C and isolated janthitrems E, F and G.

Isolation and detection: Ethanol or chloroform extracts of cultures were evaporated to dryness and subjected to preparative TLC on silica gel repeatedly which results in the separation of janthitrems (A, B and C) as pale-yellow coloured gums (Gallagher *et al.,* 1980a). Janthitrems E, F and G were isolated as colourless amorphous solids. Lauren and Gallagher (1982) developed a reversed phase HPLC to analyse janthitrems

10.4.3. Aflatrem

Biosynthesis: The diterpenoid part of the aflatrem is biosynthetically closely related to penitrem, but the biosynthesis of aflatrem involves an additional isoprenylation at C- 4 of the indole group (Steyn and Vleggaar, 1985).

Production: The toxin is intracellular. Strains can be grown on rice, corn, millets and potatoes. Large amount of toxin can be produced by using the medium containing defatted milk, dehydrated, mashed potatoes, sucrose and agar. Fourteen days incubation was optimum (Wilson and Wilson, 1964).

Isolation and detection: Extraction was carried out with diethylether and column chromatography (silica gel and florosil) to remove polar impurities. Toxin was eluted with diethylether-cyclohexane. Microneedle crystals were obtained from the toxin containing fraction. Crystaline aflatrem can be obtained by recrystalization with ethylacetate.

10.4.4. Paxilline

Biosynthesis: The toxin contains an indole nucleus linked to a diterpenoid unit. Tanabe (1977) proposed a biosynthetic scheme for this toxin.

Production: Paxilline can be produced on shredded wheat difco mycological broth and yeast extract-sucrose medium, fifteen days incubation at 25°C was optimum (Cole *et al.,* 1974a).

Isolation and detection: Fungal culture was homogenized with hot choloform and separated by column chromatography on silica gel. The metabolite was recrystalized from acetone. Cockrum *et al.* (1979) extracted with acetone and purified the toxin by chromatography (Florisil) and crystalized from diethylether.

10.4.5. Paspaline, Paspalicine and Paspalinine

Biosynthesis: Acklin *et al.* (1977) studied the biosynthesis of paspaline and related compounds. These metabolites originated through a combination of indole diterpene units.

Production: The production medium consists of sorbitol, succinic acid, potassium dihydrogen phosphate, magnesium sulphate, iron (II) sulphate and zinc sulphate. Ten days incubation at 24°C was optimum. A large scale fermentation was carried out by Kobel *et al.* (1964).

Isolation and detection: The dried mycelium of *C. paspali* was extracted with pentane, diethylether and chloroform. Paspaline is obtained by precipitating pentane fraction with methanol and crystalising it. The mother liquors were combined with diethylether and chloroform extracts and chromatographed on Kieselgel in chloroform. The richest fraction was re-chromatographed to get paspalicine (Fehr and Acklin, 1966). Gallagher *et al.* (1980a) obtained paspaline from methanol after repeated chromatography of the butyleacetate extract from dried mycelium on silica gel.

10.4.6. Paspalitrems A and B

Biosynthesis: The biosynthesis of paspalitrems involves isoprenylation at C-5 of the indole moiety (Steyn and Vleggaar, 1985).

Production, Isolation and detection: Sclerotia of *Claviceps paspali* were extracted with chloroform and evaporated to dryness and partitioned between cyclohexane and 80% aqueous methanol.

The biologically active methanol fraction was evaporated on silica gel column. Two fractions were further separated by column, TLC respectively. As a result three compounds paspalinine, paspalitrems A and B were isolated.

10.4.7. Territrems

Production, Isolation and detection: Cultures of *A. terreus* grown on rice for 14 days at 28°C yielded territrems (Ling *et al.,* 1995). Mouldy rice was extracted with chloroform and applied to a column consisting of silica gel with anhydrous sodium sulphate on the top. Territrems were eluted with a mixture of chloroform-acetone (4:1) and the fractions were monitored by TLC on silica gel using toluene-ethylacetate-formic acid (5:4:1) or benzene-ethylacetate (7:3). Territrems A and B can be separated by column chromatography on silica gel using benzeneethylacetate (5:35). Territrems B crystalizes from the chloroform in the form of colourless neddles (Ling *et al.,* 1979)

10.4.8 Fumitremorgin

Biosynthesis: Tryptophan is a precusor for fumitremorgins A and B and forms the dixopiperazine ring with proline (Yamazaki *et al.,* 1974).

Production, Isolation and detection: The production of fumitremorgins by *A. fumigatus* in a basal liquid medium depends on the addition of tryptophan (Yamazaki *et al.,* 1974). These toxins were isolated from ethylacetate extracts of the toxic strains by column chromatography on silica gel using benezene-acetone (4:1) as elute.

10.5. NATURAL INCIDENCE

Paspalin was isolated from infected *Paspalum dilatatum,* Paspalitrem A was first isolated from sclerotia of *Claviceps paspali.* Natural occurrence of the paspalinine was detected from livestock. Paspalum straggers was reported from United States, Newzeland, Australia, South Africa, Portugal and Italy (Mantle *et al.,* 1977). The tremorgenic disease syndrome resulting from the ingestion of toxic common coastal Bermuda grass by cattle was first reported in Oklahoma (Whitehair *et al.,* 1951). Rye grass stragger was reported from United States, Australia, England and Newzeland (Gallagher *et al.,* 1980). Fletcher and Harhey (1981) and Gallagher *et al.* (1980a) gave detailed report on the etiology of this disease. The fumitremorgins were isolated from a strain of *A. flavus* growing on rice (Yamazaki *et al.,* 1974). Verruculogen was isolated from groundnut (Cole *et al.,* 1972).

Penitrem A was reported from the contaminated feed, responsible for outbreaks of neurological disorder and deaths in sheep and horses (Wilson *et al.,* 1968). Penitrem has also been reported from cheese, walnuts and mouldy bread (Arp and Richard, 1979; Richard *et al.,* 1981; Hocking *et al.,* 1988). Cole *et al.* (1983) reported human intoxication resulted from the consumption of beer contaminated with *P. crustosum.* Recently Uzal *et al.* (1992) reported a tremorgenic syndrome in sheep in north east of Argentina. The syndrome was caused by tremorgenic mycotoxins produced by species of *Penicillium* on natural pastures. Natural incidence of these toxins in different foods and feeds from different parts of the world are précised in Table 10.3

10.6. STRUCTURE AND ACTIVITY RELATIONSHIP

Most of the tremorgenic toxins contain an indole moiety, in modified form as a structural feature (Steyn and Vleggaar, 1985). In penitrems, the tertiary hydroxyl group at C-22 is a prerequisite for its activity. The tertiary hydroxyl group at C- 11 is associated with the tremorgenic activity of the paspalinine. Paspalitrem A and B are the respective 3-methyl-2-butenyl and 3-hydroxy-3-methyl-1-butenyl derivatives of paspalinine. Both posses tremorgenic properties (Steyn and Vleggaar, 1985). Territrems A, B and C are all tremorgenic (Ling *et al.,* 1984). Modification in their A ring do not substantially affect their activity. However, modification in the E-ring of territrem B, results in the formation of an additional ring which greatly reduces the tremorgenic activity of territrem (Peng *et al.,* 1985)

10.7. BIOLOGICAL ACTIVITY

Aflatrem induces trembling, convulsion and even causes death. Purified aflatrem, when given orally or injected intraperitoneally in small quantity induced several behavioural changes in mice, including pronounced trembling.

Guinea pigs and rats were also susceptible to this toxin (Wilson and Wilson, 1964; Wilson, 1971; Gallagher and Wilson, 1978). Territrems exhibited acute toxicity and inhibited the growth of 1st instar larvae. They also inhibited cholinesterase activity (Dowd *et al.,* 1992).

Paspaline and Paspalicine are non-tremorgenic (Cole and Cox, 1981), where as paspalinine administered orally causes tremors in cockereles, mice and domestic animals (Gallagher *et al.,* 1980a). The LD_{50} values of both paspalitrems were lower than 14.0 mg kg^{-1} in mice. Clinical signs were tremors, diarrhoea, rough hair coat and hypersensitivity to sound stimuli. Gross clinical signs were similar to those produced by paxilline (Cole *et al.,* 1977). Paspalines and Paspalitrem A were responsible for the neurological disorder of the farm livestock in New Zealand. Paxilline induced death in cockerels at the high dosage of 10 mg per cockerel (Cole *et al.,* 1974).

Penitrems produce a unique neurotoxic syndrome in vertebrates (Maas *et al.,* 1981). In mice, penitrem A produces tremors at 250 ug kg-1 and its LD_{50} is 1.50 ms kg^{-1} (Cole and Cox, 1981). Various toxicological and pharmacological studies were carried out by Stern (1971), Hayes *et al.* (1977), Peterson *et al.* (1997), Mantle *et al.* (1978) and Sobotka *et al.* (1978). Hayes and Hood (1978) reported the toxicity of penitrems to dogs. Penitrem A was toxic to sheep and other animals (Cole and Dorner, 1985). Penitrem B was active against gram (+) positive bacteria (Surekha, 1990). Penitrem B was responsible for inhibition of mitosis. The mitotic index decreased with the increase in toxin concentration and exposure duration. At the higher concentration of toxin, cytokinesis inhibition and formation of binucleated to multinucleated cells was observed (Surekha, 1990). The inhibitory effect may be due to the blockage of DNA synthesis. PenitremA causes tremors, limb weakness, ataxia and convulsions (Maas *et al.,* 1981). Penitrem A acts by interfering with amino acid neurotransmitter release mechanisms (Norris *et al.,* 1980).

Verruculogin produces severe tremors when orally administered to mice or 1-day old cockerels (Fayos *et al.,* 1974). Verruculogen induced tremor in mice by loosing an inhibitory function of aminobutyric acid in central nervous system (Cole and Cox, 1981). The effective dose required to induce tremor in mice and cockerels is 0.39 mg and 0.33 mg per kg respectively (Fayos *et al.,* 1974). Fink Gremmels (1990) discussed the effect of verruculogen on neurotransmitter release mechanism in isolated neuroblastoma cells.

Verruculotoxin induced ataxia, prostration and complete lack of muscular coordination in one-day-old cockerels. Animals when administered with lethal doses showed mortality after 4-6 hrs (Cole *et al.,* 1975; Field *et al.,* 1978). Janthitrems injected intraperitoneally (200 mg in 0.1 ml propylene glycol) induced tremors and led to hypersensitivity to sound and touch in mice (Gallagher *et al.,* 1980). The LD50 values for territrems A, B and C in mice were 17.6, 9.06 and 6.28 mg per kg. respectively (Ling *et al.,* 1984).

Fumitremorgins A and B cause perceptible tremors in mice when administered intraperitoneally at 1 mg per mouse. Cole and Cox (1981) reported 70% mortality at 5 mg dose. The neurotoxic effects of this toxin was reported by Yamazaki *et al.* (1979, 1983).

10.8. PENITREMS

Wilson *et al.* (1968) for the first time reported the production of tremorgenic mycotoxin, penitrem A by *Penicillium cyclopium* which was responsible for the outbreak of neurological disorders in sheep and horses.

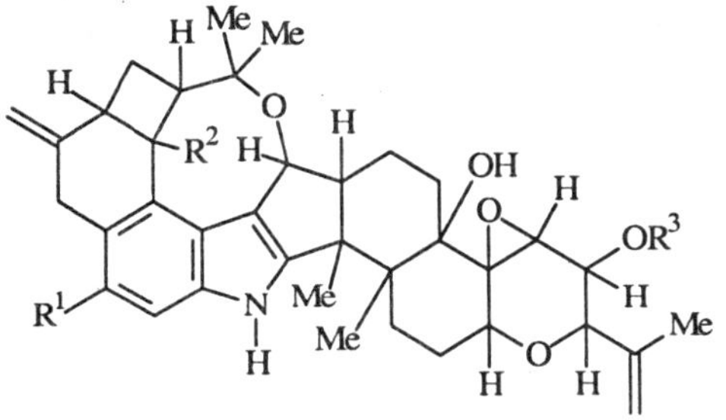

Penitrem A : R^1 = Cl, R^2 = OH, R^3 = H
Penitrem B : R^1 = H, R^2 = H, R^3 = H
Penitrem E : R^1 = H, R^2 = OH, R^3 = H
Penitrem F : R^1 = Cl, R^2 = H, R^3 = H
25-O-Acetyl Penitrem A : R^1 = Cl, R^2 = OH, R^3 = Ac

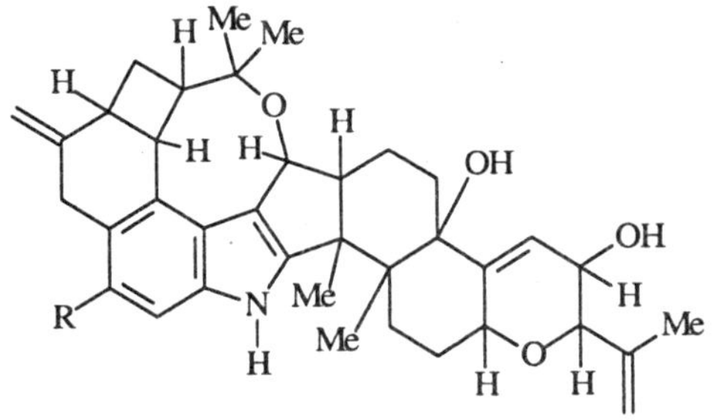

Penitrem C : R = Cl
Penitrem D : R = H

Fig. 10.8.1 Structure of penetrims

10.8.1. Physico-chemical characteristics

These compounds have a complex ring structure comprising indole and isoprenoid derivatives and differ in their functional group. Richard *et al.* (1986) reviewed excellently the structure, production, detection and biological effects of penitrems (Fig. 10.8.1).

10.8.2. Organisms

Several species of *Penicillium* (*P. aurantiogriseum, P. cyclopium, P. canescens, P. clavigerum, P. crutosum, P. communae, P. granulatum, P. janthenellum* and *P. palitans*) are reported to produce this group of mycotoxins (Ciegler, 1969; Pitt, 1970; Wagner *et al.*, 1984; Payne and Walls,

1984; El–Banna and Leistner, 1988 and El–Banna *et al.* (1987). Jesus *et al.* (1981) elucidated the structure of pentirems A-F.

Penitrems are reported to induce neurotic syndromes in vertebrates (Maas *et al.,* 1981), tremors in mice (Cole and Cox, 1981), toxic to dogs (Peterson *et al.,* 1978; Jortner *et al.,* 1986 and Hocking *et al.,* 1988) and toxic to sheep and other animals (Cole and Dorner, 1985; Laws *et al.,* 1987). Cavangah *et al.* (1998) reported that penitrem acts on cerebellum of brain of rat. Neurotic effect of penitrems was attributed to the interference of neuro-transmittance mechanism of amino acid (Norris *et al.,* 1980). Selela *et al.* (1991) reported the tremors in guinea pigs under the influence of penitrem A. Penitrem A was toxic to Heliotheszea (Dowd *et al.,* 1988). Weiser *et al.* (1988) recorded Parkinson like disease symptoms due to ingestion of penitrem A.

10.8.3. Isolation and Detection

Penitrems are elaborated in good amount in yeast extract sucrose (YES) and Skim, Milk, Yeast extract (SMKY) media. Penitrems are intracellular mycotoxins and are extracted with diethyl ether. Chloroform: methanol (95:5) or Chloroform: Acetone (93:7) or hexane and ethylacetate (6:4) are generally employed for separation of penitrems by TLC. Ferric chloride (2%) in butanol gives grey coloured spot which changes to green colour and heating at 120°C changes to variable green to blue (R_f 0.4). These give purple coloured spot when sprayed with cerium sulphate in ammonium sulphate. Hou *et al.* (1971) have developed method for quantitative estimation of penitrem.

10.8.4. Natural incidence

Penitrem A has been reported to contaminate cheese, walnuts and bread (Arp and Richard 1979; Richard and Arp 1979; Richard *et al.,* 1981; Arp and Richard, 1981 and Hocking *et al.,* 1988). Cole *et al.* (1983) reported the intoxication and tremors in man who consumed *Penicillium crustosum* contaminated beer. Jelinek *et al.* (1989) reviewed the update status of natural incidence of penitrems in foods and feeds. Hosae *et al.* (1990) could detect penitrem A, B, D, E and F in *Penicillium puberulum* infected bread in Tokyo city.

10.8.5. Production

Penicillium puberulum produced maximum penitrem A during its growth in SMKY and YES medium, while *P. aurantiogriseum* preferred YES medium followed by Richard's medium for production of penitrem B. El Banna *et al.* (1987) reported that *P. crustosum* produced good amount of penitrem in 2% skim milk, potato extract sucrose medium. When a temperature of 25-30°C was optimum for production of penitrem A by *P. puberulum, P. aurantiogriseum* preferred 20-25°C. *P. puberulum* produced penitrem A in the pH range of 5.0 – 7.0, while *P. aurantiogriseurn* opted pH 5.5-6.5. *P. puberulum* and *P. aurantiogriseum* produced maximum amount of penitrem A and penitrem B respectively during their growth in medium containing D-mannose and D-xylose. However, these fungi differed in their preference to glucose and sucrose respectively. Both the species opted potassium nitrate and sodium nitrate as nitrogen source for penitrem A production. Penitrem B production by *P. aurantioguseum* was maximum when C:N ratio in the medium was 3: 1. Further its production was stimulated by the presence of yeast extract, peptone, beef extract and malt extract. Most of the growth regulators stimulated penitrem B production only at low concentration (<40 μg/ml).

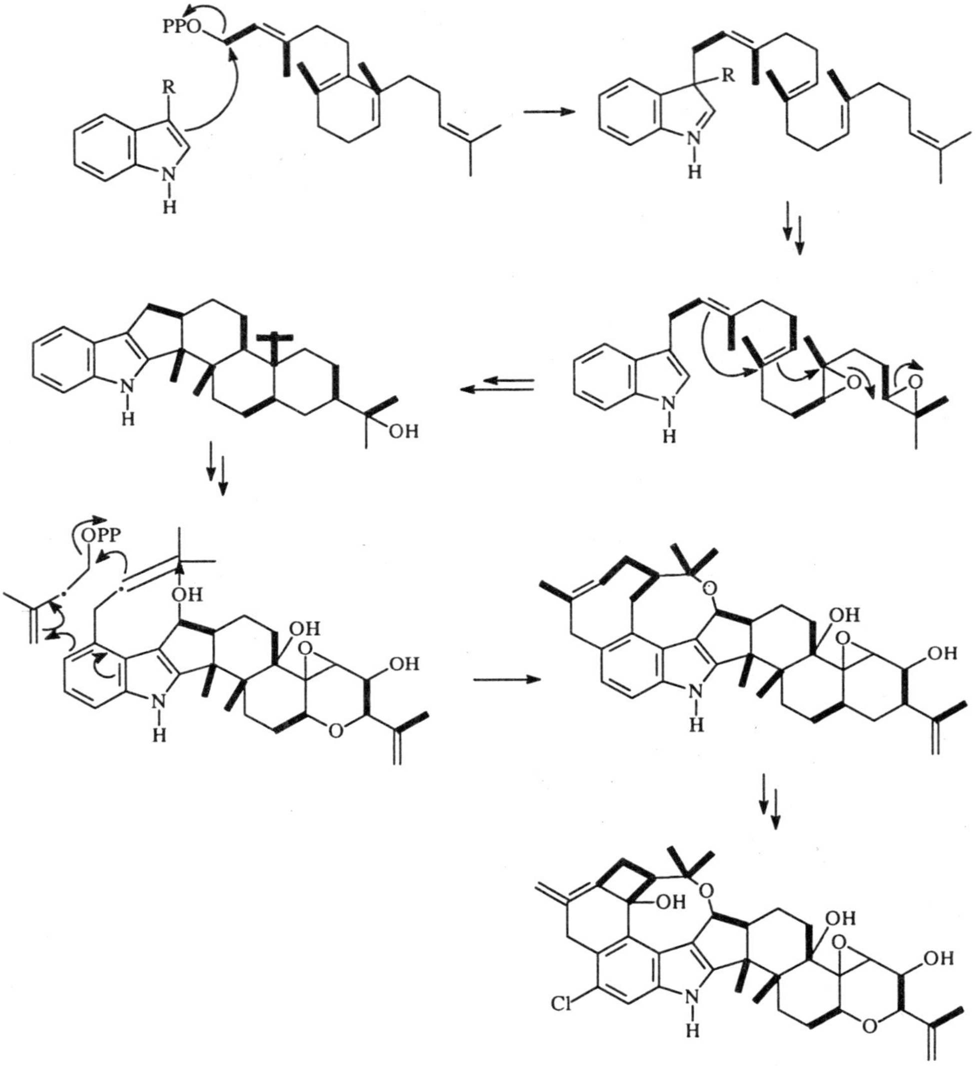

Fig. 10.1.2. **Proposed biosynthetic pathway in the formation of penitrem A.**

10.8.6. Biosynthesis

Laws and Mantle (1989) discussed elegantly the problems associated with the biosynthetic studies of penitrem. Hosae *et al.* (1990) suggested indole diterpenoid as possible precursor of penitrem. Mantle and Penn (1989) reported the stimulation of penetrim A synthesis by the presence of paxilline. Smith *et al.* (1989) suggested indole-oxocane as an intermediate in biosynthesis of penitrerm A. Further, Hesseltine *et al.* (1998) suggested oxacane fused indole as an intermediate for Penitrerm A biosynthesis (Fig. 10.1.2).

Dejesus *et al.* (1980) proposed biosynthetic pathway of penetrim A in *Penicillium crustosum.*

10.8.7. Biological activity

The culture filterates of *P. puberulum* were toxic to seed germination of wheat which increased with the increase in age of the fungus. The elongation of shoot and root were also inhibited which increased with the age of the fungus. When *P. puberulum* was inoculated to wheat seeds, the inhibition of seed germination was more than 50%. Similarly when seeds were treated with spore suspension of *P. puberulum,* the seed germination was only less than 50%. Similarly treating the seed with crude penitrem A solution, resulted in increased seed germination inhibition.

Penitrem A produced by *P. puberlum* caused growth inhibition of *Drechslera rostrata* and *Colletotrichum gloeosporioides* which increased with the increase in the concentration of the compound. *Fusarium oxysporum* was comparatively more resistant to this mycotoxin. *Escheritia coli* followed by *Micrococcus* sp., *Proteus vulgaris* and *Pseudomonas aeruginosa* were inhibited significantly by penitrem A. The degree of inhibition increased with the increase in penitrem A concentration. However, it failed to inhibit the growth of *Klebsiella pneumoniae* and *Staphylococcus aureus.* Penitrem B was toxic to *Streptococcus albus* and *Bacillus polymyxa* and was responsible for total inhibition of growth. However, this toxin was ineffective against Gram (-)ve bacteria.

Penitrem B was responsible for inhibition of mitosis which increased with the increase in Penitrem B concentration (Table 10.8.1). Chromosome breakages and extreme fragmentation were most common abnormalities. Unorganised metaphase, chromosome fragmentation followed by laggards, irregular distribution of chromosomes was some other aberrations. Inhibition of cytokinesis was another cytotoxic effect which resulted in the formation of binucleate and multinucleate cells.

Penitrems are reported to act on central nervous system resulting in the trembling and paralytic effect. Selala *et al.* (1989) suggested that penitrem acts on aminobutryic acid receptor site. Hukovic and Mulabegovic (1986) reported that penitrem A was responsible for increasing the extrinsic nerve stimulation. Susheela *et al.* (1986) recorded a disorder in glucose metabolism due to long time exposure of rats to penitrem A.

Pentirem A caused significant change in the serum and blood composition of the chick. Serum iron, calcium and protein decrease progressively with the increased dose of penitrem A. Serum albumins were reduced to a significant level (Table 10.8.2). Similarly serum cholesterol and serum urea also decreased with the increasing dose of penitrem A. On the other hand, total serum bilurubins, direct serum bilurubins, SGPT and SGOT increased progressively under the influence of penitrem A. The serum amylase activity also increased in the presence of penitrem A. Blood iron, blood haemoglobin, and blood sugars decreased under the influence of penitrem A. When significant decrease in the WBC was recorded, RBC count increased considerably. The gradual decrease in the body weight was also recorded. Percentage of mortality also increased significantly under the influence of penitrem A.

Table 10.8.1 Effect of Penitrem B on dividing cells of *Allium sativum*

				Metaphase		Anaphase		Telophase		
Concentration (µg/ml)	Duration (h0	Mitotin index (%)	Stickiness (%)	Breakages (%)	Extreme fragmentation	Fragmenttation (%)	Laggards	Fragments	Binucleate	Total
Control	-	13.26	-	-	-	-	-	-	-	-
300.00	1.00	11.14	-	1.28	1.43	1.23	-	0.46	-	4.03
	2.00	9.26	-	1.26	1.45	1.42	-	0.48	-	5.00
	3.00	8.01	0.15	2.21	2.63	2.10	-	0.72	-	7.72
	4.00	6.04	0.36	3.19	3.62	4.26	0.43	0.81	0.30	13.86
600.00	1.00	8.82	0.58	2.21	2.28	1.68	0.44	0.53	0.32	7.21
	2.00	7.93	0.92	2.86	2.94	2.24	0.53	0.68	0.60	8.96
	3.00	7.02	1.00	3.04	3.03	2.93	0.56	1.02	1.20	12.36
	4.00	5.80	1.23	3.83	5.02	5.42	0.82	1.08	1.40	18.10
900.00	1.00	6.00	1.02	2.50	2.63	2.00	0.60	0.92	1.02	8.85
	2.00	5.62	1.06	2.76	3.00	2.68	0.81	1.12	1.10	10.86
	3.00	5.00	1.85	3.12	3.54	3.01	0.88	1.26	2.23	14.25
	4.00	4.63	2.00	4.38	6.00	7.82	0.96	1.82	3.56	25.13
1200.00	1.00	4.26	1.20	3.52	3.55	3.00	1.00	2.00	2.40	13.03
	2.00	3.21	2.10	4.16	4.28	3.62	1.12	2.43	3.20	18.12
	3.00	3.21	2.32	5.13	5.39	4.01	1.16	3.21	4.10	22.14
	4.00	2.05	2.30	6.36	9.88	8.86	1.23	5.00	6.05	36.38

Table 10.8.2. Effect of penitrem A on haematological and biochemical composition of serum and blood of chick

Biochemical / Haematological constituents	Control	Concentration (mg/ml)	
		100	200
Serum Iron (µg/ml)	12.00	10.50	8.00
Serum Calcium (mg/ml)	60.10	42.25	40.00
Serum proteins (mg/100 ml)	1.62	0.68	0.54
Serum Albumins (O.D.)	0.66	0.05	0.04

Serum Cholesterol (mg/ml)	3.40	2.20	1.60
Serum urea (mg %)	58.10	43.50	28.25
Total serum bilirubins (mg %)	0.25	0.65	1.15
Direct serum bilirubins (mg %)	0.08	0.35	0.50
SGPT (Units/ml)	18.50	30.00	39.00
SGOT (Units/ml)	56.00	78.50	120.60
Serum amylase (Units/ml)	44.10	52.35	59.45
Blood iron (mg/ml)	33.25	19.15	12.20
Blood Haemoglobin (g/100ml)	11.50	6.00	2.50
Blood sugars (mg/ml)	27.50	15.40	7.60
WBC/mm^3 (in lakhs)	17.20	10.20	4.30
RBC/mm^3 (in lakhs)	2.36	10.50	13.95
Average body weight	75.60	65.40	60.20
Mortality (in percentage)	5.50	22.00	47.00

Penitrem A caused the decrease in the proteins of brain and brain, while in kidney and liver decrease was marginal (Table 10.8.3). Glycogen content of brain, heart, kidney and liver decreased progressively with the increase in the dose of penitrem A. The cholesterol content of liver also decreased significantly in the presence of penitrem A.

Table 10.8.3. Effect of penitrem A on biochemical composition of different organs of chicks

Constituents	Conc. (µg/ml)	Organs			
		Brain	Heart	Kidney	Liver
Proteins (mg/ml)	C	270.40	315.70	362.05	296.00
	100	120.20	190.60	320.40	255.50
	200	80.30	90.60	265.00	234.50
Glycogen (mg/g)	C	0.18	0.13	0.43	0.16
	100	0.90	0.09	0.20	0.11
	200	0.01	0.04	0.07	0.05
Cholesterol (mg/g)	C	--	--	--	0.28
	100	--	--	--	0.16
	200	--	--	--	0.06

There was marked loss in body weight of chicks by 7th day under the influence of penitrem A. About 60% mortality was observed within 2 hours of injection of tremorgenic mycotoxin. Toxin injected, chick exhibited drawn back, head trembling and extended legs within half an hour of treatment.

Chick became weak and showed, diarrhoea, shivering and feed refusal, other symptoms include dull behaviour, closing of eyes and sleepy eyes. The average body weight of chicks decreased considerably under the influence of penitrem A. Gross post mortem examination of liver, kidney, heart, brain and spleen revealed the following major pathological conditions.

Liver showed extensive degenerative changes of the hepatic cells with the presence of little hemorrhages in the Sinusoids. Mononuclear infiltration was also found in many areas. Degeneration of hepatic cells was a common feature. It also showed diffusive fatty degenerative changes and necrosis of liver parenchyma with bile duct hyperplasia (Hepatitis) and presence of hemorrhages. Per cellular necrosis of hepatic cells with congestion and marked degenerative changes observed were congestion of glomerulus, extensive hyalinization and necrosis of proximal and distal convolute tubules and of glomerulus with presence of hemorrhages. Presence of extensive necrosis with infiltration of mononuclear cells in brain was another pathological examination of heart of penitrem A. Extensive gliosis was also observed. Post mortem examination of heart of penitrem A given chick, revealed congestion, hemorrhages with the degenerative changes of the myocardial cells. Loss of cross striations and degenerative changes in myocardial muscle fibers. Mononuclear infiltration was also seen in between cordial muscle fibers and it is called myocarditis.

10.8.8. Management

Exploitation of aniline methoxy ethanol and ethyl acetate (Surekha and Reddy 1992), spices (Surekha and Reddy, 1992a), plant extracts (Surekha and Reddy, 1992), γ-radiation (Surekha and Reddy, 1991) and *Trichoderma viride* for the control of *Penicillium aurantiogriseum* infestation and penitrem B production have been suggested. Control of growth of *P. puberulum* and penitrem A elaboration can be successfully checked by employing volatile compounds like propionic acid, acetic acid and formic acid, leaf extracts of eucalyptus and its oil, and garlic could also check the production of penitrem A. Antioxidant like butylated hydroxyanisole was also effective in checking the growth and penitrem A production by *P. puberulum*. Cloves and coriander also inhibited the penitrem A production. Of different food prese-rvatives, propionic acid, sodium acetate and sodium metabisulphate could inhibit the growth and penitrem A production. *Pseudomonas aeurginosa* was an effective biocontrol agent of penitrem A.

Though there are studies dealing with the epidemology of disease, only limited information is available on the control of fungal infestation and penitrim production in foods, feeds and fodders.

Many of the volatile compounds such as aniline, ethylacetate, methoxyethanol, acetone and iso-amyl alcohol inhibited penitrem B production to a maximum extent. Food preservatives such as acetone, sodium metabisulphate and propionic acid were responsible for total inhibition of penitrem B production. Surekha and Reddy (1991a) reported the inhibition of penitrem B production effectively by cloves even at very low concentration. Black pepper, thymol and garlic were also effective inhibitors of penitrem B production by *P. aurantiogriseum*. O-nitrophenol, P-nitrophenol, P-chloronitrophenol and pentachlorophenol were responsible for total inhibition of penitrem B production

Surekha and Reddy (1991a) reported that x-rays at the dosage of 400kg was effective in checking penitrem B production. Moulds like *Trichoderma viride, P. griseofuluum, Aspergillus niger* and *Scytalidium state* of *Hendersonula toruloides* were responsible for total inhibition of the growth and penitrem B production by *P. aurantiogriseum*. However, more detailed investigations on the methods of control of penitrems are desirable.

CHAPTER – 11

PATULIN

Patulin, (4-hydroxy-4H furo [3,2-c] Pyran-2(6h)-one), a polyketide lactone though reported simultaneously by Brikinshaw *et al.* (1944) and Anslow *et al.* (1943) from culture of *Penicillium patulum* and *P. expansum* respectively as an antibiotic, the real credit goes to Chain *et al.* (1942) who have isolated this compound as claviformin from cultures of *Penicillium claviformine.* It has been described under several names such as clavatin (Bergd, 1944), clavacin (Katzman, 1944), expansin (Osterhass, 1946), penciden (Atkinson, 1942) and mycin, lenoopin and tercimin (Cole and Cox, 1981). It was soon realized the fact that its antiboitic properties are out weight by its toxicity to animal system including man has now given way to concern over its frequent presence in the environment under varied circumstances (Scott and Bullerman 1975). Cole and Cox (1981), Turnor and Aldrige (1983) and Engel and Tenber (1984) and Speijers (2004) have excellently reviewed various aspects of this toxin.

11.1. PHYSICO-CHEMICAL CHARACTERISITICS

Patulin is a highly reactive β-unsaturated lactone of fungal origin. It is known as 4-hydroxy-4H fUFo (3,2-c) Pyran - 2 (6H) - one and has empirical formula $C_7H_5O_5$ (Fig. 11.1). It is a crystalline compound with a melting point about 110-111°C. It is neutral substance soluble in water, ethanol, acetone, ethyl acetate, diethyl ether and chloroform and insoluble in benzene and light petrolium. Patulin optically inactive, inspite of its asymmetric carbon atom in the structure. It can be recrystlised preferably from benzene. Patulin reduces warm Fehling's solution and decolonizes potassium permanganate solution. Patulin is unstable in the presence of alkali. Boiling of patulin in $2NH_3SO_4$ for 6 hours will result in the decomposition. It is active for several weeks in buffers of pH 3.3-6.3 but is inactivated at pH 6.8. Patulin is appreciably stable in apple and grape juice. Sulphur dioxide used as a preservative sometime for juices causes rapid disappearance of patulin. It is also unstable in wheat flour, ground nut, corn under alkaline conditions. The presence of cysteine and glutathione might account for the liability of patulin in certain food stuffs. A fresh garden soil (pH 9.0) is reported to destroy patulin, but it is stable in acidic soils. Patulin has UV maxima at 275 nm (E=166,600), at 276 nm in ethanol and 278 nm after thin layer of chromatography on Silica gel plates. IR spectra of toxin at 3390, 1768 and 1745C shoulder, cm^{-1} in nujol mull and in KBr at 3360 (brod), 1765 and 1740. Patulin can be removed from apple juice by treatment with certain types of active carbons (Leggott *et al.*, 2001).

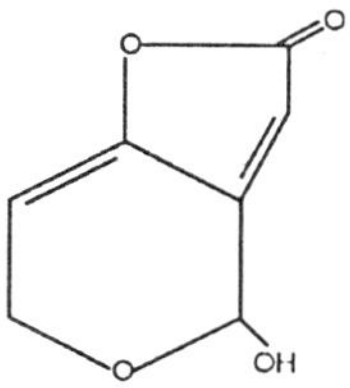

Fig. 11.1. Patulin

Patulin is produced by several species of *Aspergillus* and *Penicillium* which are versatile in their capacity to colonize different substrates. Further, patulin is elaborated under wide range of environmental conditions. Variety of biological effects of patulin may pose a serious threat to the health of man. In general if the producers adhere to the tolerance limits set and maintain good manufacture practices, there is no appreciable health risk from patulin.

11.2. ORGANISMS

In view of broad spectrum antibacterial activity of patulin intensive search for organisms producing similar substance lead to the isolation of this compound from a large number of fungi. Florey *et al.* (1944) and Kent and Heatley (1945) have isolated this compound from cultures of *Asperqillus qiganteus* and *A. calvatus* and *A. terrcus* respectively.

Similarly *Penicillium patulum* (Kent and Heatlay, 1945), *P. expansum* (Osterhass, 1946), *P. leucopus* (Umezava *et al.,* 1947), *P. divergen* and *P. equinum* (Bharat and Mair, 1948), *P. melini* (Abraham and Flory, 1949), *P. griseofulvum* (Simonarta and Dehatowar, 1957), *P. cyclopium* (Efimento and Yakimov, 1961), *P. lepidosum* (Myrchink, 1967), *Paecilomyces variotii* (Escoula, 1975), *Byssoclamys fulva (Gymnoascus* spp.) (Ciegler, 1977), *B. nivea* (Rice, 1997), and *Penicillium roqueforti* (1979) are reported to produce patulin. The malt feed which was reported to be responsible for the death of nearly 100 cattle in Japan was reported to contain patulin producing *P. urticae* (1954). Land and Hult (1987) reported that *Penicillium expansum* inhabiting wood synthesized patulin. Aran and Eke (1987) recorded production of patulin by an isolate of *Penicillium aurantiogriseum.* Sleiman *et al.* (1989) reported the production of patulin by variety of micromycetes. *P. cyaneo- fulvum* and *P. granulatum (P. divergens)* are also reported to produce patulin (Betina, 1989).

11.3. BIOSYNTHESIS

Czapek-Dox and Raulin - Thom medium were originally used for patulin producing *Penicillia* and *Aspergilli* (70 mg/l. of yield). Higher yields (700mg/lt) of patulin were obtained in complex media such as yeast extract (*P. patulum*) and potato glucose broth (1200 - 1700 mg/l.1) (Engel *et al.,* 1975). Norstadt and MCalla (1969) have reported that *P. urticae* produced good amount of patulin on potato glucose broth.

The patulin pathway is one of the best characterized pathway in fungal secondary metabolism. The first intermediate of the patulin pathway is 6-methyl salicyclic acid, which is formed from the condensation of acetyl-Co A with 3 malonyl units by *P. patulum* (Betina, 1989). Dimroth *et al.* (1970) isolated and characterized methylsalicylic acid synthetase complex and proposed reaction scheme leading to synthesis of 6 methyl salicylic acid.

Biosynthetic pathway from 6-methyl salicylic acid to patulin and enzymes mediated each step is as follows (Murphy and Lymen, 1975). 6-methylsalicylic acid-m-cresol-m-hydroxy-benzyl alchol-gentisyl alcohol-gentisaldehyde - patulin.

Turner and Aldridge (1983) reported that biosynthesis of patulin from the polyketide derived 6-methyl salicylic acid via gentisaldehyde. Murphy *et al.* (1974) and Forrester and Gaucher (1972) have detailed the pathway of patulin synthesis by *P. patulum* and *P. urticae* respectively by employing ^{3}H and ^{14}C - labeled precursor. It is detailed in Fig. 11.2. In addition, cell free systems from these organisms were used to demonstrate the conversion of m-cresolto 2, 5 dihydroxy to leave and m-hydroxy benzyl alcohol to m-hydroxy benzaldehyde and gentisyl alcohol to gentisaldehyde. Sekiguchi and Gaucher (1979) have shed light on the stages beyond gentisaldehyde to two pigments, phyllostine and iso epoxydon were isolated from mutant JI, which are interconvertable by a specific alcohol dehydrogenase in the producing organisms. To accomodate the new metabolites as pathway intermediates. Gaucher postulates a monoxygenase mediated conversion of gentisaldehyde to isoepoxydon or pilyllostine - a ring opening rearrangement to give the 7-membered lactone (a route) acyclic and cyclic forms of patulin. If the rearrangement of phyllostine may proceed via route b and the hydride shift necessary to convert the iso-patulin to the patulin skeleton will take place at a later stage. Zamir (1980) and Sekiguchi (1973) gave excellent account of biosynthesis of patulin. Grootwassink and Gaucher (1980) and Newnay and Gaucher (1982) discussed various enzymes involved in biosynthesis of patulin.

Girisham *et al.* (1989) reported that *A. terreus* produces a good amount of patulin in mineral liquid medium. Glucose and tyrosine were the best carb on and nitrogen sources respectively. *A. terreus* produced maximum amount of patulin at incubation temperature of 25°C - 30°C and pH 5.5. *P. expansum, P. patulum* and *A. clavatus* are reported to produce patulin in the temperature range of 0 to 24°C, 4-31°C and 12-24°C respectively. Minimum water activity for patulin production by these species was 0.99, 0.95 and 0.99 respectively.

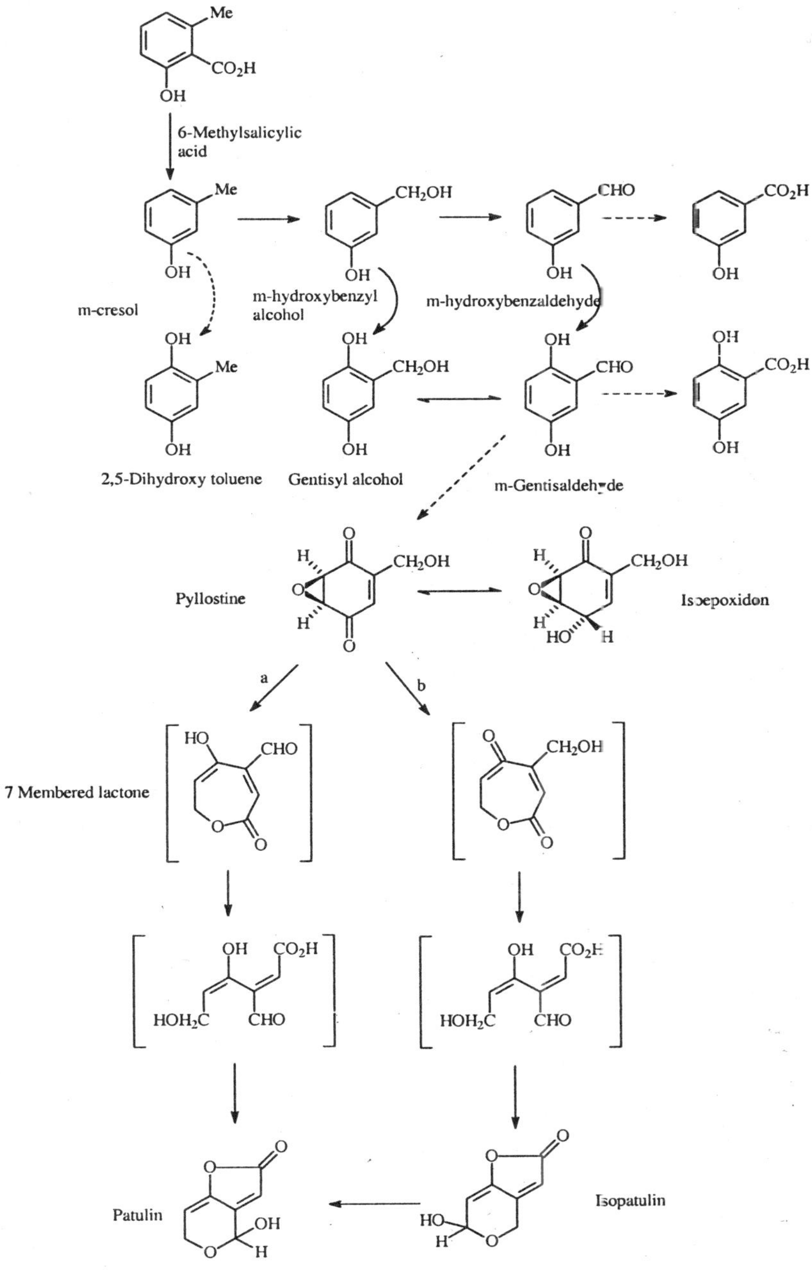

Fig. 11.2. Biosynthetic pathway of Patulin

11.4. ISOLATION AND DETECTION

Czapek-Dox and Raulin - Thom are the most commonly employed media for patulin production by different species of *Penicillium and Aspergillus.* Glucose was the best carbon source and incubation temperature of 20-25°C was optimum for patulin production (Betina, 1985). High yields may be obtained with *P. patulum* in YES (2% yeast and 15% Sucrose) for 15-20 days at 27°C (Engel *et al.,* 1975) from culture broth. Patulin extracted with analytical grade ethylacetate twice or thrice and extracts were concentrated (5 ml) on water bath and analysed for the presence of patulin. Finely ground grain is extracted with 250 ml acetonitrile hexane (4 : 1) for 30 min. Extracted filtrate (200 ml) mixed with 50 ml of hexane and evaporator to dryness in flash evaporator at 60°C up to 1 ml and used for direct detection of patulin (Pohland and Allen, 1970)

11.4.1. Chemical Methods

Chromatographic methods are reported to be more suitable for analysis of patulin. It can be divided into four types.

(i) Paper Chromatography (PC)

(ii) Thin Layer Chromatography (TLC)

(iii) High Performance Liquid Chromatography (HPLC)

(iv) Gas Liquid Chromatography (GLC)

11.4.2. Paper Chromatography

Adsorption chromatography with a Whatman filter paper No. 1 as the adsorbent with various solvent systems is being widely practiced. Spots were identified with the help of different spray reagents

11.4.3. Thin Layer Chromatography

The adsorption chromatography with a silica gel as the adsorbant and thickness of the silica gel usually is 0.25 mm. After spotting, the samples are developed in different suitable solvent spray systems and the patulin was identified on the basis of flourescence visibility in colour with the help of different spray reagents (Table 11.1). Subramanian (1982) reported the quantitative estimation of patulin calorimetrically by using 2% phenyl hydrazine hydrochloride.

Table 11.1. Detection of Patulin by different spray reagents on thin layer chromatography

Spray reagent	Detection unit	Reaction conditions	Colour product
P-Anisaldehyde	0.1	130°C 8-20 min	Visible : Reddish UV : Yellow
1% aqueous N-methyl benzothiazolone-2-hydrazone	0.06	--	Visible : Brownish
O-Uianisidine-acetic acid	0.2	80°C	Visible : Yellow-brown
0.5% aqueous 2-hydro gone 2,3-dihydro-3-methylbenzo-thiazolene-HCl	0.05	130°C 15 min.	Visible : Yellow UV Yellow-Brown
Diazotized P-nitro aniline	--	--	Visible : Brown Orange
Aqueous $KMNO_4$	--	--	Visible : Pink
1% Methanolic diphenylboric acid,aminoethyl ester,	0.1	105°C 15 min.	UV : Blue
0.5% aqueous 3-methyl-2-benzoth-iazolinone hydrazone hydrochloride	0.01	130°C 15 min.	UV : Yellow
2% Phenylhydrazene hedrochloride	5	100°C 5 min.	Visible : Yellow
3-Methyl-2-benzo-thiozolinone (MBTH)		130°C 15 min.	Visible : Yellow

Table 11.2. Rf values of Patulin in different solvent system on TLC (Silica gel.G.)

Solvent System	R_f	Reference
Chloroform; Methanol (3:1)	0.65	Schepertz *et al.* (1972)
Chloroform; Methanol (4:1)	0.72	
Chloroform; Methanol (85:15)	0.51	Schepertz *et al.* (1972)
Chloroform; Methanol (95:5)	0.40	Scott and Kennedy (1971)
Chloroform; Acetone (9:1)	0.25	Durackova *et al.* (1976)
Chloroform; Isobytylmethyl Ketone (4:1)	0.20	Durackova *et al.* (1976)
Chloroform; Acetic acid (9:1)	0.32	Forester and Gaucher (1972)
Chloroform; Acetic acid; Diethyl ether (17:3:1)	0.30	Durackava *et al.* (1976)
Benzene; tetrahydrofuren (8:2)	0.58	Peteo *et al.* (1972)
Benzene; Methanol; Acetic acid (24:2:1), (90:5:5)	0.21 0.34	Pohland and Allen (1970)
Hexane; Ether (1:3)	0.47	Pefo *et al.* (1972)
n-Butanol;Acetic acid-water (4:1:4)	0.78	Durackova *et al.* (1976)
Toluene; Ethyl acetate:formic acid (5:4:1)	0.50	Scott and Kennedy (1971)

11.4.4. Gas Liquid Chromatography (GLC)

Pohland *et al.* (1970) reported GLC of patulin, ethyl ether, and acetate and chloroacetate derivatives have to be formed and GLC conditions for detection of these patulin derivatives are summarized in Table 11.3.

Table 11.3. GLC of Patulin Derivatives

Parameter	Derivative		
	Acetate	Chloroacetate	Silyl ether
Column	6 ft. x 4 mm I.D., glass coil	6 ft. x 4 mm I.D., glass coil	10 ft. x 4 mm I.D., glass coil
Column packing	3% JXR on Gas Chrom W	3% JXR on Gas Chrom W	1% SE-30 on Gas Chrom Q
Temperature (°C)	110	130	110
Electrometer sensitivity	3×10^{-1}	3×10^{-11}	3×10^{-11}
Hydrocarbon number	15.2	17.2	--
Detector	Hydrogen flame	Hydrogen flame	Hydrogen flame
Detection limit (ng of patulin)	Ca. 60	Ca. 40	Ca. 100

11.4.5. High Performance Liquid Chromatography

HPLC has been successfully applied for the quantitative determination of patulin. Separation of patulin on partisil 100 DS or Spherisorb 50 DS (Stray, 1978) with water as the mobile phase is achieved with a detection limit of 1 µg of patulin. Similarly Ware *et al.* (1975) reported the separation of Silica gel columns with isooctane: methylene chloride; methanol (80:15:1) and isooctane: diethylether; acetic acid (750:250:5) respectively with detection limit of 11 µg/l^{-1}.

Adam (1980) and Altamyer *et al.* (1982) reported the separation of patulin on Silica gel columns with diisoprophyl ether: dichloromethane: acetic acid (95:5:0.1) and chloroform: methanol: acetic acid (94:4:2) respectively as solvents with a detection limits of 5 µg/l^{-1} (Table 11.4).

Kubachi (1985) reported simple, sensitive and reliable method for the determination of patulin in apple juice by HPLC. Jimenez *et al.* (1988) described the detection and quantification of patulin by HPLC. Lehmann and Wald (1990) described the improved method of extraction and HPLC determination of patulin. Priest and Light (1990) discussed the application of HPLC to quantification of metabolites and enzymes of patulin pathway from *Penicillium patulum.*

Table 11.4. Methods for the Determination of Patulin in Apple Juice

Extraction	Cleanup	Quantitation	Detection (μg/kg)	Ref.
Shake vigorously for 1 min with ethyl acetae (2 ml/ml)	Liquid-liquid partition	HPLC Column: Reverse-phase C_{18} (5 μm, 25 cm x 3.9 mm i.d.) Mobile phase: 0.8% Tetrahydrofuran in water Detection: UV (276 nm)	12	Branse *et al.* (1977)
Vigorously shake 50 ml sample with three 50-ml portions ethyl acetate	Silica gel column chromatography	TLC Plate: Adsorbosil-5 Solvent: 5:4:1 Toluene/ethyl acetate/ 90% formic acid Derivatization: Spray plate with 0.5% 3-methyl-2-benzothiazolinone hydrazone hydrochloride solution; heat 15 min at 130°C Detection: Yellow-brown fluorescent spot under long-wave UV light or yellow spot in visible light		Scott (1995)
Diphasic dialysis by shaking dialysis tube containing 50 ml ethyl acetate with 25 ml apple juice in a flask overnight at 150 rev/min	Silica gel SPE cartridge	HPLC Column: Reverse-phase C_{18} (15 cm x 3.9 mm i.d.) Mobile phase: 99:1 Water/ tetrahydrofuran Detection: UV (275 nm)	1.0	Prieta *et al.* (1993)
Diphasic dialysis by shaking dialysis tube containing 75 ml ethyl acetate with 200 ml apple juice in a flask 5 hr at 120 rev/min		TLC Plate: Silica gel 60 Solvent: 70:50:50:20 Toluene/ethyl/ acetate/ chloroform/90% formic acid Derivatization: Spray with 0.5% 3-methyl-2-benzothiazolinone hydrazone hydrochloride solution; heat 15 min at 130°C Detection: Yellow-brown spot under 366 nm UV light	50	Prieta *et al.* (1992)
Shake with equal volume of ethyl acetate	Liquid-liquid partition	HPLC Column: Reverse-phase C_{18} (5 μm 25 cm x 3 mm i.d.) Mobile phase: Distilled water Detection: UV (276 nm)	5.0	Moller and Josefsson (1980)

11.4.6. Bioassay

Abedi and Scott (1969) reported £ 50 value for zebra fish *(Brachydari rerio)* larvae was 18.0 μg/ patulin. Similarly Reiss (1972) and Durackova *et ai.* (1977) reported brine shrimp larvae *(Artemia salina)* and the crutacean *Cyclops fuscus* was 100 μg/patulin/ml. However, this assay for patulin was less sensitive than other mycotoxins.

Patulin is also phytotoxic (Resis, 1975), insecticidal (Cole and Rolison, 1972) and anticrustacean (Harwing and Scott, 1971), amphibians (Reiss, 1977), fish larvae (Abed and Scott, 1969) and tracheal organ explants (Cardeithae *et al.,* 1972) may be suitable to develop bioassay for the detection of patulin.

(i) Bacterial: *Bacillus subtills, Escherichia coli, Staphylococcus aureus* and *Agrobacterium tumifaciens* were inhibited by as little as 2-5 μg/ml pure patulin (Yamakawa and Ueno, 1970). Similarly, Scott and Bullerman (1975) reported that *B. megaterium* was sensitive to patulin and can be a suitable test organism for an accurate and quantitative bioassay of the toxin. The assay was found to be sensitive to 1.7 μg patulin. The assay is simple, rapid (12-15 hr), inexpensive, accurate as well as to quantitatively measure patulin in samples of liquid media, apple juice and corn.

(ii) Fungal assay: *Pythium debaryanum* and *Phythium ultimum* were completely inhibited by 2 and 1.2 μg patulin/ml respectively (Abraham and Florey, 1949). Similarly, Reiss (1973) reported that patulin suppressed the spore germination and hyphal growth of *Aspergillus flavus, A. niger, Penicillium expansum, Neurospora crassa, Cladosporium herbarum* and *Rhizopus nigricans* by 100 μg/filter disk. Some common methods of determination of patulin in apple juice are precised in Table 11.4.

11.5. NATURAL INCIDENCE

Patulin is a potent food contaminant as *Penicillium expansum* is a common storage rot organism in apples. Patulin was involved in the spoilage of fruits by moulds (Brain *et al.,* 1956). Apples, pears, bananas, pine apples, grapes, peaches and apricot fruits with spontaneous brown-rot are reported to contain patulin (Frank *et al.,* 1977). Patulin producing fungi *(P. expansum, P. utricae* and *Byssochlamys nivea)* artificially inoculated in juices are reported to grow at temperature ranging from 5 to 25°C. Patulin has been detected in commercial samples of apples and grapes juices mostly at the time of harvest but not after fermentation into wine (Scott, 1977). It was observed that juice prepared from contaminated contained lower levels of patulin than juice prepared directly from fresh apples. The concentration of patulin after five - months of storage of juice is reported to decrease from 94 to 17.5 μg/l (Williams, 1985). Apple juice fermented by *Saccharomyces* is reported to be free from patulin. Although certain countries (Norway, Sweden and Switzerland) have action levels of 50 µg-l of patulin in apple juice. There is no interest in its control (Scott, 1975) in North America because of the lack of tumorogenic activity.

Table 11.5. Natural occurrence of patulin in fruit products during 1992-1997

Commodity	Country	Publication Year	No. of samples	Incidence	Range (µg/kg)
Apple juice	Turkey	1997	44	100%	5-75
Apple juice	Brazil	1996	73	21%	6-78
Apple juice	UK	1996	173	36%	10-184
Apple juice	UK	1995	185	35%	10-490
Apple juice	Spain	1994	100	43%	10-170
Apple juice	Austria	1994	20	85%	10-148
Apple juice	Austria	1994	17	65%	10-185
Fruit juice	Australia	1992	270	57%	5-1130
Miscellaneous juice	Australia	1992	70	26%	5-32

Bread, cookies, and biscuits infected with a suitable strain have been found to contain patulin (Reiss, 1972). Many natural substrates support patulin production including barley, malt and rice (Yamamoto,1954), wheat straw (Norstadt and McCalla, 1971), animal feed (Ukau *et al.,* 1954), soil containing roots, leaves or other residue of apple fruit trees (Borner, 1963) and in commercially available apple juice in Canada and United States (Frank *et al.,* 1976). Buchanan *et al.* (1974) recorded patulin in pears and stored fruits decayed by *Penicillium expansum.* Patulin was also detected in mouldy bread and bakery goods (Reiss, 1973). Subtilled plots in Nebroska are reported to contain patulin in soil and wheat straw residues (Norstadt and McCalla, 1969). Patulin could not be detected in raw sausage and cheese even in artificially contaminated with patulin producing fungi (Ciegler, 1977). Probably this may be due to the reaction of patulin formed with SH groups of proteins leading to chemically altered compounds with decreased toxicity (Scott and Bullerman, 1976). Nahdi and Nusrath (1985) recorded the presence of patulin in mixed feed. Lane *et al.* (1988) recorded the natural occurrence of patulin in fodder. Jelinek *et al.* (1989) reported patulin from fungal infested foods, feeds and fodders. Patulin has been found in pears, peaches, strawberries, blue berriers, cherries, apricots and grapes (Majerus and Kapp, 2002). Diementese *et al.* (2005) reported patulin in conventional and organic fruit products from Italy. It is also reported to contaminate apricots, grapes, peaches, pears, apples, olives and cereals (Askar and Siliha, 1999: De Sylos and Rodrigez Amaya, 1999; Arici, 2000; Abrunhosa *et al.*,

2001; Laidou *et al.*, 2001; Kadakal *et al.*, 2002). Occurrence of patulin in fruit products from different parts of the world during 1992-1997 are precised in Table 11.5.

11.6. BIOLOGICAL ACTIYITY

(i) Cytological: Dustin (1963) reported that patulin affects specifically on dividing cells (mitotic pioson). High percentage (12%) of polyploid cells was recorded in human leucocyte cultures treated with patulin (Kawasaki, 1972). Umeda *et al.* (1977) reported single and double strand breaks in He La cell DNA at 32 pg/ml concentrations. Ueno and Kuboto (1976) reported the carcinogenic effects of patulin that interact with DNA.

(ii) Membrane permeability: Phenylalanine and threonine uptake was inhibited by patulin (Polacco and Stands, 1977). Ueno *et al.* (1976) reported that the total and non-competitive inhibition of Na+ dependent transport of glycine into rabbit reticulocytes. Riley *et al.* (1990) recorded various changes in the functioning of cultured cell treated with patulin.

Enzymes: Patulin has been reported to inhibit the activities of co-carboxylase (Karrer and Viscontini, 1947), Urease (Reiss, 1977), *E. coli* Polymerase (Moule and Hatley, 1977) and various ATPase (Phillips and Hargesis, 1977). Ashoor and Chu (1973) described the mechanism of patulin inhibition of thiolenzymes, muscle aldolase, muscle lactic dehydrogenase (LDH) and yeast alcoholdehydrogenase (ADH). Sakthisekaran and Shanmugasundaram (1990) reported the inhibition of biosynthesis of liver aldolase.

High concentration of patulin (50mM) caused lipid peroxidation, abrupt calcium influx,extensive blebbing and total lactate dehydorgenase(LDH) release occur and these appear to be implicated in the loss of structural integrity of the membrane (Riley and Showker, 1991). Patulin impaired DNA synthesis which may be due to it ability to react with sulfhydryl groups and thereby inhibit enzymes involved in the replication of genetic material. Dirheimer (1998) reported the single and double strand breaks of DNA by patulin. Mezler (1999) demonstrated that patulin induces intermolecular protein–protein cross links. Schumacher *et al.* (2003) reported DNA interstrand crosslink's in cell free DNA treated with patulin. Patulin also reacts with terminal sulfhydryl groups of protein and dpolypeptides present in food (Fleige and Mezler, 1999). Patulin has inhibitory effect on several biochemical parameters such as ATPase, alkaline phosphatase, aldolase and hexokinase activity. Patulin was able to activate glycogen phsophorylase in the liver and increase glucose level. Sorensen *et al.* (1985) reported inhibition of phagocytosis of sheep erythrocytes. Sharma (1993) reported decreased serum immunoglobulin, reduced biogenesis of lymphocytes and reduced chemilumiscience of peritoneal leuckocytes in rabbits under the influence of patuilin.

(iv) Cellular respiration: Oxygen uptake was inhibited by 80% in guinea pig brain tissue and 50% in kidney tissue by 1 mm patulin (Andrond *et al.*, 1966). Polacco and Stands (1977) reported the inhibition of Cyanidein sensitive oxidase system operation in certain plant mitochondria.

(v) Effect on macromolecular synthesis: Umeda *et al.* (1972) have demonstrated DNA Single and double strand breaks induced by patulin in FM 3A Cells. Similarly, Kawaski *et al.* (1972) reported that the suppression of DNA synthesis in free La S3 cell treated with patulin for 2h. At a concentration of 10 μg/ml^{-1} and 50 μg/ml^{-1} of patulin causes single and double-strand DNA breaks in *Escherichia coli* respectively (Lee and Roschentholer, 1987). *Bacillus megaterium* also induces temperature-sensitive prophage at the same concentration of *E. coli. in vitro* system patulin induces DNA repair synthesis and inhibits DNA replication. Similarly, *in vivo* induction of DNA repair *synthesis* and DNA strand breaks was reported (Lee and Roschentholer, 1987). Inhibition of RNA synthesis by patulin was reported by Moule and Hatey (1977). *In vitro* studies revealed that patulin specifically impairs the activity of RNA polymerase at the initiation step of transcription.

Patulin directly impairs protein synthesizing system and inhibits amino acid incorporation (Moule *et al.,* 1978). Patulin also interfers with oxidative pathways (Moule, 1984). Lee and Roschentholer (1987) reported the single strand scissions of DNA in the presence of reducing agents and cupric ions. Patulin has been reported to be carcinogen when subcutaneously injected to rats and acutely and chronically toxic to chicks producing hemorrhages in digestive tract (Smith and Ross, 1991).

11.7.1 Animal diseases

Patulin is moderately toxic to rats and mice (Ciegler, 1977). Rats showed comparable oral response with death occurring within 4 days. The pathological changes were oedematous with hemorrhaging capillary damage in liver, spleen and kidney and oedemas of the brain (Scott and Bullerman, 1975). Dickens and Jones (1961) have reported the carcinogenicity of patulin to male rats due to subcutaneous treatment. Similarly, a subcutaneous, twice-weekly at 0.2 mg application in rats produced malignant tumours at the site of injection (Yamakawa and Ueno, 1970).

Devaraj and Devaraj (1987) recorded decrease of total lipids, phospholipids and triglycerides, while cholesterol level increased. Speijers *et al.* (1988) reported the toxicity of patulin to kidney and gastro-intestinal duct at subacute toxicity. Sorenson *et al.* (1985) recorded the toxicity of patulin for alveolar macrophages. Patulin also caused increased blood pressure in anaesthetised cats (4-8/mg dose) and non-anaesthetised rabbits (0.05-5.0/mg dose).

Low levels of patulin (1-2 p.g) produced significant teratogenic response (35 to 45%) when injected into 4 day-old chicken embryo, resulted in the malformed feet ankles and beaks execephally and exophthalina (Ciegler, 1977). Mintzlaff and Christ (1971) reported liver lesions in chicks given daily oral doses of 0.2 mg patulin for 6 weeks. Devaraj *et al.* (1986) recorded the inhibition of Na-K dependent ATPase in kidney and intestine.

Jocquet *et al.* (1963) have reported the outbreak of disease in cattle due to the presence of toxic *A. clavatus.* Similarly, Moreau and Moreau (1960) have recorded pulmonary oedema and congestion in one of the cows died due to toxic wheat in Japan. Obzazhel (1987) described the symptoms of disease caused to Swine due to ingestion of patulin. The symptoms include vomitting, salviation, anorexia, weight loss and leukocytosis.

11.7.2 Phytotoxicity

Patulin exhibited phytotoxic activity and caused inhibition of seed germination and wilting of plants. Patulin contaminated soil or sand, also reduced the leaf length of wheat, maturity of root and straw weights (Norstadt and McCalla, 1971). Patulin is reported to induce nitrate reductase activity in cauliflower and white mustard (Afridi, 1962) and increased pigment in red beet root discs (Andrand *et al.,* 1966).

11.7.3. Health hazards in man

The main source of patulin in the human diet is apple juice, home made fruits and fruits juices from naturally infected fruits. The natural incidence of patulin contamination up to 84% has been reported by Scott *et al.*,(1986). Its natural occurrence, toxicity and carcinogenicity make it a potential health hazard to man (IARC, 1986). Frayssinet (1980) tried to correlate the incidence of oesophageal cancer in parts of Western France and the presence of patulin in Normandy Cider.

Patulin could be effective in the treatment of the common cold. Patulin produced some irritation in the eyes of patients with the later disease (Altmayer *et al.,* 1982). Intense tissue swelling, in addition to epidermal and dermal injury, was also observed within 48h in pach test carried out with patulin (Dalton, 1952).

Patulin was irritating to the stomach when given orally as tablets (DeRosnay *et al.,* 1952). Nausea and vomiting are reported in human due to patulin (Freerksen and Bonicke, 1951). Devaraj *et al.* (1986) feel that patulin is a diabetogenic lactone.

11.7. PHARMACOKINETICS

Dailey *et al.* (1977) recorded the excretion of patulin in urine and faeces over a period of 7d after a single oral dose. Only low levels (1-2%) appeared to expired CO_2 or remained in the carcass. Most of the patulin was excreted through urine within 24h and faecal excretion was essentially completed by 48h. Red blood cells were found to be major repository for bound activity of patulin 1-7d after treatment followed in order by other blood rich organs as the spleen, kidney, lung and liver.

CHAPTER – 12

CYTOTOXIC MYCOTOXINS - CYTOCHALASINS

The cytochalasins are secondary metabolites of fungi that have attracted the attention of several workers owing to the unique effects produced on mammalian cells in tissue culture. These are being produced by species of *Metarrhizium, Zygosporium, Penicillium, Phomopsis, Phoma, Chalara, Diplodia, Chaetomium* and *Helminthosporium*. The cytochalasins can be subgrouped into cytochalasins, *zygosporins* and chaetoglobosins. Structurally they are either phenyl or indolyl substituted polyhydroisoindoles fused to variable sized macrocyclic ring. Structures, fungal sources and characteristics of different cytochalasins are illustrated and précised in Table 12.1. These toxins exhibit acute toxicity or and cytotoxic effects and induce characteristic morphological and cytological alternations in exposed cells namely a cesation of cell movement, generalized contraction resulting in flattening of the cell and the inhibition of cytokinesis often accompanied by polynucleation. The name cytochalasin is derived from the Greek cytos (cell), chalasin (relaxation) to describe the effects of these toxins on cell motility. Though cytochalasin B has been studied extensively because of its ready availability, but studies on other cyctochalasins has been extended recently.

These toxins mainly affect microfilament associated cellular contractile mechanisms and the cytoskeletal related regulation of the plasmamembrane function. Inhibition of macromolecular synthesis is considered to be secondary. Cytochalasin B is an extremely potent inhibitor of glucose transport.

Only limited number of reports in the literature has dealt with cytochalasins occurance in human food stuff. However, cytochalasin E was extracted from cultures of *Aspergillus clavatus* isolated from moldy rice implicated in human fatality in Thialand (Glindukon *et al.*, 1976). Cytochalasin A has been recovered from *Phoma exigua* isolated from potato (Scott *et al.*, 1978), while cytochalasin B from *P. exigua* and *Hormicium* sp from potato and tomato respectively. Cytochalasin H has been isolated from *Phomopsis* sp isolated from pecans (Wells *et al.*, 1976), kodocytochalasins 1 and 2 from *Phomopsis paspalli* infecting kodomillet and Chaetoglobosin C from *Penicillium aurantio-virens* isolated from weevil damaged pecans. Springer *et al.* (1976),

Natori (1977) have excellently reviewed information available on cytochalasins in relation to their role as mycotoxins

Binder *et al.* (1973) developed a systematic nomenclature of these mycotoxins based on the following rules.

1. The skeleton of the perhydroisoindole unit including macrocycle and the C-substituents of the isoindole system except for the phenyl and the indole-3-yl rings is named as cytochalasin. The number of atoms in the macrocycle including the atoms common to it and the isoindole ring is indicated by a number in square brackets preceeding the name cytochalasin.
2. The absolute configuration of the isoindole system is established and does not require specification. For the substituents in the macrocycle, chirality may be designated by use of symbols (R) and (S).
3. The clockwise numbering system indicated in the formulae 1-4 is used for designating the positions of the substituents on the macrocycle and additional functional groups attached to the isoindole system.
4. Oxygen atoms which are part of the macrocycle are designated by the prefix oxa with appropriate locant.

Prefixes E and Z respectively, indicate trans and cis geometry of olefinic double bonds according to the IUPAC (1974) recommdnations. Tamm and Tananbaum (1978) discussed different aspects of toxicity of these compounds on animals, microbes and plants.

Ceasation of movements at the cell surface as a consequence of alternations in the microfilament-plasmamembrane complex include such effects as generalized cellular contraction, retraction of surface microvilli, zeriosis (Blebbing) and aborization.

The first two members were isolated and characterized independently in Switzerland and Great Britain (Alridge *et al.*, 1967). Compounds isolated from culture filtrates of *Phoma* and *Helminthosporium dematioideum* were called cytochalasins A and B respectively. Phomin being identical with cytochalasin B and dehydrophomin with cytochalasin A (Alridge *et al.*, 1967).

The isolation of zygosporin A from cultures of *Zygosporium masonii* proved to be identical with cytochalasin D from *Metarrhizium anisopliae* (Alridge and Turner, 1969) was the beginning of the discovery of zygosporins.

In chemical and cytoxicity studies of toxigenic fungi, strains of *Chaetomium globosum* and *C. cochiliodes* were noticed because they produced cytotoxic metabolites causing polynucleation of HeLa cells and acute toxicity to experimental animals which led to the discovery of chaetoglobosins, a novel group of cytochalasins containing an indolyl group instead of the phenyl group as in other cytochalasins.

The aspochalasins, isolated from *Aspergillus microcysticus* (Keller-Schierlein *et al.*, 1979) contain an isopropyl group in the same position.

Table 12.1 Structure and fungi producing naturally ocuring cytochalasins

Toxin	Type structure	Macrocyclic ring substituents (R)	Source	Reference
Cytochalasin A (dehydrophomin)	I-B	$—O\overset{\overset{O}{\|}}{C}CH = CH\overset{\overset{O}{\|}}{C}CH_2CH_2CH_2—$	*Helminth-osorium dematioideum, Phoma sp.*	Aldridge *et al.* (1967)
Cytochalasin B (phomin)	I-B	$—O\overset{\overset{O}{\|}}{C}CH = CH\underset{HO\ \ H}{C}CH_2CH_2CH_2—$	*Phoma sp. H. dematioideum Hormiscium* sp.	Rothweiler and Tamm, (1966), Aldridge *et al.* (1966), Aldridge *et al.* (1967)
Cytochalasin C	I-A	$—\underset{AcO\ \ H}{C}CH = CH\underset{H_3C\ \ OH}{C}\overset{\overset{O}{\|}}{C}—$	*Metarrhizium anisopliae*	Aldrige and Turner (1969)
Cytochalasin D	I-B	$—\underset{AcO\ \ H}{C}CH = CH\underset{H_3C\ \ OH}{C}\overset{\overset{O}{\|}}{C}—$	*M. anisopliae Zygosporium masonii*	Aldrige and Turner (1969) Minato and Matsumoto (1970)
Cytochalasin E	I-D	$—O\overset{\overset{O}{\|}}{C}OCH = CH\underset{H_3C\ \ OH}{C}\overset{\overset{O}{\|}}{C}—$	*Aspergillus clavatus Rosellinia necatrix*	Buchi *et al.* (1973) Aldridge *et al.* (1973).
Cytochalasin F	I-D	$—O\overset{\overset{O}{\|}}{C}CH = CH\underset{HO\ \ H}{C}CH_2CH_2CH_2—$	*H. dematioideum*	Buchi *et al.* (1973) Aldridge *et al.* (1973).
Cytochalasin G	II-D	$—O\overset{\overset{O}{\|}}{C}CH_2CH_2\overset{\overset{O}{\|}}{C}CH_2—$	*Nigrosabulum* sp.	Cameron *et al.* (1974)

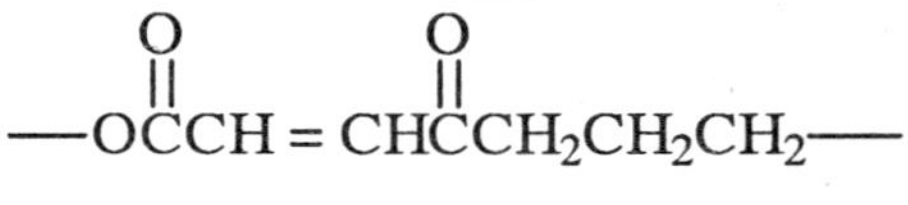

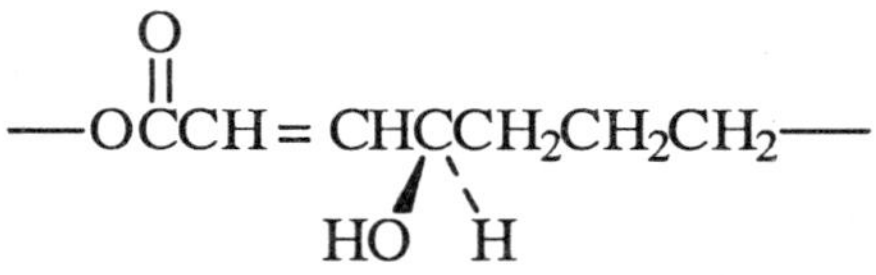

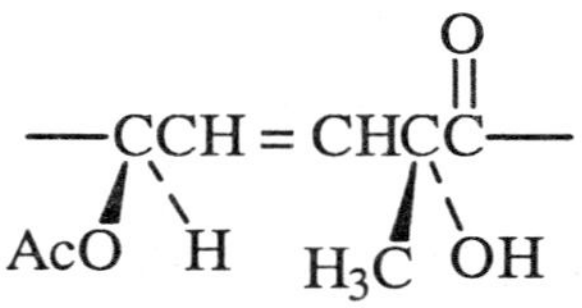

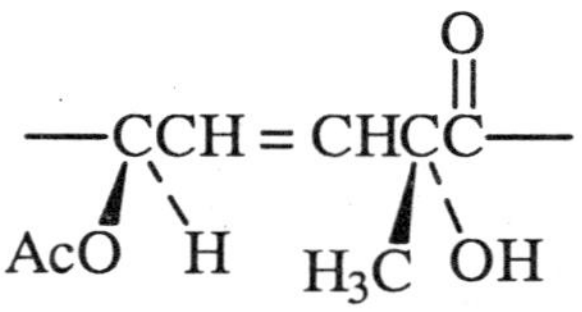

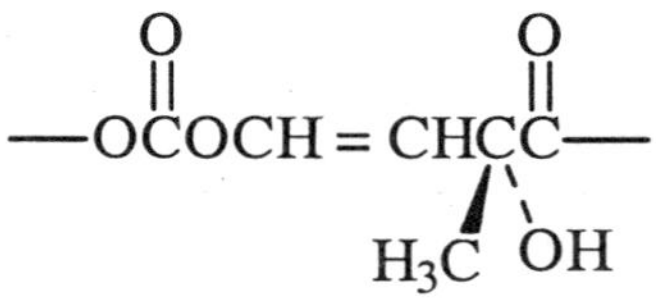

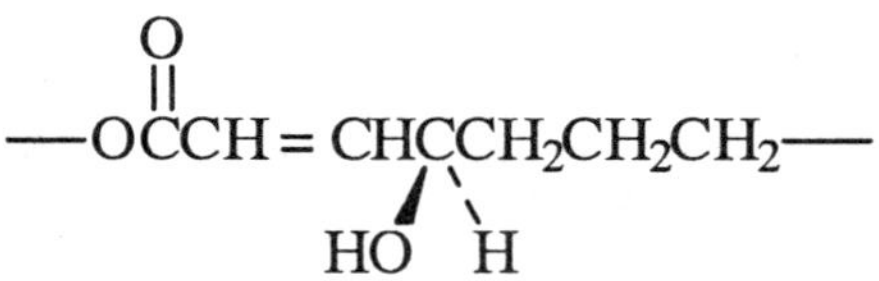

Kodo-cytochalasin-1 (cytochalasin H)	I-B	—C(OAc)(H)CH = CHC(HO)(CH$_2$)C(=O)CH$_2$—	*Phomopsis paspalli*	Mc Millan *et al.* (1977)
Kodo-cytochalasin-2 (cytochalasin J)	I-B	—C(HO)(H)CH = CHC(HO)(CH$_2$)C(=O)CH$_2$—	*P. paspalli*	Patwardhan *et al.* (1974)
Cytochalasin	K	$C_{32}H_{37}NO_6$	*Chalara microspora*	
Cytochalasin	L	$C_{32}H_{37}NO_7$	*Chalara microspora*	
Cytochalasin	M	$C_{30}H_{37}NO_6$	*Chalara microspora*	
Zygosporin D	I-B	—C(HO)(H)CH = CHC(H$_3$C)(OH)C(=O)—	*Z. masonii*	Minato and Katayama (1970)
Zygosporin E	I-B	—C(HO)(H)CH = CHCH(CH$_3$)C(=O)—	*Z. masonii*	Minato and Katayama (1970)
Zygosporin F	I-E	—C(AcO)(H)CH = CHC(H$_3$C)(OH)C(=O)—	*Z. masonii*	Minato and Katayama (1970)
Zygosporin G	I-C	—C(AcO)(H)CH = CHC(H$_3$C)(OH)C(=O)—	*Z. masonii*	Minato and Katayama (1970)
Deoxaphomin	I-B	—C(=O)CH = CHC(HO)(H)CH$_2$CH$_2$CH$_2$—	*Phoma* sp.	Binder and Tama (1973)
Proxiphomin	I-C	—C(=O)CH = CHCH$_2$CH$_2$CH$_2$CH$_2$—	*Phoma* sp.	Binder and Tama (1973)

Protophomin	I-F	$-\overset{\overset{O}{\|}}{C}CH=CHCH_2CH_2CH_2CH_2-$	*Phoma* sp.	Binder and Tama (1973)
Chaetoglobosin A	II-D	$-\overset{\overset{O}{\|}}{C}CH=CHC\underset{HO\;\;H}{C}C=CH-$	*Chaetomium globosum*	Silveston *et al.* (1976)
Chaetoglobosin B	II-A	$-\overset{\overset{O}{\|}}{C}CH=CH\overset{\overset{O}{\|}}{C}\underset{HO\;\;H}{C}\overset{\overset{CH_3}{\|}}{C}=CH-$	*C. globosum*	Sekita *et al.* (1973)
Chaetoglobosin C	II-D	$-\overset{\overset{O}{\|}}{C}CH_2CH_2\overset{\overset{O}{\|}}{C}\overset{\overset{O}{\|}}{C}\underset{\underset{CH_3}{\|}}{C}=CH-$	*C. globosum* *C. cochliodes* *Penicillium aurantio-virens*	Sekita *et al.* (1976), Natori (1977), Springer *et al.* (1976).
Chaetoglobosin D	II-B	$-\overset{\overset{O}{\|}}{C}CH=CH\overset{\overset{O}{\|}}{C}\underset{HO\;\;H}{C}\overset{\overset{CH_3}{\|}}{C}=CH-$	*C. globosum*	Sekita *et al.* (1976)
Chaetoglobosin E	II-A	$-\overset{\overset{O}{\|}}{C}CH_2CH_2\underset{HO\;H}{C}\overset{\overset{O}{\|}}{C}\underset{\underset{CH_3}{\|}}{C}=CH-$	*C. globosum*	Sekita *et al.* (1977)
Chaetoglobosin F	II-D	$-\overset{\overset{O}{\|}}{C}CH_2CH_2\underset{HO\;H}{C}\overset{\overset{O}{\|}}{C}\underset{\underset{CH_3}{\|}}{C}=CH-$	*C. globosum*	Sekita *et al.* (1976)
Chaetoglobosin G	II-A	$-\overset{\overset{O}{\|}}{C}CH_2CH_2\overset{\overset{O}{\|}}{C}\overset{\overset{O}{\|}}{C}\underset{\underset{CH_3}{\|}}{C}=CH-$	*C. globosum*	Sekita *et al.* (1977)
Chaetoglobosin J	II-C	$-\overset{\overset{O}{\|}}{C}CH=CH\overset{\overset{O}{\|}}{C}\underset{HO\;\;H}{C}C=CH-$	*C. globosum*	Sekita *et al.* (1977)
Chaetoglobosin K		$C_{34}H_{40}N_2O_5$	*Diplodia macrospoora*	

Acetylchaetoglobosin A	II-D	$-\overset{O}{\overset{\|}{C}}CH=CH\overset{O}{\overset{\|}{C}}\underset{AcO\ H}{C}\overset{CH_3}{\overset{\|}{C}}=CH-$	*C. globosum*	Natori (1977)
Apochalasin A		$C_{24}H_{33}NO_4$	*A. microa-plicus*	Keller-Schierlein and Kupfer (1979)
Apochalasin B		$C_{24}H_{33}NO_4$	*A. microa-plicus*	Keller-Schierlein and Kupfer (1979)
Apochalasin C		$C_{24}H_{35}NO_4$	*A. microa-plicus*	Keller-Schierlein and Kupfer (1979)
Apochalasin D		$C_{24}H_{35}NO_6$	*A. microa-plicus*	Keller-Schierlein and Kupfer (1979)
Engleromycin		$C_{28}H_{35}NO_6$	*Engeromyces goetzer*	Petersen *et al.* (1980)

Many aspects of the chemistry, biosynthesis and biological effects of cytochalasins have been described in a book published by Tanenbaum (1978). Between 1978 and 1981, the number of cytochalasins has increased from 24 to 36.

This system of numbering allows the assignment of the same locants to that part of the molecule which is common to the hitherto known metabolites. So far the following four types of naturally occurring basic skeletons are known (Fig 12.1). Because of the length of the systematic names, trivial names will be used in this chapter.

1. (11) cytochalasin;
2. (13) cytochalasin;
3. 24-oxa-(14) cytochalasin;
4. 21, 23-dioxa-(13) cytochalasin

12.1. ISOLATION AND DETECTION

12.1.1. Cytochalasin A and B

These are elaborated by *Phoma exigua, Helminthosporium dematioideum* and *Hormiscium* sp. from which cytochalasin B was isolated (Prbela *et al.*,

1975). Production and purification methods are similar to those employed for production and purification of deoxaphomin.

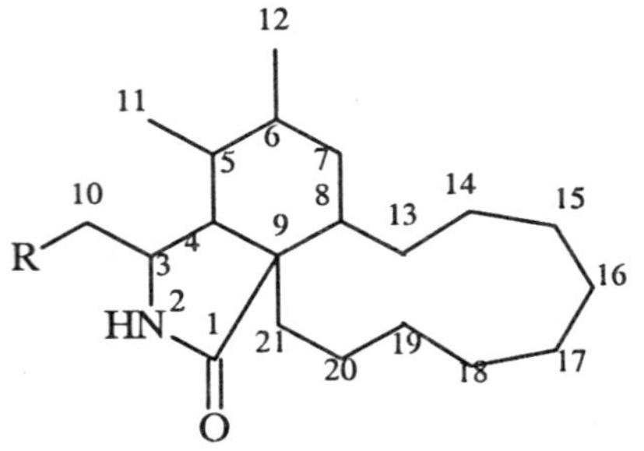

1 R = Phenyl
Indol-3-yl
Isopropyl

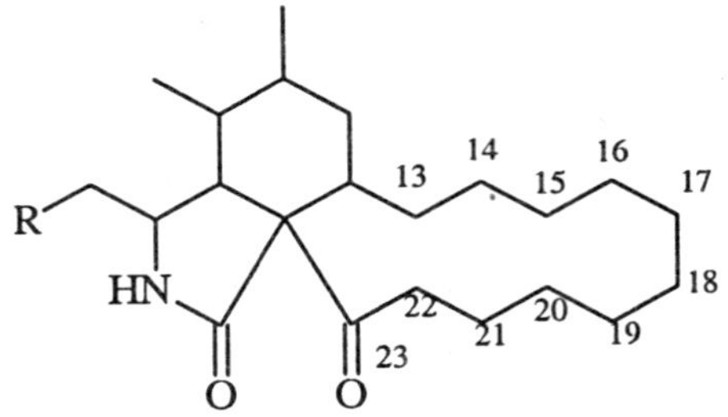

2 R = Phenyl
Indol-3-yl

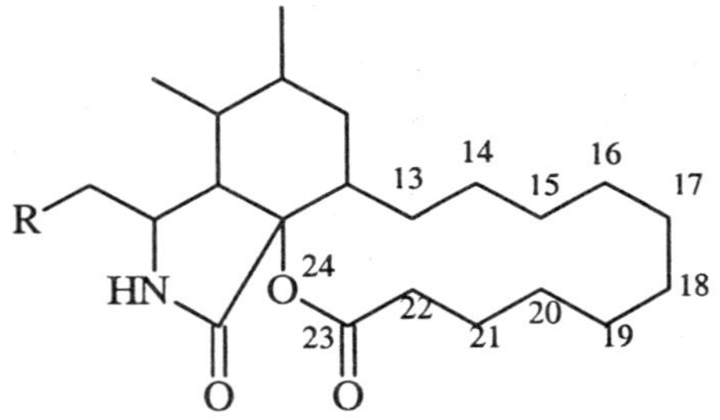

3 R = Phenyl

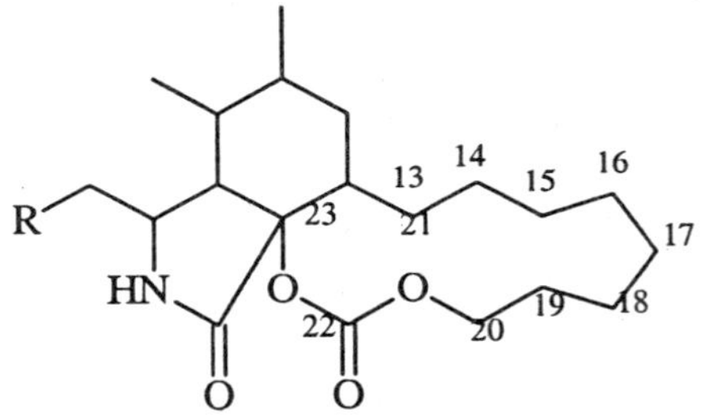

4 R = Phenyl

Fig.12.1. Structures of different groups of cytochalasins.

The TLC of cytochalasins A and B, together with other cytochalasins in several solvent systems, is characterized. TLC on silica gel G of several derivatives of cytochalasin B was carried out by Rothweiler and Tamm (1970). Chloroform, containing 0.5-20% of methanol, was mostly used as the solvent system. Detection was effected with iodine vapour.

12.1.2. Cytochalasins C and D

Metarrhizium anisopliae, *Zygosporium masonii* and *Z. mycophilum* are the producers of these mycotoxins. Biosynthetic experiments with cytochalasin D using growing cultures of *Z. masonii* demonstrated that the building blocks are one of phenylalanine, three units of methionine and acetate units. The biosynthesis of cytochalasin D was discussed by Tamm and Tanenbaum (1978).

Aldrige and Turner (1969) separated cytochalasins C and D on silica gel G using chloroform-methanol-formic acid (90:5:5). The plates were sprayed with 5% ethanolic sulphuric acid and heated to ca. 110°C for a few minutes. Cytochalasin C gave an orange and cytochalasin D a yellow fluorescence under UV light.

TLC of cytochalasin D was carried out on silica gel GF_{254} with benzene ethylacetate (7:3) or chloroform : methanol (9:1) (or both) and the k spots were rendered visible by spraying with concentrated sulphuric acid followed by heating at Ca. 180°C (Hayakawa *et al.*, 1973).

Padhye *et al.* (1975) published R_f values of seven cytochalasins (including C and D) in eleven solvent systems using three reagents for detection: ethanolic sulphuric acid, vanillin-phosphoric acid and Dragendroff's reagent (Table 12.2). Cytochalasins C and D had almost identical R_F values in most of the solvent systems used. They could be readily distinguished from each other. However, by spraying with ethanolic sulphuric acid and observing their fluorescence under UV light, cytochalasin C giving a dull orange colour and cytochalasin D a weak yellow colour. Durackova *et al.* (1976) gave additional information on the TLC behaviour of cytochalasins A, B and D in eight solvent systems and their detection with p-anisaldehyde (Table 12.3).

Table 12.2. R_F x 100 values of cytochalasins on silica gel F plates[2]

Cytochalasin	Solvent system*										
	1	2	3	4	5	6	7	8	9	10	11
A	63	53	62	39	36	0	93	67	79	95	56
B	58	38	53	36	30	0	92	63	78	95	47
C	46	34	44	23	26	0	90	59	78	92	30
D	45	32	42	19t**	20t	0	89	58	76	93	34
H	63	55	64	46	50	0	87	59	80	92	60
J	44	38	56	34	35	0	75	55	72	90	46
E	79	57	63	48	40	0	90	62	82	96	71t

* Solvent systems : 1 = chloroform-methanol (95:5); 2 = chloroform-methanol formic acid (95:5:5); 3 = chloroform-diethylamine (90:10); 4 = diisopropyl ether-ethyl acetate (90:10); 5 = cyclohexane-ethyl acetate-diethylamine (60:30:10); 6 = benzene-chloroform (50:50); 7 = n-butanol-formic acid-water (80:10:10); 8 = benzene-methanol (70:30); 9 isopropanol; 10 = acetone; 11 = benzene-acetone (70:30). ** t = Tailing.

12.1.3. Cytochalasin E

21, 23 Dioxa cytochalasins is produced by *Rosellinia necatrix* and *Aspergillus clavatus*. Cytochalasin E, crystallized from acetone-hexane, m.p. 206-208°C with decomposition; $[\alpha]_D^{25}$ = -25.6° (methanol); M = 495; $C_{28}H_{33}$

NO_5; IR absorptions typical of γ-lactams. X-ray data have been published by Buchi *et al.* (1973)

Table 12.3. R_f x 100 values of cytochalasins on (a) silica gel plates and (b) silufol sheets with P-anisaldehyde.

Cytochalasin	Solvent system*								Detection**	
	A	B	C	D	E	F	G	H	1	2
A (a)	41	58	25	85	17	33	46	99	Pale beige	Blue
(b)	34	42	23	83	20	31	44	95		
B (a)	28	46	13	78	8	16	24	98	Violet	Blue
(b)	34	34	13	74	7	14	26	94		
D (a)	19	24	10	75	0	8	14	90	Beige	Orange
(b)	32	27	10	74	4	9	20	85		

*Solvent systems : A = benzene-methanol-acetic acid (24:2:1); B = toluene-ethyl acetate-90% formic acid (6:3:1); C = benzene-ethanol (95:5); D = chloroform-methanol (4:1); E = chloroform-isobutyl methyl ketone (4:1); F = chloroform-acetate (9:1); G = chloroform-acetic acid-diethyl ether (17:1:3); H = n-butanol-acetic acid-water (4:1:4, upper layer).
*Detection; 1 = p-anisaldehyde; 2 = UV at 366 nm after p-anisaldehyde.

The procedure developed for the production of cytochalasin E and two tremorgenic mycotoxins (tryptoquivaline and tryptoquivalone) by *A. clavatus* is as follows (Demain *et al.*, 1976). Spores from a soil cultures maintaind at 4°C were inoculated on to Czapek solution agar, enriched with 0.5% yeast extract and 0.5% casamino acids (Difco Lab.), in a 120ml dilution bottle. The incubation lasted 4-6 days at 30°C. Spores were suspended in 100 ml of 0.01% (w/v) sodium lauryl sulphate, and 10ml of the spore suspension were added to each 2.8 wide-mouthed Fernbach flask. The Fernback flasks had received 300 g of pearled barley and 150 ml of warm water. They were then autoclaved, cooled inoculated and incubated on a rotary shaker at 200 rpm (ca. 5.08 cm diameter orbit) for 12 days at 30°C.

The barley was extracted twice, each with 500ml of methylene chloride per flask. After the first addition of methylene chloride, the suspension was allowed to soak over night. The flask contents were then filtered through glass-wool and the grains were blended in a warning blender for 60 sec with the second 500-ml portion of methylene chloride. The homogenate was filtered through cheese cloth and the solids were discarded. After combining the two extracts, anhydrous magnesium sulphate was added in order to remove trace amounts of water; the suspension was then filtered on a Buchner funnel with No. 595 filter paper (Schleicher & Schuell) and with either celite filter aid No. 503 (Johns-Manville) or Hyflo Super Cel. After evaporating the filtrate to dryness in a flask evaporator, the oily residue was precipitated by pouring it slowly into ice-cooled light petrolium and placing it over night in an ice-bath. The crude toxin fraction was recovered by filtration under vacuum on a medium-pore, sintered-glass filter and allowed to dry in the hood before being weighed. To remove trace amounts of light petroleum before determination of the weight of the fraction, the vial was stored overnight under negative pressure. From 1 kg of barley, approximately 35 mg of cytochalasin E and 80 mg of combined tremorgens were isolated in pure form. This yield represents an improvement over the original production of 10 mg of cytochalasin E and 11 mg of tremorgens per kilogram of glutinous rice. The tremorgens formed from this strain usually occur in a 3:1 ratio of tryptoquivaline to tryptoquivalone (Demain *et al.*, 1976).

12.1.4. Cytochalasin F

It is a minor metabolite of *Helminthosporium dematioideum* and isomeric with cytochalasin B. It was isolated by Aldridge *et al.* (1967) who assumed that cytochalasin F is a double-bond isomer of cytochalasin E. Later, their structures were revised (Aldridge *et al.*, 1973). At present, cytochalasin F is the only representative of this type of cytochalasin.

12.1.5. Cytochalasin G

This compound was isolated from an unidentified *Nigrosabulim* sp (Cameron *et al.*, 1974) and *Pseudoeurotium zonatum* (Probst and Tamm, 1981).

For stationary cultures of *Metarrhizium anisopliae* a fortified Raulin-Thom medium of the following composition (%, w/v): tartaric acid 0.266, monoammonium tartrate 0.266, $(NH_4)_2$ HPO_4 0.040, K_2CO_3 0.040, 3 $MgCO_3$. $Mg(OH)_2.3H_2O$ 0.027, $(NH_4)_2SO_4$ 0.016, $ZnSO_4.7H_2O$ 0.0042, $FeSO_4.7H_2O$ 0.0042, glucose 2.5 and yeast extract 0.010 was employed. Before sterilization, the pH was adjusted to 6.0 with 6N potassium hydroxide solution. Cultivation was carried out for 13 days.

Hayakawa *et al.* (1973) obtained cytochalasin B from shaken culture of *Zygosporium masonii* in potato-sucrose medium. A large-scale fertmention of *Z. masonii* was used to produce cytochalasin D and zygosporins D, E, F and G (Hayakawa *et al.*, 1973). In 500 ml flask containing 100 ml of a pre-seed medium (30 g of glucose, 20g of peptone and 5g of sodium chloride in 1 l of tap water, pH adjusted to 6 prior to sterilization) was employed to grow at 28°C for 96 h on a reciprocating shaker.

The culture filtrate of initial pH 3.4 was adjusted to pH 7.0 and extracted with chloroform (3 x 3 l). The solvent was removed to give a brown semi-solid residue (2.2 g), which was combined with similar extracts to give a total of 7.47 g. The extract was absorbed on silica gel (75 g) by evaporation of an acetone solution and placed on the top of a silica gel column (300 g) made up in benzene. Elution with benzene containing increasing proportion of chloroform gave gummy fractions (3 g), which were discarded. Elution with chloroform gave a series of solid fractions, the earlier of which were a single compound and combined (2.0 g) and recrystallised from acetone to give cytochalasin C.

From a shake-flask fermentation (Hayakawa *et al.,* 1973), cytochalasin D was isolated as follows. A 25 g amount of celite was added to culture broth and filtered to obtain 55g of mycelium (including the celite) and 860 ml of the filtrate. The mycelium was extracted with methanol (2 x 200 ml) and the extract was evaporated in vacuo. The residue was dissolved in ethyl acetate, washed with water, 2 N sodium carbonate solution and water, dried over sodium sulphate and evaporated at 50-60°C in vaccum. The residue gave cytochalasin D.

The producing strain of *Nigrosabulum* sp. (Cameron *et al.*, 1974) was grown as a surface culture for 28 days in a medium adjusted to pH 5.5 and containing (g l^{-1}) cerelose (50), $NaNO_3$ (2), KH_2PO_4 (1), $MgSO_4.7H_2O$ (0.5), KCl (0.5), $FeSO_4$ (0.01), oxoid yeast extract (1) and 0.1% of a minor element concentrate.

Pseudoeurotium zonatum was grown as a surface culture in a medium containing (g per litre of demineralized water) cerelose (50), $NaNO_3$ (2), KH_2PO_4 (1), $MgSO_4.7H_2O$ (0.5), KCl (0.5), yeast extract (1) and 1 ml of minor element solution pH 5.3-5.7 (non-adjusted). The sterilized medium was inoculated with a spore suspension and the flasks, each containing 1 l of the medium, were incubated for 25 days at 27°C.

The culture filtrate (65 l) was extracted at pH 6.8 with chloroform (2 x 18 l) to afford an amorphous solid (5 g), part of which (3.8 g) was dissolved in the two-phase system toluene-methanol-chloroform-water (40.5:40.5:9.5:9.5) (100 ml). The mixture was placed ion equal portions in the first four separators of Gallenkamp automatic counter current apparatus programmed to perform 120 solvent transfers with a partition period of 1 min and a settling time of 10 min. Additional lower-solvent phase (12.5 ml) was placed to each of the remaining separators and upper-solvent phase was placed in the reservoir. At completion the contents of separators 56-80 were pooled and concentrated in vacuo to *ca.* 50 ml, then extracted with ethylacetate (3 x 100 ml) to afford a solid (0.81 g), which was recrystallized from acetone-ethylacetate. Two further crystallizations of the product from methanol yielded cytochalasin G (350 mg) (Cameron *et al.*, 1974).

Probst and Tamm (1981) isolated Cytochalasin G from both the mycelium and the medium of *Pseudoeurotium zonatum*, were extracted twice with ethylacetate. The organic layer was dried over sodium sulphate and evaporated in vacuo at 40°C. The dark-brown extract was chromatographed on silica gel (100:1) using increasing amount of methanol in methylene chloride. From the fractions containing 1.25% of methanol, cytochalasin G was isolated by crystallization from acetone-hexane.

12.1.6. Cytochalasins H and J

Cytochalasins H and J were originally named kodo-cytochalasin-1 and kodo-cytochalasin-2 respectively and reported to be produced by *Phomopsis paspalli* (Patwardhan *et al.*, 1974). *Phomopsis paspalli* was cultivated in large conical flasks on 2% potato-dextrose-agar medium for 3 weeks (Patwardhan *et al.*, 1973). The entire material including the medium was agitated with an amount of 95% ethanol sufficient to cover the entire material and the mixture was left at 20°C for 4 days. The aqueous alcoholic extract was drained off and evaporated under reduced pressure. The residue (65 g) was mixed with sodium sulphate (300 g) and extracted by cold percolated, with the following sequence of solvents to furnish various extracts; light petroleum (124 mg) (gum), diethyl ether (585 mg) (yellow sticky mass), ethyl acetate (1.7 g) (yellow solid) and acetone (618 mg, gum).

TLC and preparative TLC were carried out on silica gel layers containing 15% gypsum. TLC of the diethyl ether extract in chloroform-methanol (9:1) and spraying with concentrated sulphuric acid gave four spots with R_f 0.56 (yellow, minor) and 0.32 (red, major). The red spots correspond to kodo-cytochalasin-1 (cytochalasin H) and kodo-cytochalasin-2 (cytochalasin J).

12.1.7. Cytochalasins K, L and M are produced by *Chalara microspora* (Fex, 1981)

The fungus was grown in a stationary culture and both the medium and mycelium were extracted with ethylacetate. The cytochalasins K, L and M were isolated using reversed-phase chromatography.

12.1.8. Zygosporins D, E, F and G

These four zygosporins were isolated as minor metabolites of *Zygosporium masonii* (Minato *et al.*, 1973). The shaken flask and large-scale fermentations of *Z. masonii* were used for production of zygosporin D, E, F and G as described for cytochalasin D (Hayakawa *et al.*, 1973).

The residues A and B obtained after isolation of cytochalasin D were combined and chromatographed on silica gel (830 g). Fraction 1, eluted with benzene, benzene-chloroform (9:1, 7:3 and 5:5) and chloroform, contained an oil (42 g). Fraction 2, eluted with chloroform (25 g) was crystallized from ethyl acetate to give a crystalline product (C) (270 mg) and a paste (D) (24 g). The latter was chromatographed on alumina (690 g) to give an oil (22 g), eluted with benzene and then chloroform, and a paste (E) (1.2 g), eluted with chloroform-methanol (20:1, 9:1 and 8:2). Fraction 3, eluted from the silica gel column with chloroform-methanol (9:1) was dissolved in light petroleum and a precipitate (F) (340 mg) was collected. The crystalline product C was separated into cytochalasin D (R_F 0.40) (150 g) and zygosporin E (R_F 0.48) (51 mg) by preparative TLC using toluene-methanol (10:1) as the solvent system. The precipitate F was separated into cytochalasin D (R_F 0.50) (135 mg) and zygosporin D (R_F 0.40) (32 mg) by preparative TLC using chloroform-methanol (10:1). In addition to preparative TLC used in the isolation of zygosporins D, E, F and G as described earlier, TLC was used to characterize the degradation products and derivatives of the four zygosporins.

12.1.9. Aspochalasins A, B, C and D

Aspergillus microcysticus (Keller-Schiertein and Kupfer, 1979) produces these mycotoxins. Originally, one of these compounds (asposterol), was isolated as an antibacterial and antifungal antibiotic. Five years later, new fermentation extracts were chromatographed; the antibiotic (now named aspochalasin B) and its inactive co-metabolites, aspochalasin A, C and D, were isolated and the structures of the four compounds were elucidated.

Two hundred ml medium containing (g l^{-1}) glucose (50), L(+)-tartaric acid (3), diammonium D-tartrate (3), $(NH_4)_2HPO_4$ (0.4), K_2CO_3 (0.4), magnesium hydroxide carbonate (0.3), $FeSO_4.7H_2O$ (0.05), $ZnSO_4.7H_2O$ (0.05) and demineralized water; pH 3.7 before and after sterilization was inoculated with 2 ml of a conidial suspension (9 x 10^7 ml^{-1}) and incubated on a rotary shaker at 27°C for 48 h. The culture filtrate was extracted repeatedly with about 1/20 volume of ethyl acetate (Heberle *et al.*, 1974).

The crude extract (5.5 g) was dissolved in a small volume of acetic acid ester and light petroleum was added until no more precipitate was formed. The precipitate (3 g) was filtered and contained mostly aspochalasins C and D with trace amounts of aspochalasin B. The filtrate was evaporated and the residue (2.3 g) contained aspochalasins A and B. The precipitate was chromatographed in a cold room (at about 4°C) on a column of Kieselgel 60 (125 g), using benzene-ethyl acetate and ethyl acetate alone. Aspochalasin B (250 mg), almost pure aspochalasin C (150 mg) and pure aspochalasin D (1.231 g) were obtained. Further purification was achieved by repeated chromatography and crystallization from ethyl acetate. The residue from the evaporated filtrate was similarly handled, yielding a fraction (687 mg) containing aspochalasin A and B, followed by a fraction containing aspochalasin C (202 mg) and crude aspochalasin D (450 mg). A separation of the fraction that contained cytochalasins A and B on 120 g of Kieselgel at 4°C yielded 62 mg of pure aspochalasin A and 670 mg of pure aspochalasin B (Keller-Schierlein and Kupfer, 1979).

TLC of the aspochalasins was performed on Kieselgel 60-Fertigplatten F_{254} and the spots were revealed by spraying with 50% sulphuric acid and heating to 200°C, with iodine vapour or from the fluorescence under UV light (Keller Schierlein and Kupfer, 1979).

R_f of values 0.35, 0.27 and 0.54 were obtained for aspochalasins C, D and B, respectively (in ethyl acetate, blue fluorescence). In chloroform-methyl acetate (4:1), aspochalasins A and B had R_f values of 0.53 and 0.26, respectively. In the same paper, TLC data for derivatives and degradation products of aspochalasins are given.

12.1.10. Deoxaphomin, proxiphomin and protophomin

These toxins are produced by *Phoma* sp. (strain S 298) and *Phoma exigua* (Tamm and Tancubaum, 1978). Deoxaphomin, proxiphomin and protophomin were isolated as minor metabolites from cultures of *Phoma* sp., strain S 298. The main metabolites of this strain are cytochalasin A (20-dehydrophomin) and cytochalasin B (phomin).

Phoma sp., strain S 298, was grown in static culture in a medium containing (g l^{-1}) glucose (20), peptone (2), yeast extract (2), malt extract (2), KH_2PO_4 (2), $MgSO_4.7H_2O$ (2) and demineralized water. Incubation in Erlenmeyer flasks was carried out at 18°C for 11 days.

The culture medium was separated from the mycelium by filtration and the mycelium was washed three times with water. Ten literes of the filtrate were extracted four times with 8:1 of ethyl acetate and the extract was washed three times with mycelial washings. The extract was dried over calcium chloride, filtered and evaporated in vacuo. The dark-brown crude extract was dissolved in methylene chloride methanol (9:1) and filtered through alumina (activity II). From this partially purified extract, phomin (cytochalasin B) was obtained by fraction crystallized in alumina or Kieselgel. Phomin was eluted from alumina (activity II) with methylene chloride-methanol (98:2) and from Kieselgel with methylene chloride-methanol (98:2). Dehydrophomin, which moved slightly more quickly, was eluted with methylene chloride or methylene chloride-methanol (99:1).

Deoxaphomin was isolated as follows. A 100^{-1} volume of a culture was homogenized and extracted four times with 70 1 of ethyl acetate. A mixture of silicone oil and propylene glycol was used as an anti-foaming agent. The extract was evaporated and the residue (110 g) was separated between light petroleum and methanol-water (9:1). The non-polar fraction was stored and the polar fraction was purified by filtration through alumina (activity II, 200 g) in methylene chloride-methanol (9:1). Methylene chloride was added to the extract (about 35 g) and phomin was precipitated quantitatively. Crystallization from acetone gave *ca.* 4.5 g of phomin, which, on TLC (methylene chloride-methanol, 95:5), showed a polar strongly UV-active impurity. Repeated crystallization of phomin from acetone yielded 4 g of pure phomin. The mother liquors from both crystallizations were combined (ca. 1 g) and the impurity (deoxaphomin) was separated on four preparative plates (20 x 20 cm, 2 mm) with chloroform-acetone (3:1). The crude product was further purified using four preparative separations (chloroform: acetone, 3:1; twice with chloroform-acetone-formic acid, 90:5:5; chloroformacetone, 3:1). The substance was extracted with chloroform (1:1) and the extracts were checked for purity by TLC (chloroform-acetone, 3:1; chloroform-acetone-formic acid, 90:5:5). Amorphous deoxaphomin (96 mg) was obtained.

The following procedures were used to isolate proxiphomin and protophomin (Binder and Tamm, 1973). The residue after isolation of phomin and deoxaphomin (35 g)[29] was chromatographed on 1 kg of Kieselgel. Fractions 7-9 (1 1 each), eluted with methylene chloride-methanol (9:1), contained several non-polar components. Fractions 7-9 were combined and chromatographed again on 100 g of Kieselgel. From fractions 17-26 (methylene chloride, 100ml each), 180 mg of crude proxiphimin resulted and fractions 60-68 (methylene chloride-methanol, 98:2) contained protophomin and several impurities.

As proxiphomin contained two impurities (TLC) with methylene chloride-methanol, 98:2), 60 mg of the crude preparation were purified using preparative TLC (20 x 20 cm, 1 mm) with methylene chloride-methanol. Extraction of the main zone with chloroform-acetone (4:1) resulted in 55 mg of chromatographically pure proxiphomin (TLC) with methylene chloride-methanol, 98:2, and methylene chloride-ethyl acetate, 9:1).

The protophomin-containing fractions (60-68) (59 mg) were chromatographed on layers of 20 x 20 x 0.02 cm, yielding 29 mg of crude protophomin, which was submitted to further purification by TLC, twice with chloroform-acetone-formic acid (96:2:2) and once with methylene chloride-methanol (98:2). Extraction of the zones with chloroform-acetone (3:1) yielded 19 mg of almost pure protophomin, which, after crystallization from acetone, gave 15 mg of pure protophomin.

12.1.11. Chaetoglobosins and acetyl Chaetoglobosins

In this family of cytochalasins, twelve compounds have been isolated and characterized, namely chaetoglobosins A-K and three 19-0-acetylchaetoglobosins.

Producing organisms

Except for chaetoglobosins K and L, the members of this family of mycotoxins have been isolated from strains of *Chaetomium globosum*

(Springer *et al.*, 1976) Chaetoglobosins A-E were also isolated from strains of *C. molliilium*, *C. rectum*, *C. subaffine* and *C. cochlioides* (Nogawa *et al.*, 1979). Chaetoglobosins K and L are produced by *Diplodia macrospora* (Probst and Tamm, 1982). Chaetoglobosin C was also isolated from *Penicillium aurantio-virens* (Springer *et al.*, 1976). The chaetoglobosins are characterized in the same order as given in Table 12.1.

The structures of Chaetoglobosins A and B were verified and the relative stereochemistry was established for chaetoglobosin A by X-ray analysis.

Udagawa *et al.* (1979) prepared a mycelial inoculum of each strain of *Chaetomium* spp. by dissecting agar surface of 14-day-old cultures grown on oatmeal agar or potato dextrose agar slants at 23-25°C. Each fungus was grown in 1-1 Roux flasks on a medium composed of 200 g of polished rice and 20 ml of sterile water per litre flask. Prior to its sterilization the rice was well washed and soaked with tap water. After inoculation, the flasks were incubated in a static condition at 23-25°C for 21 days and in some instances at 30°C for 14 days.

Production of 19-0-acetyl derivatives of Chaetoglobosins, A, B and D by *C. globosum* (strain Lederle H-124) was as follows. The microorganism was grown in shake cultures in a medium containing (g 1^{-1}) molasses (20), glucose (10) and bactopeptone (5) in demineralized water; pH adjusted to 7.5 with sodium hydroxide before sterilization. The sterilized medium was inoculated with a spore suspension. The 500 ml Erlenmeyer flasks, each containing 250 ml of medium, were shaken at 28°C for 11 days.

To produce Chaetoglobosin C, *Penicillium aurantio-virens* was grown on shredded wheat medium. Chaetoglobosin K-producing *Diplodia macrospora* was cultured on potato dextrose agar slants at approximately 26°C for 10 days and was then maintained at 5°C until transferred to Fernbach flasks (2.8 1), each containing 100g of shredded wheat, 200 ml of Difco mycological broth (pH 4.8), 2% yeast extract and 20% sucrose (Kirksey and Cole, 1974) for production of the toxin. Inoculated flasks were incubated in the laboratory at about 24°C for 19 days.

To produce chaetoglobosins K and L, *Diplodia macrospora* was grown in standing cultures in ten 1-1 Erlenmeyer flasks. Each flask contained 40 g of shredded wheat, 80 ml of Difco mycological broth (pH 4.8), 16 g of sucrose and 1.6 g of yeast extract. The flasks were sterilized and inoculated with a suspension of spores of *D. macrospora* and incubated at 27°C for 22 days.

In screening *Chaetomium* spp. cultures for chaetoglobosin production, the moulded rice grains were extracted with methylene chloride twice at room temperature and the combined solution was evaporated. The residual rice grains were extracted with ethyl acetate twice at room temperature and the solution was evaporated. The extracts were examined by cytotoxicity tests and TLC.

The mycelium of *P. aurantio-virens* was extracted with chloroform and, after concentration, subjected to a series of elutions on a silica gel column. Toxicity was found in the ethylacetate fraction (oral, 1-day-old cockerels). A second column employing gradient elution (chloroform-ethyl acetate) was used for further purification. The crude toxin was crystallized from ethyl acetate to give small needles of Chaetoglobosin C.

From submerged cultures of *C. globosum*, the isolation was as follows (Prohst and Tamm, 1981). The same volume of ethanol was added to the culture broth, the mixture kept over night at 4°C and the mycelium separated by filtration. The ethanol was then evaporated in vacuo at 40-50°C and the aqueous layer extracted four times with methylene chloride. The organic layer was dried over sodium sulphate and evaporated in vacuo at 40°C. The brown residue was dissolved in benzene and kept for 3 days at room temperature. The crystallized mixture of Chaetoglobosin A and 19-0-acetylchaetoglobosin A was separated by crystallizaton and chromatographed on silica gel (100:1) using increasing amounts of methanol in methylene chloride. The benzene solution was evaporated in vacuo at 40°C and the residue was chromatographed on silica gel (100:1) in the same manner. The isolated yields, referred to a 1-1 scale, were 200-700 mg for 19-0-Acetylchaetoglobosin A, 150-600 mg for chaetoglobosin A, 70-80 mg for chaetoglobosin C, 10 mg for 19-0-acetylchaetoglobosin D and 6 mg for 19-0-acetylchaetoglobosin B.

Chaetoglobosin K was isolated from growing culture of *Diplodia macrospora* with the help of 300 ml of acetone (Cutler *et al.*, 1980). The mycelium and substrate were macerated with a homogenizer and the suspension was strained through cheese cloth to remove the pulp. The liquid was filtered through Whatman No. 1 paper and the clarified filtrate was reduced in vacuum at 50°C to yield the aqueous phase. This was extracted twice with ethyl acetate, each volume of the solvent being equal to twice that of the aqueous portion. The combined extracts were dried over sodium sulphate, reduced to a small volume in vacuo, then placed on a silica gel (70-230 mesh) chromatographic column (100 x 9 cm I.D.) that had been slurry packed in benzene, followed by stepwise elution with 1/1 each of benzene, diethyl ether, ethyl acetate, acetone and methanol. Individual solvents drained to the top of the silica gel before addition of the next sequential solvent. The acetone fraction, exhibiting biological activity, was further fractionated by first reducing it to a small volume in vacuo at 50°C and placing it on a silica gel (70-230 mesh) column (50 x 4.0 cm I.D.) in benzene. The 250 ml of benzene were added to the top and immediately a linear gradient of benzene (a further 1 1) to acetone (1 1) was used for elution.

Fractions of 20 ml were collected, evaporated to approximately 2 ml and further examined by TLC and bioassay.

Chaetoglobosins K and L were isolated from static cultures of *Diplodia macrospora* with the help of 150 ml of acetone were added. After transferring mycelium and substrate into a 5-1 beaker, 1.5-1 of 70% aqueous acetone (Probst and Tamm, 1982) were added and the suspension was macerated with a homogenizer and filtered through a layer of celite. After washing the residue with acetone, the organic solvent in the filtrate was removed in vacuo. The remaining aqueous phase was extracted four times, each with an equal volume of methylene chloride. The dried extracts (sodium sulphate) were evaporated in vacuo and the resulting crude product (red-brown oil) was dissolved in 90% aqueous methanol and extracted three times, each with an equal volume of light petroleum. The combined light petroleum extracts were dried (sodium sulphate) and evaporated in vacuo. The remaining water-methanol phase was diluted with water, the methanol removed in vacuo and the aqueous phase extracted three times, each with an equal volume of methylene chloride, and a further three times, each with an equal volume of ethyl acetate. The extracts were dried separately (sodium sulphate) and evaporated in vacuum. TLC of the three extracts with methylene chloride-methanol (95:5) as solvent showed that only the methylene chloride extract contained compounds with a positive reaction for phenols and indoles after spraying with 5% solutions of ammonium cerium (IV) nitrate in acetone and of hydroxylammonium chloride in 80% aqueous acetone. The light petroleum and ethylacetate extracts were discarded. The methylene chloride extract (ca. 2 g of red-brown oil) was chromatographed on 200 g of silica gel using increasing amounts of methanol in methylene chloride. The fractions eluted with methylene chloride-methanol (97:3) contained Chaetoglobosin K. In the following fractions eluted with methylene chloride-methanol (95:5), chaetoglobosin L and diplosporin were present.

For the separation of Chaetoglobosins K and L (Probst and Tamm, 1982) on TLC, CH_2Cl_2: methanol (95:5) and toluene-ethylacetate-formic acid (5:4:1) served as solvents. The fractions from the column chromatography containing the chaetoglobosins K and L were purified by preparative TLC on silica gel with toluene-ethyl acetate-formic acid (5:4:1).

Chaetoglobosins A-E were chromatographed on silica gel F_{254} using benzene-ethylacetate (1:1) and benzene-chloroform-methanol (10:10:3) as the solvents. Metabolites were detected by UV irradiation at 254 and 365 and by spraying with Ehrlich's reagent (5% in dilute hydrochloric acid) and colouration after heating (Udgawa *et al.*, 1979).

The R_f value of Chaetoglobosin K was 0.53-0.56 on silica gel 60 TLC plates developed with toluene-ethylacetate-formic acid (5:4:1) and it was observed as a dark-spot under short wave UV light (Cutler *et al.*, 1980).

Decreasing R_f values of the five Chaetoglobosins on the same sorbent and methylene chloride-methanol (95:5), showing increasing polarity from left to right: 19-0-acetylchaetoglobosin A > chaetoglobosin C > 19-0-acetylchaetoglobosin B > 19-0-acetylchaetoglobosin D > chaetoglobosin A (Probst and Tamm, 1981).

Natural incidence: Except sporadic reports of incidence of these mycotoxins, no detailed investigations are available.

12.2. BIOLOGICAL ACTIVITY

These mycotoxins exhibit variety toxicities, such as cytotoxicity, mutagenicity, teratogenocity and cytophysiological effects including metabolism and macromolecular synthesis, interacts with contractile system and alters membrane structure and function.

Considerable variation exists in toxicity of different cytochalasins towards different microorganisms. Cytochalasin A was more toxic to *E. coli,* while cytochalasin B and D were least toxic. Similarly cytochalasin A was potent fungicide, while protozoa and *Euglena* were unaffected by these toxins. The cytotoxicity of cytochalasins, Chaetoglobosins and zygosporins are précised in Table 12.4.

Table 12.4. Cytotoxicity and acute toxicity of the cytochalasins

	Cytotoxicity	Animal Toxicity					
Toxin	ID_{50} (µg/ml)	Species	Sex	Age or Weight	Route	LD_{50} (mg/kg)	Reference
Cytochalasin A	3.2	--	--	--	--	--	Umeda *et al.* (1975)
Cytochalasin B	1.0-3.2	--	--	--	--	--	Umeda *et al.* (1975)
Cytochalasin C	<0.32	Mouse (Swiss albino, DS)	M	18-20g	i.p.	>100	Umeda *et al.* (1975)
	0.12						Minato *et al.* (1973)
Cytochalasin D	0.32-1.0	Mouse (Swiss albino, DS)	M	18-20g	i.p.	10	
	0.15						
	0.89	Mouse	--	--	s.c.	1.85	
					p.o	36.0	
		Mouse (Swiss Webster)	F	32g	i.p.	2.1	Shepard and Greenaway (1977)

		Mouse (BALB/c)	F	22g	i.p	2.2	Shepard and Greenaway (1977)
		Mouse (C578L/6J)	F	29g	i.p.	2.2	Shepard and Greenaway (1977)
Cytochalasin E	--	Mouse (Carworth)	M	21g	i.p.	4.60	Glinsukon (1974)
			F	20g	i.p.	4.90	
		Guinea pig (Hartley)	--	Wean-ling	i.p.	0.5-1.5	Glinsukon (1974)
		Rat (Fisher)	M.F	1dy; 5g	i.p.	0.98	Glinsukon (1974)
			M	69g	i.p.	2.60	
			F	71g	i.p.	2.61	
			M	70g	p.o	9.10	
			M	80g	i.t.	1.30	
Cytochlasin H	--	Chicken (DeKlb)	M	1day	p.o.	12.5	
Kodo-cytochalasin	--	Mouse	--	--	--	<2	
Zygosporin D	0.4	Mouse (Swiss albino, Ds)	M	18-20g	i.p.	10	Minato *et al.* (1973)
Zygosporin E	0.4	--	--	--	--	--	Minato *et al.* (1973)
Zygosporin F	≥10	--	--	--	--	--	Minato *et al.* (1973)
Zygosporin G	0.68	Mouse (Swiss albino, Ds)	M	18-20g	i.p.	750	Minato *et al.* (1973)
Chaetoglobosin A	3.2-10	Mouse (DDD)	M	5weeks	s.c.	6.5	Umeda *et al.* (1975)
			F	5weeks	s.c.	17.8	Umeda *et al.* (1975)
		Mouse (DDD)	M.F	5weeks	Po. >400		Umeda *et al.* (1975)
		Rat (Wistar)	M.F	5weeks	i.p. 2.0		Umeda *et al.* (1975)
			M.F	5weeks	p.o.	>400	Umeda *et al.* (1975)
Chaetoglobosin B	3.2-10	--	--	--	--	--	Umeda *et al.* (1975)
Chaetoglobosin C	10-32	--	--	--	--	--	Umeda *et al.* (1975)

Chaetoglobosin D	3.2	--	--	--	--	--	Umeda *et al.* (1975)
Chaetoglobosin E	3.2	--	--	--	--	--	Umeda *et al.* (1975)
Chaetoglobosin F	10	--	--	--	--	--	Umeda *et al.* (1975)
Chaetoglobosin G	3	--	--	--	--	--	Sekita *et al.* (1977)
Chaetoglobosin J	3	--	--	--	--	--	Sekita *et al.* (1977)

a. The concentration of toxin producing an actual or approximated inhibition of growth in Hela cell cultures; b. LD_{100} values; c. The degree of inhibition was not stated.

12.2.1. Cytokinesis

Cytochalasins are reported to inhibit cell division or cytokinesis (Schroeder, 1978) without affecting nuclear division which results in the formation of binucleate cells.

12.2.2. Locomotion and cytoplasmic streaming

Cytochalasin B was a rapid and reversible inhibitor of cellular locomotion and ruffling activity at low concentration but did not affect flagellar or ciliary movement at higher concentrations (Carter, 1997). Locomotion was totally inhibited in human, horse and rabit leukocytes after 1 hr exposure to cytochalasin B at 2.5 µg/ml concentration. Inhibition of locomotion by cytochalasin B was not dependent upon mucopolysaccharide synthesis in chicken embryo heart cells. Cytoplasmic streaming in plant cells was also inhibited (Thomas *et al.*, 1978). No ultrastructural alteration in microfilaments of *Nitella* cells was observed during inhibition of streaming.

12.2.3. Mutagenicity and teratogenicity

Cytochalasin B is reported to cause long multicentric chromosomes (Oneill, 1972). Human lymphocytes exposed to cytochalasin H developed a variety of chromosomal aberrations in 25-50% cells including tetraploidy and premature chromosomal condensation, extension and construction (Mulkeherkar *et al.*, 1978). Cytochalasins also exhibit teratogenicity. Cytochalasin B caused neural tube closure in chicken embryo which includes everted neural tubes, absence of optic vessels and general retardation of neuralation. Increased foetal resonation was also noted. Exencephally was the predominant defect along with other abnormalities including hypognathia and coxial skeletal defects.

12.2.4. Cytophysiological effects

Interestingly large numbers of cytophysiological processes are altered by cytochalasins. Most of the information is based on *in vitro* system and their relevance to animal toxicity can only be summarized. Subacute cytochalasin exposure may produce a spectrum of subtle but profound effects in animals. Cytochalasin action is rapid with effects noted within the first few minutes after exposure to the toxin and even the reversibility of action is also fast after cytochalasin removal suggesting non-covalent interaction of toxin to cellular receptor molecules.

12.2.5. Attachment and adhesion

The cytochalasins affect the attachment of suspended cells to a solid surface and the substrate adhesivity of these cells once attached. Attachment and the concomitant process of cell spreading are mediated by the cellular skeletal and contractile systems. Once a cell comes to rest on the substrate, microspikes or filopodial extensions arise from the cortex around the circumference of the base and contact with and attach to substratum. At this stage cell spreading occurs which involves the extension of flat lamella of peripheral cytoplasm from the central apex of the cell which terminates in a raised brim in a configuration. Similar to that of a Mexican hat 30-60 min after initial setting of the cell which gets inhibited and gets dislodged by cytochalasin.

12.2.6. Morphogenesis

The effects of cytochalasins on morphogenesis and fertilization in different plant and animal experimental system have been studied and they are summarized in Table 12.5. Cytochalasin B was effective in blocking axonal growth in culture chicken embryo dorsal root gansha.

12.2.7. Endocytosis

The endocytosis is a process in which extracellular fluid particles are incorporated by the cell via invagination of plasmamembrane and the resultant formation of endocytic vacuoles or vesicles. There are three types of endocytosis which are characterized by the size of the particles ingested or the vesicles formed.

1. **Phagocytosis**: Ingestion of particles more than 1 μm in diameter and is performed by macrophages and polymorphonuclear leucocytes and produces vesicles 200 – 700 nm indicator.

2. **Macroendocytosis:** is also noted in macrophages and certain specialized epithelial cells and produces vesicles 200–700 nm in diameter.
3. **Micropinocytosis:** is responsible for protein transport in vascular endothelial cells and forms microvesicles of 60-70 nm in diameter.

Macropinocytosis was inhibited by cytochalasin B, while micropinocytosis was insensitive. The effects of cytochalasis on endocytosis are précised in Table 12.6.

Table 12.5. Effect of cytochalasins on fertilization and morphogenesis

Biological systems	Dosage/duration	Response	Reference
Brown alga eggs and embryos (*Pelvetia fastigiata*)	100 μm CB/ml; 30 min 4 μg CB/ml; 14 days	Inhibition of photopolarization in eggs Inhibition of apical hair formation	Nelson and Joffe (1973)
Polyporus biennis	20μg CB/ml; 3 hr	Irregular hyphae growth; inhibition of initiation of clamp connection formation	Betina *et al.* (1975)
Neurospora crassa	50μg CD/ml; time not given	Stimulation of hyphae branching	Zismek and Epel (1975)
Surf clam eggs (*Spisula solidissima*)	0.5-5μg CB/ml; 10 min	Prevention of polyspermy block after fertilization	Goldsomho *et al.* (1977)
Echiuroid worm eggs and sperm (*Urechis caupo*)	1-2μM CB;30 sec-10min	Inhibition of sperm poenetration and formation of sperm penetration cone	Raff (1972)
Marine snail embryos (*Ilyanassa obsolete*)	1-10μg CB/ml; 1.5-4hr	Prevention of polar lobe formation	Arnold and Hilliams-Arnold (1974)
Squid embryos (*Loligo pealei*)	0.2-2.0μgCB/ml; 10-30min	Blastodisc formation and polar body extrusion inhibited in porecleavage embryos; generalized anamalies of organ placement and differentiation in 3- to 4- day embryos	Lash *et al.* (1973)
Ascidian larvae (*ten species*)	0.25-10μg CB/ml; 90 sec-4hr	Inhibition of tail resporation in metamorphosis via disruption of apical filaments and porevention of caudal epidermis contraction	Luchlel *et al.* (1976)

Frog embryos (*Xenopus laevis*)	7.5μg CB/ml; 4-10min	Regression of furrow formation during cleavag and interference with cell junction development	
Newt embryos (*Triturus alpestris*)	2-10μg CB/ml; 0-100min	Regression of first cleavage furrow	Solman *et al.* (1976)
Chicken embryos	10μg CB/ml	Inhibition of neurulation and separation of neural folds with concomitant disturbances of micro-filament arrays	Karfunkel (1972)
Chicken embryos	5μg CB/ml; 3hr	Prevention of nuclear migration in neural ells inhibition of estrogen-induced tubular gland formation and regression of presently existing gland cells and associated microfilaments	Messler and Auclair (1974)
Chicken embryo spinal cord and dorsal root ganglia cell cultures	7-10μg CB/ml; 0-24hr	Cessation of axon elongation and alteration of microfilament network	Yamada *et al.* (1970)
Chicken oviduct explants	5 μg CB/ml; 3 hr	Prevention of nuclear migration in neutral cells	Messler and Allclai (1974)
	30-50 μg CB/ml 6 h4	Inhibition of estrogen induced tubular gland formation and regression of presently existing gland cells and associated microfilament	Yamada *et al.* (1970).
Chicken embryo duodenal explants	0.1μg CB/ml; 15 min-6 hr	Formation of elongated branched microvilli and alteration of their spatial distribution	Burgers and Grey (1974)
Chicken embryo duodenal explants	1.0μg CB/ml; 24-36hr	Prevention of previllous ridge formation and disruption of microfilament bundles	Burgers (1975)
Chicken embryo cardiac muscle	20-100μM CB; 45min-4hr	Disruption of preexisting myofibrils and inhibition of new myofibril formation; loss of Z-babds and disarrayed myosin filaments	Manasek *et al.* (1972)

Chicken embryo limb bud mesodermal cells	2.5μg CB/ml; 24-72hr	Irreversible inhibition of chondrogenesis. Independent of reduction in porotein synthesis or glucose uptake	Parker *et al.* (1978)
Mouse oocytes	5μg CB/ml; 5-16hr	Arrest of oocyte meiosis evidenced by inhibition of polar body emission and loss of homologous chromosome separation	Wassarman *et al.* (1976)
Mouse eggs and spermatozoa	5-10 μg CB/ml; 1-7hr	Reduction of fertilization rate; induction of triploidy in fertilized eggs	Neimierko and Kumar (1976)
Preimplantation mouse embryos	0.25-10μg CB/ml; 0-24hr	Inhibition of morula to blast-ocyst transformation; decrea-se of trophoblast out growths noted within 3 to 5 min	Granholm and Brenner (1976)
Mouse embryos	10μg CB/ml; 3-8.5hr	Induction of tetraploidy and 2n/4n mosaic embryos	Tarkowski *et al.* (1977)
Mouse embryos	10μg CB/ml; 2hr	Abnormal structure and development of neuroepithelial cells	Webster and Langman (1978)
Mouse embryo salivary glad epithelium cultures	7-10μg CB/ml; 18-72hr	Loss of normal branching morphogenesis noted within 6 to 10 hr; microfi-lament disorganization	Spooner and Wessells (1972)
Mouse skeletalmuscle explants	0.25-2μg CD/ml; 0-7 days	Increased contractility, retraction and segmental constriction of myotubes after short exposures; pro-longed treatment resulted in formation of leptomeric bodies and myofibril disorientation	Miranda and Godman (1973)

12.2.8. Exocytosis

Exocytosis is a process by which cells secrete substances into the extracellular space and is reported to be inhibited by cytochalasins.

12.2.9. Immunological functions

Several immunological important processes are altered by cytochalasin treatment. For the most part, the nature of the affected mechanisms is unknown although membrane surface properties and other factors controlling cellular aggregation are undoubtedly implicated. The reversible inhibition of chemotoxis by cytochalasin B has been reported for human rabbit leucocytes responding to bacterial cells.

12.2.10. Viral infectibility

Cytochalasin B and C reversibly enhance the infectivity of isolated HSV-DNA from 3-6 folds in rabbit kidney cells, udeliving the role of the viral proteins play in modulating host cell functions. An additive effect on viral infectivity was observed. Cytochalasin D exerted additive effect on viral infectivity.

Both cytochalasin B and dihydrocytochalasin totally inhibited Ca-dependent blood clot retraction in platelet rich human plasma. Cytochalasin B also caused osmotic fragility and increased deformability of human erythrocyctes, water movement across food bladders was promoted by the action of vasopressin or cyclic AMP. These also caused profound disruption of fluid balance mechanisms in rodents by cytochalasin F or Chaetoglobosin A (Ohtsubo *et al.*, 1978). Inhibition of fluid transport was also accompanied by alternations in the electrical properties of the membrane by cytochalasin B. Neurotransmission was also blocked through cat cerrical sympathetic ganglia by high concentration of cytochalasin B (Nakazato and Douglas, 1973). Low levels of cytochalasin B caused a rapid nearly total loss of end plate activity at nerve muscle contracts in cultured chicken embryo muscle cells and spinal cord explants (Rubin *et al.*, 1976). Cytochalasins also blocked the fast axoplastric transport of radio labelled proteins in rat sciatic nerve (Crooks and Mc Clure, 1972). Cytochatasins completely inhibited *in vitro* ovulation in ovary fragments of frogs (Schrocder, 1973) and fish (Schroedar and Pendergross, 1976). Cytochalasin B also caused a reversible dose-dependent blockage of MSH or cyclic AMP induced darkening in frog skin which was correlatable with a drastic reduction in microfilaments of the dendritic process of melanocytes (Malawista, 1971).

12.2.11. Metabolism and macromolecular synthesis

Cytochalasin B at 5 mg/ml concentration inhibited the glucose incorporated to glycogen in chicken embryo myoblasts. Similarly it inhibited the glycolysis. At high concentration cytochalasin B (100 μM) caused total inhibition of ATP synthesis in mitochondria probably due to rapid dose dependent inhibition of ATP and Mg^{++} induced mitochondrial contraction. Cytochalasin D and E caused 50% activation of superoxide generation (Nakagawara *et al.*, 1976). There is also an inhibition of nucleic acid and mucopolysaccharide synthesis by cytochalasins. Cytochalasins inhibited cytokinesis but did not affect immediately nuclear division with the result that bi, tri or multinucleated cells were produced. Cytochalasin B has stimulated orotic acid incorporation into mouse 3T3 fibroblast RNA by four fold which may be attributed to increased transport of the precursor (Cohn *et al.*, 1972). Inspite of inhibition of variety of processes of cells like glucose and glucosamine uptake (Cohn *et al.*, 1972), morphological alterations (Cohn *et al.*, 1972), lectin induced agglutination (Kaneko *et al.*, 1975), cell sorting (Shepro *et al.*, 1969), cytokinesis (Estensen, 1971) antiviral interferon activity (Bourgeade and Chaney, 1971) lectin induced agglutination (Kaneko *et al.*, 1975), virus replication (Dix and Courtney, 1976), locomotion (Spooner *et al.*, 1972), ruffling membrane activity (Zigmond and Hirsch, 1972) and cytoplasmic streaming (Bradley, 1973), protein synthesis remained unaffected. Cytochalasin B was potent inhibitor of glucosamine uptake (Plagemann and Erbe, 1973). Cytochalasin may also interfere with cellular metabolism and function at points other than plasmamembrane and contractile microfilaments. Mitochondrial and nuclear processes also may be directly affected. Cytochalasin B altered some as yet undefined intermediary step in the hormone-receptor pathway. Cytochalasins affect microfilament structure *in vivo* and do interact with various components of contractile systems *in vitro* (Lazarides, 1976).

Cytoplasmic gelation in HeLa cell (Weihing, 1976) and *Acanthamoeba* extract (Pollard, 1976) was also inhibited by cytochalasin A, B and D. The disaggregation of actin from a high molecular weight protein fraction was implicated in this process (Weihling, 1976). Cytochalasin A disrupted microfilament in mouse 3T3 cells produced actin-containing rod elements in addition to the aggregate central actin masses (Rathket *et al.*, 1977).

12.2.12. Alterations in membrane structure and function

Cytochalasins mainly affect plasmamembrane at cellular level. These mainly causes alterations in the spatial arrangement and surface accessibility of specific membrane components including antigen, lectin, and cell surface polypeptide hormone receptors and inhibition of membrane transport functions, particularly hexose transpot system. Cytochalasins interact with high affinity membrane binding sites atleast one of which appear to be intimately associated with glucose carrier. Cytochalasin B altered the availability of surface sialic acid groups. Cytochalasin B blocked the ATP induced recovery of cell mobility in rat erythrocytes previously impaired by X-ray irradiation (Salto *et al.,* 1973). Further, they concluded that Cytochalasin B interfered with contractile proteins in the membrane.

12.2.13. Structure activity relationships

At present information on structure activity relationship of only cytochalasins A to E is available (Table 12.1) practically no information on zygosporins and cheatoglobosins is available. Further with the data of only 2 or 3 congers, no unequivocal statement of relative potency are possible to make table 12.7. Comparative studies on CB and dehydro–CB in inhibiting number of morphological and motile processes in several biological systems including zeiosis and arborization, locomotion, cytokinesis and protoplasmic streaming revealed equal potency inspite of the fact that dihyro-CB was ineffective in inhibiting sugar transport. CA us unique among cytochalasins in that it behaves as a sulfhydryl reagent. Further, biological effects of CA are reversible. These effectiveness may be prevented if the system in pretreated with thiols, such as glutathione, cysteine or dithiothreitol. Structure activity relationships at t present information on structure activity relationship of only cytochalasins A to E is available. Practically no information on zygosporins and chaetoglobosins is available. Further with the data of ony 2 or 3 congers no uniquelocal statements of relative corporative potency are possible to make (Table 12.7). Studies of on CB in inhibiting number of morphological and mortile process in several biological systems including zeiosin and arborization, locomotion, cytokinesis and protoplasmic streaming revealed equal potency inspite of the fact that dihydro-CB was ineffective in inhibiting sugar transport. CA is unique among cytochalasins in that it behaves as a sulfhydryl reagent. Further biological effects of CA ae not reversible. These effects may be prevented if the system is pretreated with thiols, such as glutathione, cysteine or dithrothreitol.

Table 12.6. Effect Cytochalasins on Exocytosis

Biological system	Secretory product	Conditions	Response	Reference
Human leukocytes	Lysosomal enzymes (β-glucuronidase marker)	5μg CB/ml; 15 min	An approximate 35% increase in enzyme release after zymosan particle challenge	Zuber *et al.* (1973)
Rabbit leukocytes	Lysosomal enzymes (β-glucuronidase marker)	0.5-10μg CB/ml; 15-30 min	Enhanced release of enzymes following stimulation with retinal or rabbit globulin coated latex particles and antigen antibody complexes; enzymes not rele-	Hawkins (1973)

			ased from isolated secretory granules	
Mouse peritoneal macrophages	Lysosomal enzymes (β-glucuronidase and (β-galactosidase markers)	2-10μg CB/ml; 5-72hr	Time and dose-dependent release of enzymes; effect noted after 24-hr exposure to 10μg CB/ml	Davis *et al.* (1973)
Guinea pig peritoneal leukocytes	Lysosomal enzymes (β-glucuronidase marker)	2μg CB/ml; 30min	An approximate 45% increase in enzyme release without particle challenge	Tou and Stjomhom (1974)
Human blood platelets	Lysosomal enzymes (β-glucuronidase marker) N-acetyl-glucosam-inidase)	1-50μg CB/ml; 10 min	Enzyme release stimulated by CB only after collagen induction	Haslam *et al.* (1975)
Mouse perit-oneal macrop-hages (Thiglyco-llate-stimulated)	Elastase and lysozyme	4μg CB/ml; 24-72hr	An approximate 60% inhibition of clastase secretion after 48hr; no effect on lysozyme	Gordon and Werb (1976)
Rat parotid tissue Slices	α-Amylase	52μM CB; 1 hr	Stimulated α-amy-lase release at low cyclic AMP levels (<0.3mM); and high cyclic AMP (1.0mM) induced amylase release	Butcher and Goldman (1974)
Water mold Achyliya ambisexualis	Cellulase	2-20μM CB; 1hr	Reversible dose-dependent inhi-bition of cellulase secretion; variable effects with CB	Thomas *et al.* (1974)
Rabbit and human fibro-blast cultures	Collagenase	1-10μg CB/ ml; 2-8 days	Enhanced collag-enase activity in medium	Harris *et al.* (1975)
Rat liver slices	Plasma proteins (albumin)	5-50 μM CB; 3 hr	No significant inh-ibition of plasma protein release	Redmann *et al.* (1975)
Rat pancreas pieces	Chymotrypsinogen and lipase	0.1-10μg CB/ml; 1hr	Optimum stimu-lation of enzyme release at 0.5-2 μg CB/ml; 1hr	Banduin *et al.* (1975)

Guinea pig hypogastric nerve-vas differences preparation	Dopamine-β-hydroxylase	3μg CB/ml; 75mins	Blockage of electrically induced enzyme release; little effect on basal enzyme ecretion	Tou *et al.* (1974)
Rat lymph node cells	Lymphotoxin	0.05-5μg CB/ml; 3days	Dose-dependent inhibition of ovalbumin-stimulated lymphotoxin production	Yoshinaga *et al.* (1972)
Mouse plasmacytoma Cells	1g	100μM CB; 4hr	No effect on Ig release	Parkhouse and Allison (1972)
Mouse spleen cells	1g antibodies	3-300μg CB/ml; 18hr	No effect on antibody secretion toward sheep erythrocytes	Teplitz *et al.* (1974)
Mouse spleen cells and viral-infected BHK Cells	Interferon	0.6-12.5μg CB/ml or 0.12-10μg CD/ml; 8hr	Complete inhibition of interferon production at 10μg CB/ml or 0.37μg CD/ml	Ito *et al.* (1976)
B. Hormones				
Rat pancreas islets	Insulin	10μg CB/ml; 90 min	Increased insulin release following glucose induction; no effect on basal insulin secretion without glucose stimulation; effects reversible	Orci *et al.* (1972)
Rat pancreas islets	Insulin	1-30μg CD or 15 μg CB/ml; 30-60 min	An approximate doubling of CD-enhanced glucose-induced insulin release at 5 μg CD/ml; CB, but not CD, blocked alloxan inhibition of glucose-induced insulin release	Mc Daniel *et al.* (1975)
Rat pancreas islets	Insulin	10μg CB/ml; 90 min	Stimulation of glyceraldehydes-leucine-, and glucose-induced insulin secretion	Levy *et al.* (1976).

Mouse thyroid glands	Thyroid hormones	0.1-10μg CB/ml; 5hr	Dose-dependent increase in basal thyroid secretion at ≥ 1.0 μg CB/ml; total inhibition of thyroid stimulating hormone (TSH)- induced secretion at 3.0 μg CB/ml	Williams and Wolff (1971)
Dog thyroid slices	Thyroid hormones	0.1-30μg CB/ml; 2-6hr	Complete inhibition of cyclic AMP and TSH-induced thyroid secretion at 1 μg CB/ml; effects reversible.	Neve *et al.* (1972)
Bovine parathyroid pieces	Parathyroid hormones	5μg CB/ml; 30min-3hr	Enhancement of hormone secretion (approximately 50%) 1hr after treatment; additive effect with retinal induction of secretion	Chertow *et al.* (1974)
Bovine pituitary Gland slices	Growth hormone	10 μg CB/ml; 195 min	Inhibition of prostaglandin induced stimulation of hormone release; no effect with CB alone	Cooper *et al.* (1972)
Rat pituitary Glands	Prolactin and growth hormone	10μg CB/ml; 6 – 7 hr	Decreased secretion of prolactin in female rat pituitary glands and growthy hormone in male and female rat glands.	Maclead *et al.* (1973)
Frog and rat pituitary glands	Melanophore stimulating hormone (MSH)	0.01 – 100 μg CS/ml 2-6 hr	Time-dependent, reversible inhibition (up to 60%) of MSH release; no inhibition of release from frog glands in the poresence of glucose	Bower and Hadley (1975)

C. Neurotransmitters

Guinea pig hypogastric nerve-vas deferens preparation	Norepinephrine	3µg CB/ml; 75 min	Inhibition of electrically induced transmitter release; little effect on basal secretion in unstimulated poreparations	Thoa *et al.* (1972)
Rat cervical ganglia	Acetylcholine	0.5-30µg CB/ml; 1hr	Dose-dependent inhibition of electrically stimulated transmitter release; half-maximal inhibition at 1.5µg CB/ml	Pumplin and McClure (1974)
Human blood platelets	5-hydroxytryptamine	1-50µg CB/ml; 10 min	Potentiation of release only after collagen fiber induction	Haslam *et al.* (1975)
Rat brain synaptosomes	Noradrenaline, dopamine and α-aminobutyric acid (GABA)	0.15mM CB; 15-30 min	Inhibition of K^+-stimulated dopamine (30%) and noradrenaline (90%) release; no effect on GABA release	Nicklas and Berl (1974)

D. Other Substances

Pollen antibody-treated human skin	Histamine	2-20µM CB: 105 min	Inhibition of skin histamine release after pollen antigen exposure	Yamamoto *et al.* (1973)
Serum-sensitized rat peritoneal mast cells	Histamine	1-100µM CB or 1-50 µM CA; 30-90 min	Total inhibition of histamine release with 10 µM CB	Orr *et al.* (1972)
Serum-sensitized human lungs	0.01-5µg/ml CA or CB; 15 min	Antigen-induced stimulation of histamine release with CA treatment; variable effects with CB; CA effects not reversible		Orange (1975)

Human leukocytes	Histamine	0.1-10μM CB or CD; 0.1-100μM CA; 30min	Equal enhancement of complement induced histamine release by CB and CD; no effect with CA	Hook and Siraganian (1977)
Mouse BALB/3T3 cells	Prostaglandins	0.1-5μg CB/ml or 0.3-5μg CA/ml; 10-120 min	Dose-dependent inhibition of thrombin-, bradykinin, and serumstimulated prostaglandin release by CB; CA approximately one third as active	Pong and Devine (1976)

Table 12.7: Comparative Potency of Cytochalasin Congeners

Biological effect of process	Biological system	Response[a]	Potency series[b]	References
Anti bacterial activity	*Escherichia coli: Bacillus subtilis*	Enhancement	CA>CD>(CB)	Betina and Micekova (1972)
Anti fungal activity	*Saccharomyces cerevisiae: Botrytis cinerea*	Enhancement	CA>(CB,CD)	Betina and Micekova (1972)
Acute toxicity	Mouse	Enhancement	CE ~ = CD>CC	Glinsukon *et al.* (1974)
Cytotoxicity	HeLa cells	Enhancement	CC ~=CD>CB~=CA	Umeda *et al.* (1975)
Teratogenicity	Chicken embryo explants	Enhancement	CD>CA>CB>CE	Greenway and Plus (1977)
Zeiosis and arborization	Mouse fibroblasts (3T3)	Enhancement	CE>CD>CB = dihydro-CB	Atlas and Lin (1928)
Enucleation	Mouse mammary tumor cells (MCH-C18)	Enhancement	CB = dihydro-CB	Atlas and Lin (1978)
Cytokinesis	Mouse embryo cell line (Rose)	Inhibition	CE>CA>CD>CB	Greenway and Plus (1977)
	Mouse fibroblasts (3T3)	Inhibition	CB ~ = dihydro-CB	Atlas and Lin (1978)

	Fertilized sea urchin eggs	Inhibition	CB = dihydro-CB	Lin *et al.* (1998)
Membrane ruffing activity	Moiuse fibroblasts (3T3)	Inhibition	CE>CD>CB = dihydro-CB	Lin *et al.* (1998)
Cytoplasmic streaming	*Elodea leaves*	Inhibition	CB = Dihydro-CB	Lin *et al.* (1998)
Adhesion	Ehrilich ascites cells	Inhibition	CA>CB>CE	Lin *et al.* (1998)
Axonal outgrowth	Chicken embryo dorsal root ganglia	Inhibition	CB = dihydro-CB	
Histamine release	Rat peritoneal mast cells	Inhibition	CA>CB	Orr *et al.* (1972)
	Human lung tissue	Enhancement	CA>CB	Orange (1975)
Insulin secretion	Rat pancreatic islets	En	CD = CB	Williams and Wolff (1971)
Lymphcyte-mediated cytotoxicity	Immune mouse spleen lymphocytes	In	CA = CB	Bubbers and Henny (1975)
Rosette formation	Rabbit alveolar macrophages and complement-coated human erythrocytes	In	CA>CE>CD>CB	
Mitogen-induced lymphocyte activation	Human peripheral blood lymphocytes	En	CA = CD = CE >>CB> (dihydro-CB, γ-lactone CB)	Greene *et al.* (1976)
Viral infectivity	Poliovirus and Vero cells	En	CD>CB	Ponem (1977)
	Herpes simplex virus and rabbit kidney cells	In	CD>CB	Farber and Eberle (1976)
Blood clot retraction	Human platelet-rich plasma	In	CB = dihydro-CB	Henney (1978)
Superoxide generation	Guinea pig polymor-phonuclear leukocytes	En	CE = CD>CC>(CA, CB)	Nakagawara *et al.* (1976)
Polymerization of G-actin	Rabbit muscle	In	CA > CB	
Insulin binding to	Human llymph-	In	CA = CB = CD	

cell surface	ocyte cell line (1M-9)			
Human growth hormone binding to cell surface	Human lymphocyte cell line (1M-9)	In	CA>CB	
Glucose transport	Human crythrocyte ghosts	In	CB>>CA> (CC, CD,CE, dihydro-Cb, γ-lactone CB)	Lin *et al.* (1998)
	Mouse fibroblasts (3T3)	In	CB> (dihydro-CB)	Atlas and Lin (1978)
Binding to high affinity membrane Site 1 (glucose carrier)	Human crythrocyte ghosts	En	CB>>CA> (CC,CD,CE dihydro-CB, γ-lactone CB)	
Binding to high affinity membrane site 11 (nonglucose carrier site)	Human crythrocyte ghosts	En	CE>CA>CB>dihydro-CB> γ-lactone CB	
Binding to high affinity memb-rane sites (nong-lucose carrier sites)	Human crythrocyte ghosts	En	CE>CD> dihydro-CB ~ = CB	
Binding to membrane receptors	Human peripheral blood lymphocytes and leukocytes	En	CE > CA > CD>CD	
	Mouse fibroblasts (3T3)	En	CD>CB ~ = dihydro-CB	Atlas and Lin (1978)

a. En = enhancement; in = inhibition.

b. Congeners listed in parentheses were inactive under the conditions employed

CHAPTER – 13

TERATOGENIC MYCOTOXINS

13.1. ZEARALENONE

Zearalenone (ZEN), a phenolic resorcylic acid lactone, was first isolated and characterized in 1962 from cultures of *F. graminearum* obtained from corn and implicated in a field outbreak of hyperesterogenism in pigs in the United States by Stob *et al.* (1962). Zearalenone is a nearly universal corn contaminant and often occurs in the same sample with trichothecenes (Lud *et al.*, 1990). Despite its dissimilarity to steroid compounds, ZEN produces potent hyperesterogenic responses in susceptable animals which is unique among known mycotoxins (Mitterbauer *et al.*, 2003). Chekowski (1998), (2000) and Alldrick (2004), Gaumy *et al.,* (2001) and Agag (2004) have excellently reviewed different aspects of incidence and toxicity of Zearalenone

13.1.1. Physico-Chemical characteristics

Zearalenone, chemical name is [6 (10-hydroxy-6-oxo-trans-1-undecenyl)-β-resorcylic acid lactone]. Its empirical formula is $C_{18}H_{22}O_{54}$ with molecular weight of 318 (FDA, 1979). ZEN structure was first elucidated by Urry *et al.* (1966).The name Zearalenone is derived partly from the generic name of the host plant infected by *Fusarium* (Zea) and partly from its chemistry (resorcylic acid lactone, en=double bond at C-12 and 1 one = ketone (Urry *et al.*, 1966). It is a white crystalline compound, that exhibits blue-green fluorescence when excited by long wave length UV light (360 nm) and a more intense green fluorescence when excited with short wave UV light (260 nm). Zearalenol (α-Zearalenol) is an hydroxylated derivative of ZEN due to ZEN reductases present in animal tissues (Diekman and Green, 1992).

In methanol, UV absorption maxima occur at 236 nm, 274 nm and 316 nm. Maximum fluorescence in ethanol occurs with irradiation at 450 nm. Solubility in water is about 20 μg/ml. Zearalenone is slightly soluble in hexane and progressively more so in benzene, acetonitrile, methylene chloride, methanol, ethanol and acetone. It is also soluble in aqueous alkali. Structure of ZEN and its derivatives are presented in Fig.13.1 and some synthetic analogues of ZEN are depicted in Fig. 13.2.

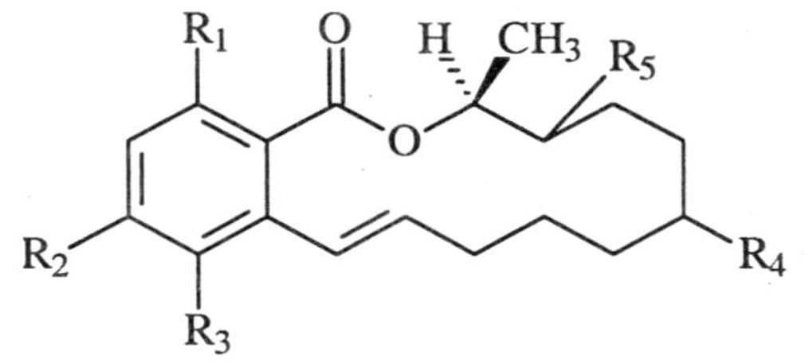

R_1	R_2	R_3	R_4	R_5	Cl' et C_2	Nom
OH	OH	H	O	H_2	CH=CH	Zearalenone
OH	OH	H	OH	H_2	CH_2=CH_2	α et β-zearalonol
OH	OH	H	OH	H_2	CH=CH	α et β-zearalenol
H	H	H	H_2	H_2	CH_2=CH_2	dideoxyzearalane
OH	OH	H	H_2	O	CH_2=CH_2	tetrahydrozearalenol
O-CH_3	O-CH_3	H	O	O	CH=CH	dimethylzearalenone
O-CH_3	OH	H	O	O	CH=CH	o-methykzearalenone
OH	O-CH_3	H	O	O	CH=CH	p-methylzearalenone
H	H	H	O	O	CH_2=CH_2	dideoxyxzearalenone
H	OH	H	O	O	CH_2=CH_2	2-deoxyzearalenone
OH	H	H	O	O	CH_2=CH_2	4-deoxyzearalenone
OH	OH	OH	O	O	CH_2=CH_2	5-hydroxyzearalenone
OH	OH	H	O	O	CH_2=CH_2	Zearalanone
OH	OH	H	H_2	H_2	CH_2=CH_2	Zearalenol
OH	OH	H	O	OH	CH=CH	8=hydrozearalenone
OH	OH	H	O	H	CH=CH	7-dehydrozearalenone
OH	OH	HC=O	O	H_2	CH=CH	5-formylzearalenone
OH	OH	H	OH	OH	CH=CH	6,8-dihydroxyzearalene

Fig 13.1. Structure of zearalenone and its derivatives

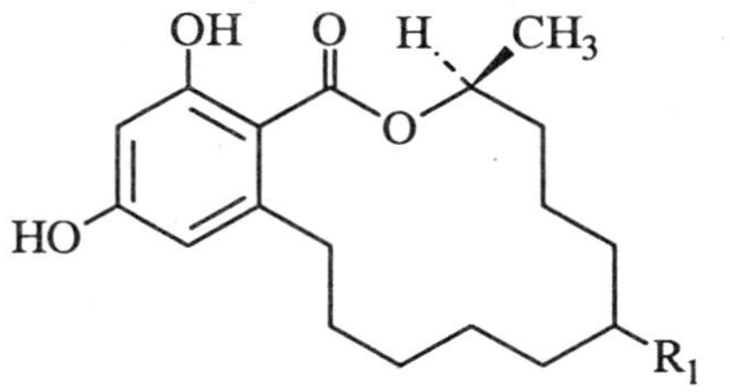

Zearalenone R_1 = O; Zearalene R_1 = H; Zearalenol R_1 = OH

Fig. 13.2 : Synthetic zearalenone analogs

13.1.2. Producing organisms

The most important producer of ZEN is *F. graminearum*, which frequently is referred to incorrectly in the literature as *F. roseum* (Marasas, 1991). Isolates of *F. graminearum* that produce large amounts of ZEN in culture have been obtained from mouldy corn implicated in porcine hyperestrogenism in the United States and elsewhere (Vesonder *et al.*, 1981). Isolates of *F. graminearum* from scabby barley and wheat in Japan (Ichnoe *et al.*, 1983) have shown to produce ZEN in culture and have also been obtained from many other agricultural commodities (Windels *et al.*, 1989; Vincelli and Parker, 1999). Some other species of *Fusarium* that produce ZEN are *F. culmorum* (Ramakrishna *et al.*, 1989; Snijder *et al.*, 1990) and *F. crookwellense* (Vesonder *et al.*, 1991; Sydenham *et al.*, 1991). *F. sporotrichioides* and *F. equiseti* (Marasas, 1991) and *F. semitectum* (Weidenborner, 2001). *F. moniliforme*, *F. tricinctum* and *F. oxysporum* are also reported to produce ZEN (Agag, 2004).These fungal species occurred naturally in corn, barley, animal feeds, hay, oats, sesamum and sorghum. The natural occurrence of ZEN is favored by high humidity and low (10-15°C) temperatures (CAST, 1989 and Weidenborner, 2001).

13.1.3. Biosynthesis

The structure of zearalenones suggest that they are biosynthesized by the act as polymalonate pathway resulting in head to tail condensation of acetate units into a polyketide that requires relatively little modification to accomplish conversion to zearalenones (Fig. 13.3), Steele *et al.* (1974), Mirocha and Pathre (1979) and Shier (1998) traced the biosynthetic pathway of Zearalenone by using radiolabelled acetate and concluded that Zearalenone is synthesized by the acetate-polymalonate which leads to the formation of polyketide which can be converted with little modification to zearalenone.

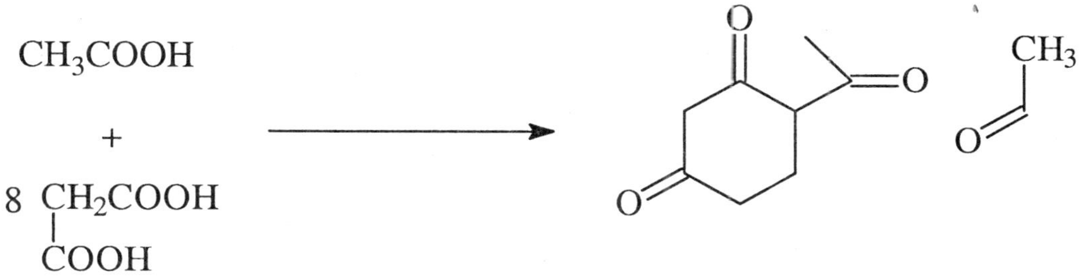

CH_3COOH + 8 CH_2COOH (with COOH) → [structure]

Fig 13.3. Head to tail condensation of acetate units

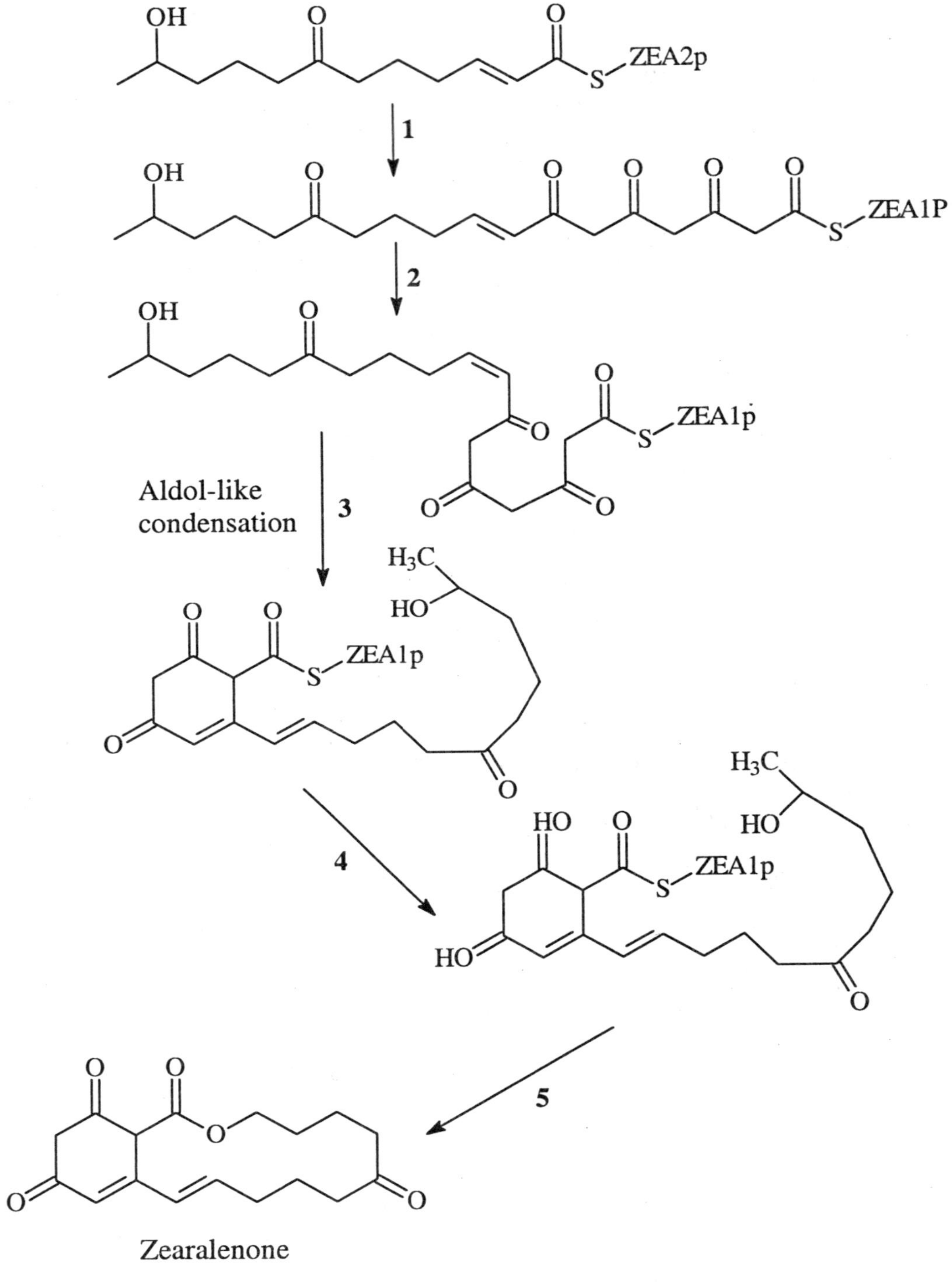

OH O O
S ZEA2p

1

OH O O O O O
S ZEA1P

2

OH O
O
O ZEA1p
S
Aldol-like condensation
3
O O

H_3C
O O HO
S ZEA1p
O
O

4

H_3C
HO O HO
S ZEA1p
HO
O

5

O O
O
O O

Zearalenone

Fig -13.4 : Biosynthetic pathway of Zearalenone

Gafoor and Trail (2006) have reviewed the biosynthetic pathways of Zearalenone. From studies of biochemical biosynthesis revealed the entire ZEA molecule would be synthesized from a single acetyl-CoA molecule and eight malonyl-CoA molecules (Fig. 13.3) as previously determined (Mirocha and Pathre, 1979). The first 10 carbon additions are catalyzed by ZEA2p, which has the reductive domains KR, DH and ER and could variably reduce the keto groups on the polyketide chain. The keto group can be partially reduced to yield an alkene or can be fully reduced to an alkane depending on the length of the carbon chain growing on the ACP domain. The remaining three rounds of C_2 additions are generated by ZEA1p, which is a nonreducing PKS and which lacks KR, ER, and DH domains. The unreduced ketones added by ZEA1p are highly reactive and form a ring that is rapidly aromatized due to enolization of the ketone groups via a nonenzymatic mechanism is for C_{11} and C_{12} to be added by ZEA1p, not ZEA2p, an alternative that remains to be tested experimentally. The two ZEA proteins may be dimerized or may act independently. Dimerization would facilitate the transfer of the product of ZEA2p to ZEA1p (Fig.13.4). In *in vitro* studies, PKSs can be promiscuous in terms of starter unit specificity (Watanabe *et al.*, 1996). If ZEA1p and ZEA2p are associated, then ZEA1p should use the correct starter unit and there by maximize the efficiency of ZEA biosynthesis.

13.1.4. Isolation and Detection

ZEN and some of its derivatives fluorescence blue-green when excited by UV radiation (360nm) and fluorescence even more intensively when irradiated at 260 nm. This property has been exploited to detect ZEN on TLC plates (Gillespic and Schenk, 1977). The current methods for the detection of Zearalenone include techniques like thin layer chromatography (William, 1984)and gas chromatography (Soares and Rodrigues Amaya, 1989; AOAC, 1995) Mass spectrometry (Onji *et al.*, 1998) and HPLC (Krska, 1998). Some of the detection methods of TLC are précised in Table 13.1.

Enzyme linked immunosorbent assays of Zearalenone using polyclonal, monoclonal and recombinant antibodies suggested by Park *et al.* (2002) proved to be good for detection of ZEN in cereals. Liquid-liquid partition (Soares and Rodrigues – Amaya, 1989; Quiroga *et al.*, 1994 and AOAC, 1995), open chromatographic columns (Kim *et al.*, 1993), chromatographic cartridges (Onji *et al.*, 1998; Ware *et al.*, 1999), multifunctional (Romer, 1994) and immunoaffinity columns (Scott and Trucksess, 1997) are some of the clean up steps employed in the determination of zearalenone.

Krska (1998) considers the use of multifunctional and immunoaffinity clean up columns for purification of zearalenone to be the best as it is a time saving. Use of immunochemical techniques such as Enzyme-Linked Immuno Sorbant Assays (ELISA) (Bennett *et al.*, 1994) and immunoaffinity columns with flourimetry have also been suggested for determination of Zearalenone by Scott and Trucksess (1997). Liquid and gas chromatographic methods which are commercially employed for ZEN analysis are as follows.

Table 13.1: Detection of Zearalenone produced by *Fusarium* with TLC

Solvent system	R_f value	Spray reagent	Detection	Reference	
				UV	Visible
Toluene: Ethyl Acetate: Formic acid (6:3:1)	0.78	$Ce(SO_4)_2$ 1% in 6N H_2SO_4 2,4, DNP $FeCl_3$ 3% in ethanol	—	brown Dark orange Pink	Grost-Allman and Steyn (1979) Mirocha *et al.* (1974) Pathre and Mirocha (1979)
Benzene: Methanol :	0.42	50% H_2SO_4 CH_3OH	Yellowish brown	—	Ishili *et al.* (1981)
Acetic acid (24 : 2: 1)		H_2SO_4 20% $AlCl_3$	charring Blue green	—	Kamimura *et al.* (1981) Kamimura *et al.* (1981)
		Anisladehyde Methanol 85 ml H_2SO_4 5 ml Glacial acetic Acid (10ml)	Faint brown		Scott *et al.* (1970)

Sample preparation: A mixture of organic solvent (acetonitrile, ethyl acetate, methanol, chloroform and acetone) with water or aqueous acidic solutions is commonly used for extracting ZEN from cereals (Schumacher *et al.*, 1997). For purification of the extract, SPE cartridges (Zollner *et al.*, 1999; Scudamore and Patel, 2000), liquid-liquid partition (Scott, 1999; Josephs *et al.,* 2001), MycoSep column (Silva and Vargas, 2001) and recently specific IA columns (Krska, 1998; Visconti and Pascale, 1998; Kruger *et al.*, 1999; Zollner *et al.*, 1999; Josephs *et al.*, 2001) have been used.

Separation and detection techniques: HPLC has been the most often used technique for ZEN analysis in cereals (Begneris *et al.*, 1986; Sydenham and Shepherd, 1996; Krarka, 1998; visconti and Pascale, 1998; Josephs *et al.*, 2001) (Table 13.2). Reversed phase separation with fluorescence detection, due to native fluorescence of the ZEN molecule, has frequently been applied. In addition to fluorescence detection, electrochemical, UV and diode array detection have been used (Sydenham and Shepherd, 1996; Schumacher *et al.*, 1997). GC techniques, mainly with MS detection, have also been utilized. With reports of simultaneous detection of different *Fusarium* toxins, and LC-MS and LC-MS/MS also appear to be a promising techniques (Schwadorf and

Muller, 1992; Ryu *et al.*, 1996; Sydenham and Shepherd, 1996; Muller *et al.*, 1998; Onji *et al.*, 1998; Rosenberg *et al.*, 1998; Zollner *et al.*, 1999; Tanaka *et al.*, 2000). Some of the methods currently used for ZEN analysis in various food stuffs are precised in Table 13.3.

Table 13.2. Some common HPLC methods

Toxin analyzed	Extraction	Clean-up	Mobile phase	Separation, detection	Reference
ZEN	ACN-water	MycoSep 224	MeOH-water	HPLC-Flu	Silva and Vargas (2001)
ZEN	CAN	IA column	ACN-water	HPLC-Flu	Schumacher *et al.* (1998)
ZEN	CAN-water	IA column	ACN-water-MeOH	HPLC-Flu	Vissconti and Pascale (1998)
ZEN (OA)	Chloroform-water-phosphoric acid	Silica-column	MeOH-water phosphoric acid	HPLC-Flu	Langseth *et al.* (1989)

ACN = acetonitrile, MeOH = methanol, HPLC-Flu = high performance liquid chromatography with fluorescence detector.

Table 13.3 Methods for the Determination of Zearalenone (and/or a and b-Zearalenol).

Cereals	Shake ground sample 30 min with 3:1 acetonitrile/water (10 ml/g)	Florisil column chromatography	HPLC Column: Silica gel (Nucleosil 50-10, 30 cm x 4 mm i.d.) Mobile phase: 50:15:2:1 90% Water-saturated chloroform/cyclohexane/acetonitrile/ethanol Detection: Fluorescence (ex: 176nm; em: 460 nm)	20	Tanaka *et al.* (1985)
Cereals	Shake ground sample 30 min with 75:25 methanol/water (10 ml/g)	Precipitation and liquid-liquid partition	HPTLC Plate:Whatmann LHP-K Solvent: 95:5 chloroform/ethanol Derivatization: Spray plate with Fast Violet B, borate buffer, and sulfuric acid solution; heat Detection: Visualization as violet spots	80	Swanson *et al.* (1984)

Cereals	Shake ground sample 30 min with 90:10 acetonitrile/4% KCl (4 ml/g)	Liquid-liquid partition	HPLC Column: Reverse-phase C_{18} (25 cm x 4.6 mm i.d.) Mobile phase: 80:20 Methanol/water Detection: Fluorescence (ex: 285 nm; em:440 nm) Derivatization: Precolumn with 0.25 M aluminum chloride in 3:1 methanol water		Metmarpki and Sludamore (1991)
Corn	Shake 50 g ground sample 30 min with 25 ml water, 25 g diatomaceous earth, and 250 ml chloroform	Silica gel column chromatography and liquid-liquid partition	TLC Plate: Silica gel GHR Solvent: 95:5 Chloroforma/alcohol Derivatization: Spray plate with 20% aluminum chloride solution and heat 5 min at 130°C Detection: Visualization as greenish-blue fluorescent spot under short-wave UV light before spraying and blue fluorescent spot under long-wave UV after spraying	<300	Scott (1995)
Corn	Shake ground sample 30 min with 90:10 acetonitrile/4% KCl (4ml/g)	Precipitation and liquid-liquid partition	TLC Plate: Silica gel 60 aluminum sheet Solvent: 90:10 Chloroform (1% ethanol)/acetone Detection: visualization of fluorescent spot under 254 nm UV light	50	Zuiroga *et al.* (1994)
Corn	Shake ground sample 30 min with 3:1 acetonitrile/water (4 ml/g)	Liquid-liquid partition and Florisil column chromatography	HPLC Column: Reverse-phase C_{18} (5 μm, 15 cm x 4.6 mm i.d.) Mobile phase: 70:30 Methanol/water Detection: Fluorescence (ex: 236 nm; em: 418nm)	2.0	Kang *et al.* (1994)

Powdered milk	Reconstitute 9.6 g in water; filter; pass through immuno-affinity column	Immunoaffinity column chromatography	ELISA Format: Direct competitive ELISA in microtiter wells Determination: Enzyme immunoassay reader	5.0	Tanaka *et al.* (1985)
Barley	Stir ground sample 15 min with 60:40 methanol/wter (5 ml/g); add zearalenone 6'-oxime as internal standared	Liquid-liquid partition and piperidino-hydro-xypropyl-Sephadex LH-20 column chromatography	HPLC Column: Reverse-phase C_{18} (5μm, 25cm x 4.6mm i.d) Mobile phase: 11:10 Acetonitrile/0.3% sodium acetate (pH 3.2) Detection: Fluorescence (ex: 274nm; em: 446 nm)	1.0	Tanaka *et al.*(1993)
Corn	Shake 50 g ground sample 15 min with 20 ml water, 25 g diatomaceous earth, and 250 ml chloroform	Liquid-liquid partition	HPLC Column: Bondapak Reverse-phase C_{18} (25 cm x 4.6 mm i.d.) Mobile phase: 1:0:1.6:2.0 Methanol/ acetonitrile/ water Detection: Fluorescence (ex: 236nm; em: 418 nm)	>50	Scott (1995)
Corn	Shake ground sample 30 min with 10:1 dichloromethane/water (5.5 ml/g) made basic with cupric carbonate	Liquid-liquid partition	HPLC Column: Reverse-phase C_{18} (25 cm x 4.6 mm i.d.) Mobile phase: 80:20 Methanol/water Detection: Fluorescence (ex: 280nm; em: 465 nm)	1.5-2.2	Chang and De Vires (1984)
Cereals	Shake ground sample 2 hr with ethyl acetate (6.7 ml/g)	Liquid-liquid partition	GC Column: Fused-zsilica capillary (25 m x 0.25 mm i.d.) Carrier gas: Helium Temp.profile: 180^{0}C; 15^{0}C/min to 285^{0}C Derivatization: Trimethylsilylation with N,)-bis(trimethylsilyl) trifluoroacetamide Detection: Electron impact with ion trap	1.0	Schadorf and Muller (1992)

Milk	Basify milk (250 ml) with ammonium hydroxide and shake 1 min with acetonitrile (70 ml)	Liquid-liquid extraction and aminopropyl SPE column	HPLC Column: Reverse-phase C_{18} (5μm, 25cm x 4.6mm i.d) Mobile phase: 61:35:4 Methanol/ water/ acetonitrile Detection: Fluorescence (ex: 236 nm; em: 470nm)	0.2- 2.0	Scott and Lawrance (1988)
Corn and wheat	Shake ground sample 3 min with 70:30 methanol/water (10 ml/g)	Filtration	ELISA (Agri-Screen Kit for Zearalenone, Neogen Corp., Lansing, MI) Format: Direct competitive ELISA in microtiter wells Determination: Visual or spectrophotometric comparison of color change compared with 500 ng/g control.		Benett *et al.* (1994)

13.1.5. Natural Incidence

The natural occurrence of ZEN is favored by high humidity and low temperature (10-15°C) (CAST, 1989). Weidenborner (2001) feels that a temperature between 12 to 14°C are more favorable for production of ZEN but production may also takes place at temperature below 10°C and even below freezing.

F. graminearum produced ZEA on maize grain at a_w 90-0.91 over a 12 week incubation period at 25-30°C. Optimum conditions for ZEA production were at >0.98 a_w. However, it requires a minimum of 22 to 25% moisture to grow on cereal grains (Jacobsen *et al.*, 1993). The fungus survives best between crops of corn and small grains in residue on the soil surface (Vincelli and Parker, 1999). The spores of this fungus spreads by rain splash or air movement to corn silk where they germinate and colonize on the silks during wet weather. By growing down the silk channel, the fungus is able to attack developing ear and cause ear rot. A reddish - brown discoloration of the kernels progresses from the tip downwards and effect ears. A whitish red mold growth may present on the highly affected ears (Vincelli and Parker, 1999). High levels of ZEN in cereals specially maize accumulate during storage mature *Fusarium* infected grains that have not significantly dried because of wet weather at harvest or in grains that were stored wet (moisture >22%). Variety of food grains and other foods including corn products

(Williams, 1985), corn beer (Lovelace and Nyathi, 1977), wheat flour (Tanaka *et al.*, 1985) and walnuts (Jemmali and Mazerand, 1980) are reported to be contaminated by Zearalenone. The presence of ZEN in whole plants and parts of maize used for silage making in Germany (Oldenburg, 1995) were contaminated by ZEN concentrations up to 300 μg/kg which mainly accumulated at the end of the ripening process with subsequent contamination of the silage. *Fusarium* species may produce ZEN and some other toxins (Christensen *et al.*, 1988). Abnormally high concentrations of both nivalenol and ZEN have been recorded in Japanese barley samples and in maize in New Zealand (Placent *et al.*, 1999). Zearalenone has also been shown to be potential contaminant of both amaranth (Brester *et al.*, 1998) and black pepper (Dimic and Skinjar, 1995). Okoye (1987) reported zearlenone in the rangeof 90-4600 μg/liter of beer in different parts of Africa, while beers sold in Korea and Canada no zearalenone could be detected (Scott *et al.*, 1993; Shim *et al.*, 1999). Occurrence of zearalenone in cereals and animal feed in different countries during 1996-1998 is summarized in Table 13.4.

Table 13.4. Natural incidence of zearalenone in cereals and animal feeds during 1996-1998 in different parts of the world

Commodity	Country	Publication Year	No. of samples	Incidence	Range (μg/kg)
Corn feeds	UK	1998	67	52%	20-1800
Cereals	Egypt	1997	135	18%	5-45
Poultry feeds	Argentina	1997	300	1%	30-280
Wheat	Germany	1997	445	31%	1-8036
Barley	Germany	1997	240	22%	2-311
Corn	Canada	1997	126	69%	5-647
Barley	Canada	1997	210	1%	4-21
Wheat	Switzerland	1996	154	16%	10-121
Wheat	Switzerland	1996	123	4%	10-18
Mouldy corn	USA	1996	98	67%	100-400
Corn	Argenitna	1996	2271	30%	5-2000
Wheat	Bulgaria	1996	140	69%	10-120

World wide occurrence of ZEN in cereals and its co-occurrence along with trichothecenes particularly deoxynivalenol has been reported by Tanaka and Ueno (1977) from all over the world. Natural incidence of ZEN in different agricultural commodities has also been reported from different parts of the world (Jemmali and Mazerand, 1980; Thiel *et al.*, 1982; Tanaka *et al.*, 1985). Large number of reports are also available about the natural occurrence of

ZEN in corn and corn based foods in Brazil (Sabina and Rodrigues-Amaya, 1993; Nordin and Luchese, 1998), Argentina (Reshik *et al.*, 1996; Solovcy *et al.*, 1999), Uruguay (Pincire and Silva, 1997), Korea (Kim *et al.*, 1993; Park *et al.*, 2002), Canada (Scott, 1977), UK (Scudamore *et al.*, 1998), Italy (Vissconti and Pascale, 1998) and Indonesia (Ali *et al.*, 1998). Sydenham and Shepherd (1996) recorded ZEN in 1-69% of samples with a contamination range of 4-8036 μg/kg. FAO (1998) considers contaminated cereals as the main source of ZEN in Nordic countries (Erikson and Alexander, 1998). Trigo-Stockli *et al.* (1996) revealed that the level of ZEN was more in the bran and low in flour indicating the ZEN exists on the surface of the grains. ZEN production has been reported on grains in the field, during harvest, commercial grain processing and/or subsequently during storage of any food or feed stuff containing the grain (Weidenborner, 2001). ZEN has also been detected in barley and malt (Park *et al.*, 1992), pecans (Schroeder and Storey, 1976), tomatoes (Sphan *et al.*, 1977) and bananas (Jimenez *et al.*, 1992).

13.1.6. Biological Activity

The toxicity of zearalenone is unique among known mycotoxins inspite of its dissimilarity to steroid compounds. It produces potent hyperesterogenic a response in susceptible animals (Columbe, 1993). Zearalenone was classified as a weak oestrogen having anabolic properties (FDA, 1979). Swines are the most commonly affected animals, cattle, poultry and laboratory rodents are also affected but to a lesser degree (CAST, 1989; Diekman and Green, 1992). The biological activity of zearalenone appears to be due to its hormonal activity in regulating perithecia formation during the sexual stage of the life cycle of *F. graminearum* (Wolf and Mirocha 1973). Low concentrations of ZEN (0.1 to 1.0 μg) increased the number of perithecia by nearly 100%; where as higher levels (> 10 μg) almost totally blocked the formation of perithecia in 4 day old cultures (Mirocha *et al.*, 1977). Zearalenone shows predominantly oestrogenic properties that are manifested in the form of hyperesterogenism in female swine, cattle and sheep causing severe reproductive problems (Jemmali, 1987). Esterogenicity includes atrophy of seminal vesicles and testis, squamous, metaplasmia of prostrate gland, osteropersis, myelofibrosis of bone marrow, cytoplasmic vacuolization of adrenal glands, hyperkeratosis of vagina and endometrial hyperplasia. ZEN caused androgenic response in male turkey poults (Olsen *et al.*, 1986). Young male pig exposed to ZEN undergoes symptoms of feminization (Newborne, 1987). Transmission of ZEN or its metabolites into milk appears to be minimal (Prelusky *et al.*, 1990).

Estrogenic disturbances are also suspected in sheep. Abortion, mild vulvo vaginitis and hypertrophy of the uterus are recorded (Radostits *et al.*, 1994). Reduction in the incidence of ovulation and infertilization were seen in ewes dosed with ZEN for 10 days before bating (Smith *et al.*, 1990). In the outbreak of ZEN mycotoxicosis in two farms in Northern Queensland most prepubertal gilts in the herds displayed enlarged testis and signs of oestrus such as red ,swollen vulvas. In several cases both rectal and vaginal prolapses occurred . In one of the farms 25 pigs died.

Zearlaenone and its metabolites possess estrogenic activity in pigs, cattle and sheep (D Mello *et al.*, 1999) and therefore, it is possible that this toxin may play a role in hormonal balance and mammary cancer in regions with high zearlaenone ingestion (Pitt, 2000). According to Etienne and Dourmad (1994) zearalenone reduces embryonic survival and sometimes fortal weight when administered above threshold level. It may also affect the uterus by decreasing LH and progesterone secretion and by altering the morphology of uterine tissues.

Smith *et al.* (1990) observed exposure of ewes to ZEN caused decline in ovulation rate, cycle length and an increase in duration of menstrus but did not affect pregnancy rate or embryonic loss. Bailey *et al.* (1976) recorded decreased growth rate, food intake, fertility, resorptions, still births, abortion and bone malformations in fueturesres in rats due to intake of ZEN at 10 mg/kg. Egg production decreased in hen due to consumption of ZEN (Branton *et al.*, 1989). ZEN (22 mg/kg) in the diet caused alterations in the reproductive tract of swine such as in the uterus and affects follicular and embryo development (zearalenone affects reporduction of pigs most seriously (Tiemann and Danicke, 2007).

According to Marasas and Nelson (1987) dietary levels of ZEN as low as 1 μg/g cause clinical signs of hyperestrogenism in pigs. Several alterations of immunological parameters were recorded at high ZEN concentration, *in vitro* (Eriksen and Alexander, 1998). Lang *et al.* (1992) recorded degeneration of embryos due to progressive disorganization of the embryonic disk. Marin *et al.* (1996) recorded increase in interleukin (1L)-2 and IL-5 in murine thymosa cell line EL-4 in the presence of ZEN suggests its immunotoxicity. In Newzealand ZEN in pasteure is a recognized cause of infertility in sheep (Towers and Sprosen, 1992). Its acute toxicity is low and LD_{50} (mg/kg bw) for oral and intraperitoneal administration are 2000 and >500 respectively in the mouse, >4000 > 10,000 > 5500 in the rat and >15,000 for oral administration in chicks. Coulumbe (1993) and Hussain and Brasal (2001) feel that swine is highly sensitive to Zearalenone. Mirocha *et al.* (1971) and Mirocha and Christensen (1974) gave detailed account of toxicity of ZEN. Several of closely related metabolites of Zearalenone produced by *Fusarium* spp also possess similar properties, although a few have been proven to occur naturally. Garies (1993) reported that swine fed by a diet containing 50 mg/kg of pure ZEN suffered abortion and stillbirths, while levels above 10 mg/kg reduced the litter size and reduced the weight of piglets. Trial feeding of female pigs revealed that a concentration of 0.25 mg/kg or even less, produced distinct redness and swelling of the vulva, slight swelling of the mamma and numerous vesicular follicles and some cystic follicles on the ovarian (Bauer *et al.*, 1987). ZEN decreased the serum level of oestrogen in animals (Ito and Ohstudo, 1994).

Coulombe and Roter (1993) feel that sexually mature boars are relatively insensitive to ZEN contamination. He also discussed at length biological action of ZEN in different animal systems. In mice ZEN affects ovary DNA and exhibited mitotoxicity (Aida *et al.*, 2001). Kuiper-Goodman (1991) recorded increased incidence of pituitary adenomas in both male and female animals with progression to malignancy with the consumption of ZEN. Gimeno and Qunitanilla (1983) reported the natural outbreak of ZEN mycotoxicosis in horse. Symptoms in exposed mares were enlarged edematous vulva, prolapsed vagina, over sized uterus and internal hemorrhage.

ZEN at concentration of 2.0 to 5.6 and 1.5 ppm in a feed given to swine was reported to swine induce hyperesterogenism (Mirocha and Christenson, 1974). Pigs showing clinical signs of hyperesterogenism was reported to be fed by maize containing 8 ppm ZEN (Blaney *et al.*, 1984). α-Zearalenone is three times more estrogenic than β-Zearalenone based on a rat uteropic assay (Hagler *et al.*, 1979). Elghor (1994) and Hassanane *et al.* (2000) have reported that the increase in cytotoxic effect of ZEN was usually associated with decrease in number of cells with chromosomal aberrations.

Prelusky *et al.* (1994) feel that ZEN may cause immunotoxicity. ZEN has also been considered as a possible agent that caused pubertal changes among young children of Puerto Rico (Saeniz and Rodrigues, 1984). Coppock *et al.* (1990) observed that ZEN about 750 ppb and 500 ppb of DON caused poor consumption of feed, depressed milk production, diarrhea and total reproductive failure. Towers *et al.* (1995) and Sporsen and Towers (1995) and Smith *et al.* (1995) feel that measurement of urinary zinc can successfully give the ZEN intake. In sheep ZEN causes lower conception, reduced ovulation, increased twinning rates and a 10-20% decline in fertility of ewes. In dairy cattle reproductive problems arise due to ZEN at 400 ppb. Kuiper-Goodman *et al.* (1987) concluded that pure ZEN is less toxic than naturally occurring contaminated cereals indicating the presence of additional toxic substances in matrix.

Presence of ZEN in red meat and poultry may cause pre-mature thelarke in Italy and premature thelarke pubersche, gynocomastia and precocious pseudo puberty in Puerto Rico (Hayes, 1994). Carcinogenic activity of ZEN still remains controversial (Tomaszewski *et al.*, 1998; Hilakivi *et al.*, 1999). Resistance of ruminants to ZEN is attributed to its degradation by resident microorganisms.

Infertility, reduced milk production, and hyperesterogenism in cows have been reported to be associated with ZEN (Diekman and Green, 1992).

Reduced feed intake, vaginitis, vaginal secretions, poor reproductive performance and mammary gland enlargement in virgin heifer (Jones *et al.*, 1994)) are some of the toxic syndromes of ZEN. Schoental (1983) described 'cold cow syndrome' characterized by decreased body temperature, diarrhea and swelling of the vulvae and other symptoms due to excess intake of ZEN through feed

Although swine have been found to be the most sensitive domesticated animal to zearalenone, calves have been reported to show earlier sexual maturity. Dairy cows have been reported to have vaginitis, prolonged oestrus and infertility (Palti, 1978) and sheep are reported to become sterile (Towers and Sprosen, 1992).

The oestrogenicity of zearalenone caused atrophy of seminal vericles and teses, squamous metaplasia of prostrate gland, osteoperosis, myelofibrosis of bonremarrow, cytoplasmic vacuolization of adrenal gland, hyperkeratosis of vagina and endometrial hyperplasia. Makela *et al.* (1995) feel that zearalenone is one of the most potent natural xenoestrogens. In New Zealand zearalenone in Pasteur is a recognized as a cause of infertility in sheep (Towers and Sprosen, (1992).

13.1.7. Mechanism of Action

ZEN and some of its metabolites have been shown to competitively bind to estrogen receptors (ER). Binding to specific receptor has been demonstrated in uterus, mammary glands, liver and hypothalamus of different species. Effects of Zearalenone are similar to Zearalenol. Zearalenone is generally considered to produce estrogenic effects 5-10 times greater than those of ZEN (Jacobsen *et al.* 1993). α-Zearalenol is 10 times more estrogenic activity than ZEN, where as β-isomer is considered less than that of ZEN (Weidenborne, 2001). α-Zearalenol showed comparatively stronger binding than β-zearalenol. The relative binding affinities to rat uterine cytoplasmic receptor for ZEN and derivatives were α-zearalenol > zearalenol > β-zearalenol > ZEA > β-zearalenol (Eriksen and Alexander, 1998). ZEN was shown to bind and activate both ERα and ERβ in cells transected with ERα and ERβ (Fig. 13.5). For ERα, ZEA was found to be a full antagonist and for ERβ to it's a mixed antagonist (Kuiper and Goodman 1998). In blood ZEN and zearalenol bind to human sex hormone-binding globulin to some extent (Eriksen and Alexander, 1998).

Hayes (1994) feels that ZEN interacts with estrogen receptors, translocation of receptor –ZEN compete to the nucleus, combination with chromatin receptors, selective RNA transcription leading to biochemical effects including increased water and lowered liquid content in muscle and increased permeability of the uterus to glucose, RNA, RNA protein precursors. It also induces biphasic changes in the luteinizing hormone but not the follicle stimulating hormone in sperm (Bongiovanni, 1983). Estrogenic activity of ZEN and several of its derivatives is due to depression of goanadotrophin levels (Radostits *et al.*, 1994).

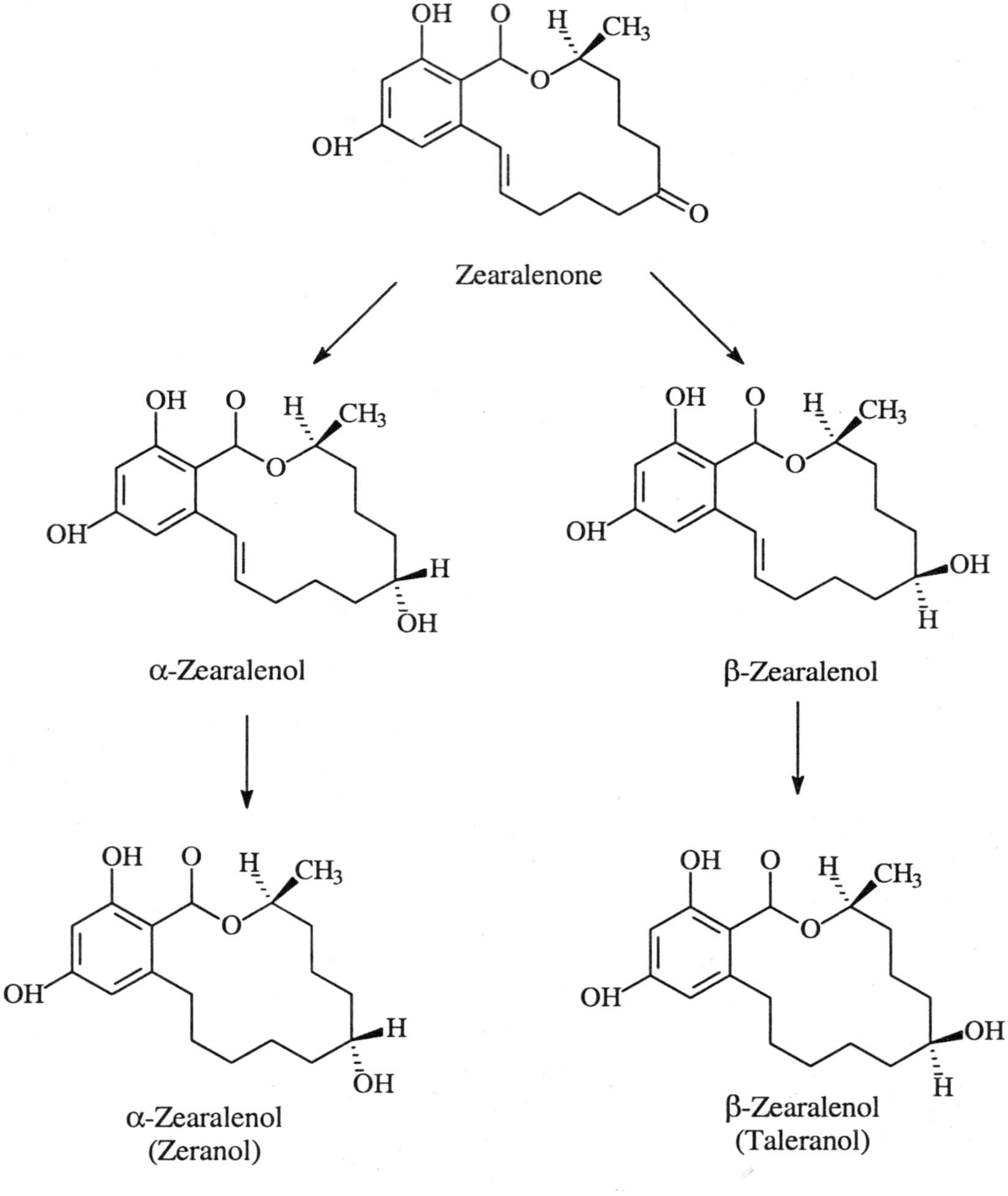

Fig-13.5 : Metabolic derivatives of Zearalenone

ZEN competes with estuarial 11B (E2) in binding to cytosolic estrogen receptors, activates the expression of initial protein and increases RNA polymerase I and II activities in nuclei (Boyd and Wittliff, 1978; Kawabata *et al.*, 1982). A number of factors including sex, and species of animal, environmental factors, nutritional status and mycotoxin synergism etc can either directly or indirectly modulate toxic effects.

Pharmacokinetics

Studies on pharmacokinetics and metabolism indicate that ZEN is fairly, rapidly adsorbed by following oral administration and can be metabolized by intestinal tissue in pigs and possibly in humans during its absorption with the formation of α and β-zearalenol, which are subsequently conjugate with glucouronic acid. Quantitatively glucouranide conjugates with the parental compound and α-zearalenol predominates in these species (WHO, 2000). Differences between species in the metabolism of ZEN were recorded. In the rat, most of the ZEN is found free zearalenone or glucouranide conjugate thereof while only small amount of zearalenol and their conjugates were formed (Kuiper-Goodman *et al.*, 1987; JECFA, 2000).

Broiler chicks and a lying hens are affected very little by dietary ZEN even when massive doses are given (Romer, 1990; Jacobsen *et al.*, 1993; Lesson *et al.*, 1995). Feed containing 300 ppm of ZEN in chicks resulted in increased weight of bursa of Fabricus, increased comb weight and increased number of cysts in the genital tract (Meronuck *et al.*, 1970). ZEN is rapidly adsorbed after oral administration is extensively absorbed by tissues in rats, rabbits and humans (WHO, 2000). Biehl *et al.* (1993) recorded about 80-85% of ZEN is absorbed by pig tissues. Kuiper–Goodman *et al.* (1987) could provide evidence for deposition of ZEN in adipose tissues. ZEN and its metabolites are excreted mainly in the bile in most animal species except rabbits in which urine is the main route. In pigs 45% ZEN is recorded in urine during first 48 hrs, 22% was recorded in feaces (Biehl *et al.*, 1993). In chickens ZEN is distributed chiefly to the liver and gall bladder (Mirocha *et al.*, 1982) and excreted in the form of ZEN or the derivatives. In laying hens most is excreted in faeces but residues may occur in yolk (Dailey *et al.*, 1980)

Biliary excretion which enters hepatic circulation occurs in rats and mice, while urinary excretion predominates in rabbits. Urinary excretion is also the main route of excretion in pigs inspite of the demonstrated enter-hepatic circulation of ZEN owing to high degree of reabsorption in the gut. Urinary excretion of ZEN is also significant in human (Kuiper-Goodman *et al.*, 1987). Zearalenone is metabolized in rat liver by one of the two pathways. The principal pathway is glucuronidation and the lesser pathway is reduction to an isomer of zearalenol.

In humans and pigs, ZEN was found mainly as glucouranide conjugates of ZEN and Zearalenone in urine with 24 hrs of uptake. Turkeys metabolize ZEN to Zearalenone, while chickens tend to produce equal amounts of ZEN to highly active Zearalenone (Wyatt, 1991). The amount of detectable ZEN in animal tissues depends on the contamination of feed, treatment of animals with ZEN or ά-Zearalenol, duration of exposure to the toxin, the persistence of ZEN in the animal and species variation (WHO, 2000). Chicken fed with 100 mg/kg Zen for 8 days, contained 59-103 mg/kg ZEN in muscle and up to 681 mg/kg in liver (Mirocha *et al.,* 1982). A dose of 544.5 mg/day given to cow for 21 days yielded maximum concentration of Zearalenone of only 2.5 mg ZEN, 3mg/L ZEN in mice (Prelusky *et al.*, 1994). Eggs accumulated metabolites of ZEN in the yolks even after 94% of the dose had been eliminated in excreta (Dailey *et al.*, 1980). Both bovine and ovine rumen fluid rapidly reduce ZEN to the more readily excreted α and β-zearalenone (Kiesseling *et al.*, 1984). Grosse *et al.* (1997) observed that vitamin A C and E reduced DNA aducts in kidney and liver by 70-90% in mice exposed to zearalenone.

13.1.8. Management

Usual strategies both at pre-harvest and post-harvest stage in the management of mycotoxins are also apply in true spirit in the management of ZEN Factors associated with the inhibition of ZEN production in the field are complex and genetics associated with such properties would be equally more complex. Increased understanding of the nature, origin, formation of ZEN may contribute to its control of contamination of foods and feeds.

Methods aimed at preventing or reducing level of ZEN contamination can be broadly classified as preventive or curative. Under this, there have been five basic approaches (i) prevention of the initial growth of the *F. graminearum* and subsequent production of ZEN (ii) detection of ZEN in food and selective removal of contaminated portion which, however, reduces the toxin only to a negligible level, (iii) inactivation or destruction of the ZEN by chemical, physical and biological means, (iv) development of effective and economic methods for controlling the mould growth and subsequent production of ZEN and (v) utilization of genetic resource.

Wet milling of corn reveals that ZEN concentrates in the gluten (48-56%) none is found in the starch. On the other hand, dry milling results in high concentration of toxin in the germ and feed fraction. During bread making 34-40% of ZEN in wheat flour is lost and in making instant noodles 48-62% is lost and in preparing biscuits 16-27% is lost (Scott, 1984). Fermentation of corn results as ZEN free ethanol found in solids used for animal feed (Bennet *et al.*, 1981).

Addition of non-nutritive sorbents or bacteria, yeast and modified yeast cells capable of reducing the bioavailability of ZEN are assuming increasing interest and are critically reviewed by Galvano *et al.* (2001). Eight *Lactobacilli* and three *Leuconostoc* species could transform ZEN to α-zearlaenol. Similarly *Streptococcus* and *Enterococcus* could bind ZEN to the extent of 49% and could help to decrease its toxicity in animal (Niderkorn *et al.*, 2007).

A lactone hydrolase was found to be responsible for detoxification of ZEN (Takahashi-Ando *et al.*, 2004). Bauer *et al.* (1987) detoxified ZEN by employing calcium hydroxide monomethylamine. Grosse *et al.* (1997) reported that vitamins A, C and E reduced DNA- ZEN adduct formation in kidney and liver of mice. Vitamin E was able to prevent genotoxicity of ZEN (Ghedira *et al.*, 1998). Modified mannanoligosaccharide derived from the cell wall of *S. cerevisiae* was reported to bind ZEN. Trenholm *et al.* (1996) reported that cholestyramine was capable of binding efficiently ZEN.

Gut flora of pig is reported to degrade ZEN (Kollarczik *et al.*, 1994). Similarly Megharaj *et al.* (1997) reported that mixed culture of bacteria can remove ZEN from culture medium. El-Mezami *et al.* (2002) reported that *Lactobacillus rhamnosus* GG and *Lactobacillus rhamnosus* LC705 removed the levels of ZEN and its main derivative of ά-zearalenol in liquid media to a significant level.

The bioavailability of zearalenone can be decreased with the help of probiotic strains such as rumen microbes and intestinal microflora (EL-Mezami *et al.*, 1999; EL-Mezami *et al.*, 2002). Gamma and electron beam irradation at 30K Gray resulted in total elimination of species of *Aspergillus*, *Penicillium* and *Fusarium* (Dovidio *et al.*, 2007). Kale *et al.* (2006) reported that yeast culture and activated charcoal are good adsorbents as they could adsorb ZEN to the extent of 81% and 79% respectively and can be applied in detoxification of zearalenone in foods and feeds.

CHAPTER - 14

Gastrointestinal mycotoxins

Trichothecenes

Trichothecenes are a class of fungal secondary metabolites characterised by the tetracyclic sesqueterpene 12, 13-epoxy- A' -trichothecene skeleton. Freeman and Morrison (1949) were the first to isolate one of these compounds as an antibiotic, trichothecene, from culture of *Trichothecium roseum*. Another trichothecene called trichodermin was isolated as an antifungal metabolite of *Trichoderma viride*. However, the structure of these compounds was elucidated by Godtfredsen *et al.* (1967). Subsequently several trichothecenes were added to this list (Betina, 1989). There are over 140 compounds of this group produced by different species of *Cephalosporium, Cylindrocarpon, Fusarium, Myrothecium, Stachybotrys, Trichoderma, Trichothecium* and *Verticimonosporium* which are generally associated with the toxicity of moudly cereal grains (Table 14.1). However, natural incidence of only deoxynivalenol (DON), nivalenol (NV), T -2 toxin, HT-2 toxin and diacetoxyscirpenol (DAS) have been reported. The structural elucidation of triepoxide trichothecene, baccharin, has been reported from a Brazilian plant, *Beaccharis megapotamica* (Kupchan *et al.,* 1976). Because of the wide variety of toxic trichothecenes as well as ubiquitous nature of trichothecenes producing fungi, these toxins are recognised as the most important class of mycotoxins in the world (Ueno, 1983). The trichothecene compounds are usually crystalline compounds. The biological activity of these compounds are attributed to the 12, 13-epoxytrichothecene. The structural alterations result in the modification of biological response (Ueno *et al.,* 1977). The other common structural feature of these compounds is 9, 10 double bond. Beasley *et al.* (1989), Miller and Treholm (1994), Jarvis *et al.* (1995), Chu (1997) and D mello *et al.* (1999) while discussing various aspects of trichothecenes, stressed the need for more work specially in respect of screening methods, pharmacokinetics and therapy of trichothecene mycotoxicosis.

Table 14.1. Trichothecenes toxins produced by different Fungi

Name of the fungus	Name of the toxin
F. acuminatum	DAS, HT-2, T-2, NS
F. anguioides	DAS
F. avenaceum	DAS
F. bacteridioides	DAS
F. culmorum	CN, DAS, DON, DON-m, Zen4, 15-Diacetoxy-3 7-dihydroxy-12, 13-epoxy en-8-one
F. diversisporum	DAS
F. episphaeria	NV-d, Fu-X, Zen
F. equiseti	DAS, DON, Fu-X, NS, NV, NV-d, T-2, Zen
F. lateritium	4, 15-diacetoxy trichothec-3, 7-diol 4, 15-diacetoxy trichothecene-3,7,8-trio DON, DAS, NS, T-2
F. graminearum	DON, 3-ADON, Fu-X, NV
F. nivale	CN, DON, Fu-X, NV, NV-d, T-2, Zen, NV m4β, 15-diacetoxy-3,7-dihydroxy-12,13 epoxy trichothec-9-8-one
F. oxysporum	DAS, Fu-X, DON, T-2, Zen
F. moniliforme	DAS, T-2, Zen
F. poae	Acetyl T-2 toxin, DAS, HT-2, NS, T-2
F. roseum	4-AOS, DON, DON-d, DON-m, DAS, DDON, Fu-X, HT-2, AOS, NS-m, NV, NV-d, T-2, SC, Zen
F. rigidiusculum	DAS, NS, T-2
F. sambucinum	DAS
F. scirpi	DAS, DON, NV-d, T-2, HT-2 4, 15, 15-triacetocy trichothecene-3, 7-diol
F. solani	DAS, HT-2, NS, T-2, Zen
F. semitectum	DON, DAS, NS, NV, Fu-X
F. sporotrichioides var poae	HT-2, NS, T-2
F. sporotrichioides	4-acetoxy trichothec 3,8,15-triol 4,8,Diacetoxy trichothec 3,15-diol, DAS, HT-2, T-2
F. sulphureum	4 AOS, $C_{21}H_{28}O_8$ trichothecene, AOS, 3-Acetyl-DON, DAS, HT-2, T-2
F. tricinctum	Acetyl T-2 toxin,

	$C_{19}H_{26}O_{28}$ trichothecene,
	DAS
	4,8 Diacetoxy 12,13-epoxy trichothec-9-one-3a, 15-diol
	HT-2, T-2, NS, NS-m, T-2T, Zen
	HT-2, T-2, NS, NS-m, T-2T, Zen
Myrothecium roridum	Rr-A,B,C, Vr-A, 2DVr-A
Mverrucaria	Vr-A,B,J,K,
	Rr-A,H,
	Trichoverrol-A,B, Trichoverrin-A,B
Trichoderma viride	TD, PB, GL, VD
T. lingnorum	T-2, TD
Trichothecium roseum	12,13-Epoxy trichothec-9-enc
	CT, TCT, TT, TC-d
	Resenolactone, Resolactone
Cylindrocarpon spp.	Rr-E, H, E-Iso-Rr-E,
	Err-H, DERr-H,
Verticimonosporum diffractum	Vertisporin
Cephalosporium	
Crotocinigenum	CT
Stachybotrys atra	St-C, D.E,F,G,H,Rr-E,Vr-J

Diacetoxyscirpenol=DAS; Diacetyl calonectrin-DAC; Deoxynivalenol=DON; Deoxynivalenol diacetate = DON-d; Deoxynivalenol monoacetate=DON-m; 3, Acetyldeoxynivalenol = 3ADON; HT-2 toxin=HT-2; T-2 toxin=T-2; Neosolaniol=NS; Zearalenone = Zen; Calonectrin=CN; Fusarenon-X=Fu-X; Neosolaniol monoactate= NS-M; Nivalenol=NV; Nivalenol monoacetate=NV-m; Nivalenol diacetate= NV-d; Scirpentriol=SC; T-2 Tetraol=T-2T; Acetyldeoxynivalenol=3-Acetyl-DON; Acetoxyscripenol=AOS; 4-Acetoxyscirpenol=4AOS; Roridin=Rr-A,B,C,D,E,H; Epoxyroridin-H=Err-H; Diepoxyroridin-H=DERR-H; Epoxy-isororidin-E=E-Iso-Rr-E; Verrucarin=Vr-A,B,J,J; 2-Dehydroverrucarin = A=DVr-A; Trichodermin=TD; Pachybasin=PB; Gliotoxin=GL; Viride=VD; Crotocin=CT; Trichothecolone=TCT; Trichothecin=TT; Trichodiol=TC-d; Satratoxin=St-C, D,E,F,G,H; 3,15 Diacetyl deoxynivalenol=DON.

14.1. PHYSICO-CHEMCIAL CHARACTERISTICS

Trichothecenes are known to possess a basic, tetracyclic, sesquiterpenoid structure which includes a six member oxygen containing ring, an epoxide in the 12,13 position and an olefinic bond in the 9,10 position. The trichothecenes possess oxygen containing substituents at one or more of 3, 4, 7, 8 and 15C position. These substituents can be esterified into hydroxyl, Keto (Position 8 only) or epoxide (Position 7, 8, 9, 10) groups or combination of these. Since most of the naturally occuring trichothecenes contain an epoxide ring and double bond at C-12, 13 and C-9, 10 respectively, these are characterised as 12, 13-epoxytrichothecene. Greenhalgh *et al.* (1989) reported another group of trichothecenes called apotrichothecenes which are characterised by a transfused AIB system as opposed to the cis fused system of trichothecenes. Both classes of compounds are derived from farnesyl pyrophosphate but differ in cyclitization of trichodiene. An oxidative intermediate other than trichodiol or trichotriol is suggested for the formation of apotrichothecenes.

On the basis of their chemical structure, the trichothecenes can be divided into two groups of compounds.

1. The alcoholic derivatives of the trichothecene and their simple ester.

2. More complex macrocyclic di or tri-ester

According to Ueno *et al.* (1985) trichothecene mycotoxins are mainly classified into four groups, on the basis of chemical structure.

Type-A: Hydroxyl or acyl-substituted trichothecene characterised by and represented by T -2 and diacetoxyscirpenol (DAS).

Type-B: 8-Keto trichothecenes characterised by carboxyl function at C-8 and represented by nivalenol and deoxynivalenol.

Type-C: 7-8 Epoxy trichothecenes characterised by second epoxy function at C-8 or C-9, 10 and represented by crotocin and crotocol.

Type-D: Verrucarins and roridins characterised by a macrocyclic ring between C-4 and C-5 represented by verrucarins, roridins and satratoxins.

Busby and Wogan (1981) suggested a 5th group for 7, 8-epoxy roridins (epoxy isororidin E, epoxy roridin H and diepoxy roridin H).

Tamm and Tori (1984) classified these compounds into 4 groups based on O_2 atom (Fig. 14.1) was subsequently supported and followed by Betina (1989).

I. Trichothecenes without a O_2 atom at the C-8 position, eg trichodermin.

II. Trichothecenes with O_2 functions at other than ketone at the C-8 position, eg T -2, neosolaniol.

III. Trichothecenes with a ketone at C-8 position, eg trichothecin, nivalenol, fusarenon-X.

IV. Macrocyclic trichothecene (Fig. 14.2) eg. Verrucarins, roridins, satratoxins and Baccharinoids.

TRICHOTHECENES WITHOUT AN OXYGEN FUNCTION AT THE C-8 POSITION

TRICHODERMOL
(RORIDON C)

$R_1 = H, R_2 = H, R_3 = H; R_4 = H$

TRICHODERMIN

$R_1 = OH; R_2 = H; R_3 = OH; R_4 = H$

SCIPDENTRIOL

$R_1 = H; R_2 = OCCH{=}C(CH_3)CH_2CH_2OH$ (trans)

$R_3 = 0$

OCCH=CH. CH=CH. CH(OH) CH(OH) CH_3

$R_4 = H$

DIACETOXY SCIRPENOL

$R_1 = H; R_2 = OCOCHE; R_3{=}OH; R_4{=}OH$

TRICHOTHECENES WITH AN OXYGEN FUNCTION OTHER THAN KETONE

NEOSOLANIOL

$R_1 = OH; R_2{=}H; R_3{=}H; R_4{=}OH; R_5{=}H$

HT-2 TOXIN

$R_1{=}CCOCH_3; R_2{=}H; R_3{=}OH; R_4{=}OCOCH_3;$
$R_5{=}OH$

ACETYL T-2 TOXIN

$R_1{=}OCOCH_2CH(CH_3)_1; R_2{=}H; R_3\text{-}OCHOCH_3$
$R_4{=}OCOCH_3; R_5{=}OH$

CROTOCIN

$R_1{=}OCOCH; R_2{=}OH; R_3{=}OCOCH_3$
$R_4{=}OCOCH_3; R_5{=}OH$

TRICHOTHECENES WITH A KETONE AT THE C-8 POSITION

TRICHOTHECIN

$R_1{=}H; R_2{=}H; R_3{=}OH; R_4{=}H$

DEOXYNIVALENOL

$R_1{=}H; R_2{=}H; R_3{=}OCOCH{=}CHCH_3$ (CIS)

NIVALENOL

$R_1{=}OH; R_2{=}OH; R_3{=}H; R_4{=}OCOCH_3$

FUSARENON-X

$R_1{=}OH; R_2{=}OH; R_3{=}OH; R_4{=}OH$

Fig. 14.1 Structure of some non-macrocylic trichothecenes

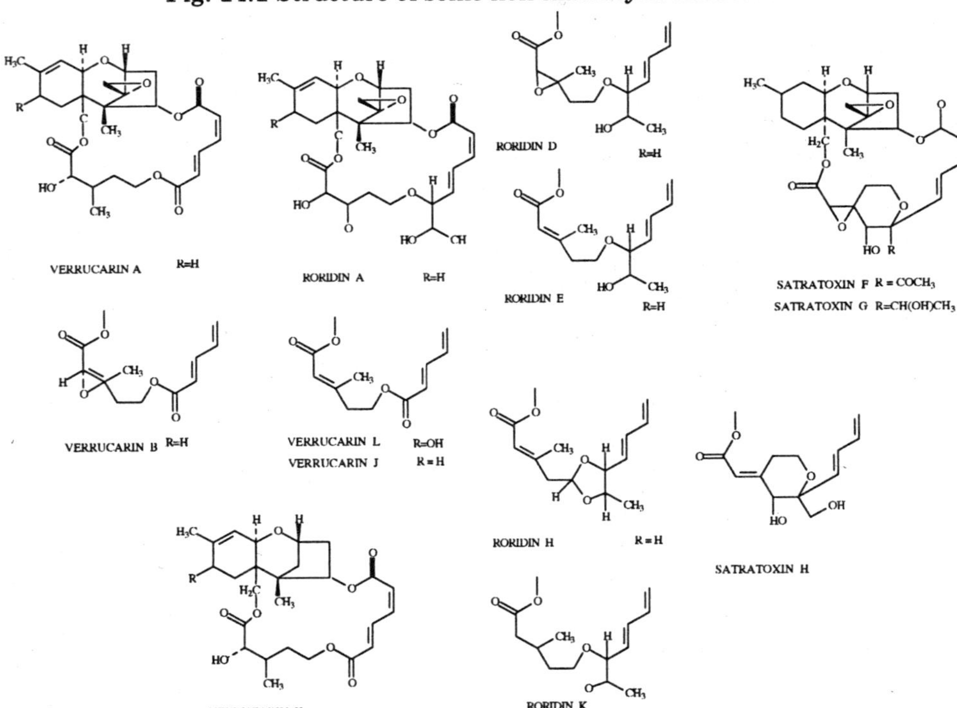

Fig. 14.2. Structure of macrocyclic trichothecens

Buck and Cole (1991) have excellently discussed the natural occurrence, biochemical mechanism of action, toxicity, clinical symptoms, lesions, diagnostic and therapeutic consideration of trichothecenes.

Trichothecenes are formed through a condensation of three mevalonate units to give farnesyl pyrophosphate and through its cyclization via the intermediate trichodiene. Turner and Aldridge (1983) have proposed the following scheme for the biosynthesis of trichothecenes mainly based on feeding experiments with *Trichothecium roseum* and isolation of trichodiene. The trans farnesyl pyrophosphate isomerizes to cis, trans farnesyl pyrophosphate and thus loses a mevalonoid 5-pro-S-hydrogen atom and 5-pro-R-hydrogen at C-I of farnesyl pyrophosphate is retained at C-II of the trichothecene. During the cyclization of the cis, trans farnesyl pyrophosphate to give trichodiene, a 1, 5 hydride shift from C-6 to C-2 of the trichothecene skeleton takes place. Finally the O_2 atoms are introduced into trichodiene molecule with retention of configuration. Several key enzymes in the biosynthetic pathway of trichothecenes have been identified. Atleast six genes (Tri 1 to Tri 6) involved in biosynthesis of trichothecenes have been cloned and these genes are clustered together on a chromosome (Proctor *et al.,* 2000)

Desjardins *et al.* (1986) have investigated the biosynthesis of T-2 toxin by *Fusarium sporotrichioides.* The oxygen molecules incorporated into T -2 toxin are there at C-4 and C-15 and isovalerate at C-8 were derived from H_2O and six at C-1, 12-13 epoxide, hydroxyl groups at C-3, C-4 and C-15.

Baldwin *et al.* (1986) attributed the multiplicity of trichothecenes to the specificity of hydroxylase and esterase of these fungi to the substrates. They have demonstrated biotransformation of 3-acetyl deoxynivalenol and deoxynivalenol molecule into 15-acetyl deoxynivalenol, 3, 15 diacetyldeoxy-nivalenol and fusarenon-X.

The macrocyclic systems of verrucarins and roridins contain a C6 unit derived from mevalonic acid (verrucarinic acid) and a straight chain C6 or C unit derived from acetate (Turner and Aldridge, 1983). Jarvis *et al.* (1983) provided evidence suggesting that incomplete macrocycle trichodermadie-nediols A and B, trichoverrols A and B and trichoverrins A are precursors of the macrocyclic trichothecenes. Similarly baccharinoids are detected in *Bacclaris Megapotamica* and *B. cordifolia* infected with *Myrothecium* are resultant of biotransformation of roridins of the fungus (Jarvis *et al.,* 1981 and 1983).

Production of trichothecene mycotoxins are reported to be determined by both genetic and environmental factors. Ueno *et al.* (1975), Miller *et al.* (1983) and Miller and Greenhalgh (1985) demonstrated that the production of fusarenon-X, neosolaniol, T–2 toxin and deoxynivalenol are influenced by nutrient status of the medium. The mycotoxin production by *Fusarium graminearum* could be regulated by complementation between two mutant types (Bu'Lock *et al.,* 1986). Biosynthesis of trichothecene has been very well reviewed by Tamm and Breitenstein (1980), Jarvis *et al.* (1983), Turner and Aldridge (1983) and Moss (1984).

The family of trichothecene mycotoxins is important because its members exhibit acute and chronic toxicity. They are also subjects of great concern as possible carcinogens, mutagens and genotoxic agents (Ueno, 1983). Some members of this family possess antibiotic, phytotoxic, cytotoxic, cancerostatic, antileukemic, immunosuppressive, antiviral and insecticidal activity.

The important structural features that are known to affect biological activities of trichothecenes are (1) The presence of the 12, 13-epoxy ring (2) Presence of hydroxyl or other groups at appropriate positions on the trichothecene nucleus (3) The structure and position of a side chain (4) The presence of a second epoxy ring on the trichothecene nucleus and (5) The presence of a macrocycle ring.

Trichothecenes are reported to vary in their toxicity (Ueno, 1983). Schappert and Khachatourians (1986) compared the effect of 8 trichothecenes on yeasts *(Kluvveromyces gracilis* and *Saccharomyces cerevisiae).* Verrucarin A inhibited the growth of *K. gracilis* where as other 7 trichothecenes did not. All 8 trichothecenes inhibited mitochondrial function. GU *et al.* (1991) suggested that temperature may affect the toxicity of trichothecenes to the yeast cells by regulating the composition of yeast cell membrane. The order of toxicity was as follows:

Verrucarin A> Roridin A> T-2 toxin> DAS > HT-2 toxin>Verrucarin > T-2 triol > T-2 toxin.

15-acetoxyscirpenol, HT –2 toxin, acetyl T –2 toxins and neosolaniol were inhibiting the growth of *Chlorella pyrenoidosa* at 1 mg, where as verrucarol, T-2 tetraol, nivalenol, fusarenon-x, deoxynivalenol and 3acetyl dehydroxynivalenol were inactive. Esterification at C-15 appears to be important for the growth inhibitory activity. Inhibition of protein synthesis appears to be involved in the toxicity. The presence of 12, 13 epoxide (oxinane) ring and C-9, C-10 double bond are basic requirements for toxic effects and biological activity. The hydrogenation opening of epoxide ring results in non-toxic compounds. Yoshizawa *et al.* (1985) feel that animals are able to metabolize some trichothecenes into inactive diepoxy derivatives. Lymphotoxicity of 8-Keto trichothecenes decreases according to the C-4 substituents as order follows: acetyl > hydroxyl > hydrogen (Forsell and Pestka, 1985). The structure and position of a side chain affects the antifungal and cytotoxic properties of trichothecenes (Claridge and Schmitz, 1978). Repeated deacetylation or hydroxylation of T-2 toxin significantly reduces immunotoxicity of its metabolites.

The reduction of C-9 and C-10 bonds to give the corresponding dihydro-derivative consistently lead to the diminished cytotoxicity (Grove and Mortimer, 1969). Acetylation of some, but not all other hydroxyl substituents generally lead to the increased toxicity (Ueno, 1977). Introduction of hydroxyl group at appropriate position on the trichodthecene skeleton often serves to increase the biological activity. The presence and position of second epoxide in the trichothecene nucleus produces profound effect on bioactivity of these compounds. A second expoxide at C-7, C-8 reduces and at C-9, C-10 increases their activity (Jarvis *et al.*, 1983). Bergmann and Yagen (1989) have elegantly discussed the structure and activity relationship of trichothecenes on toxicity to brain of rat.

14.2. ISOLATION AND DETECTION

Methods of isolation and detection of the trichothecenes can be divided into two categories.

1. Physico-chemical Methods
2. Biological Methods

Many of the biological tests are highly sensitive but lack specificity, while physico-chemical methods are generally specific but lack sensitivity.

(i) Physico-chemical Methods: Chromatographic methods are more suitable for analysis of trichothecene mycotoxins. Most of the trichothecenes do not contain functional groups which would permit the use of spectrophotometric methods. Thin Layer Chromatography (TLC), which affords simple and rapid detection, HPLC as well as Gas Liquid Chromatography (GLC) have been widely employed for the qualitative and quantitative analysis of trichothecene mycotoxins (Table 14.5).

(ii) Thin Layer chromatography: Most of the studies reported are on adsorption chromatography with a silica gel as the adsorbent. The thickness of the thin layer is usually 0.25 mm. various solvent systems are being employed as developing agents are listed in Table 14.2. Clean-up method is essential before the samples are spotted on TLC.

(iii) Clean-up Procedures: Mycotoxins are extracted with methanol or acetonitrile or aqueous mixed solvents from food or feed samples. Fats are removed 3 to 4 times with n-hexane or petroleum ether. Trichothecenes are extracted with chloroform from the original extracts. Chloroform extracts are subjected to column chromatography for purification and isolation of trichothecene toxins. The cleaned up extracts are separated on TLC plates and developed with any of the solvent systems mentioned in Table 14.2

Table 14.2. R_f value of different trichothecene mycotoxins

Mycotoxin	A	B	C	D	E	F
TYPE-A						
Scirpentriol	--	--	--	--	0.14	--
Diacetoxyscirpenol	0.36	--	0.50	0.14	0.28	--
T-2 toxin	0.40	--	0.52	0.16	0.44	--
Trichodermin	0.67	--	--	0.51	0.57	
Neosolaniol	--	--	0.29	--	0.15	--
HT-2 toxin	0.09	00	0.21	0.03	0.13	--
TYPE-B						
Trichothecin	0.66	--	--	0.53	0.58	--
Nivalenol	--	--	0.03	--	--	--
Fuserenon-X	--	--	0.29	--	0.25	--
Diacetoxynivalenol	--	--	0.44	--	--	--
Dexynivalenol	--	--	0.20	--	--	--
TYPE-C						
Verrucarin-A	0.28	0.59	--	0.47	--	--
Verrucarin-B	0.47	0.69	--	0.63	--	--
Verrucarin-J	0.59	--	--	0.64	--	--
Roridin-A	0.18	0.21	--	0.20	--	--
Roridin-D	0.21	--	--	0.29	--	--
Roridin-E	0.40	--	--	0.35	--	0.50
Roridin-H	0.59	--	--	0.72	--	--
TYPE-D						
Crotocol	0.25	--	--	0.07	0.30	--
Crotocin	0.65	--	0.65	0.34	0.40	--
Satratoxin-C	--	--	--	--	--	0.95
Satratoxin-F	--	--	--	--	--	0.42
Satratoxin-G	--	--	--	--	--	0.28
Satratoxin-H	--	--	--	--	--	0.20

Solvent systems: A = Chloroform, Methanol (98:2); B = Chloroform, Methanol (97:3); C = Chloroform, acetone (3:2); D = Benzene, Tetrahydrofuran (85:15); E = Ethyl acetate, Toluene (3:1); F = Iso propanol, Chloroform (3:97)

(iv) Detection: Thin Layer Chromatography is being extensively employed in the purification and detection of trichothecene toxins. Hence some procedures are necessary to detect or measure spots by colour/ flourescene (Table 14.3) after the samples are developed with suitable solvents and they will be detected and identified on the basis of flourescenece/colour

Table 14.3. Spray reagents for detection of different trichothecene mycotoxins

Spray reagent	DAS	T-2	HT-2	NS	NV	Fu-X	DON	SA	VER	RO
Sulphuric Acid (F)	Blue Green Char-ring	Bright Red	Sky blue	Green, sky slue	Bright red sky blue	Bright red sky blue	Charring, blue green	-	-	-
p-anisaldehyde (V)	Red	-	-	-	Yellow	-	Yellow	-	-	-
2,4 DNP	-	Orange	-	-	-	-	-	-	-	-
4(p-nitrobenzyl)	Bluish	Bluish	-	-	Bluish	-	Bluish	-	-	-
Pyridine (V)	Violet	Violet	-	-	Violet	-	Violet	-	-	-
Nicotinamide-2-acetylpyridine (F)	Bluish	Bluish	-	-	Bluish	-	Bluish	-	-	-
Chromotrophic acid (V)	Borwn	Purple	-	-	-	-	-	-	-	-
Phloroglucinol (VB)	Pink	Pink	Pink	Pink	Pink	Pink	Pink	Pink	Pink	Pink
Antimonium	Blue	-	-	-	-	-	-	-	-	-
Trichloride (F)	Purple	-	-	-	-	-	-	-	-	-
$AlCl_3$(F)	-	-	-	Bluish	Bluish	Bluish	-	-	-	-
0.7% Fast (V) Violet Acetic acid	Blue	-	-	-	-	Purple	-	-	-	-
Pyridine (V)	-	-	-	-	-	Blue	-	-	-	-
Phosphomolybdic acid	Blue	-	-	-	-	-	-	-	-	-
Resorsinol (Acidic)	-	-	-	-	-	-	-	Red	-	-

Kamimura *et al.* (1981) developed simulataneous detection of different *Fusarium* mycotoxins by sequential spraying plates with alcoholic $AlCl_3$, H_2SO_4 and 2,4-DNP on a TLC developed in chloroform: methanol (97:3). By this method, Zen, NV, DON, Fu-X, T-2, HT-2, DAS, NS, Mn and Bu could be detected by their characteristic fluorescence/visual colour.

Rao *et al.* (1985) reported phloroglucinol-HCl spray reagent to react with the allyl groups of the 12, 13 epoxy trichothecene giving a pink coloured complex. Further, it was found to obeying Beer's law thus enabling to estimate trichothecenes quantitatively. Ramakrishna and Bhat (1987) have made comparative assessment of seven spray reagents for the detection of four naturally occurring trichothecene toxins (DON, NV, T-2 and DAS). Ramakrishna *et al.* (1987) developed a minocolumn technique for the detection of deoxynivalenol in agricultural commodities. Still some more methods of determination of DON are precised in Table 14.4.

Table 14.4 Methods for the Determination of Deoxynivalenol and various trichothecenes

Food	Extraction	Cleanup	Quantitation	Detection (µg/kg)	Ref.
Corn, wheat barley	Shake 25 g ground sample with 125 ml 80:20 chloroform/ ethanol plus 10 ml water	Silica gel column chromatography	GC Column: CP-sil 5 fused silica capillary (0.14 µm film, 50 m x 0.32 mm i.d.) Carrier gas: Helium Detection:Electron capture Temp.profile: 50°C for 2 min; 40°C min to 160°C; 2°C/min to 220°C Derivatization: Treat with heptafluoro-butyric acid anhydride with 4-dimethylaminopyridine catalyst	12	Blancy *et al.* (1988)
Milk		Precipitation; centrifugation; filtration; liquid-liquid extraction with (Extrelut column (EM Science); charcoal alumina column chromatography	HPLC Column: Reverse-phase C_{18} (5 µm 25 cm x 4.6 mm i.d.) Mobile phase: 96:4 Water / acetonitrile Detection: UV (220 nm; range 0.0001 AUF)	5	Uudat-hala *et al.* (1994)
Wheat	Shake ground sample 30 min with 84:16 acetonitrile/water (4 ml/g)	Charcoal-alumina-celite column chromatography	HPTLC Plate: Silica gel (Whatman LHP-K) Solvent: 80:10:10 Chloroform/acetone /2-propanol Detection: Visualization and densitometry under long-wave UV light Derivatization: Spray with methanolic aluminum chloride	40	Fernandez *et al.* (1994)

Cereals	Shake 50 g ground sample 30 min with 250 ml 80:20 ethyl acetate/ acetonitrile and 10 ml water	Liquid-liquid partition and Florisil SPE column chromatography	GC Column: OV-1701 capillary (0.21 μm film, 25 m x 0.32 mm i.d.) Carrier gas: Helium Detection:Electron capture Derivatization: Silylation with TriSil/TBT Temp.profile: 60°C for 2 min; 40°C min to 150°C; 5°C/min to 250°C hold for10 min	Moller and Gustavsson (1992)
Corn	Shake ground sample 30 min with 3:1 acetonitrile/water (4 ml/g)	Liquid-liquid partition and Florisil column chromatography	GC-MS Column: DB-5 fused silica capillary (0.25 μm film, 30 m x 0.25 mm i.d.) Carrier gas: Helium Detection: mass spectrometry with selected ion monitoring Derivatization: trimethylsilylation (TMS) Temp.profile: 120°C for 5 min; 5°C/min to 270°C	Kang *et al.* (1994)
Grains	Blend ground sample with methanol (20 ml/g)	Liquid-liquid partition and silica SPE column chromatography	GC-MS Column: DB-5 fused silica capillary (0.25 μm film, 30 m x 0.25 mm i.d.) Carrier gas: Helium Detection: Negative ion chemical ionization mass spectrometry Derivatization: Treat with heptafluorobutyrylimidazole (HFBI) Temp.profile: 150C for 1 min; 10°C/ min to 250°C; 25°C/min to 300°C/ hold for10 min	Krishn-amurthy *et al.* (1996)

Wheat	Blend ground sample 5 min with 90:10 methanol/4% aqueous KCl (6 ml/g)	Precipitation, liquid-liquid partition, and column chromatography with mixture of charcoal, alumina, and diatomaceous earth	GC Column: BD-255 capillary (0.25 µm film, 15 m x 0.33 mm i.d.) Carrier gas: Hydrogen Detection: Flame ionization Derivatization: Treat with HFBI Temp. profile: 100°C for 1 min; 30°C/ min to 155°C; 5°C/min to 220°C	100- 500	Furlong and Soares (1995)
Grains	Shake ground sample 30 min with 50:50 methanol/water (4 ml/g)	Partition chromatography and silica SPE column chromatography	GC Column: SE-30 capillary (0.25 µm film, 25 m x 0.32 mm i.d.) Carrier gas: Helium Detection: Electron capture Derivatization: Treat with HFBI Temp.profile: 180-220°C at 4°C/ min; 15°C / min to 250°C; 250°C for 5 min	200	Sydenham and Thiel (1989)
Cereal grains	Blend ground sample 5 min with 50:50 methanol/water (5 ml/g)	Precipitation, liquid- liquid partition, and silica gel and cyano SPE column chromatography	GC Column: DB-5 fused silica capillary (0.25 µm film, 30 m x 0.32 mm i.d.) Carrier gas: Helium Detection: Electron capture Derivatization: Treat with HFBI Temp.profile: 75° for 2 min; 5°C/ min to 225°C; 5°C /min to 250°C; 30°C/min to 280°C	50-100	Cohen and Lapointe (1984)

(v) Gas Liquid Chromatography: Sylvia *et al.* (1986) have described high-performance liquid chromatography method of detection of trichothecene. Swanson *et al.* (1987), Bata *et al.* (1983), Richardson *et al.* (1984), Yagen *et al.* (1985) and Rood *et al.* (1986) have developed gas liquid chromatography methods for detection of trichothecene mycotoxins. Lattanzio *et al.* (2008) determined different trichotheces in cereal based products with the help of liquid–chromatography tandem mass spectrometry. Different methods of detection of difference trichothecenes are precised in Table 14.4.

Biological Methods

(i) Skin toxicity Test: Chang *et al.* (1974) suggested a rabbit bioassay test for screening larger number of field samples of corn for T-2 toxin. Similarly, Hayes and Schiefer (1979) estimated T-2 and DAS toxins quantitatively by injecting rats and rabbits subcutnaeously.

(ii) Vomiting-Test: In non-macrocyclic trichothecenes (*Fusarium* toxins) causing emesis and vomiting are common symptoms. Type A (T -2, HT -2, DAS) and Type B trichothecenes (NV, DON, Fu-X) when injected to one day old duckling, vomiting takes place at the intervals of 2-3 min.

(iii) Cytotoxicity Test: Eppley (1974) described cytotoxicity test for detection of trichothecene toxins using brine shrimp larvae. A modified petri dish is filled with artificial sea water and with brine shrimp eggs. Petri dishes thus prepared are incubated at 20-38°C for 30-48 hours. The resultant larvae were removed from the dish with syringe or pipett and incubated in hatching medium and the concentration of larvae is adjusted to 20-30/test. Samples to be tested were dissolved in chloroform at concentration of 0.1-1.0 mg/ml and desired quantities of solutions are added to the centre of each depression in the spot dish. Then the dish is covered with glass and incubated over night. Numbers of dead shrimp were counted under low power of wide field microscope. Muller (1984) developed a technique to detect T -2 trichothecene in feed stuffs by employing seeds of pea and pollen of tobacco.

(iv) Biochemical assay: Ueno (1983) has developed biochemical assay by employing rabbit reticulocyte. Locke-Ringer's solution (0.6 ml), 0.2 ml reticulocyte suspension, 0.1 ml of test solution was pre-incubated at 37°C for 15 min. Subsequently, 0.1 ml of leucine (14°C) was added and incubated for 40 min. At the end of incbation, the reaction was stopped by addition of 1.0 ml of 10% perchloric acid. The radioactivity of the acid insoluble fraction was counted.

Foster *et al.* (1975) developed another biochemical assay by employing glutathione. The 12, 13 epoxy trichothecenes react with glutathione and form 12-13 epoxy trichothecene glutathione complex. The glutathione-epoxide transferase catalyses this reaction. The residual glutathione is estimated photometrically. This method allowed the detection of 1.0 μg of T -2 toxin.

(v) Radio-Immunoassay: Chu *et al.* (1979) reported a radioim-muno-assay method for detection of trichothecene. In this method T -2 toxin is first converted to T-2 hemisuccinate by reacting the toxin with succinic anhydride in presence of pyridine resulting in the T-2 hemisuccinate. The resulted T-2-hemisuccinate was coupled with bovine serum albumin by using water soluble carbodimide (EDPC). Then the bovine serum albumin- T-2 hemisu-ccinate was mixed, with adjuvant and mixed emulsion was injected intrade-rmally to rabbit. Bleeding and booster injections were made at appropriate times after the initial immunization and the antisera were prepared with 33.3% saturated $(NH_4)_2$ SO_4. By employing antisera and (3H) T -2 toxin, the radioimmunoassay was developed. The sensitivity of sthe assay for T-2 toxin detection was in the range of 1-20 μg per assay.

Contamination of different cereals by some trichothecenes such as T-2 toxin, vomitoxin, trichothecin, HT-2 toxin, nivalenol, deoxynivalenol and diacetoxyscirpenol have been reported from different parts of the world (Vesonder and Hesseltine, 1981; Frisvad, 1983; Bata *et al.,* 1983; Karppanen *et al.,* 1985; Tanaka *et al.,* 1988; Ranfft *et al.,* 1990 Trenhom *et al.,* 1991 and Lancova *et al.* (2008). However, from India Rukmini and Bhat (1978) have reported the occurrence of T-2 toxin in sorghum grains contaminated with head moulds. Ghosal *et al.* (1977) have reported the T-2 toxin, diacetoxys-cirpenol and 12, 13 epoxy trichothecene in safflower seeds infested with *Fusarium oxysporum.*

Varieties of trichothecene producing fungi were associated with maize seed samples collected from different parts of Andhra Pradesh. Species of *Fusarium* were dominant and occurred with considerable percentage of incid-ence. Samples collected from Karimnagar were with highest percentage of *Fusarial* infestation, while samples collected from Nellore were with least percentage of *Fusarium* incidence. *M. roridum* was associated with seed samples collected from Warangal, Karimnagar and Nalgonda, while *T. viride, T. roseum* and *S. atra* were associated with seed samples collected from Nellore in addition to above 3 districts. However, the incidence of these fungi was high in samples collected from Karimnagar, while samples collected from Nellore supported least number of trichothecene producing fungi. The incide-nce of species of *Aspergillus* and *Penicillium* was considerably high, which may be one of the reasons for limited appearance of trichothecene producing fungi.

Out of different species of *Fusarium* isolated from maize, 50% of strains were toxigenic and produced one or other trichothecene toxins. Six strains each produced zearalenone and T-2 toxin. Four strains each produced deoxynivalenol and diacetoxyscirpenol, two strains each produced fusarenon-

X and neosolaniol. Krishna Reddy *et al.* (1987) have reported the incidence of toxigenic species of *Fusarium* in both pre-harvest and post-harvest maize. Sixty six per cent of isolates *of M. roridum* were toxigenic and elaborated roridin A and E. Similarly three out of five isolates of *S. atra* were toxigenic and produced satratoxin. Two out of four isolates of *T. viride* were toxigenic and elaborated trichodermin. Seventy five per cent of *T. roseum* isolates were toxigenic and elaborated trichothecin and only one isolate produced crotocin.

Five species of *Fusarium* were isolated from samples of sorghum collected from six districts of Andhra Pradesh. The incidence of these fungi was highest, while that of *T. viride* was least. *M. roridum* could be detected in samples collected from Nellore, Chittoor and Warangal. *S. atra* could be detected in samples collected from Prakasam and Visakhapatnam. *T. roseum* was also detected in samples collected from Prakasam.

Out of eight isolates of *Fusarium* spp., six were positive for one or other trichothecenes. Five isolates produced zearalenone, while four isolates elaborated diacetoxyscirpenol. T-2 toxin was elaborated by three isolates. Nivalenol and deoxynivalenol were elaborated by two strains each. Fusarenon-X and neosolaniol were elaborated by one isolate each screened. Both the isolates *of M. roridum* elaborated roridin A and E. Two out of three isolates *of T. viride* were positive for trichodermin. One out of two isolates of *T. roseum* elaborated trichothecin.

Out of 10 samples of sorghum analysed for the presence of trichothecene mycotoxins, only two samples contained detectable amount of diaceto-xyscirpenol and zearalenone. Similarly out of 19 samples of pearlmillet, 3 samples contained zearalenone, deoxynivalenol and roridins. Only one sample out of 15 samples of maize was contaminated by zearalenone. Aflatoxins, citrinin and cyclopiazonic acid were other mycotoxins detected in maize.

None of the 8 samples of rice contained trichothecenes but 4 samples contained other toxins. Only in two samples out of 42 samples of poultry feed contained detectable amount of T-2 and satratoxin. In one sample of broken rice contained fusarenon-X and deoxynivalenol. None of the four samples of fish meal contained trichothecene mycotoxins. When 6 samples of legume mixture were analysed for the presence of trichothecenes, only one sample contained T-2 and roridin. One mixed feed sample also contained zearalenone and diacetoxyscirpenol. None of the 29 samples of rice fodder contained trichothecene mycotoxins, while two out of 18 samples of maize fodder were contaminated by zearalenone and diacetoxyscirpenol. Two samples of ground nut fodder were found to contain trichothecin and roridin.

The incidence of trichothecene producing fungi in pearl millet was marginal and It was dominated by species of *Aspergillus* and *Penicillium*. The incidence of *Fusarium* species was highest in pearl millet samples collected from Warangal followed by Nalgonda. Similarly the percentage of incidence was also quite high in samples collected from Visakhapatnam. The percentage incidence of these fungi in samples collected from Chittoor and Nellore was least. *M. roridum* could be detected only in samples collected from Nellore. *T. viride* and *T. roseum* could not be detected in pearl millet samples collected from Nellore and Visakhapatnam. *T. roseum* could be detected in samples collected from Warangal and Nalgonda. The percentage of incidence of these fungi was quite low.

Out of ten isolates of species of *Fusarium* screened seven were toxigenic and produced one or other mycotoxins. Five isolates were positive for zearalenone, while four isolates each elaborated diacetoxyscirpenol and T-2 toxin. Nivalenol was elaborated by three isolates. Deoxynivalenol and HT-2 toxin were elaborated by two isolates each. Only one isolate of *M. roridum* elaborated roridin A. Similarly one isolate of *S. atra* elaborated satratoxin. Two out of three isolates *of T. viride were* positive for trichodermin, while only one isolate of *T. roseum* elaborated trichothecin.

It is also evident that the incidence of trichothecene producing fungi varied with type of pulse samples and place of collection. Three species of *Fusarium* were isolated from samples of blackgram collected from Nellore, Krishna and East Godavari. *M. roridum, T. viride* and *S. atra* could be isolated from samples collected from Nellore, East Godavari and Chittoor respectively. In general, incidence of these fungi was comparatively low, while that of species of *Penicillium and Aspergillus* were dominant in the spermosphere of blackgram. It is also evident that out of eight isolates of *Fusarium* species, six were toxigenic. Five isolates elaborated zearalenone, while diacetoxyscirpenol was elaborated by three isolates only. Fusarenon-X and T -2 toxin was elaborated by two isolates each. Nivalenol was elaborated by one isolate of *Fusarium.* Deoxynivalenol and HT-2 toxin could not be detected in any of the cultures of *Fusarium* screened. Only one isolate of *M. roridum* elaborated roridin A and E; while *S. atra* elaborated satratoxin. Out of two isolates of *T. viride,* only one elaborated trichodermin.

The incidence of trichothecene producing fungi in redgram was limited and is dominated by species of *Aspergillus* and *Penicillium.* Species of *Fusarium* could be detected in samples collected from Karimnagar, while *M. roridum* and *T. viride* could be detected in samples collected from Khammam and Warangal respectively. *S. atra* could not be detected in the spermosphere of red gram. Two out of three isolates each of *Fusarium* from the spermosphere of redgram screened, elaborated zearalenone and deoxynivalenol. Only one strain each of *Fusarium* elaborated diacetoxyscirpenol, HT -2 toxin and T -2 toxin. Only one isolate of *M. roridum* elaborated roridin A.

The incidence of trichothecene producing fungi in greengram was found to be limited and was dominated by species of *Aspergillus* and *Penicillium.* Trichothecene producing fungi could be isolated in samples collected from

Nellore and Chittoor, while *M. roridum* and *S. atra* could not be detected in any of the samples collected. *T. viride* could be traced in samples collected from Nellore. Out of four isolates of *Fusarium* three were toxigenic and elaborated one or other trichothecene. Two were positive for zearalenone. Only one strain each elaborated deoxynivalenol, diacetoxyscirpenol, HT -2 toxin and T-2 toxin. *T. viride* elaborated trichodermin.

Different vegetable seeds were colonized by different species of Fusarium. Percentage of incidence of *Fusarium* was high in bhendi seeds followed by bean, tomato and chilli in descending order. The seeds of tomato were colonized by *M. roridum*. Similarly *T. viride* and *T. roseum* could be detected on seeds of bean. *S. atra* could be detected on methi, chilli and bean seeds. Species of *Aspergillus* and *Penicillium* dominated over trichothecene producing fungi in the spermosphere of these vegetables. Only one species of *Fusarium* could be detected.

Only one species of *Fusarium* and *S. atra* could be detected on oil seeds screened. Incidence of *Fusarium* was highest in sesamum. *S. atra* could not be detected in ground nut seeds. The incidence of these fungi was very low in sunflower seeds.

Poultry feeds contain energy sources besides proteins, vitamins and minerals. These feeds are made from a large variety of ingredients like oil seed cakes, cereals, millets and their brans, fish meal etc. Poultry feed used in most parts of the country have about 50% of maize and 25%, ground nut cake which are very good substrates for mould growth and mycotoxins specially aflatoxin production. Feed which is one of the key factors for successful poultry production is prone to mould infestation ditectly due to faulty storage or indirectly by using the contaminated ingredients in feed preparation. Fungi besides causing discoloration of feed and depleting the nutrients may also produce toxic substanes which are potential health hazardous to animals and in turn to man (Wyatt, 1979; Patterson and Roberts, 1980; Osweller and Trampel, 1985; Gabal *et al.*, 1986). Stachybotryotoxicosis of poultry due to *S. atra* results in the loss of an appetite, reduced growth rate and watery discharge from the necrotic lesions in the mouth (Ivanics *et al.*, 1990). The estrogenic syndrome in swine, poultry and dairy cattle due to ingestion of mould infested feed is well documented (Gimeno and Quintanilla, 1981).

The natural occurrence of trichothecenes in cattle feed and corn have been reported by Hsu *et al.* (1972). Prelusky *et al.* (1984) have reported the non-transmission of deoxynivalenol to the milk following oral administration of the toxin to dairy cows which then attributed to the endogenous ruminant microorganisms. Though incidence of different mycotoxin producing fungi and mycotoxin production in different agricultural commodities has been reported by several workers (Betina, 1984, Gelderblom *et al.*, 1984;

Pepeljnjak and Cventic, 1984; Abbas *et al.*, 1986; Blaney *et al.*, 1986; Kim and Roh, 1986; Mislivec *et al.*, 1987; Krishna Reddy *et al.*, 1987), comparatively very little information is available with poultry feeds (Krishna Reddy *et al.*, 1987; Widiastuti *et al.*, 1988 and Bekes, 1988).

Mycotoxin contaminated feed has caused serious economic loss of which the turkey X disease (Sargeant *et al.*, 1961) is a well known example. Incidence of mycotoxicoses in poultry due to ingestion of contaminated feed has been reported from different parts of India (Shankaran *et al.*, 1970; Choudry *et al.*, 1971;, Neelakantan *et al.*, 1978). Still many times they go unnoticed as proper reporting is not being done. Though there are several studies dealing with mycoflora of poultry feeds and mycotoxicoses of poultry from different countries (Stoloff *et at.*, 1976; Mirocha *et at.*, 1976; Burditt *et ai.*, 1983; Westlke and Dutton, 1985; Morcho Romo and Fernandez, 1986; Boikov and Khasanova, 1986; Moran *et al.*, 1987; Ewaidah, 1988), comparatively very little is known about the mycoflora of poultry feeds in India (Katoch *et al.*, 1975; Neelakantan *et al.*, 1978; Reddy *et al.*, 1984; Sudarshan Singh, 1987; Krishna Reddy *et al.*, 1987).

Almost all the poultry feed samples collected from different regions of Andhra Pradesh contained one or other trichothecene producing fungi. Species of *Fusarium* could not be detected in samples collected from Krishna district, while species of *Fusarium* occurred with highest percentage of incidence in samples collected from Ranga Reddy and least percentage of incidences of *Fusarium* spp in samples collected from Karimnagar. *M. roridum* could not be detected in samples collected from Mahabubnagar, Visakhapatnam, Nellore, Vijayanagaram, Krishna and Ranga Reddy districts. Similarly *T. viride* could not be detected in samples collected from Mahabubnagar, Guntur, Nellore, Vijayanagaram and Cuddapah. *T. roseum could* be detected in samples collected from Mahabubnagar, Guntur, Vijayanagaram, Krishna and Chittoor, while *S. atra* could be traced in samples of Visakhapatnam, Guntur, Vijayanagaram, Ranga Reddy, Cuddapah and Chittoor. However, the percentage of incidence of *Aspergillus* was highest and trichothecene producing fungi formed minor proportion of the fungal population.

Out of 52 strains of *Fusarium* screened 36 were positive for one or other trichothecene group of mycotoxins. Fifteen isolates were positive for zearalenone, while 12 isolates elaborated deoxynivalenol. Eight isolates were positive for nivalenol. Neosolaniol was elaborated by five isolates. Four isolates elaborated HT -2 toxin. Fifteen isolates of *Fusarium,* elaborated T -2 toxin. Diacetoxyscirpenol was elaborated by ten isolates of *Fusarium.* Fusarenon-X was elaborated by five isolates. Six out of eight isolates of *M. roridum* elaborated roridin and four isolates could elaborate verrucarin. Six out of nine strains of *S. atra* elaborated satratoxin, while eight out of ten strains of *T. viride* elaborated trichodermin. Similarly seven out of ten strains of *T. roseum* elaborated trichothecin.

Almost all the cattle feed samples collected from different regions of Andhra Pradesh contained one or other trichothecene producing fungi. Species of *Fusarium* could not be detected in samples collected from Guntur. Only one cattle feed sample collected from Karimnagar supported the growth of *M. roridum. T. roseum* could be detected in samples collected from Nellore only, while *S. atra* could be detected in samples collected from Visakhapatnam and Chittoor. *T. viride* could be detected in samples collected from Guntur, East Godavari and Chittoor. The cattle feed was dominated by species of *Aspergillus* and *Penicillium.*

Out of 21 isolates of *Fusarium* screened 19 were toxigenic and produced one or other trichothecene toxins. Thirteen strains were positive for zearalenone, while eight isolates elaborated deoxynivalenol. Nivalenol and diacetoxyscirpenol were elaborated by two and three isolates each. Five isolates produced fusarenon-X. Three out of four isolates of *M. roridum* were positive for roridins. Five out of seven isolates of *S. atra* elaborated satratoxins. Six out of eight strains of *T. viride* elaborated trichodermin. While four out of 6 strains of *T. roseum* elaborated trichothecin.

Though percentage of incidence of trichothecene producing fungi was low, one or other species was spotted in all the samples of ground nut cake. Species of *Fusarium* could not be detected in samples collected from Visakhapatnam, while *T. roseum* could be detected only in samples collected from Khammam. Two species of *Myrothecium* could be recorded in samples collected from Cuddapah and Karimnagar, *S. atra* could be detected in samples collected from Cuddapah, Karimnagar, Visakhapatnam and Chittoor. *T. viride* could not be detected in samples collected from Nellore, Visakhapatnam and Warangal. It is clear from the present investigations that the trichothecene producing fungi formed a low percentage of population than other fungi.

Ten out of fourteen isolates of *Fusarium* were toxigenic and elaborated one or other trichothecene toxins. Eight were positive for zearalenone, while six isolates elaborated T-2 toxin. Nivalenol and HT-2 toxin were elaborated by two isolates each. Four strains produced deoxynivalenol, while five isolates elaborated diacetoxyscir-penol. Three out of four isolates *of M. roridum* elaborated roridins. Four out of five isolates of *S. atra* elaborated satratoxins. Similarly three out of 5 isolates of *T. viride* elaborated trichodermin. Only one isolate of *T. roseum* elaborated trichothecin.

Sesamum cake samples collected from different regions of Andhra Pradesh supported the growth of trichothecene producing fungi with low percentage of incidence. Species of *Fusarium* could not be detected in samples collected from Chittoor. Sesamum cakes collected from Chittoor supported the growth of *S. atra*. *M. roridum* could not be detected in samples collected from Krishna. *T. roseum* could be detected in samples collected from Nellore only. *T. viride* could be detected in samples collected from Guntur and Chittoor. Incidence of trichothecene producing fungi was low when compared to that of species of *Aspergillus* and *Penicillium*.

Out of six isolates of *Fusarium,* four were toxigenic and elaborated one or other trichothecene toxins. When three each strains elaborated zearalenone, deoxynivalenol, diacetoxyscirpenol and T -2 toxin, HT-2 toxin and fusarenon-X were produced by one isolate each. Only one isolate of *S. atra* elaborated satratoxin, while *T. roseum* elaborated trichothecin. Out of 3 isolates of *M. roridum,* two were positive for roridins. Only one of the two isolates of *T. viride* elaborated trichodermin.

Straw of rice, maize, sorghum and stylograss left after harvesting of the grains, is used as fodder for livestock. Though these are poor in nutritive value, containing about 3% protein and 40% total digestable nutrients (TDN), these straws along with small quantities of protein supplements are used as a sole feed. These are the main bulk of fodders for livestock throughout the country. It is the usual practice in the country side to store the straw under open conditions. Tropical conditions, harvesting practices, unseasonal rains, high moisture content of straw during baling and open storage not only promote fungal proliferation but also encourage to elaborate many mycotoxins.

Dairy cattle exposed to many fungal spores from fodders may often develop sub-clinic disease (Austwick, 1962). Fungi from mouldy fodders may cause infections, provoke allergic responses in sensitized subjects or poison with toxic metabolites (Lacey; 1975). Ainsworth and Austwick (1973) have isolated fungi from outbreaks of mycotic abortion in cattle including species which are not commonly pathogenic, but most common in mouldy fodders. Stachybotryotoxicosis of horse, cattle, sheep and swine is reported to be due to satratoxins elaborated by a species of *Stachybotrys*. Many animal diseases like stomatitis, thrombocytopenia, gross'necropsy are reported to be due to toxic substances elaborated by moulds proliferating in feeds and fodders (Fortuskny *et al.,* 1959). Fescue foot of cattle and sheep is reported to be due to T -2 toxin. Animal disease of cracks at the junction of hoofs and on the skin or on their rear legs is reported due to mycotoxins in pasture grass (Yates, 1971).

Many fungi found in fodders produce toxic metabolites, but a few of their toxins have been identified in fodders on which they grow (Kady and Heath, 1982; Bekes, 1988; Kordic *et al.,* 1990). Fungi may also cause animal diseases indirectly in ways that mimic mycotoxicoses. Fungal growth may also deplete the nutritive value and can alter the availability of micronutrients (Broster, 1973). Besides these, there are several reports of mycotoxicoses and death of livestock (Wiseman *et al.,* 1973; Harwingand Munro, 1975; Knight, 1981;

Heathcote, 1984; Deng *et al.,* 1984; Manickam *et al.,* 1985, Ruhr, 1987) due to the ingestion of mould infested fodder from different parts of the world. However, there are only a few scattered reports of such diseases of livestock from India. Further, very few reports of mould infestation of fodder (Rajendran *et al.,* 1975; Surekha and Reddy, 1990) from India are available.

The percentage of incidence of *Fusarium* species was highest in rice fodder samples collected from Ananthapur followed by Ranga Reddy and it was least in samples collected from Nellore. Species of *Myrothecium* could not be detected in samples collected from Ranga Reddy, Ananthapur, Nellore and Khammam. Similarly *T. roseum* could not be detected in samples collected from Ranga Reddy, Visakhapatnam and Nellore. *T. viride* and *S. atra* could not be found in samples collected from Chittoor, Warangal, Ranga Reddy and Khammam respectively.

Out of 18 strains of *Fusarium* screened for mycotoxin production, 11 strains were toxigenic and produced one or other mycotoxins. Zearalenone was elaborated by 9 strains, while diacetoxyscirpenol was elaborated by 7 strains. Fusarenon-X, nivalenol and neosolaniol were elaborated by 2 strains each. Similarly deoxynivalenol and HT-2 toxin were elaborated by six and four isolates respectively. Out of three each isolates of *M. roridum* and *M. verrucaria,* two each were positive for production of roridins and verrucarins repectively. Both the isolates of *S. atra* were toxigenic and elaborated satratoxin. Species of *Fusarium* were common in groundnut fodder collected from different regions of Andhra Pradesh. *T. roseum* was detected only in samples collected from Chittoor and Khammam. *S. atra* could not be detected in samples collected from Nellore and Cuddapah. Species of *Myrothecium* could be recorded in samples collected from Warangal, Nellore artd Cuddapah. The incidence of *T. viride* was common and could be spotted in all the samples collected from different regions of Andhra Pradesh.

Out of thirteen isolates of *Fusarium,* eight were positive for production of mycotoxins. Zearalenone was elaborated by 5 isolates. Nivalenol, fusarenon-X and diacetoxyscirpenol were produced by 2 isolates each. Neosolaniol was produced by one isolate, while deoxynivalenol was produced by three isolates. Only one isolate of *M. verrucaria* elaborated verrucarin, while isolate of *S. atra* could produce satratoxin.

The incidence of species of *Fusarium* was high in sorghum fodder samples collected from Ranga Reddy and Visakhapatnam. *M. roridum* could be detecteed with low percentage of incidence from Visakhapatnam and Karimnagar. Similarly *T. viride* and *T. roseum* were associated with samples collected from Visakhapatnam and Prakasam respectively. *S. atra* could be traced in sample collected from Ranga Reddy and Karimnagar. The incidence of *Fusarium* species was highest, while other trichothecene producing fungi were limited in their population. Similarly percentage of frequency and abundance was more with species of *Fusarium*. Out of eight isolates of *Fusarium,* five were positive for production of one or other mycotoxins. Zearalenone and nivalenol were elaborated by three isolates each. HT-2 toxin, diacetoxyscirpenol and deoxynivalenol were elaborated by one isolate each. Both the strains of *M. roridum* were toxigenic. Both the strains of *S. atra* were positive for satratoxin. Most of the trichothecene producing fungi were capable of elaborating trichothecene toxins.

Species of *Fusarium* formed a dominant elements among the fungi associated with stylograss fodder. The percentage of incidence of *Fusarium* was highest in samples collected from Ranga Reddy and least in samples collected from Chittoor. Species of *Myrothecium* could be detected in samples collected from Visakhapatnam and Nellore, while *T. roseum* and *S. atra* were traced in samples collected from Nellore, *T. viride* could not be detected in samples collected from Visakhapatnam. The percentage of frequency and abundance were also more for species of *Fusarium,* while rest of the trichothecene producing fungi fluctuated in their population.

Out of ten isolates of *Fusarium,* six were positive for one or other mycotoxins. Zearalenone was elaborated by 4 strains. While Diacetoxyscirpenol was elaborated by 2 isolates. Similarly three isolates elaborated neosolaniol, while only one isolate elaborated deoxynivalenol. Fusarenon-X and HT -2 toxin were detected in culture filterates of different species of *Fusarium* understudy. Both the strains of *M. roridum* produced roridins. Only one isolate of *S. atra* elaborated satratoxins.

All the species of *Fusarium* except *F. merismoides, F. scirpi, F. tricinectum* and *F. udum* which failed to produce zearalenone, produced one or other trichothecene mycotoxins. Nivalenol was produced by *F. equiseti, F. moniliforme* var *intermedium, F. oxysporum, F. semitectum* and *F. scirpi,* while *F. semitectum, F. solani, F. oxysporum, F. moniliforme* var *intermedium* and *F.moniliforme* produced deoxynivalenol. Similarly fusarenon-X was produced by *F.moniliforme, F. moniliforme* var. *intermedium, F. oxysporum, F. solani, F. semitectum, F. scirpi* and *F. udum, F. equiseti, F. graminearum, F. moniliforme, F. moniliforme* var *intermedium, F. oxysporum, F. solani, F. semitectum, F. palladoroseum* and *F. tricinctum* elaborated neosolaniol. *F. equiseti, F. moniliforme, F. moniliforme* var *intermedium, F. solani, F. semitectum* and *F. udum* produced HT-2 toxin. *F. culmorum, F.moniliforme* var *intermedium, F. merismoides, F. tricinctum* and *F. udum* failed to produce T -2 toxin. *F. solani* elaborated solaniol only. *F. oxysporum* could elaborate javanicin. Diacetoxyscirpenol was elaborated by *F. acuminatum, F. equiseti, F. moniliforme, F. moniliforme* var *intermedium, F. mersmoides, F. oxysporum, F. solani, F. semitectum, F. scirpi, F. pallidoroseum* and *F. tricinctum,* while *F. semitectum* could produce butenolide.

Most of the species of *Fusarium* produced more than one trichothecene mycotoxins. Thus species of *Fusarium* may pose a great threat to the livestock and indirectly to the health of man. Production of more than one trichothecene mycotoxin by a single species of *Fusarium* may be attributed to transformation reactions that take place in the surrounding environment of growing culture. Betina (1989) also attributed the production of multiple trichothecenes by species of *Fusarium* to the acetylation and deacetylation reactions that take place in the external environment of the fungus.

14.3. FACTORS INFLUENCING

High incidence of fungi producing trichothecenes and production of these substances in various agricultural commodities are reported from various corners of the world. Hsu *et al.* (1972) have reported the natural occurrence of trichothecenes in cattle feed and corn. Greenway and Puis (1976) and Mirocha *et al.* (1977) have also reported the presence of trichothecenes in barley and maize, commercial pellets and mixed feed in British Columbia, Canada and USA. Yoshizawa (1983) and Vesonder *et al.* (1978) have demonstrated heavy contamination of corn and barley with trichothecenes in Japan and USA respectively. Jemmali *et al.* (1978) and Szigeti (1976) have reported the natural contamination of barley seeds by trichothecene producing fungi from France and Hungary respectively. *Stachybotrys* strains which are reported to elaborate macrocyclic trichothecene toxins (satratoxins G and H, roridin E and verrucarin J) were isolated from straw in connection with field outbreaks of apparent stachybotryosis in Hungary and Czechoslovokia (Harrach *et al.*, 1983; Bata *et al.*, 1985). Scott *et al.* (1985) and Abbas *et. al.* (1986) have reported the contamination of wheat and other grains with trichothecene producing fungi in Canada and USA respectively. These toxins have also been reported to occur naturally from other parts of the world such as Yugoslavia (Nowar, 1975), North America (Vesonder, 1983), Portugal (Gimeno, 1983), Canada (Scott, 1983), Finland (Hintika, 1983), Denmark (Palle, and Bendicte, 1983), Britain (Patterson, 1983), Italy (Giovani, 1983), South Africa (Marasas and Krick, 1983), Taiwan (Tseng, 1983), Germany (Gedeck and Bauer, 1983), Hungary (Harrach *et al.*, 1983), and Czechoslovokia (Betina and Binovska, 1979). Bhat *et al.* (1976), Rukmini and Bhat (1978), Bhat (1983) and Bhavanishanker and Shanta (1987) reported the natural occurrence of differenet trichothecenes in different agricultural commodities from India.

Abbas *et al.* (1985) and Scott *et al.* (1985) reported that there is no appreciable loss of DON in wheat during its cleaning, milling and baking. Scott (1984) elegantly discussed different aspects of contamination of these toxins in various agricultural commodities. Prelusky *et al.* (1984) reported the non-transmission of DON to the milk following oral administration of the toxin to dairy cows which they attributed to the endogenous ruminant microorganisms. According to Swanson *et al.* (1987), T-2, DAS and DON have least effect on ruminent animals as their toxins are likely to be biotransformed by endogenous microorganisms in the elementary tract of ruminants.

Sorenson *et al.* (1987) have reported satratoxin H in all samples and satratoxin G and trichoverrol A and B in conidia of some samples of *Stachybotrys atra* and visualized the potential health hazards of inhalation of aerosols containing these conidia. Dashek *et al.* (1985) have elegantly discussed the possible warfare agents in relation to the "Yellow Rain and Disease" in South-East Asia.

In India there are only a few scattered reports of occurrence of trichothecene toxins. Tripathy *et al.* (1967) have reported the death of 22 birds out of a flock of 236 Rhode Island red pellets in the Orissa state poultry breeding farm. Sharma and Singh (1971) and Rajendran *et al.* (1975) have reported the disease diagnosed as stachybotryotoxicosis, which involved 160 cattle with a 40 per cent mortality, was attributed to the presence of *S. atra* in straw samples. Rukmini and Bhat (1978) have reported the occurrence of T-2 in sorghum grains contaminated with head moulds. Ghosal *et al.* (1977) have reported the T-2, DAS and 12, 13. epoxytrichothecene in safflower seeds infected with *Fusarium oxysporum.* Vora (1978) in his survey of different parts of the northern India reported the incidence of various species of *Fusarium.* Bhat and Tulpule (1983) discussed the trichothecene problem in India. Chakrabarti and Ghosal (1986) have reported the occurrence of free and conjugated 12, 13-epoxytrichothecenes in banana fruits. Bhat *et al.* (1989a) have attributed the outbreak of mycotoxicosis in Kashmir Valley to the contamination of wheat with deoxynivalenol. Some of the factors influencing the incidence of trichothecene in foods and feeds are discussed.

(i) Relative Humidity (RH)

Of various factors influencing biodeterioration and mycotoxin elaboration relative humidity (RH) of a storage place is primary importance. Storage fungi grow at moisture content which is in equilibrium with relative humidity ranging from 65-70 to 85-90%. Each microorganism has a minimum moisture level for the invasion and proliferation, failing which deterioration is not possible.

Though influence of humidity on biodeterioration of different food grains by storage fungi has been studied (Diener and Davis, 1967; Christensen and Kaufmann, 1969; Christensen, 1973; Arumugham and Shanmugavelu, 1977; Khan and Siddiqui, 1980; Perez *et al.,* 1982; Reddy and Reddy, 1982; Dange and Patel, 1984; Belcheva *et al.,* 1985; Kochler *et al.,* 1985; Kingsland, 1986; Girisham and Reddy, 1986; Hareeshvardhan Rao, 1988), limited information is available on deterioration of feeds and fodders by trichothe-cene producing fungi (Bacon *et al.,* 1973; Jones and Hamilton, 1985; Kurbatskaya and Trostanetskii, 1987).

Proteins, FAA, reducing sugars, total nitrogen. Phenols increased due to fungal infestation. The increase in their constituents, however, varied with the fungus. *F. pallidoroseum* elaborated increasing amount of trichothecene toxins with the increase of RH of incubation chamber. Similarly *M. roridum* could elaborate roridins, which was maximum in feed incubated at RH 90 and 100. *S. atra* also produced maximum amount of satratoxin at RH 90 and above. *T. roseum* was a poor producer of trichothecin. However, the amount of toxin increased with the increase in relative humidity of storage chamber. In general humidity of 80% and above was conducive for mould infestation and production of different trichothecene mycotoxins.

(ii) Temperature

Trichothecene producing fungi were inactivated to significant extent at an incubation temperature of 40°C. All the five fungi under study were more active at 25°C and biodeteriorating activity decreased both at lower and higher incubation temperature. Ptoteins, FAA, reducing sugars, total nitrogen and total phenols of cattle feed increased due to fungal infestation. The increase in proteins was more significant in *T. viride* inoculated feed followed by *F. pallidoroseum.* On the other hand, increase in protein content was minimum in *T. roseum* inoculated feed. Increase of FAA content was more in *S. atra* inoculated cattle feed and minimum in *T. viride* infested feed. Reducing sugars (RS) increase was more in *F. pallidoroseum* inoculated feed. However, RS content varied both with the fungus and incubation temperature. The variation in RS content may be attributed to their assimilation by infesting fungus or polysaccharide hydrolysing activity of the fungus. FFA content decreased due to fungal infestation. The decrease was more significant in *T. roseum* inoculated feed. The decrease in FFA may be attributed to the assimilation of FFA in feed by the inoculated fungus. On the other hand, increase of FFA was recorded in different crop seeds infested by storage fungi (Girisham, 1986; Krishna Reddy and Reddy, 1988; Hareeshvardhan Rao and Reddy, 1988). Free fatty acid content was generally taken as an index for viability of the seed. The difference may be due to the fact that fatty acids in feed are already in a hydrolysed state and the infesting fungus assimilated the same as a source of energy. On the other hand, FFA present in the viable seed are held in a polymerised state and are in storage form. Naturally the infecting fungus attacked these stored lipids, fats and released fatty acids by producing lipase.

The cattle feed suffered weight loss due to fungal infestation. The weight loss of feed caused by different fungi varied with the incubation temperature.

The weight loss was maximum in *S. atra* inoculated feed at incubation temperature of 25°C but at 20°C it was more in *F. pallidoroseum* inoculated feed. On the other hand, at 15°C weight loss was more in *S. atra* inoculated feed. Weight loss of cattle feed decreased significantly at incubation temperature above 30°C. However, *F. pallidoroseum* and *T. viride* were versatile and remained active over a temperature range of 15 to 40°C. On the other hand, *M. roridum* was more sensitive to temperature and it was active only between temperature of 20°C and 35°C. Toxin production by *F. pallidoroseum* was maximum at incubation temperature of 25°C and decreased to significant extent at 15°C and 40°C. A gradual decrease in quantity of toxin production was observed with the increase of incubation temperature. *M. roridum* elaborated roridins at all incubation temperatures tried and it was maximum at 25°C and decreased slightly at 20°C and 15°C. On the other hand, the decrease was gradual at elevated temperature (30°C and 35°C). Almost similar behaviour was recorded with *S. atra*. *T. roseum* produced minimum amount of trichotheceme at 25°C and toxin production decreased at 30°C as well as at low temperature. On the other hand, the decrease in toxin production was more pronounced at 35°C and 40°C. Similarly *T. viride* also elaborated trichodermin in higher quantity at 25°C and the quantity of toxin decreased gradually at lower and higher incubation temperatures. However, the decrease in trichothecene was very significant at 40°C of incubation temperature.

(iii) Water activity (a_w)

The growth and metabolism of microorganisms demands the presence .of water in an available form. The most useful measurement of the availability of water is water activity (a_w). Of various environmental factors determining the mould growth, moisture content, temperature and humidity are of paramount importance. Microorganisms have an absolute demand for water and without water no growth can occur. The exact amount of water needed for the growth of microorganisms varies. Technically this water requirement is best expressed in terms of available water or water activity (a_w). The a_w for pure water would be 1.0 and hence a_w is always less than 1.0.

Each microorganism has a maximal, optimal and minimal aw for growth. As the a_w decreases below the optimum level, various changes occur in growth and metabolism of an organism, such as lengthening of lag phase of growth, decrease in the rate of growth and decrease in the amount of cell substances synthesized. These changes vary with the organism and with the solute employed to reduce a_w. Factors that may, affect a_w requirements of microorganisms include (i) the type of solute employed to reduce a_w, (ii) nutritive value of the culture medium, (iii) temperature, (iv) oxygen supply, (v) pH and (vi) inhibitors. The highest being for bacteria (0.97) and lowest for moulds. Yeasts occupy intermediate stage. The reduction of a_w below the optimum for a mould delays spore germination and reduces the rate of growth. The lower the a_w the less is the substratum available to the growth of moulds (Corry, 1978).

A large number of workers including, Northolt *et al.* (1979), Northolt and Van Egmond (1981), Northolt and Bullerman (1982), Roland and Beuchat (1984), Magan *et al.* (1984), Richard *et al.* (1985) and Patterson and Damoglou (1986) have studied the effect of a_w on growth of microorganisms. However, studies on the effect of water activity on growth and metabolism of trichothcenes producing fungi and toxin production in foods feeds is very limited (Bacon *et al.,* 1973; Kurbatskaya and Trostanetskii, 1987; Surekha and Reddy 1990).

The fungi trichothcenes producing were sensitive to changes in a_w. Fungal activity was maximum at a_w 0.92 and above, while the activity was minimum at a_w 0.75. Protein content of feed increased due to fungal infestation. The increase was significant at aw 0.92 and above and minimum at lower aw 0.75. The increase was maximum in *T. viride* and *S. atra* infested feed and minimum in *M. roridum* infested feed. *F. pallidoroseum T. viride* infestation of feed at aw 0.75, caused marginal increase in proteins. Free amino acids, reducing sugars, phenols and total nitrogen content of feed increased due to fungal infestation. The increase was more pronounced in feed with high aw (0.95) and minimum in feed with low aw (0.1 Total nitrogen of feed in *M. roridum, S. atra* and *T. roseum* infested feed was only marginal, while in rest of the two species it was significant. However, *T. viride* was an exception as at a_w 0.75, the increase in reducing sugars total nitrogen was marginal only. Interestingly free fatty acids content decreased due to fungal infestation. This may be due to the fact that it as such may be rich in free fatty acids and infesting fungi must have utilized. Thus resulting in the decrease instead of increase as recorded many of the stored food grains (Laxma Reddy and Reddy, 1992).

The free fatty acid content was more in feed incubated at a_w 0. than in feed incubated at a_w 0.95. The cattle feed suffered significant weight loss due to fungal infestation. However, the degree of weight loss varied with the fungus and a_w of the feed. *T. viride* was more aggressive and caused maximum weight loss, specially at higher a_w. On the other hand *F. pallidoroseum* was inactivated to a considerable extent at a_w 0.87 and below.

All the present fungi elaborated one or other trichothecene mycotoxins. The fungi exhibited wide range of tolerance to a_w. All the fusaria elaborated significant amount of toxin at all aw tried. However, *T. viride* elaborated only marginal amount of toxin at a_w 0.75. *S. atra* and *M. roridum* were efficient in production of satratoxin and roridin respectively and their production was maximum at a_w 0.95 and decreased significantly below this water activity.

iv. Competitive Environment

The fungi usually grow in a mixed population on a variety of agricultural commodities and responsible either for their deterioration resulting in the decreased nutrient quality or elaborate certain mycotoxins which are health hazardous to man. Fungi growing on the surface of the substratum interact closely and are bound to influence the growth and activities of other organism. This phenomenon was demonstrated by Lillehoj *et al.* (1970). They have also reported the inhibition of elaboration of different mycotoxins by fungi studied by them in the presence of certain fungi. One of the many functions attributed to mycotoxins is their ability to serve as "Survival factors" for the producing organisms. Control of mycotoxins has been a subject of intensive study during the last two decades. Prevention, removal and detoxification of moulds and mycotoxins from feeds and foods are three major strategies which can be adopted in order to declare the feed and food stuffs out of danger for human and animal consumption.

The interaction between *S. atra* and other toxigenic fungi in a common culture medium was somewhat different, when *A. ochraceus* and *S. atra* were grown in a mixed culture, *S. atra* was totally inhibited for production of satratoxin, while *A ochraceus* remained unaffected for production of ochratoxin A. The interaction of rest of the fungi was almost the same to that observed in U shaped culture vessel. The inhibitory effect of *A. flavus, A. terreus and A. nidulans* was more pronounced than observed in 'U' shaped culture vessel. While, only marginal increase in inhibition of satratoxin production was observed in the presence of *P. crustosum, P. aurantiogriseum and P. griseofulvum.* The inhibitory effect of sister fungi on *S. atra* for production of satratoxin could overcome significantly when other fungi were inoculated 5 days later to *S. atra*. Shantha (1991) feels that the counteracting ability of different fungi depends on various factors such as sequence of infestation, humidity and pH of the substratum.

It is clear that the competitive environment and interaction of serveral factors will influence the particular biological process. *S. atra* which was inhibited significantly by other toxigenic fungi could over come when it was inoculated 5 days in advance. Similarly the close interaction is also another factor which affect toxin production. The degree of inhibition of satratoxin production by *S. atra* in a mixed culture was comparatively more than what was observed in U shape culture vessel.

The biomass production by *A. nidulans, P. crustosum, P. griseofulvum* and *P. aurantiogriseum* was only little more when they were grown along with *S. atra.* However, biomass production by *A. ochraceus* decreased in the presence of *S. atra.* Similarly biomass production by *A. flavus* was inhibited in the presence *of S. atra.* The biomass production by *S. atra* in combination with other fungi significantly decreased, when *S. atra* was inoculated 5 days in advance of other fungi. U shaped culture vessel could not give correct assessment in the simultaneous interaction of fungi. On the other hand, the mixed culture may facilitate the close interaction of fungi but each may not grow independently and show activity freely. Further, the biomass in a mixed culture can not give assessment of individual fungus.

M. roridum was inhibited only marginally in the presence of *A. flavus, A. ochraceus* and *P. aurantiogriseum,* while *A. terreus* was moderate in itsinhibitory effect. *A. flavus, A. ochraceus* and *P. crustosum* remained unaffected in the presence *of M. roridum* for production of aflatoxin, ochratoxin A and penitrem A respectively, Mild inhibition of *P. aurantiogriseum, P. griseofulvum, A. nidulans* and *A. terreus* for production of penitrem B, CPA, sterigmatocystin and patulin and terreic acid respectively was recorded in the presence of *M. roridum.*

M. roridum when grown along with other fungi such as in the presence of *A. terreus, A. ochraceus* and *P. crustosum* in mixed culture was inhibited for production of roridins to a significant level. The other fungi were also affected in the presence of *M. roridum* for production of respective mycotoxin. Inhibitory effect of other fungi on roridin production by *M. roridum* could be over come to a great extent, when *M. roridum* was inoculated 5 days in advance of other fungi. The recovery was more significant in the presence of *A. terreus* and *P. aurantiogriseum.*

When *M. roridum* was inoculated 5 days in advance, it was affected only marginally for biomass production. However, it was more pronounced by the presence of *P. griseofulvum.* On the other hand, the biomass production by *A. flavus* and *P. griseofulvum* were affected significantly. Biomass production by rest of the fungi understudy remained unaffected.

The presence of other toxigenic fungi affected both biomass and trichothecin production by *T. roseum. T. roseum* was inhibited significantly in the presence of *A. flavus, A. ochraceus* and *P. aurantiogriseum,* while rests of the fungi were marginal in their inhibitory activity. *A. flavus, A. ochraceus* and *P. crustosum* remained uneffected in the presence of *T. roseum* for production of aflatoxin, ochratoxin A and penitrem A respectively. Mild inhibition of *P. aurantiogriseum, P. griseofulvum, A.nidulans* and *A. terreus* for production of penitrem B, CPA, sterigmatocystin and patulin and terreic acid respectively was recorded in the presence of *T. roseum.* In mixed culture *T. roseum* was inhibited for production of trichothecin to a significant level. Other fungi were also affected in the presence of *T. roseum* for production of respective mycotoxins. When *T. roseum* was inoculated 5 days in advance of other fungi, inhibitory effect of other fungi on trichothecin production by *T. roseum* could over come to a great extent. On the other hand, other fungi were affected significantly for production of respective toxins.

The biomass production by *A. nidulans, P. crustosum, P. griseofulvum* and *P. aurantiogriseum* was only little more when they were grown along with *T. roseum*. However, biomass production by *A. ochraceus* decreased in the presence of *T. roseum*. Similarly, biomass production by *A. flavus* was inhibited by the presence *of T. roseum*. The biomass production by *T. roseum* in combination with *A. flavus* or *A. terreus* significantly decreased when *T. roseum* was inoculated 5 days in advance of other fungi. The biomass production by *P. crustosum, P. aurantiogriseum* and *P. griseofulvum* was also decreased, when *T. roseum* was inoculated 5 days earlier to other fungi.

The presence of other toxigenic fungi affected both biomass and trichodermin production by *T. viride*. The degree of inhibition was more in the presence of *A. flavus* and *A. ochraceus* and least in the presence of *A. nidulans* and *P. griseofulvum,* when U shaped culture vessel was employed. On the other hand, production of aflatoxin and ochratoxin A by *A. flavus* and *A. ochraceus* respectively were not affected in the presence of *T. viride,* while *P. crustosum, P. aurantiogriseum* and *P. griseofulvum* were inhibited in the presence of *T. viride* for producction of penitrem A, penitrem B and cyclopiazonic acid respectively. In dual culture *A. flavus, A. nidulans, A. ochraceus* and *P. aurantiogriseum* inhibited the production of trichodermin by *T. viride* to a significant level. The inhibition of trichodermin production by *T. viride* in the presence of rest of the fungi was only marginal. Aflatoxin, sterigmatocystin and ochratoxin A were inhibited significantly and could be produced only in trace amounts in the presence of *T. viride*. Rest of the fungi were also inhibited but to a lesser extent. The inoculation of *T. viride* 5 days in advance of other fungi, adversely affected the other fungi for production of respective toxins.

The biomass production by *T. viride* was inhibited in the presence of other fungi only marginally. In dual culture the biomass yield was more than in the culture of individual fungus. The inoculation of *T. viride* 5 days in advance of other fungi did not make much difference in biomass production.

In the presence of other fungi production of DAS by *F. pallidoroseum* was inhibited significantly. The degree of inhibition was more in the presence of *A. flavus* and *P. griseofulvum* and least in the presence of *P. crustosum,* when U shaped culture vessel was employed. On the other hand, production of aflatoxin and ochratoxin A by *A. flavus* and *A. orchraceus* respectively was not affected in the presence of *F. pallidoroseum,* while *A. nidulans and A. terreus* were marginally inhibited. *P. crustosum, P. aurantiogriseum* and *P. griseofulvum* were inhibited considerabley for production of penitrem A, penitrem B and cyclopiazonic acid respectively in the presence of *F. pallidoroseum*. In dual culture, *F. pallidoroseum* was inhibited for production of DAS in the presence of *A. flavus, P. griseofulvum, P. crustosum* and *P. aurantiogriseum*. Interaction of *F. pallidoroseum* and other fungi *like A.*

flavus or *A. terreus* or *A. ochraceus* did not make much difference in 'u' shaped culture vessel and dual culture in the amount of toxin production by these fungi. The inoculation of *F. pallidoroseum* 5 days in advance of other fungi, adversely affected the toxin production by other fungi both qualitatively and quantitatively.

Biomass production by *F. pallidoroseum* was least in the presence of other fungi. However, presence of *P. crustosum, P. aurantiogriseum* and *P. griseofulvum* did not make much difference in biomass production by *F. pallidoroseum.* On the other hand, the biomass of *F. palladoroseum* was significantly inhibited in the presence of *A. flavus, A. terreus, A. nidulans* and *A. ochraceus.* On the other hand, *A. flavus, A. terreus* and *A. nidulans* were inhibited marginally in the presence of *F. pallidoroseum.* The biomass production in dual culture was more than what *F. pallidoroseum* could achieve individually. However, the biomass production by *P. griseofulvum* was the same, when grown individually and incombination with *F. palladoroseum.* When *F. pallidoroseum* was inoculated 5 days in advance, the biomass production by *A. ochraceus* and *P. griseofulvum* was almost the same. In rest of cases the biomass production was marginally affected.

14.4. BIOLOGICAL ACTIVITY

Generally trichothecenes do not require metabolic activation to exert their biologically important effects. The biological activity of the toxins to skin and intestinal mucosa has been attributed to direct reaction with cellular components and to the presence of reactive electrophilic epoxide functional group.

Trichothecenes induced mycotoxicoses are difficult to distinguish because they affect many organs including gastrointertinal tract and haematopoietic cardiovascular systems. Infection of foods and feeds contaminated by trichothecene mycotoxins causes many types of mycotoxicoses in humans and animals including moldy corn toxicosis, scabby wheat toxicosis (or red mold, or akakbi-byodisease or scaby barley poisoning), feed refusal and emetic syndrome (swine), fusariotoxicoses, hemorrhage syndrome and alimentary toxic aleukia. Some of the common manifestations of trichothecenes toxicity are depression of immune responses and nausea and sometimes vomitting. Because of number of closely related metabolites likely to cooccurein combination in foods and animal feeds, the toxicology is complex with both synergistic and antagonstic effects which has been discussed elegantly by Miller (1995). T2 Toxin acts synergistically with aflatoxins (Raju and Devegowda, 2000). Continous exposure to trichothecenes results in skin rashes which may proceed to necroticesions. Presence of macarocyclic trichothecenes in air-borne fungal spores may contribute to some forms of sick building syndrome (Croft *et al.*, 1986).

(ii) Nucleic acid synthesis: Nivalenol and fusarenon-X are inhibitors to DNA synthesis by 60 and 85% respectively but RNA synthesis was slightly (5-10%) affected (Ohtsubo and Saito, 1970). Similar observations were made for DAS (Liao *et al.,* 1976) and trichodermin (Gupta and Siminovitch, 1978). Ueno *et al.* (1971) feel that these trichothecenes disturb the ligase reactions of DNA Synthesis. Trichothecenes affect RNA synthesis only secondarily.

(iii) Protein Synthesis: All 12, 13-epoxytrichothecenes are potent inhibitors of protein synthesis on eukaryotic ribosomes (Gale *et al.,* 1981). The inhibitory effect of trichothecenes depend apart from the presence of 12, 13-epoxy- ring, type of double bond, the number and position of hydroxy groups and the type of esterifying acids. Thus little or no substituent at C_4 makes trichothecene a initiation inhibitor (verrucarin, fusarenon-X, T-2). As ester group only at C_4 position makes the compound a chain elongatil on termination inhibitor with another ester (trichodermin, DON)

(iv) Other biochemical processes: Phosphate uptake and lipid synthesis was reported to be inhibited by fusarenon-X (Chiba *et al.,* 1972). Fusarenon-X, T-2 and neosolaniol inhibit creatine phosphokinase. Lactate and alcohol dehydrogenase activity was reduced by treatment with fusarenon-X (Ueno and Matsumoto, 1975). Kobayashi *et al.* (1989) have induced callus initiation in alfalfa cotyledon by treating with trichothecene mycotoxins. Sergeen *et al.* (1990) elucidated the mechanism of impairment of calcium metabolism by T-2 mycotoxin.

(v) Antibacterial activity: Many of the fungal metabolites (mycophenolic acid, penicillic acid citrinin, patulin and trichothecene) were originally considered to be antibiotics. Subsequently intensive studies on isolation from contaminated material and their toxicity to animal system led to these grouping *of* the compounds under mycotoxins. Trichothecene do not possess strong antimicrobial activity. They are active only at higher concentration. These toxins are reported to be active against both gram (+) ve and gram (-)ve bacteria. Satratoxins and roridins were active only against gram (+) bacteria (Buckelew *et al.,* 1972). However, roridins were active even against gram (-) ve bacteria at higher concentration. DAS was reported to be inhibitory to *Escherichia coli, Bacillus subtilis* and *Micrococcus pyogenes* (Buckelwe *et al.,* 1972).

Trichothecene toxins were active against both gram (+) ve and gram (-) ve bacteria. However, fusarial toxins failed to inhibit the growth of *E. coli,* while satratoxin was ineffective in checking the growth of *Proteus vulgaris* and *Klebsiella* sp. Roridin was most active against all the bacteria under study. Further, *Streptococcus albus, B. megaterium* and *B. subtilis* were inhibited to a significant level, while inhibition of growth of *Proteus vulgaris, E. coli* and *Klebsiella* sp. was marginal. However, fusarial toxins were comparatively mild in their toxicity towards bacteria under study.Hence some of these bacteria may be exploited for bioassay agents.

(vi) Cytoxicity: Trichothecenes exhibit a wide range of cytotoxic symptoms. Failure of cytokinesis is a characteristic feature. Satratoxins are higly toxic, while trichothecenes are least toxic (trichothecenol and crotocin). These compounds are also reported to be as Karyorrhexis in a manner similar to radiomimetri injury (Saito and Ohtsubo, 1973). Grove and Mortimer (1969) reported the diacetoxyscirpenol (DAS) was highly toxic to HE P_2 of human origin and BHK of baby hamster kidney. Similarly Ohtsubo *et al.* (1968) and Bodan and Zoldac (1974) have reported toxicity of nivalenol and related compounds and T-2 to He La cells and epithelial cells of pig kidney. Tanaka *et al.* (1973) while evaluating the cytotoxicity of 20 kinds of trichothecens concluded that macrocyclic trichothecenes were highly toxic followed by microcyclic type A (T-2) and type B (nivalenol). Schindler *et al.* (1974) demonstrated the breakdown of polyribosomal profile at cellular level. Roridin and Verrucarins were inhibitors of chain initiation step of protein synthesis (type B), while trichothecenes like trichodermin and trichodermol were inhibitors of chain elongation termination step (E-type). It is said that the peptidyl transferase, which is localised in 60s of eukaryotic ribosomes marked by the binding trichothecene molecule.

The trichothecene compounds have been reported to possess potent cytotoxicity to eukaryote (Ohtsubo and Saito, 1970). Eppley (1974) recommended cytotoxicity test for detection of trichothecene toxins using brine shrimp larvae. Reiss (1973) recorded inhibition of mitosis in onion roots by *Fusarium toxins.* He recorded induction of various abnormalities in the cell division. Reiss (1975) reported the mitotic inhibition and chromosomal aberrations in onion cells under the influence of diacetoxyscirpenol. T-2 and related trichothecenes have been reported to possess a potent cytotoxicity to eukaryotes (Ohtsubo and Saito, 1970)

Trichothecene toxins were responsible for decrease in mitotic index. Mitotic inhibition was more pronounced in cells treated with fusarial toxins followed by roridin and satratoxin. Mitotic index decreased with the increease in concentration of toxin (Table 14.5). Chromosome breakages, extreme fragmentation and enlargement of nuclei were most common abnormalities. Other abnormality of metaphase was unorganised orientation of chiromosomes. At anaphase, fragmentation of chromosomes was most frequent followed by laggards. At telophase bridges were most frequent followed by fragmentation and laggards (Plate 1 Figs. 1-12). The most striking effect of satra-toxin was inhibition of cytokinesis resulting in binucleate cells. When the cytotoxicity of three toxins was compared, satratoxins were more potent, in the cause of cytotoxicity as compared to roridin and fusarial toxins.

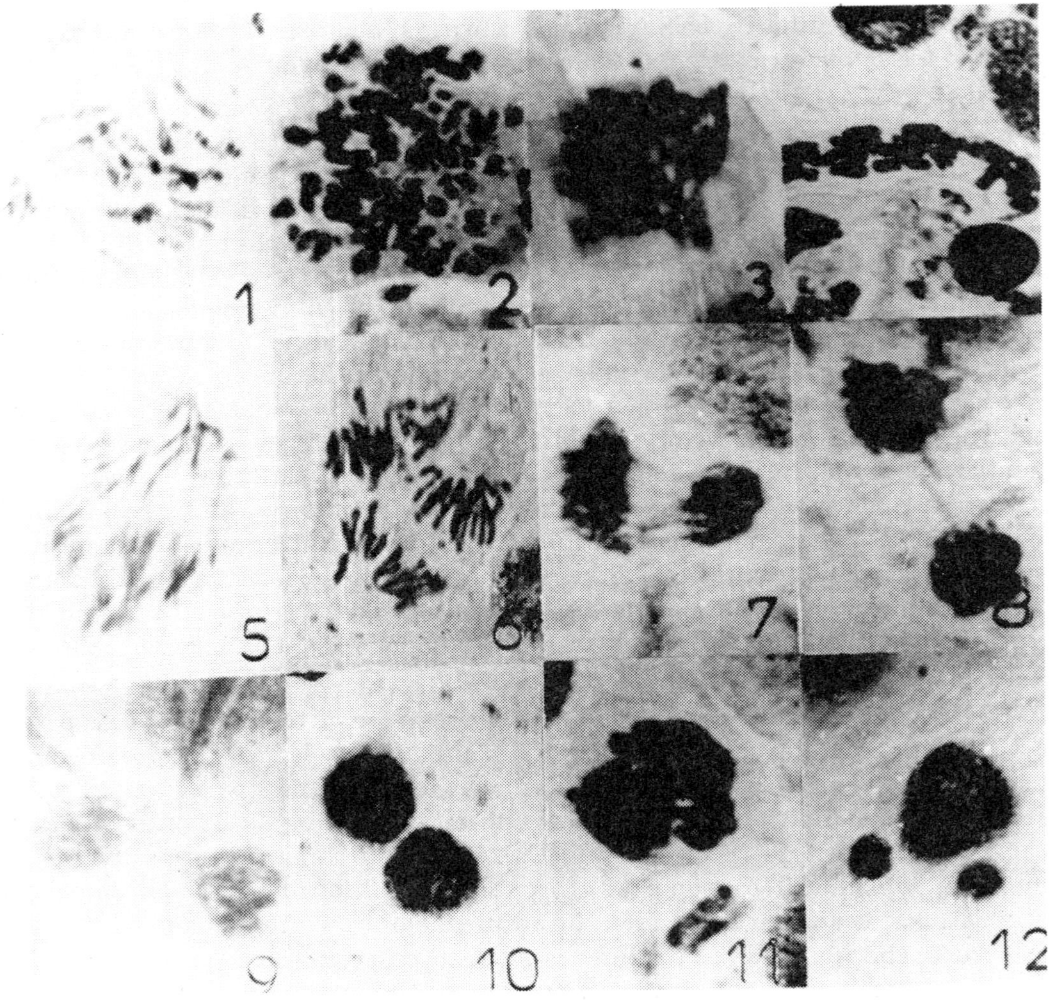

PLATE - I. Mitotic abnormalities induced by trichothecenes in meristematic root cells of *A. sativum*

Fig. 1. Chromosome fragmentation at anaphase and metaphase; Fig. 2. Multiple fragmentation or extreme fragmentation of chromosome Dl; Fig. 3. Stickiness of chromosomes at metaphase; Fig. 4. Unorientation of chromosomes; Fig. 5 & 6. Fragmentation, multipolar orientation of chromosomes at anaphase; Fig. 7. Multiple bridges of chromosomes; Fig. 8. Stickiness, laggards at late anaphase or early telophase; Fig. 9. Sticky bridge at late anaphase; Fig. 10 Binucleate cell; Fig. 11. Sticky trinucleate cell; Fig. 12. Cell showing two micro nuclei

According to Berek *et al.* (2001) T_2 toxin almost completely suppressed IL-2 and IL-5 production at ng/ml or more. Marin *et al.* (1996) reported that T_2 toxin inhibited cytokine prodution and proliferation.Deoxynivlenol modulated clellar and humoral immune function by altering nitrogens induced proliferative response of murinelymphocytes, viability of spelnic lymphocytes and host resistance (Meky *et al.*, 2001). Berk *et al.* (2001) found that nivalenol and deoxynivalenol had an inhibitory effect on mitogen induced blast formationa and natural killer reactions. T_2, diacetoxyscirpenol, nivalenol and deoxynivalenol are reported to inhibit effectivley mitogen induced lymphocyte proliferation and immunoglobulin production in human lymphocyte in a dose dependent manner (Thuvander *et al.*, 1996).

Table 14.5. Effect of trichothecene toxins on dividing cells of *A. sativum*

Cytological feature (in percentage)	Satratoxin	Roridin	Fusarial
Mitotic index	11.04	12.15	12.55
Prophase			
Enlargement of nuclei	4.95	4.38	3.29
Metaphase			
Stickiness	5.91	4.39	4.34
Fragmentation	4.78	3.34	1.08
Anaphase			
Bridges	4.77	5.51	3.19
Fragmentation	2.98	-	2.17
Laggards	2.98	6.38	1.08
Telophase			
Bridges	4.77	1.06	1.59
Fragmentation	2.98	4.62	6.38
Laggards	-	4.62	-
Cytokinesis inhibition (bipolar)	4.44	-	-
Multipolar segregation	1.49	4.31	-
Micronuclei	1.49	4.31	-
Others	3.95	4.34	1.95
Total	46.50	39.47	25.07

(vii) Animal diseases: The trichothecenes produce disease symptoms such as nausea, emesis, decreased feed consumption, feed refusal, skin irritation, diarrhoea, histological changes, haemorrhage and reproductive problems. Inhibition of protein and DNA synthesis is also reported (Scott *et al.,* 1985). Epidemological surveys have revealed trichothecene mycotoxins to be common natural contaminants of cereals and fodders are etiological agents of mysterious diseases of farm animals and human beings. The various aspects of diseases caused by trichothecenes have been reviewed from time to time by Bamburg and Strong (1971) Smalley and Strong (1974), Tamm

(1977), Bamburg (1976), Tamm and Breitensteirn (1980), Ueno (1983), Tamm and Tori (1984), Betina (1989) and Beasley *et al.* (1989).

Trichothecene mycotoxins cause variety of health hazards in different domestic animals Bubien *et al.* (1989) and Fenerstein *et al.* (1989) have discussed different aspects of toxicity of trichothecenes on cardiovascular and nervous system respectively (Table 14.6).

Table 14.6. Health hazards caused in different animals due to trichothecene mycotoxins

Diseases	Organs affected	Symptoms
Epithelionecrosis	Cutaneous surfaces, muc-ocutaneous junctions mucosa of upper alimentary canal	Discomfort, changes jin ability to mediate physiologic disruption of an effective defense barrier to secondary infection. Cutaneous necrosis.
Gastroenteritis	Intestinal mucosa	Epithelial cell (crypt) necrosis, diarrhoea, vomiting, mucosal necrosis, disruption of gut epithelium by erosions and ulcerations.
Coagulatory	Hematopoietic tissues	Changes in differential serum protein contents including lowered levels of gamma, alpha, beta and beta 2 protein.
Depressed hematopoiesis	Hematopoietic tissues	Lowered PCV levels, leukopenia, hypoactive marrow, and thrombocytopenia.
Suppression of immunoligic mechanisms	Spleen, thymusand lymph	Necrosis of lymphoid tissues, reduced concentrations of IgA and IgM and some components ofcomplement 3, depressed phagocytic function of both and neutrophils.

Toxicity of trichothecenes is highest in verrucarins and roridins such as verrucarinA and roridinA, followed by hydroxyl or aceyl substituted trichothecenes, such as T-2 toxin and 8-keto trichothecenes (nivalenol) and non-macrocyclic compound of crotocin (type-b) is the least toxic trichothecene.

According to Berek *et al.* (2001) T_2 toxin almost completely suppressed IL-2 and IL-5 production at 5 mg/ml or more. Marin *et al.* (1996) reported that T-2 toxin inhibited cytokine production and proliferation. Deoxynivalenol modulated cellular and humoral immune function by altering nitrogen-induced proliferative response of murine lymphocytes, viability of spienic lymphocytes and host-resistance (Meky *et al.*, 2001). Berek *et al.* (2001) found that nivalenol and deoxynivalenol had an inhibitory effect on mitogen induced blast formation and natural killer reaction. T-2, diacetoxynivalenol, nivalenol and deoxynivalenol are reported to inhibit effectively mitogen induced lymphocyte proliferation and immunoglobulin production in human lymphocytes in a dose dependent manner (Thuvander *et al.*, 1996).

Pharmacological investigations with fusarenon-X revealed that this mycotoxin induced hypothermia (Matsuka *et al.,* 1979). Diarrhoea is one of the characteristic responses of animals administered with a sublethal dose of toxins (Matasuka and Kubota, 1982). Shivefer and Beasley (1989) have extensively discussed the effect of trichothecene mycotoxins on the digestive system and energy metabolism involving cholesterol, lipids, fats, soluble vitamins, carbohydrates, oxidative phosphorylation and glucose utilization.

(vii) Effect on chick and mice: Trichothecenes are known to be potent inhibitors of protein synthesis in eukaryotic ribosomes. Trichothecenes are also reported to inhibit DNA and RNA synthesis in mammalian cells (Ueno, 1985 and Thompson, and Wannemoches 1990). Sunija *et al.* (1987 and 1988) have reported long term experiments with feeding of T-2 toxin on the metabolic alterations in rats. In liver T-2 toxin inhibited protein synthesis, DNA, synthesis, phospholipid synthesis, microsomal NADPH, cytochrome C and NADH cytochrome b5 reductases, hepatic GOT and GPT; Serum GOT and GPT, however, were markedly increased. From these changes it is inferred that liver and its microsomal functions are altered including the inhibition of microsomal electron transport chain, thereby causing disorganisation of liver cell membranes leading to necrosis of liver cells. T-2 toxin induced liver lipid peroxidation, damages the liver membranous system which may impair the structural and functional organization of the membrane. Necrosis of liver cells has been reported to be due to enhanced lipid peroxidation (Plaa and Witsch, 1976; Sunija *et al.,* 1987). The increase in protein peroxidation could also be due to enhanced free radicals by T -2 toxin (Segal *et al.,* 1983)

Trichothecenes are rapidly absorbed from the gastrointestinal tract and distributed throughout the body (Glavits and Vany, 1988; Betina, 1989). Within the first few hours after taking trichothecene contaminated food, peak concentration of toxin appears in the blood, there after the toxin occur in the bile, gall bladder, liver, kidneys and intestine.

Vomiting and feed refusal are common toxicological features of trichothecene type of mycotoxins. Staggering toxicosis in Siberia, alimentary toxic aleukia in Russia, red mould disease in Japan, mouldy corn toxicosis in United States, stachybotryotoxicosis in Hungary, dendrodochiasis in the Central Europe and related toxicosis have been reported to be characteristic symptoms of trichothecenes. Carcinogenicity of T-2 toxin was reported by Schoental *et al.* (1979) in rat and they observed cardiovascular lesions and tumours in the digestive tract. Wyatt *et al.* (1973) and Chi *et al.* (1977) have studied extensively the mycotoxicoses of poultry. Some of the symptoms of diseases of poultry are decrease in feed consumption, egg production, thickness of egg shell, hatchability is affected adversely, uric acid accumulation, and bursa of Fabricius was reduced in weight, increased rate of liver and pancreas. There are only limited reports regarding the trichothecene mycotoxin effects on human. Trichothecenes are reported to induce alimentary toxic aleukia (ATA or septic angina) which consists of four stages (Betina, 1989). Similarly cancer of the oesophagus of man has been reported in South Africa and Northern China and in both the areas *Fusarium* was responsible. Marasas *et al.* (1979) also reported the highest known Oesophageal cancer in Africa occurs in the South Western districts of the republic of the Transkei; where as in northern districts of same country is relatively low. Administration of DAS intravenously daily dosages ranging from 5 mg to 6 mg/kg body weight caused symptoms like nausea, vomiting accompanied by hypotension, diarrhoea, system disfunction of nervous control, fever, chills, stomatitis, erythema and hair loss (Dashek *et al.,* 1985).

Free Fatty Acid and glycogen content of different vital organs of chick decreased gradually with the increased dose of *fusarial* toxins (Table 14.7). However, the shock was reduced with prolonged exposure time. On the other hand, soluble proteins increased due to *Fusarial* toxins. The increase was dose dependent. Further, prolonged exposure resulted in the decreased response. Satratoxin was responsible in decreasing soluble proteins, free Amino acid and glycogen content. These contents further decreased when dosage was increased. Proteins of liver as well as muscle decreased only marginally, while that of brain and kidney decreased significantly at higher dosage of satratoxin. Similarly FAA of heart changed little, while that of kidney decreased under the influence of satratoxin. However, the decrease of glycogen content in all vital organs was uniform.

Roridin was also responsible for decreasing free amino acid and protein content of the vital organs brain, liver and kidney decreased very significantly. When decrease in FAA was more at a dose of 25 mg/kg of toxin in brain tissue, it was at 50 mg/kg of toxin in kidney. Similarly FAA of liver also decreased. The decrease was more pronounced at a dosage of 50 mg/kg. The glycogen content of all the vital organs decreased. The decrease was progressive with the increased dosage. However, the glycogen of liver and muscle was reduced to a significant level. Glycogen content of brain, heart and kidney decreased only marginally.

Trichothecene mycotoxins exerted a significant effect on the composition of blood in chick (Table 14.6). However, the toxicity varied with the mycotoxin. *Fusarial* toxins were responsible for decreased soluble proteins, blood sugars, haemoglobin content and albumins, SGOT and SGPT increase was more pronounced at higher concentration of the T_2 toxin. However, on prolonged exposure, the shock was recovered to some extent.

Satratoxin was also responsible for decrease in soluble proteins, sugars, haemoglobin content and albumins. The decrease in these constituents was dose dependent. SGOT and SGPT increased in the presence of satratoxin.

The increase was again dose dependent. Roridins were also responsible for the decrease in soluble proteins, reducing sugars, haemoglobin content and albumin content of blood in chick. The decrease was further significant with the increase in concentration of the toxin. On the other hand, SGOT and SGPT activity increased in the presence of roridins. The increase was again dose dependent. Satratoxin affected the blood composition to a significant level. Soluble proteins, reducing sugars and haemoglobin decreased significantly. The decrease in these constituents was dose dependent. The SGOT and SGPT activity increased due to satratoxin injection. The activity of these enzymes increased with the increase in the concentration of the toxin.

Table 14.7 Effect of Trichothecenes toxins on biochemical composition of different organs of chick

Name of the toxin	Constituents (mg/g)	Concentration (mg/kg)	Brain	Heart	Organs Kidney	Liver	Muscle
Fusarial toxins	Protein	C	2.9	9.1	5.5	5.6	5.6
		50	3.4	10.4	5.8	6.0	7.2
		100	4.3	10.5	6.2	8.1	8.0
		150	4.3	10.7	6.4	9.1	8.2
	Free amino acids	C	12.96	7.3	15.36	9.36	13.1
		50	10.00	5.2	12.96	8.64	12.1
		100	11.22	4.8	12.14	8.64	11.4
		150	9.84	2.8	10.32	6.24	10.8
	Glycogen	C	0.24	0.22	0.17	1.67	0.29
		50	0.22	0.77	0.12	1.53	0.24
		100	0.17	0.14	0.12	1.45	0.22
		150	0.14	0.12	0.08	1.37	0.17
Satratoxin	Proteins	C	14.0	11.2	15.4	15.4	16.8
		200	11.2	9.8	11.2	15.4	12.6
		300	8.4	8.4	7.0	14.0	11.2
	Free amino acids	C	4.58	4.58	11.76	7.2	6.0
		200	4.08	4.08	10.8	5.3	4.5
		300	3.60	4.08	3.62	4.8	2.8
	Glycogen	C	0.29	0.24	0.29	103	0.98
		200	0.24	0.22	0.24	0.79	0.79
		300	0.20	0.17	0.22	0.71	0.66
Roridin	Proteins	C	3.7	4.1	4.7	3.3	4.0
		25	2.8	3.5	3.8	2.6	3.7
		50	2.7	3.5	3.3	2.6	3.2

Free amino acids	C	10.8	8.88	11.52	9.12	9.60
	25	5.72	7.44	9.60	8.04	6.36
	50	2.5	6.84	4.32	2.88	7.40
Glycogen	C	0.22	0.17	0.56	0.38	0.48
	25	0.22	0.14	0.42	0.11	0.14
	50	0.17	0.08	0.42	0.08	0.08

Satratoxin inhibited the synthesis of FAA, protens and glycogen in brain, heart, kidney, liver and spleen. In heart, FAA and glycogen synthesis was inhibited to only a limited extent. The inhibition of proteins systhensis was only marginal in the heart followed by liver, while the decrease of proteins was significant in kidney.

Clinical signs of intoxication were observed by reduced emotional behaviour. There was a marked loss in body weight of chicks by 7th day of receiving of satratoxin. About 30% of chicks died within 24 hours, at higher doses (300, 150, 50 mg/kg) of toxins. The head is drawn back and legs extended fully backwards. Within 3-4 hours of treatment they become weak and diarrhoea, vomiting, shivering and feed refusal. Other observations were dull behaviour, closing of eyes and sleepy eyes.

Pathological changes observed on gross postmortem examination of liver, kidney, heart and spleen revealed the major pathological conditions.

Liver: Liver showed diffuse fatty degenerative changes and necrotic areas were also seen in many places. Mononuclear infiltration was found in many areas and a few places were more intensive. Degeneration of hepatic cells was a common feature. Haemorrhage in certain areas was prominent in satratoxin treated chicks. Appearance of pykonotic nuclei was also observed. Roridin treated chicks showed a loss of emotive behaviour and loss of weight. Fatty degenerative changes and necrotic areas were observed throughout the section. Congestion of sinusoides was observed. Mononuclear infiltration was also found at a few places.

In chicks received *Fusarial* toxins (zearalenone and diacetoxyscirpenol), the liver showed necrotic areas and different grades of degenerative changes (Plate II Fig. 1-12). Mononuclear infiltration was also observed in certain areas.

Critical perusal of pathological symptoms revealed that roridin toxin is more toxic and fusarial toxins were mild and induced comparatively less pathological abnormalities. Satratoxin was intermediate in its toxicity towards the functioning of liver. Subramaniam *et al.* (1986) also observed necrosis of the liver with mild fatty changes, periportal fibrosis and fatty vacuolation in ochratoxin A treated chicks.

(a) Heart: Postmortem examination of heart of chick received satratoxin revealed loss of cross striations and degenerative changes in many cordiac muscle fibres. Mononuclear infiltration was seen in between cordiac muscle fibres. Roridin caused degenerative changes in muscle fibres at a few areas. Treatment with fusarial toxins resulted in mononuclear infiltration and degenerative changes of myocardial fibres of certain focal areas.

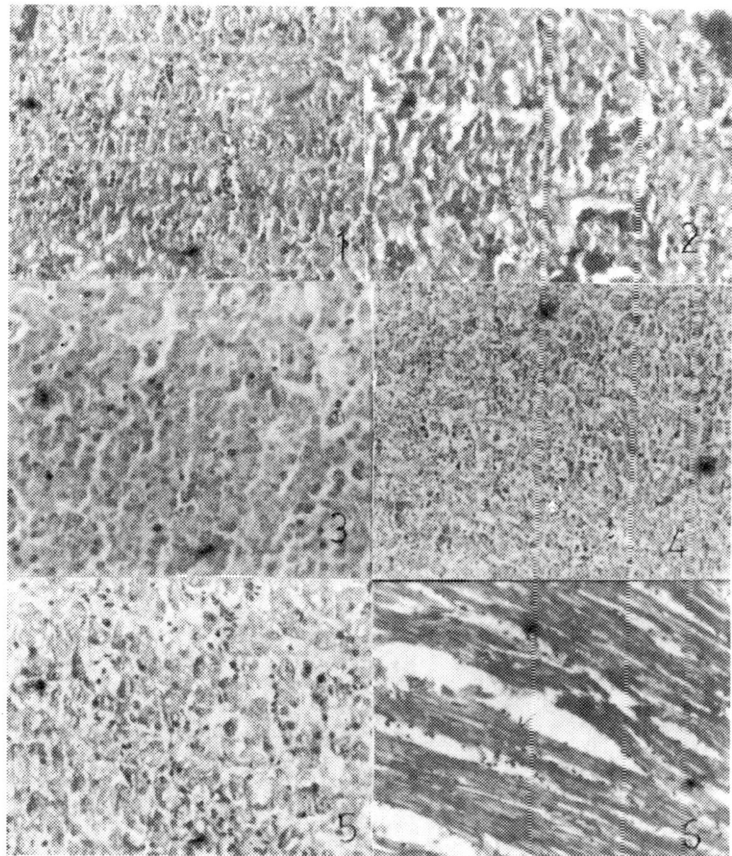

PLATE-II

Fig. 1 Microtome section ofliver showing necrotic areas and different grades of degenerative changes, H&E X 400

Fig. 2 Microtome sections of heart showing loss of cross striations, H&E 400

Fig. 3 Microtome section of kidney showing congestion of glomerulus, haemorrhage changes and degenerative changes in tubular pithelium, H&E X 400

Fig.4 Microtome sections of heart of mice showing loss of cross striations cordiac muscle fibres, H&E X 400

Fig. 5 Microtome section of kidney of mice showing a few areas of haemorrhage and tubular degeneration H&E X 400

Fig. 6 Microtome section of spleen of mice showing depletion of lymphoid follicles, H&EX 400

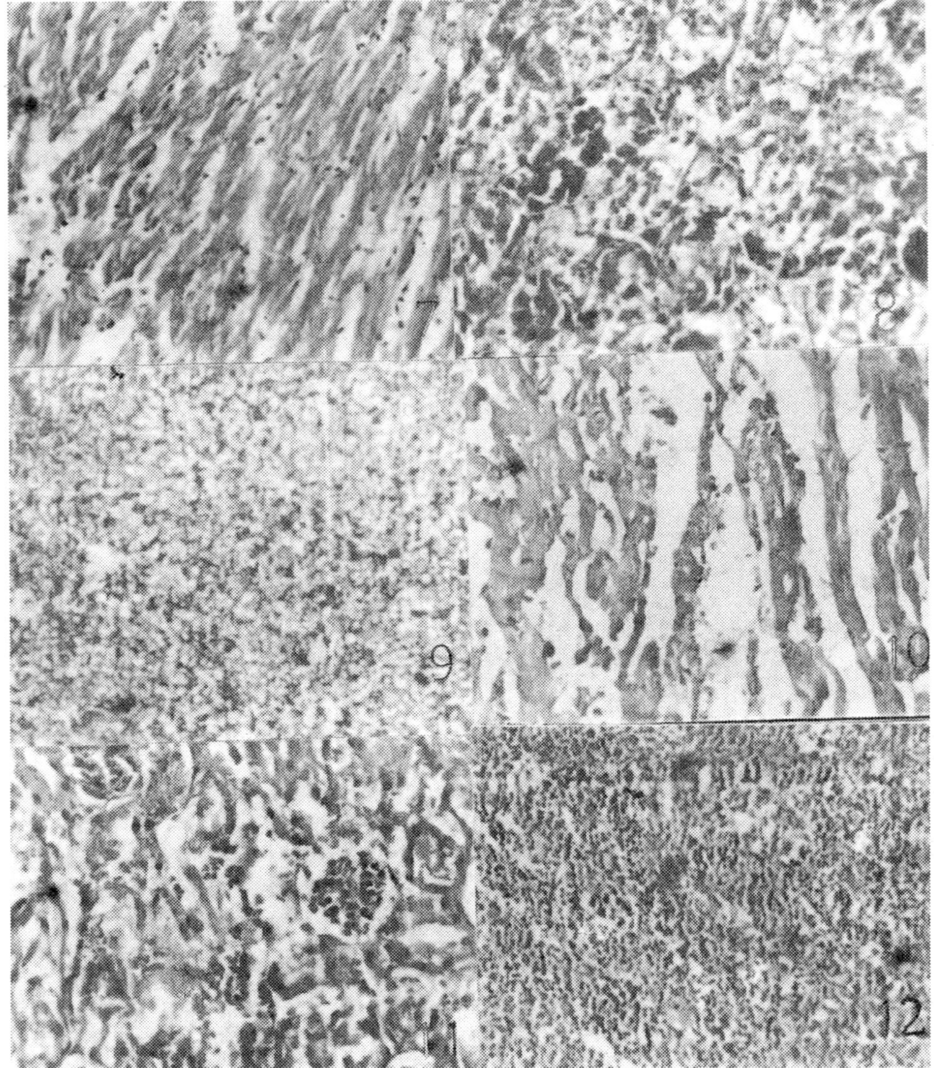

Fig. 7. Microtome section of heart showing degenerave changes of myocardial fibres of certain focal areas H&E 400

Fig. 8. Microtome section of kidney showing congestion of glomerulus, haemorrhage changes and degenerative changes in tubula epithelium. H&E X 400

Fig. 9. Microtome section of liver of mice showing fatty degenerative changes and necroces, H&E X 400

Fig. 10. Microtome section of heart of mice showing loss of cross striations cordiac muscle fibres, H&E X 400

Fig. 11. Microtome section of kidney mice showing a few areas of haemorrhage and tubular degenerative H&E X 400

Fig. 12. Microtome section of spleen mice showing depletion of lymphoid follicles, H&E X 400

(b) Kidney: Satratoxin induced congestion of glomerulus, haemorrhage changes and degenerative changes in tubular epithelium. Almost similar changes were observed in roridin and fusarial toxin treated chicks. However, the degree of toxicity varied. Roridin was highly toxic, while fusarial toxins were mild. The satratoxin was moderate in its toxicity. Dwivedi and Burns (1984) have also observed tubular nephrosis in broiler chicks given ochratoxin A.

Satratoxin receiving mice lost body weight and emotive behaviour. Liver showed fatty degenerative changes, appearance of necrotic areas and haemorrhages, while heart showed loss of cross striations in many cordiac muscle fibres. Degenerative changes and necrotic areas were also recorded in many places. Kidney showed a few areas of haemorrhage and tubular degeneration. Simultaneously degeneration of epithellium was observed. Spleen showed the depletion of lymphoid follicles and lymphoid cells

(ix) Human diseases: Further there are only a few conclusive data regarding the toxic effects of specific trichothecene mycotoxins in humans. The investigations of Soviet scientists culminated in the description of trichothecene induced alimentary toxic aleukia (ATA or septic angina). This consists of four stages 1. Local irritation of the oral mucous membranes resulting in hypermia of the oral mucosa accompanied by weakness, fever, nausea and sleep disturbances 2. Alterations such as leukopenia stage 3. Development of patechial rash on the skin spreading all over the body pharyngitis and ulcerative changes of the larynx with possible serene haemorrhagic diatherin of nasal, oral, gastric and intestinal mucosa (haemorrhagic stage) and appearance of secondary infections (recovery stage).

Administration of DAS intravenously daily dosages ranging from 5 mg to 6 mg/kg body weight caused symptoms like nausea, vomiting accompanied by myelo-suppression, hypotension, pitheli, central nervous system disfunction, fever, chills, stomatitis, erythema and hair loss (Dashek *et al.,* 1985).

Trichothecenes are also involved in the so called sick building syndrome in humans. *Stachybotrys atra* has been associated with pulmonary hemorrhage cases in the level and Ohio (Nielsen *et al.*, 2002). Mycotoxicoses caused by trichothecenes are difficult to distinguish because they affect many organs including the gastrointestinal tract and the hematopotix, nevous immune, hepatobiliary and caridiovascular systems (Table 14.9). DON has been found to be primarily responsible for outbreaks of scabby wheat toxicosis in humans are quite common in several countries but these toxicoses rarely cause death (Luo, 1988). T_2 toxin and related trichothecenes are the most immunosuppresants and cause significant lesion in the lymphnodes, spleen, thymus and bursa fabricius. The heart and pancreas are other target organs for T_2 toxin intoxication (Chu, 2000).

Table 14.8. Effect of Trichothecenes toxin on biochemical composition of blood of chick

Name of the toxin	Concentration (mg/kg)	Proteins (mg/g)	Sugars (mg/g)	Haemoglobin (in O.D)	Albumins (in O.D)	SGOT (units/ml)	SGPT (units/ml)
Fusarial toxins	C	1.0	4.7	0.11	0.1	60	56
	50	0.8	3.8	0.07	0.04	120	71
	100	0.6	2.9	0.06	0.03	152	96
	150	0.5	2.6	0.05	0.02	190	140
Satratoxin	C	1.1	8.5	0.42	0.12	20	15
	200	1.0	7.7	0.32	0.10	30	30
	300	0.8	7.2	0.25	0.05	52	36
Roridin	C	0.8	3.3	0.39	0.12	10	10
	25	0.6	2.9	0.33	0.06	23	30
	50	0.5	2.3	0.28	0.05	29	36

Table 14.9. Outbreaks of trichothecene mycotoxicoses with oral exposure

Country	Year	No. of Subjects affected/. Exposed	Clinical symptoms and signs	Onset of symptoms/ recovery	Source of exposure	Toxic fungal species	Toxin	Ref.
USSR	1932-47	------	"Alimentary toxic aleukia"		Grain	*Fusarium sporotrichoides*		CAST (2003)
			1. Hyperaemina of mucous membranes of oral cavity and pharynx, gastritis, gastroenteritis, excessive salivation, abdominal and oesophagal pain and diarrhoea	A few hours/ 2-3 days		*Fusarium poae*		
			2. Genralized indisposition, vertigo, unpleasant taste in mouth, progressive leukopni, granulocytopenia and lymphocytosis	3-4 weeks				

			3. Haemorrhagic diathesis and angina, petehial rash, catarrhl diphtheritic, gang-renous pharyn-gitis, ulcerative and gangrenous laryngitis, aphoni, asphyxia	Few days/ 2 weeks				
Japan	1956	25; 0	"Scabby grain toxicosis"-nausea, vomiting, drowsiness		Rice	*Fusrium roseum* *Fusarium nivele*		Ito *et al.* (1986)
China	1961-1985	7818; 0	"Scabby grain toxicosis"-nausea, vomiting, abdom-inal pain diarrh-oea, dizziness, headache		Corn Wheat	*Fusarium* sp.	Deoxyn-ivlenol Zeara-lenone	Li *et al.* (1981)
China	1984-1985	463/ 600;0	Nausea, vomiting, abdominl pain, diarrhoea, dizz-iness, headache	5-30 min	Corn Wheat	*Fusarium* sp.	Deoxyni-vlenol Zeara-lenone	Li *et al.* (1981)
India	1987	97/ 224;0	Mild to moderate abdominal pain, feeling of fullness imitation of	15 min-1hour/ 1-2 days	Wheat	*Fusarium* sp. *Asper-gillus flavus*	Nivalenol Deoxyni-valenol T-2 Acetyldeo-xyniva-lenol	Bhat *et al.* (1989)
China		97/ 165;0	Nausea	0-30min		*Fusarium hetero-sporum* *Fusarium gramin-earum*	T-2	Li *et al.* (1981)

(x) Effects on plants: Joffe and Palti (1977) have observed varied type of toxic effects of culture filtrates of 642 isolates of *Fusarium* on the growth of seedlings of different crop plants. Similarly, Brian *et al.* (1961) have also observed inhibitory effect of DAS on the development of roots of seedlings at a concentration of 10 mg/ml. Burmeister and Hesseltine (1970) have recorded the complete inhibition of seed germination of peas even at concentration of 4 mg/ml of trichothecene. Marasas *et al.* (1971) recorded severe wilting, necrosis and reduction in the average weight and length of the seedling when treated with T -2 toxin. Stahl *et al.* (1973) recorded inhibition of elongation of hypocotyl stem of soyabean when treated with T-2 toxin. Nummi *et al.* (1975) have developed bioassay for detection of T-2 toxin based on the observation of inhibition of amylase of germinating seeds of barley. Reiss (1973) recorded inhibition of mitosis in onion roots by *Fusarium* toxins. He also recorded induction of various abnormalities in the division. Suzuki *et al.* (1971) recorded the inhibition of growth of *Chlorella pyrenoidosa* by different *Fusarium* toxins.

(xi) As Warfare Weapons: Mirocha and Dickson (1982) and Rosen (1984) reported detection of T-2 toxin in vegetable samples and water from Kampuchia and Laos respectively. Marshall (1982) reported the T-2 toxin in blood in victims 18 days after attack. Further, trichothecenes have been detected in human blood, urine and body tissue samples from South East Asia chemical warfare victims (Mirocha *et al.*, 1983). Caldwell (1983) has discussed elegantly the possibility of trichothecenes specially T-2 toxin as a natural contaminant in food stuffs in South East Asia and called this as yellow rain.

14.5 METABOLISM AND PHARMACOKINETICS

Trichothecene metabolism and tissue distribution has not been extensively studied. The trichothecenes are rapidly absorbed from the gastrointestinal tract and distributed through out the body within the first hours after administration, peak concentration of the toxins (and/or their metabolies) appear in the blood. There after the trichothecenes occur in bile, gall bladder, liver, kidneys and intestines. Ueno *et al.* (1971) have examined the distribution of fusarenon-X in mice by using a labelled compound (H-fusarenon-X). The compound was found to accumulate in liver after 30 min. of subcutaneous injection and reached kidney and spleen after 3 hours and the toxin was excreted after 12 hours of dosing, leaving no activity of the compound in the organ. The toxin was mostly excreted in 24 hours, through urine (90%) and smaller amount through faces. Prelusky and Trenholum (1984) reported that deoxynivalenol was found in all tissues of swine fed within 0.33-1.0 hours in time except in plasma, liver and probably lung where it is excreted in 24 hrs. Swanson and Corley (1989) elegently discussed the distribution, metabolism and excretion of trichothecenes in different animals.

Ellison and Kotsonis (1974) by employing bovine and human liver homogenates have studied the metabolism of T-2 toxin. It was found that T-2 is transformed to HT-2 which resulted from the C-4 deacetylation. Nakamura *et al.* (1977) have also reported the metabolism of fusarenon-X from nivalenol. This transformation was found to be mediated by microsomal esterase of liver and brain of rats and in the liver, kidney, spleen and brain of rabbits (Ohta *et al.*, 1977).

14.6. MANAGEMENT

The problem of trichothecene mycotoxin contamination in agricultural commodities is so complex that no practical method has been proved to be effective in elimination of any of the mycotoxins completely. Factors associated with the inhibition of trichothecene production in the field are complex and genetics associated with such properties would be equally more complex. Wolf Hall *et al.* (2002) discussed different strategies to be adopted in making beverages specially beer free from fusarium mycotxins. Some of the strategies include irradiation grains and ozonization during malting, biolgocial control of moulding during malt preparation, storage practices and incorporating detoxifying genes into fermenting yeasts.

Methods of preventing or reducing level of trichothecene mycotoxin contamination can be broadly classified as preventive or curative under this, there have been five basic approaches (i) prevention of the initial growth of the moulds and subsequent production of trichothecene (ii) detection of trichothecene mycotoxins in food and selective removal of contaminated portion which, however, reduces the toxin only to a negligible level (iii) inactivation or destruction of the toxin by chemical, physical and biological means (iv) to develop effective and economic methods for controlling the mould growth and subsequent production of mycotoxins and (v) utilisation of genetic resource.

Carson and Smith (1983) by employing tritium (3H) demonstrated the prevention of T -2 absorption by intestine wall when benzonite was added to the feed. Similarly Detzler (1985) by feeding lignin to rats could prevent feed refusal and growth depression. Propionic acid prevented T-2 production by *F. tricinctum* and *F. sporotrichioides* (Muller and Thaller, 1981). Similarly calcium hydroxide mono-methylamine reduced the T-2 toxin and DAS to a non-detectable level. Bicarbonate salts are most efficient inhibitors of trichothecenes (Roinstead *et al.*, 1994). *Curtobacterium* species were able to assimilate T-2 toxin as single carbon and energy source (Veno *et al.,* 1983). A marine monoclonal antibody (CSH6) capable of neutralising the *in vitro* protein synthesis inhibitory effect of T -2 toxin on man. Gendroff *et al.* (1987) have reported a monoclonal antibody against T -2 toxin.

Cameron and Colvin (1989) reported one step procedure for the slective removal of the 12, 13 epoxide rings of diacetoxyscirpenol and deoxynivalenol. Baurer *et al.* (1982) detoxified T-2, deacetoxyscirpenol and zearlalenone by employing calcium hydroxide monomethylamine. Selenium is reported to have protective effect against trichothecene (T-2) mycotoxicosis.

Dietiry use of selenium in combination with ascorbic acid and α-tocophenol provided protection against acute toxicosis caused by trichothecenes, deoxynivalenol and T2-toxins (Rizzi *et al.*, 1995). Atroshi *et al.* (1995)

concluded that selenium, vitamin E and vitamin C act as antioxidants and free radical scavenger systems that protect the spleen and brain against membrane damage caused by T2-toxin and deoxynivalaenol.

Acetic anhydride and formic acid were effective in checking the mycelial growth of *S. atra*. Propanol, acetone and ethyl acetate were ineffective in checking the mycelial growth of *S. atra*. The growth of *S. atra* was limited in the atmosphere of vapour of isoamyl alcohol and methyl propanol. The inhibitory effect of rest of the compounds on growth and satratoxin production by *S. atra* varied.

Toxigenic strains of *Fusarium* were shown to allow high trichothecene production. While bran was demonstrated to contain biochemical inhibitors able to significantly reduced tricothecene production (Laetitia *et al.*, 2007).

(i) Effect of phenols: P-napthol, P-nitrophenol, P-cresol and P-napthol were fungitoxic and responsible for inhibition of growth and satratoxin production by *S. atra*. Catechol and methoxy hydroxy benzoate were inhibitory to satratoxin production to a significant level. Pentachlorophenol was less toxic.

(ii) Food preservatives: The preservation of high moisture corn with food preservatives has been suggested by Voelker *et al.* (1986). Singh *et al.* (1987) have also reported the use of acids for protecting moist wheat grains from microbial colonization during storage. Ammonium carbonate at 1% concnetration was effective inhibhitor of fungal contamination and mycotoxin production and still acceptable for consumption based on sensorial analysis (Sampapundo *et al.*, 2007).

Sodium metabisulphate and propionic acid were responsible for total inhibition of satratoxin production. On the other hand, sodium acetate, sodium chloride, potassium metabisulphate and p-aminobenzoic acid were ineffective in checking the growth and satratoxin production by *S. atra*. Boric acid and benzoic acid were effective in checking the satratoxin production. Benzoic acid was effective only at higher concentration. Citric acid and acetone exhibited marginal inhibitory effect on satratoxin production (Laxma Reddy and Reddy, 1992).

Physical methods of control such as (1) removal of the infected seeds or portions, (2) heat treatment, (3) absorption, filtration or density separation, (4) exposure to sunlight and other radiations and (5) milling and food processing methods. Of various types of radiations, sunlight and ionizing radiations are being extensively used in the control of fungal infestation and mycotoxin contamination (Aster *et al.*, 1986).

(iii) Gamma radiation: Gamma irradiation is an another physical method, which is known to cause injury to microorganisms and has been widely reported to prevent or delay food spoilage (Moy, 1984). The growth and satratoxin production by *S. atra* decreased gradually with the increase of dosage of gamma-radiation the amount of satratoxin production gradually decreased with the increase in the dosage of gamma-radiation. 400 KR was most effective in checking the satratoxin production by *S. atra*.

(iv) Biocontrol: Cooney *et al.* (2001) feel that colonization of *Fusarium subglutinans* would influence the production of mycotoxins by other species of *Fusarium*. *Lactobacillus Propionobacteria* degraded different trichothecene mycotxins which varied both with bacterium and trichothecene EL -Nezami *et al.* (2002).

Chapter 15

MISCELLANEOUS MYCOTOXINS

15.1. ALTERNARIA TOXINS

Alternaria is a common fungus that usually invades crops at the pre-harvest stage and under suitable conditions may lead to production of mycotoxins. Alternaria toxins have been implicated in animal and in human health disorders (Woody and Chu, 1992). The mycotoxins that occur most frequently are tenuazoic acid, alternariol monomethyl ether and alternanol. Alternuene, isoalternuene and altatoxins I and II may occur occasionally (Fig. 15.1).

15.1.1. Organisms

Alternaria alternata is a good producer of alternariols ($C_{14}H_{10}O_5$, AOH), alternariol monomethylether ($C_{15}H_{12}O_5$, AME) and alterneune ($C_{15}H_{10}O_6$). *A. solani* isolated from potatoes produce AOH, AME and ALT. *A. alternata* (Harwing *et al.*, 1978), *A. citri* (Stinson *et al.*, 1981), *A. rakuchiana* (Kinoshita *et al.*, 1972), *Phoma sorghina* (Steyn and Rabie, 1976), *A. tenuissima* (Umetsu *et al.*, 1972), an isolate of *Aspergillus physii* and member of sphaeropsidales (Kaczka *et al.*, 1964) are reported to produce tenuazoic acid under natural conditions or in culture.

15.1.2. Physico-chemical Characteristics

Tenuazoic acid is colourless, viscous oil and is a monobasic acid with pK 3.5. It is soluble in methanol and chloroform. It forms a complex with calcium, magnesium, copper, iron and nickel ions. It gives reddish-brown reaction with ethanolic ferric chloride. Alternariol and alternariol monomethyl ether crystalises from ethanol as colourless needles with melting point and decomposition at 35°C and 267°C respectively. They are soluble in most organic solvents and give a purple colour reaction with ethanolic ferric chloride and blue fluorescent under UV. Alternuene crystallizes as colourless prisms melting at 180°C and fluoresces bright-yellow under UV light.

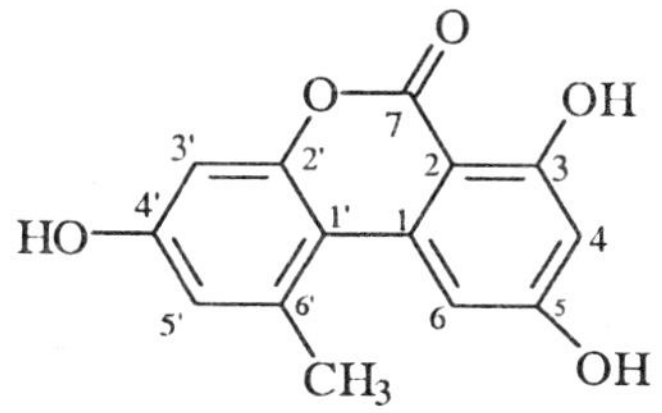

Alternariol

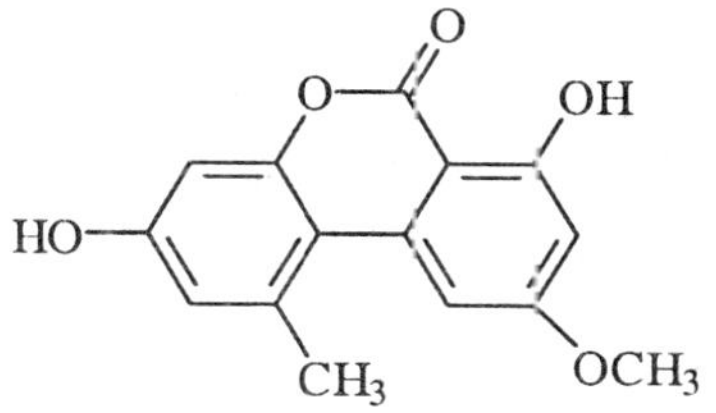

Alternarion methylether

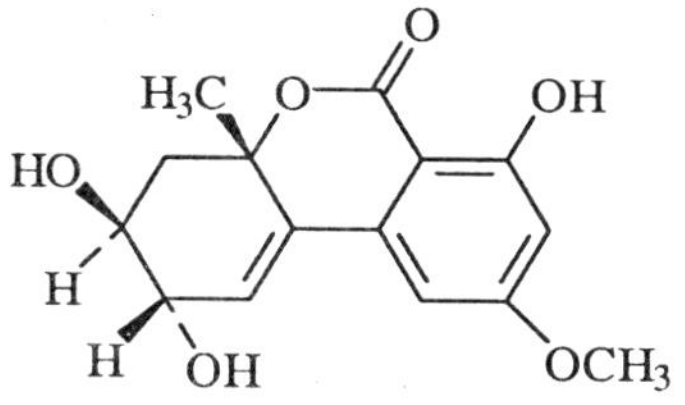

Alternuene

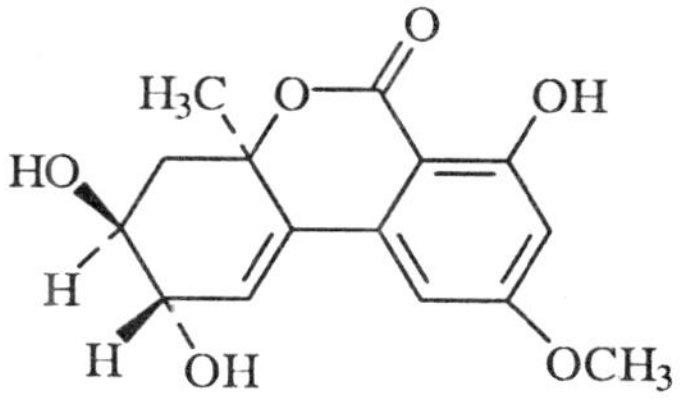

Isoaltenuene

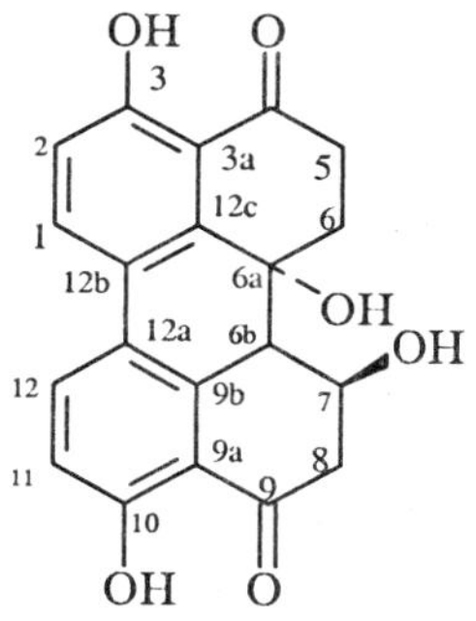

ATX-1

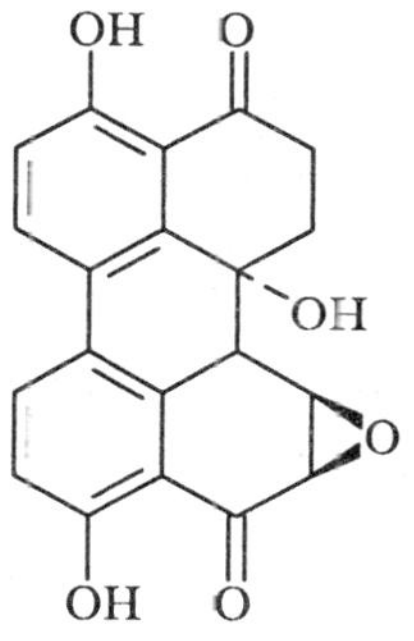

ATX-II

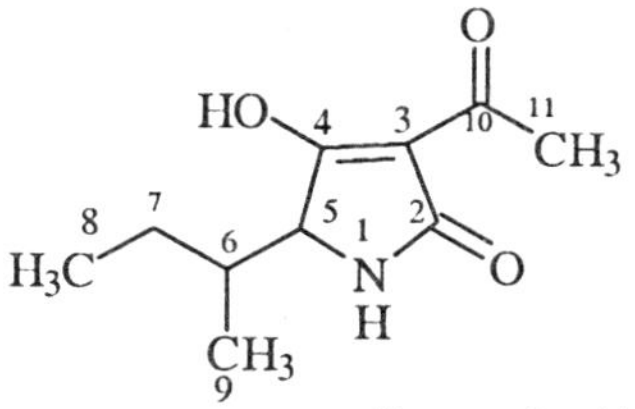

Tenuazonic acid

Fig 15.1. Structures of Alternaria toxins

15.1.3. Alternariol, alternanol and alternariol monomethylether Production

Unsupplemented milled rice at 35% moisture content is a convenient substrate for production of AME, AOH and ALT (Burroughs *et al.* 1978).

Good yields of alternariols are obtained after incubation for 14-21 days at 25°C. Good yields of alternariols are also obtained when *A. alternata* is grown on modified Czapek Dox broth (MCD) for 28 days at 25°C. Alternariols can also be recovered after 10-14 days but yields are reduced. Yeast extract-sucrose broth (YES) consisting of sucrose at 40 gl^{-1} and yeast extract at 20 $g\ l^{-1}$, is also a fairly good liquid medium for production of alternariols (Maas *et al.*, 1981). Altemoller *et al.* (2006) could synthesize alternuene and isoalternuene in 10 steps starting from quinic acid.

15.1.4. Isolation and detection

Pero *et al.* (1971) extracted rice cultures with acetone-water (7:3), then separated metabolites with a silica gel G column. AME, AOH and ALT were eluted from the column with benzene containing 5, 15 and 50% of tetrahydrofuran (THF). The progress of elution was monitored by a long wave UV light, AME in the 5% THF fraction was usually sufficiently pure to crystallize from ethanol solution after addition of water and cooling. The AOH fraction often required further purification with a silica gel column to remove a red pigment. The silica gel was washed first with chloroform and then with chloroform-methanol (97:3) to elute the red pigment. AOH was washed from the column with acetone. The ALT fraction also usually required additional purification by re-chromatography on a silica gel column eluted with 25% THF in benzene.

Stinson *et al.* (1980) homogenized rice cultures in a blender with a minimum amount of water, acidified to pH 2.0 and extracted twice with chloroform. The chloroform extracts were combined, dried over anhydrous sodium sulphate and concentrated by evaporation under nitrogen. AME, AOH and ALT were separated by column chromatography as described by Pero *et al.* (1971). This procedure could also be applied to liquid cultures.

Methanol extracts of air-dried, ground substrates were prepared in a blender by Seitz *et al.* (1975). After filtering, aqueous ammonium sulphate was added to the methanol extract and the resulting aqueous phase is extracted twice with benzene. The benzene extracts were combined, dried over anhydrous sodium sulphate, concentrated and chromatographed on a silica gel column. After addition of the sample, the column was washed first with benzene, then with 5% acetone in benzene to elute AME and finally with 4% methanol in benzene to elute AOH.

TLC is a useful purification method. Concentrated fractions from silica gel columns are streaked on to silica gel TLC plates. After developing with an appropriate solvent, the plate is viewed under short wave UV light and the band is scraped from the plate into a small fritted-glass filter and washed with ethanol or acetone to strip the metabolite from the silica.

The following extraction and clean-up procedure proved to be convenient for the routine analysis of grain sorghum and cultures of *Alternaria* on rice and other grains (Scitz *et al.*, 1926). Maas *et al.* (2000) analysed liquid and rice cultures by a similar method except that ammonium sulphate and hexane were not used. Fifty grams of grain or culture medium were homogenized with methanol in a blender for 2 min. A 30 ml aliquot of filtered methanol extract, 60ml of 20% ammonium sulphate and 30 ml of hexane were added to a 250ml separating funnel. The funnel was shaken for about 30 sec and the layers were allowed to separate. The hexane layer was discarded by aspira-ting it into a trap. The aqueous methanol-ammonium sulphate layer was extracted twice with 5ml portions of methylene chloride and the combined extracts were evaporated to dryness on a steam-bath by a slow stream of nitrogen. The residue can be dissolved in a solvent appropriate for the chromatographic method used for separation and detection.

The hexane extraction of the aqueous ammonium sulphate-methanol layer can be deleted if the metabolite level is high and the oil content low. It not only removed oil and pigments but also 29% of the AME, so AME analyses of methylene chloride extracts were adjusted upwards by a factor of 1.4. Neither AOH nor ALT was removed by the hexane.

Extracts of fruits and vegetables were analysed by homogenizing 200g of tissue in a blender, adjusting the pH to 2.0 and extracting with two 500-ml portions of chloroform (Stinson *et al.*, 1980). The extracts were combined, dried over anhydrous sodium sulphate and concentrated to 25 or 50 ml before analysis by high-performance liquid chromatography (HPLC).

AME, AOH and ALT are well resolved on pre-coated silica gel plates. Commonly used developing solvents include chloroform-acetone (88:12) and toluene-ethyl acetate-formic acid (5:4:1). Approximate R_F values in the two solvents are 0.65 and 0.7 for AME, 0.4 and 0.5 for AOH and 0.1 and 0.3 for ALT, respectively (Scott *et al.*, 1980; Chu and Bennet, 1981). The compounds are visible as blue spots under long and short wave UV light, AME and AOH being brighter under the latter. As little as 0.1 µg spot is visible. Spraying with aluminium chloride in ethanol brightens the fluorescence, particularly under long wave UV light. AME and AOH spots remained fluorescent blue after spraying with 50% ethanolic sulphuric acid, but changed to a greenish fluorescence on heating for 5 min at 100°C. Sulphanilic acid spray produces a yellow spot with ALT, a cherry-red spot with AOH and a wine-coloured spot with AME (Pero *et al.*, 1971). Care is needed to prevent mistaking AME for zearalenone as they have similar fluorescence and R_f values on TLC (Seitz *et al.*, 1975). AME, AOH and ALT can also be detected with p-anisaldehyde and iron (III) chloride spray reagents (Durakova *et al.*, 1976).

Early HPLC analyses of AME, AOH and ALT in sorghum grain and grain-based cultures involved normal-phase systems (silica columns) and isooctane –THF or light petroleum–THF with solvent programming (Seitz and Mohr, 1976). Metabolites were detected by simultaneous monitoring of fluorescence and absorbance at 350nm, with quantitation usually based on the former. Sorghum grain extracts, prepared as described above, were pre-cleaned with a small silica column if many samples were analysed. Reversed-phase HPLC is now more convenient for separation of AME, AOH and ALT.

Chu and Bennett (1981) reported the separation of *Alternaria* metabolites with an Ultrasphere-ODS C_{18} reversed-phase column (Altex Scientific, Berkeley, CA, USA) and the following solvent programme: 50% methanol in 1.5% acetic acid for 15 min, then a linear increase in methanol concentration to 65% over 30 min, followed by a linear increase to 100% over 15 min. In addition, the flow-rate was programmed: 0.5 ml min^{-1} to start, increasing to 1.0 ml min^{-1} over 1 min beginning at 14 min and an increase to 1.5 ml min^{-1} over 1 min beginning at 39 min. Elution was monitored by measuring the absorbance at 254 nm. The times at which AME, AOH and ALT eluted were 46.5, 31.7 and 17.3 min, respectively. Trimethylsilyl (TMS) derivatives of AME, AOH and ALT were resolved by gas chromatography (GC) on Gas-Chrm Q having 3 and 10% containing of OV-17 (Pero *et al.*, 1971). Different methods adapted for detection of *Alternaria* mycotoxin are précised in Table 15.1.

Table 15.1. Methods for the Determination of *Alternaria* Toxins

Food	Food	Extraction	Cleanup	Quantitation	DL (µg/kg)	Ref.
AOH, AME, ALT, ATX-I, TA	Olives	Blend 50 g ground sample 2 min with 45:40:40:1 methanol/water/ hexane/ HCl	Centrti-fugation and liquid-liquid partition	1. Two-dimensional TLC Plate: Silica gel with fluorescence indicator Solvent: (a) 88:12 Chloroform/ acetone (b)60:30:10 Toluene/ ethyl acetate/formaic acid Detection: Visuali zation under long-and short-wave UV light 2. HPLC Column: Reverse-phase C_{18} (7µm,25 cm x 4mm i.d) Mobile phase: 80:20 Methanol/water plus 300 mg $ZnSO_4.7H_2O$/L or 60:40 methanol/ water Detection: UV at 280 or 257 nm	30-200	Viscant *et al.* (1986)

AOH, AME, ALT, TA	Tomatoes and apples	Homogenize tissue, adjust pH to 2.0 with HCl, and blend with 80:20 chloroform/ echanol (2 ml/g)	Centrifugation and filtration of organic phase	HPLC Column: Silica (7.5 µm, 30 cm x 5 mm i.d.) Mobile phase: 95:5 Chloroform/ methanol Detection:UV (280 nm)		Ozeelik *et al.* (1990)
AOH, AME	Sun flower seeds	Shake homogeneized seeds with methanol (3 ml/g)	Precipitation and liquid-liquid partition	TLC Plate: Silica gel G-60 Solvent: 88:12 Chloroform/acetone		Torres *et al.* (1993)
AOH, AME, ALT	Fruits and vegetables	Homogenize tissue, adjust pH to 2.0, and extract twice with chloroform (2.5 ml/g)	Dry extract over sodium sulfate and concentrate	HPLC Column: Reverse-phase C_{18} (10µm, 30cm x 3.9mm i.d) Mobile phase: 65:35 Acetone/water Detection: UV (324 nm)		Heisler *et al.* (1980)
AOH, AME, ALT, ATX-I, ATX-II	Corn, rice, tomatoes	Blend 50 g sample with 75 ml methanol	Precipitation and liquid-liquid partition	Gradient HPLC Column: Reverse-phase C_{18} (5 µm, 12.5 cm x 4 mm i.d.) Mobile phase: 60:40 Methanol/water plus 0.1M sodium nitrate and 1 mM HNO_3 Detection: Electrochemical with dual in-series electrodes	<100	Viconti *et al.* (1991)
AME, TA	Tomatoes and tomato products	Blend 50 g for 3 min with 450 ml water and 100 ml chloroform	Silica gel column chromatography	HPLC Column: Reverse-phase C_{18} Mobile phase: 85:15 Methanol/water water with 300 mg $ZnSO_4 \cdot 7H_2O$/L Detection: UV (280 nm) for TA; Fluorescence (ex: 278; rm:389 nm) for AME	3-25	Stack *et al.* (1985)
AoH, AME, ALT, ATX-I, ATX-II	Corn, rice, tomatoes	Blend 50 g sample with 75 ml methanol	Precipitation and liquid-liquid partition	Gradient HPLC Column: Reverse phase C_{18} (5µm, 12.5cm x 4.6 mm i.d.) Mobile phase: (a) methanol (b) HPO - quidified water (pH 3.0) Gradient: 50-70% on 10	400	Palmisano *et al.* (1989)

		mm; 70-85% in 0.1 mm, Detection: Photoboid is crante for 8 min.

15.1.5. Tenuazonic acid

15.1.6 Production

Davis *et al.* (1977) produced TeA by growing *A. alternata* and *A. tenuissima* on 2% yeast extract 4% sucrose (YES) medium and on moist, delinted cotton seed for 28 days at 25°C. Selected isolates of *A. alternata* produced good yields of TeA in a 1:1 mixture of yellow corn and milled rice containing 35% moisture after 21 days incubation at 24°C (Saur *et al.*, 1978). Production of TeA by *A. alternata* in tomato and citrus fruits has also been reported (Stinson *et al.*, 1981).

15.1.7. Isolation and detection

Davis *et al.* (1977) isolated and purified TeA from YES and cotton seed cultures as follows. YES broth cultures were adjusted to pH 2.0 and extracted with chloroform. Cotton seed cultures were extracted with chloroform-methanol-formic acid (94:3:3). The organic phase was extracted with 5% sodium hydrogen carbonate solution. The aqueous phase was adjusted to pH 2 by adding dilute hydrochloric acid, and then extracted with hexane. The hexane extract was dried over anhydrous sodium sulphate and evaporated to dryness under vacuum or with a nitrogen stream. The residue was dissolved in methanol, filtered and the filtrate streaked on to 450-μm silica gel G-HR TLC plates. The plates were developed in toluene-ethyl acetate-88% formic acid (6:3:1). The TeA band (R_f 0.5-0.6) was scrapped from the plates and TeA was eluted from the silica with methanol.

Kaczka *et al.* (1964) used ethyl acetate and light petroleum (b.p. 30-60°C) in place of chloroform and hexane, respectively. TeA was purified by formation of its NN-dibenzylethylenediamine (DBED) salt. The residue from evaporation of the light petroleum extract was dissolved in Skellysolve B (essentially n-hexane) and DBED was added until there was no further precipitation. The oily precipitate was dissolved in acetone and benzene for crystallization. The DBED salt was converted into the acid by dissolving the salt in methanol and acidifying with hydrochloric acid. The solution was filtered and then evaporated to dryness under vacuum. The residue was extracted with light petroleum (b.p. 30-60°C) and the extract was evaporated under vaccum at about 30°C, leaving TeA as a viscous oil.

The solvent was evaporated and the resulting crude extract partitioned between hexane and 90% methanol. The 90% methanol fraction was chromatographed on a silica gel H column developed with chloroform-methanol (8:1:1.2). The collected fractions were analysed by TLC with chloroform-methanol (8.5:1.5). The fractions containing tenuazonates gave a dark absorbent spot at R_f 0.3 that turned red-brown when sprayed with ethanolic ferric (III) chloride. The combined fractions gave the tenuazonates as an amorphous powder that contained Mg^{2+}, Ca^{2+} and Na^+ ions.

On silica gel GHR (250-500 μm) developed in toluene-ethyl acetate-formic acid (5:4:1) or 6:3:1), TeA is usually found at R_f 0.5-0.65, TeA can be detected by spraying with ethanolic ferric (III) chloride (red-brown), 2,4-dinitrophenyl hydrazine (yellow) and 1% cerium (IV) sulphate in 6 N sulphuric acid (grey-brown) (Steyn and Rabic *et al.*, 1976). About 2.5 μg is the visual detection limit with ferric (III) chloride spray reagent.

(i) High-performance liquid chromatography: Heisler *et al.* (1980) determined TeA in fruits and vegetables by using a 30 cm x 3.9 mm I.D. μBondapak C_{18} (Waters ASSOC) column eluted with methanol-water at 2.0 ml min^{-1} (Heisler *et al.*, 1980). The retention time was about 5 min. As little as about 0.05 μg of TeA could be detected by measuring the absorbance at 278 nm.

Scott and Kanhere (1980) separated TeA from tomato paste interferences by using a μBondapak C_{18} column coated with 4-dodecyldiethylenetriamine (C_{12}-dien) and a mobile phase of methanol-water (60:40) containing 1% of ammonium acetate, 0.025% C_{12}-dien and 0.001 m zinc sulphate. The retention time, which was sensitive to small changes in composition of the mobile phase, was 7.7-10.9 min when the flow rate was 2.5 ml min^{-1}. About 5 ng of standard TeA was the detection limit by measuring the absorbance at 280 nm. When acetonitrile was used in place of methanol in the mobile phase, two isomeric forms of TeA were partially separated. An anion-exchange column also worked well for eliminating interferences from tomato paste.

(ii) Gas chromatography: The trimethylsilyl (TMS) derivative of TeA gave a single sharp peak at 11.5 min when chromatographed on 3% OV-17 with a temperature programme from 100 to 250°C at 8°C min^{-1}[42]. The column was glass (5 ft. x 2 mm I.D.) packed with the silicone coated on Gas-Chrom Q (100-200 mesh), The flow-rate of the carrier gas (nitrogen) was 25ml min^{-1}. The detection limit was about 0.1 μg. The TMS derivative was prepared by dissolving TeA or its salts in No-bis (trimethylsilyl) acetamide-trimethylchlorosilane-pyridine (6:2:9) and heating at 50°C for 10 min.

Sauer *et al.* (1978) used a similar method with 3% OV-101 and 3% OV-17 columns in an attempt to find TeA in weathered grain sorghum. The OV-101 column minimized interferences.

(iii) Spectrophotometry: Meronuck *et al.* (1972) used UV spectroscopy for the qualitative detection of TeA in water extracts of *Alternaria* cultures on corn-rice (1:1). TeA was reported as present if absorption maxima at 239 and 278 nm were observed (Table 15.1). Others used the absorption at 278 nm for the quantitative determination of TeA in cleaned up extracts and the pH dependence of the UV spectrum to confirm the presence of TeA (Davis *et al.*, 1977).

15.1.8. Natural Incidence

Alternariol (AOH) and its monomethyl ether (AME) have been found in weathered sorghum grain (Seitz *et al.*, 1975) and discoloured pecans (Schroder and cole, 1977). Only traces of altenuene (ALT) and altertoxin I (ATX-I) have been found in heavily weathered sorghum grain (Saur *et al.*, 1978); tenuazoic acid (TeA), probably the most toxic metabolite, has not been found (Meronuck *et al.*, 1972). Other than in certain fruits mentioned below, the natural occurrence of TeA has been demonstrated only in diseased tobacco leaves and in blast-diseased rice plants (Mikami *et al.*, 1971; Umetsu *et al.*, 1974). Commercial tomatoes and apples with advanced natural infections of *Alternaria*, and oranges and lemons that had been inoculated with *A. citrii* contained AME, AOH, ALT, ATX-I and TeA. AME and AOH production was highest in oranges and lowest in tomatoes; TeA was present in all the fruits and its abundance was fairly high in all except apple. Trace amounts (0.1 $\mu g\ g^{-1}$ or less) of TeA was found in several samples of commercial tomato pastes (Scott and Kanhere, 1980).

AME, AOH, ALT and TeA including ATX-1 is also a fairly common meabolite produced in culture but its structure is not yet known. Kralova *et al.* (2006) reported the incidence of altenuene, alternariol, and monomethyl ether in fiber flax and linseed from Czechoslovakia.

15.1.9. Biological activity

Alternaria toxins exhibit both acute and chronic effects. The LD_{50} values for alterniol monomethyl ether, alterniol, altenuene and altertoxin OI in mice are 400, 400, 50 and 0.2 mg/kg body weight respectively, while for tenuazonic acid 162 and 115 mg/kg body weight (by intravenous) for male and female mice respectively. *Alternaria* toxins have been implicated in animal and human health disorders (Woody and Chu, 1992). Gruber Schley and Thalmann (1988) during their study on the outbreak of suspected mycotoxicosis to the contamination of *Alternaria* in cereal samples collected from few diseased than from healthy animals. Incidents of deaths of rabbits and poultry due to the toxic action of *Alternaria* species in the fodder and feed have been reported by Wawrzkienicz *et al.* (1989). *Alternaria* sp. were also detected in cereals samples in which *Fusarium* sp. were implicated as the likely cause of an outbreak of alimentary toxic aleukia in Russia (Joffe, 1960).

AME, AOH and ALT are toxic to *Bacillus mycoides* and HeLa cells, and weakly toxic to mice when administered as a single dose intraperitoneally (Pero *et al.,* 1973). When administered subcutaneously, AOH and AME in combination (1:1) and AOH alone showed foetotoxic and teratogenic effects in mice.

TeA is highly active biologically. Its antibiotic, antiviral and antineoplastic activity and mycotoxicology have been reviewed (Harvan and Pero, 1976). Its primary modes of action in mammalian toxicology appear to be inhibition of protein synthesis (Carraxo and), interference with the release of newly formed protein from ribosomes. The toxicity of TeA also might be related in part to its ability to complex selectivity *in vivo* with trace metals (Steyn and Rabie, 1976). TeA, possibly produced mainly by *Phoma sorghina,* has been implicated as a cause of onyalia, a haematological disorder, in people living south of the Sahara in Africa (Steyn and Rabie, 1976). TeA fed for 3 weeks to young chickens at a level of 10 μg per gram of feed or incubated daily at a dose of 1.25 mg kg^{-1} caused marked lesions in various tissues (Gambrone *et al.,* 1978). Lethal effects of *Alternaria* cultures fed to rats and chicks were apparently due to TeA (Saur *et al.,* 1978).

It also exhibits antitumor, antiviral and antibacterial activity. Alternadiol and alternariol monomethyl ether showed toxicity and teratogenic effects in mice including synergistic effect when administered with a combination of the other toxins (Pero *et al.,* 1973). Most *Alternaria* toxins exhibit considerable cytotoxicity including mammalian toxicity. The alterotoxins are of particular concern because of their mutagenic activity (Stack and Prival, 1986).

15.2 SPORIDESMINS

Though the disease of farm animals, popularly known as facial eczema but better named as Pithomycosis was noticed in New Zealand at the beginning of this century. Its casual agent was established only by Percival and Thorton (1958). This toxic substance was found to be produced by *Sporidesmium bakeri* which was older name of *Pithomyces chartarum*. When the substances extracted from *P. chartarum* were given to guinea pigs and lambs, they showed liver damage characteristic of facial ecizema. Synge and White (1959) elucidated the chemical structure of Sporidesmin, a metabolite of *P. chartarum*. Taylor (1971) gave detailed account of toxicology of sporidesmin. The biological activity appears to be connected with sulfur containing ring of sporidesmin. *P. chartarum* is reported to produce several sporidesmins such as Sporidesmin A and sporidesmin B, which are less active and occur in low concentration. Sporidesmin E is present in small concentration in benzene solvent fraction, sporidesmin D and F are with open S-S-bridge are much less active. Sporidesmin C is present as a minor component. Spondesmin G prepared from sporidesmin or sporidesmin E by hydrogen polysulfide treatment, converted to sporidesmin D by sodium borohydride and to sporiden and sporidesmin E when ethereal solution is allowed to stand under fluorescent lamp. Sporidesmin H has two sulfurs, H in place of Cl on the aromatic ring Cl instead of the tertiary OH and the secondary OH as in sporidesmin. Ronaldson (1974) has reported, several modified sporidesmins containing ester group (Table 15.2.1). In conclusion, sporidesmins are produced by *P. chartarum* and have not been reported to be produced by other organisms.

15.2.1 Chemistry of Sporidesmins

Different sporidesmins produced by different fungi are depicted in Table 15.2.2.

Table 15.2.1. Physico-chemical characteristics and toxicity of sporidesmins

	Chemical Formula	Nature of compound	M.P. (in °C)	Toxicity			Reference
				B. subtilis MIC	TC (MIC)		
Sporodesmin A	$C_{18}H_{20}ClN_3O_6S_2$ (x-2, R_1=R_2OH)		179	120	3.0	0.4	Ronaldson *et al.* (1963)
B	$C_{18}H_{20}ClN_3O_5S_2$ (n-2, R_1=OH, R_2-H)	Amorphous	183	400	1.0	5.0	Ronaldson *et al.* (1963)
C	$C_{22}H_{24}ClN_3O_8S_3$		230-240	>1000	6.0	2.0	Hodges and Shanon (1965)
D	$C_{20}H_{26}ClN_3O_6S_2$	CRYST-ALLINE	110-120	-	1000	500	Jamieson *et al.* (1969)
E	$C_{18}H_{20}ClN_3O_6S_3$ (x-3, R_2=R_2=OH	CRYST-ALLINE	180=189	10	1.0	-	Rahman *et al.* (1969)
F	$C_{19}H_{22}ClN_3O_6S$	Amorphous	144-148	-	>1000	-	Jamieson *et al.* (1969)
G	$C_{18}H_{20}ClN_3O_6S_4$ (x-4, R_1=R_2-OH)		148-153	-	2.00	-	Francis *et al.* (1972)
H	Has 2S, H in place of Cl on the aromatic ring, Cl instead of tertiary OH and secondary OH			-	10.0	0	Taylor (1967)

Table 15.2.2. Organisms producing other sporidesmins in culture

Organism	Place of isolation	Mycotoxin produced	Yield (μg/ml)	Reference
Arachniotus aureus	--	Aranotin acetylar-anotin apoaranotin	--	Nagariagan *et al.* (1968)
Aspergillus terreus	--	Gliotoxin acetylaranotin	--	Miller *et al.* (1968)
Chaetomium cochliodes	--	Chetomin	7-15	Geiger *et al.* (1944)
Chaetomium globosum	Zea mays	Chetomin	10-20	Christensen *et al.* (1966)
Chaetomium minutum		Chaetocin	19	Hauser *et al.* (1970)
Gliocladium fumbriatum	--	Gliotoxin	50-80	Brian and Hemiming (1945)
Aspergillus fumigatus	--	Gliotoxin	20-80	Betina and Barut (1964)
Aspergillus sp.	--	Gliotoxin	--	Stanly *et al.* (1964)
Penicillium obscurum	--	Gliotoxin	--	Muller *et al.* (1945)
P. cinerascens	Soil-Cumberland, United Kingdom	Gliotoxin	--	Bracken and Raistreck (1947)
P. terlikowskii	--	Isodehydrogliotoxin dehydrogliotoxin glitoxin, glitoxin acetate	272	Johnson *et al.* (1953)
P. terlikowskii	--	Dehydrogliotoxin, gliotoxin	180	Lowe *et al.* (1966)
Penicillium sp.	Bath, Maine, U.S.A.; Anvers, Belgium	Gliotoxin	--	Betina *et al.* (1968)
Pithomyces chartarum	High sporing sector from plate exposed at Hamilton, New Zealand.	Sporidesmin sporidesmin B sporidesmin E sporidesmin G	0.3 0.03 0.03 0.0003	Percival and Thornton (1958)
Pithomyces chartarum	Debris of Paspalum dilatatum – Mount Albert, Auckland, New Zealand	Sporidesmin	1.5	Dingley (1962)
Verticillium sp.	Basidiocarp of Coltricia cinnamomea	Verticillin-A	24	Katagir *et al.* (1970)

15.2.2 Natural Incidence

Clare and Gumbley (1962) reported sporidesmin in spores of *P. chartarum*, while smith *et al.* (1961) reported it to be the cause of facial eczema.

15.3.3 Biological activity

Sporidesmins exhibit varied toxicity, towards animals. Though several sporidesmins (B, D, E, F) are produced, it is sporidesmin which contributes to the toxicity. The natural mycotoxicosis occurs commonly in both sheep and cattle in New Zealand and less often in Australia. It occurs rarely in sheep in South Africa. Photo dynamic disease of cattle in okaloma had all the characteristics of pithomycotoxicosis. Sporidesmin E is reported to cause capsules disease in sheep. Responses of different animal to sporidesmins are précised in Table 15.2.3.

Table 15.2.3 Effect of sporidesmins on different animals

Species	Time period of dosage (days)	Total dose (mg/kg)	Typical responses
Cattle	3	3.0	100% mortality within 3-5 days 50%
Calves	3	0.8	photosensitized; icterus; low mortality
Gravid cows	3	1.0	Marked weight loss; 50% photosensitized, icterus with some mortalities; typical serious field outbreaks
	3	0.5	Transient weight loss; liver lesions, mainly asymptomatic
Sheep	3	3.0	100% mortality within 4 days Severe illn-ess;
	3	1.0	80% photosensitized; 70% mortaility
	3	1.0	Asymptomatic; trivial lesions
Chickens	Continuous (5.0 mg/kg)	30 (maximum)	100% mortality within 6 days; lesions of liver
	Continuous (0.5 mg/kg)	7 (maximum)	100% motality within 14 days; lesions of liver and gizzard
Rabbits	1	1.2	50% mortality within 14 days; severe lesions of bile ducts and liver; icterus
	21	1-2	Hepatomegaly, marked liver fibrosis at 28 days; some mortalities
Rats	Continuous (0.2 mg/day)	4-8	Severe bile duct and liver lesions; icterus; pleural effusions and ascites; mortality 2-4 months
Mice	Continuous (0.2 mg/day)	250-300	Average survival time 35 days; bile duct and liver lesions; icterus; anemia; some pleural effusions and ascites
	Continuous (0.1 mg/day)	200-250	Average survival time 57 days; lesions as above

15.3 STERIGMATOCYSTIN

Sterigmatocystin is a toxic metabolite of mainly *Aspergillus nidulans* that is closely related in structure to aflatoxins. It consists of xanthane nucleus attached to a bifuran structure (Fig. 15.3.1). Sterigmatocystin has been found in mouldy grain, green coffee beans and cheese. Since analytical methods for determination of sterigmatocystin at its low concentration was not available, it remained unnoticed in foods when it is elaborated in low concentration.

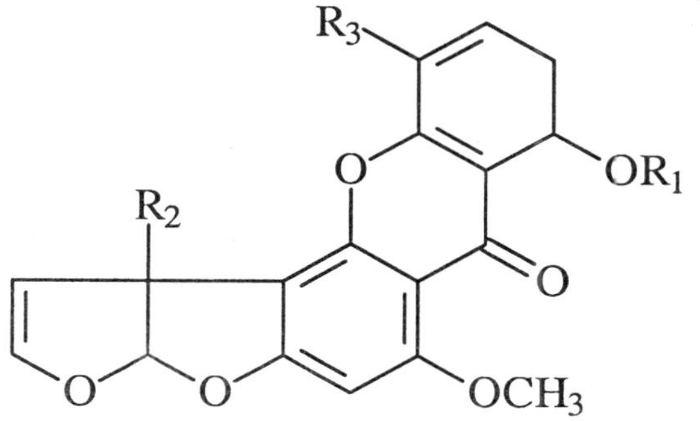

Sterigmatocystin

	R_1	R_2	R_3
Sterigmatocystin	H	H	H
O-methyl sterigmatocystin	CH_3	H	H
Aspertoxin	CH_3	OH	H
5-Methyl sterigmatocystin	H	H	OCH_3

Fig 15.3.1. Structure of sterigmatocystin and its derivatives

15.3.1. Physico-chemical characteristics

Sterigmatocystin crystallizes as pale yellow needles and is readily soluble in methanol, ethanol, acetonitrile, benzene and chloroform. Its structural formula is $C_{18}H_{12}O_6$. Its molecular weight is 324 and melting point is 246°C and has absorption maxima in benzene is 15200 nm. It reacts with hot ethanolic KOH and is melthylated by methylsulphate and methyliodide, methanol or ethanol in acid produces dihydroethoxy sterigmatocystin.

15.3.2. Organisms

It is mainly produced by *Aspergillus nidulans* and *A. versicolar*. Reports of Smith and Ross (1991) that it is produced by *A. ustus* and certain members of *A. glaucus* group can consider to be misidentification (Frisvad and Samsol, 1991).

15.3. 3. Isolation and Detection

Table 15.3.1. Methods for determination of Sterigmato Cystin.

	Food	Extraction	Cleanup	Quantitation	Detection (μg/kg)	Ref.
Sterig-mato-cystin	Barley wheat	Shake 50 g ground sample 30 min with 180 ml acetonitrile and 20 ml 4% KCl solution	Liquid-liquid partition and silica gel column chromatography if necessary	TLC Plate: Adsorbosil-1 Solvent: 90:5:5 Benzene/ methanol/ acetic acid Detection: Visualization as bright yellow fluorescent spot under short-wave UV light Derivatization: Spray plate with 20% aluminum chloride solution and heat 10 min at 80°C		Scott (1995)
Sterig-matocy-stin	Cheese	Blend sample 3 min with 85:15 acetonitrile/4% KCl (5.5 ml/g)	Liquid-liquid partition and cupric carb-onatediatoma-ceous earth (1:2) column chromato-graphy	TLC Plate: Prelocated Sil G-25HR Solvent: 85:10:5 Benzene/methanol/ acetic acid Detection: Visualization or densitometry under long-wave UV light Derivatization: Spray plate with 15% aluminum chloride solution and heat 5 min at 110°C; spray with 18:82 silicone/anhydrous ether and heat 3 min at 110°C	2.0-5.0	Francis *et al.* (1988)

4. Natural incidence: Sterigmatocystin is reported in cereals, rice, various nuts, green coffee beans and cheese (Davis and Diener, 1987 and Pitt and Hocking, 1997).

5. Biological activity: Toxic effects of Sterigmatocystins are much the same as those of aflatoxin B_1. Chronic symptoms include induction of hepatomas in rats, pulmonary tumors in mice and renal lesions and alterations in the liver and kidneys of African green monkeys. Rats fed 5-10 mg/kg of sterigmatocystin for two years showed a 90% incidence of liver tumors (Ohtsubo *et al.,* 1978). Wannematcher *et al.* (1991) have studied acute toxicity, carcinogenicity and metabolism of strerigmatocystin and have compared with those of aflatoxin and several other hepatoxic mycotoxins). Sterigmatocystin is reported to cause bloody diarrhoea, loss of milk production and in some cases caused the death of cattle which ingested about 8mg/kg (Vesonder and Horn, 1985). Richard and Thurston (1986) are of the opinion that streigmatocystin is a potential to cause human cancer.

15.4. SLAFRAMINE

Slaframine an intracellular metabolite of *Rhizoctonia leguminicola,* an incitent of black rot of red clover. The ingestion of good amount of red clover weather as pasture, having or silage by cattle, horses or sheep resulted in the extensive salivation or slobber. This is also accompanied by lacrimation, diarrhea, frequent urination and urorexia. Later Byers and Broquist (1960) could establish that fraction that is responsible for salivation could be extracted by water or alcohol. Aust (1966) isolated salivation factor from zonate leaf spot of red clover as picrate and characterized it as slaframine (Slafra to slover). Because of basic nature of salivery factor along with its physiological action Aust *et al.* (1966) proposed the trivial name Slaframine. Free base of Slaframine is clear oil which undergoes rapid degradation by oxidation in air (Aust *et al.*, 1973). Numerous salts can be prepared including the dipicrate, hydrochlorite, acetate and other (Brondquist and Syner, 1971).

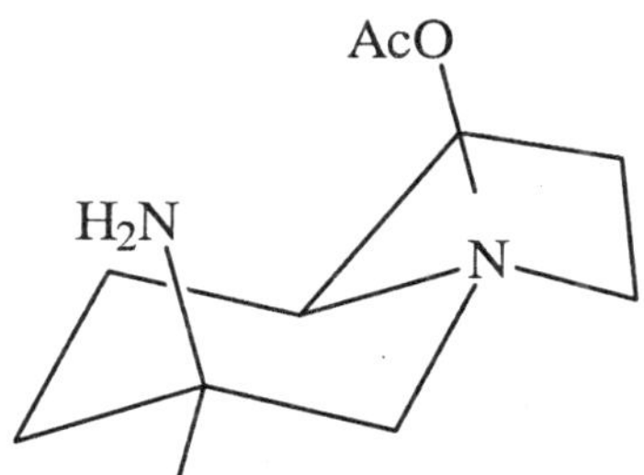

Fig.15.4.1. Stereochemistry of slaframine (1S, 6S, 8aS) 1-acetoxy-6-amino Octa hydroindolizine

Slaframine is isolated as dipicrate, $C_{10}H_8N_2O_2.2C_6H_3N_3O_7$, m.p. is 178-184 and molecular weight of 198. The structure of slaframine can be depicted as (1S, 6S, 8aS)-acetoxy-6-aminooctahydraindolizine (Fig 15.4.1)

15.4.1.Extraction and detection

Crump *et al.* (1967) used a defined medium containing 20 g soybean meal, 20 g dextrose, 5g $CaCO_3$ 5g corn steep liquor in one litter of distilled water. Increase in produce slaframine by addition of casamino acids (Difco) to the basal medium. Production of slaframine decreased drastically when growth temperature was raised to above 25°C. HCl gradient elution on Dowex so ion exchange columns replaced by silic acid and florisil column chromatography pooled fractions were crystallized as the slaframine dipicrate salts from 20% ethanol

The concentration of slaframine in culture or feed samples can most readily measured by pig bioassay technique. Various degrees of response rating systems based on the intensity of salivation. The slaframine can be concentrated or deionised from test samples by an initial extraction with chloroform at neutral pH and then again at pH 10. However, slaframine gets hydrolysed above pH 10.0. Byers and Broquist (1960) reported the extraction of slaframine from forage with hot ethanol, the ethanol evaporated and the residue redissolved in hot water. The extracts at pH 2.0 when chromatographed on paper (butanol – acetic acid) gave a spot (R_f 0.12) which produce a pink colour with Dragendorff's reagent for alkaloids. Amt (1974) estimated slaframine by GLC. Aust *et al.* (1968) used butanol : acetic acid : water (4:1:1) and Dragendorff's reagent to visualize slaframine.

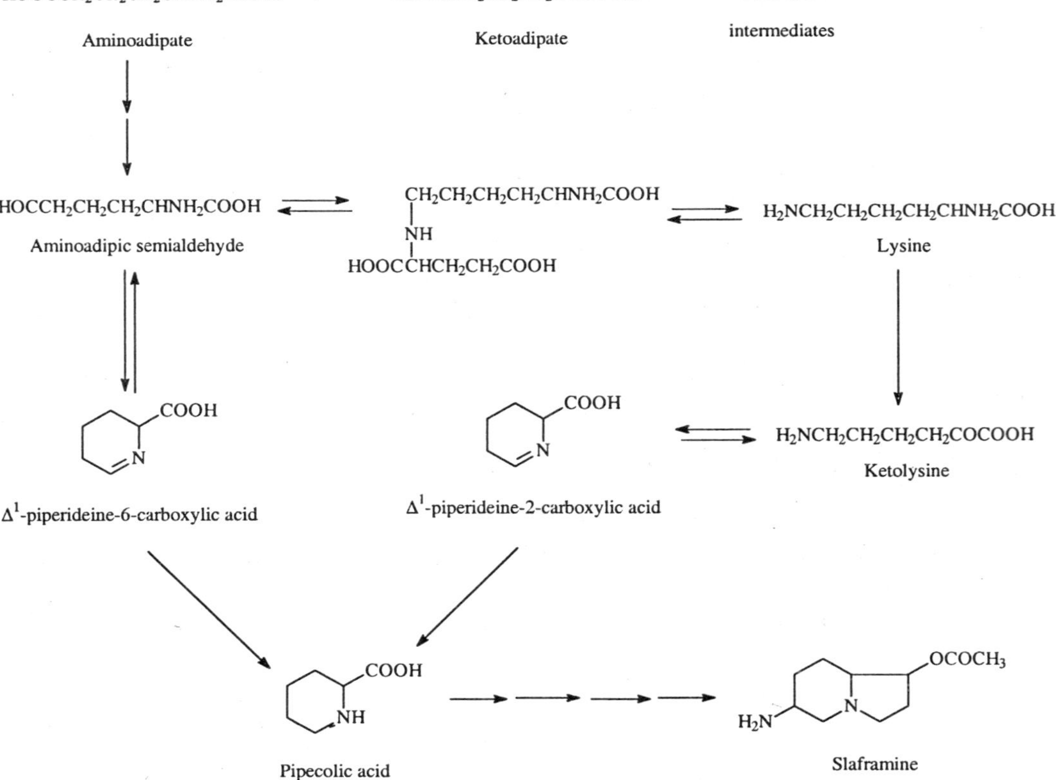

Fig. 15.4.2: Biosynthetic pathway of slaframine

15.4.2. Biosynthesis

Systemic studies using labeled (^{14}C) amino acid to growing cultures of *R. leguminicola* revealed that lysine was involved in the biogenesis of slaframine as follows.

Lysine-2-(keto—aminohexanoic acid Δ^1-piperideine-2-carboxylic acid
→ L-pipecolic acid

Guengerich and Broquist (1973) proposed a tentative scheme of biosynthesis of slaframine as lysine-pipecolic acid – keto octahydro indolizine → 1-hydroxyocta hydroindolizine → 1,6-dihydroxy octahydro indolizine → 1-hydroxy-6-amino octahydro indolizine → 1,6-dihydroxy octahydroindolizine → 1-hydroxy-6-amino octahydroindolizine → slaframine (Fig. 15.4.2)

It is clear that entire L-lysine is involved in the synthesis of slaframine and octahydroindozoline ring is formed from pipecolic acid by amination and esterification.

15.4.3. Biological activity

It is a salivation inducer by two mechanism (1) cholinergic system causing viscous secretions with the release of acetylcholine and the andrenagnic system causing a watery type of saliva with the release of adrenalin. It is not a cholin esterase inhibitor. Aust *et al.* (1968) feels that liver is involved in the formation of slaframine metabolite. Slaframine caused decrease in respiration rate, heart beat rates and body temperature. Such response was also observed in rats. Prolonged salivation in field resulted in the decreased milk yield. According to Crump *et al.* (1963) suffocation accompanied by severe emphysema is responsible for death of animal. Administring of slaframine in pregnant rats caused uterine hemorrhage, foetal abortion and frequently maternal death.

References

Abarca, M. L., F. Accensi, J. Cano and F. J. Cabanes 2004 *Antitoxic van Leeuwenhoek* **86:**33-49.

Abarca, M. L., M. R. Bragulat, G. Sustello and F. J. Cabanes 1994 *Appl. Envir. Microb.* **60:** 2650-2652.

Abouzied, M. M. and J. J. Pestka 1994 *J. AOAC.* **77:** 495-500.

Abbas, H. K. and C. J. Mirocha 1988 *Mycotoxin Res.* **4** : 67-75.

Abbas, H. K., C. J. Mirocha and J. Tuite 1986 *Appl. Envir. Microb.* **51:** 841-843.

Abbas, H. K., C. J. Mirocha, R. J. Pawlosky and D. J. Pusch 1985 *Appl. Envi. Microb.* **50**: 482-486.

Abbas, H. K. and C. M. Ocams 1995 *Plant Dis.* **79:** 642.

Abbas, H. K., C. M. Ocams, W. Xie, C. J. Mirocha and W. T. Shier 1995 *Plant Dis.* **79**: 968.

Abbas, H. K., R. D. Cart, W. Wright, C. J. Xie, S. L. Mirocha, T. J. Richard, G. L. Dvonak, Sciumbato and W. J. Shier 1999 *Mycopathologia* **147:** 97-104.

Abbas, H. K. and R. T. Riley 1996 *Toxicon.* 34: 133-136.

Abbas, H. K., R. J. Smeda, B. C. Gerwick and W. Shier 2000 *J. Natural Toxins* **9:** 85-100.

Abbas, H. K., R. N. Paul, C. D. Boyette and S. O. Duke 1992 *Can. J. Bot.* **70:** 1824-1833.

Abbas, H. K., W. P. Williams, G. L. Windham, H. C. Pringle, W. P. Xie and W. T. Shier 2002 *J. Agric. and Food Chem.* **50** : 5246-5254.

Abdel -Wahhab, M. A., A. M. Hassan, H.A. Amer and K. M. Naguib 2004 *J. Appl. Toxicol.* **24:** 469-474.

Abdelhamid, A.M., E.A. Sadik and E.A.Fayzalla 1985 *Acta Phytopathology Acadamiae Scientiarum Hungariae* **20:** 309-320.

Abdelhamid, A.M., I. IL-Shawaf, S.A. El-Ayoyy, M. M. Ali and T. Gamil 1990 *Archives of Animal Nutrition* **40:** 517-537.

Abedi, Z. H. and P. M. Scott 1969 *J. AOAC.* **52:** 963.

Aboul-Enein, H.Y., O.B. Kutluk, G. Altiokka and M. Tuncel 2002 *Biome .Chromatogr.* **16:** 470-474.

Abraham, E. P. and H. W. Florey 1949 In Antibiotics (eds H. W. Florey, E. Chain, N. G. Heatley, M. A. Jennings, A. G. Sanders, E. P. Abraham and M. E. Florey) Vol. I, Oxford University Press, London, pp. 273-355.

Abrunhora, L., R. R. Paterson, Z. Kozakiewicz, N. Lima and M. Arici 2000 *Ernahrung* **24:** 257-259.

Abrunhora, L., R. Serra and A. Venancio 2001 *J. Agric. Food Chem.* **50:**7493-7496.

Acharya, K. T. 1986 *J. Oil Technol. Assoc. India* **18:** 96-107.

Acklin, W., F. Weibel and D. Erigoni 1977 *China.* **63.**

Adam, R. 1980 *Deutsche Lebensmittel-rundschall* **765:** 123-125.

Adav, S.S. and S.P. Godinwar 1997 *Pharma Toxicol. and Endccry.* **118:** 185-189.

Afridi, M. M. 1962 *Hind. Antibiot. Bull.* **5:** 51.

Agag, B. I. 2004 *Ass. Univ. Bull. Environ. Res.* **7:** 159-176.

Ahmed, S. K. and P. L. Singh 1991 *Food Addi. and Contam.* **8:** 723-730.

Aidia, E., L. Makawy, S. M. Hassanane, A. N. El-Sayed and B. D. Alla 2001 *Reprod. Nutr. Dev.* **41:** 79-89.

Ainsworth, G.C. and P.K.C. Austwick 1973 In Fungal diseases of Animals (Commonwealth Bureau of Animal health, Review Series, No. 6), 2nd ed. Farhham Royal; Common Wealth Agricultural Bureaux.

Aish, J. L., E. H. Rippen, T. Barlow and S. J. Hattersley 2004 In Mycotoxins in Food Detection and Control (eds. N. Magan and M. Olgen), Wood Head Published Lim. pp. 308-338.

Akiyama, H., M. Miyahara, M. Toyoda and Y. Saito 1995 *J. Food Hyg. Soc. Japan* **36:**77-81.

Akiyama, H., Y. Kikuchi, N. Narita, M. Suzuki, Y. Goda, K. Takatori, M. Ichinoe and M. Toyoda 2000 *J. Food Hyg. Soc. Japan* **41:** 30-37.

Alam, M. M., M. Nagase, T. Yoshizawa and N. Sakato 2000 *Biotechnol. Biochem.* **64**:210-213.

Albert, J. F., W.C.A. Gelderblom, P. G. Thiel, W. F. O. Marasas, D. J. Van Schalkywyk and Y. Behred 1990 *Appl. Envir. Microb.* **56** : 1729-1733.

Aldridge, D. C., B. F. Burrows and W. B. Turner 1973 *J. Chem. Soc. Chem. Commun.* 148-149.

Aldridge, D. C., J. J. Armstrong, R. N. Speake and W. B. Tuner 1966 *J. Chem. Soc. Chem. Commun.* 26 - 27.

Aldridge, D. C., J. J. Armstrong, R. N. Speake and W. B. Turner 1967 *J. Chem. Soc. Commun.* 1667 - 1676.

Aldridge, D. C. and W. B. Turner 1969 *J. Chem. Soc. Commun.* pp. 923-923.

Aldrige, D. C. and W. B. Turner 1969 *J. Antibiotics* **22:** 170-176.

Alldrick, A. J. 1996 *Food Addi. and Contam.* **13**: 27-28.

Alldrick, A. J. 2004 *Appl. Biol.* **68:** 139-146.

Ali, N., A. Y. Sarjana and J. Yashizawa 1998 *Food Addi. and Contam.* **15:** 377-384.

Altmayer, B., K. W. Eichborn and R. Plapp 1982 *Z. Bebensun Unters Forsch* **175:** 172-174.

Aman, I. M. 1994 *Chem. Microbiol. Technol.* Lebson **16**: 184.

Andrand, G., J. Conquelet, M. Dorel and P. Trochen 1966 *CRS. Soc. Biol. Fil.* **160:** 325.

Anisa Athas, M., H. S. Prakash and H.S. Shetty 1988 *Indian J. Microbiol.* **28:** 125-127.

Anjana Singh 1983 In Mycotoxins in Food and Feed (eds. K. S. Bilgrami, T. S. Prasad and K. K. Sinha), Allied Press, Bhagalpur, pp. 55-68.

Anonymous 1969 In Indian standard speculation for edible groundnut flour expeller presented and solvent extraction pp.4684 - 4875.

Anonymous 1980 Food Technology, Tamilnadu Agric. Univ. Coimbatore.

Anonymous 1986 In IARC monograph on the evaluation of the carcinogenic risk of chemicals to humans. **40:** 67-82.

Anonymous 1992 Information bulletin No. 7 Kenya Agricultural Research Institute, Nairobi. pp. 4

Ansari, A. A. and A. K. Shrivastava 1991 *Food Addi. and Contam.* **7:** 815-820.

Anslow, W. K., H.Raistrick and G. Smith 1943 *J. Soc. Chem. Ind.* **62:** 236.

AOAC, 1995 Official methods of Analysis of *AOAC*, Internal 16th ed. (ed. M.D. Gaithersberg), *AOAC*. pp. 44-45.

Apisman, J. W. 2001 *Environ. Health Perspect.* **109:** 245-249.

Apleton, J. A. and M. R. Tansey 1977 *Mycologia* **687:** 882-885.

Aran, N. and D. Eke 1987 *Food Microbiol.* **4**: 101-104.

Arai, M. and T. Hibino 1983 *Cancer Lett.* **17**: 281-287.

Arai, T. and M. Otmo 1969 *Ann. Rept. Inst. Food Microbiol.* Chiba Univ. **22**: 81.

Arici, M. 2000 *Ernahrung* **24:** 257-259.

Arnold, J. M. and L. D. Hilliams1974 *J. Embryol. Exp. Morphol.* **31:**1-25

Arp, L. H. and J. L. Richard 1979 *J. Amer. Vet. Med. Ass.***85:** 556-565.

Arp, L. H. and J. L. Richard 1981 *Mycopathologia* **73:** 109-113.

Arumugham, S. and K. G. Shanmugavelu 1977 *Seed Res.* **5:** 25-31.

Ashoor, S. H. and F. S. Chu 1973 *Food Cosmet. Toxicol.* **11:** 617-624.

Askar, A. and H. Siliha 1999 *Fruit Processing* **9**: 74-77.

Atkinson, N. 1942 *J. Expol. Biol. Med. Sci.* **20:** 284-287.

Athapol, N. and T. D. Cardona 1991 In: Mycotxin Prevention and Control in Food Grains (eds. R. L. A. S. Frio, P. A. Hjcks and J. V. Lozare) Sangkok, pp. 207-215.

Atlas, S. T. and S. Lin 1978 *J. Cell Biol.* **6**: 360.

Atroshi, F., A. Rizza, I. Biese, M. Salonen, L. A. Lindberg and H. Saloniemi 1995 *J. Animal Physiology and Animal Nutrition* **74:** 157-164.

Atroshi, F., A. Rizza, T. Westermark and T. Ali-Velimas 1998 *Pharmacol. Res.* **38:** 289-295.

Atroshi, F., A. Rizza, T. Westermark and T. Ali-Velimas 2002 *Toxicology* **180:**151-167.

Augustine, B., M. Parvathi and Indira Kalyanasundram 1984 *J. Food Sci. and Technol.* India **21**: 312-316.

Aust, S. D., H. P. Broquist and K. L. Rinehart Jr. 1966 *J. Am. Chem. Soc.* **88**: 29-79.

Aust, S. D., H. P. Broquist and K. L. Rinehart Jr. 1968 *Biotechnol. Bioeng.* **10**: 403-412.

Austwick, P. K. C. 1962 Recent Progress in Microbiology **8:** 644-651.

Avantaggiato, G., R. Havenaar and A. Visconti 2005 *Food and Chem. Toxic.* **42:** 817-824.

Azcona-Olivera, J. I., M. M. Abouzied, R. D. Platter and J. J. Pestka 1992 *J. Agri. Food Chem.* **40 :** l531-1534.

Azcona-Olivera, J. I., M. M. Abouzied, R. D. Plattner, W. P. Norred and J. J. Pestka 1992a *Appl. Envir. Microb.* **58:** 169-173.

Aziz, N. H. and L. A. E. Moussa 1997 *Food Microbiology* **41:** 150-154.

Bacigaupo, M.A., A. Ius, G. Meroni, M. Dovis and E. Petruzzelli 1994 *Analyst* **119**: 2813-2815.

Bacon, C. W. and D. M. Hinton 2002 *Biological Control* **23:** 274-284.

Bacon, C.W., I.E. Vates, D.M. Hinton, F. Meredith, W. T. Allaben, J.R. Bucher and P. C. Howard 2001 *Envir. Health Perspec.* **109:** 325-332.

Bacon, C. W., J. G. Sweeney, J. D. Robbins and D. Burdick 1973 *Appl. Microbiol.* **26:**155-160.

Bailey, D.E., G.E. Cox, K. Morgneidge and J. Paylor 1976 *Toxicol. Appl. Pharmacol.* **37:** 144-155.

Balachandran, C. and K. R. Parthasarathy 1996 *Mycopathologia* **133:** 159-162.

Balachandran, C. and R. Ramakrishanan 1987 *Indian J. Poultry Sci.* **22:**105-108.

Baldwin, N.C.P., B.W. Bycroft, P.M. Dewick and J. Gilbert 1986 *Z. Naturforsch.* **410:** 845.

Ballarini, G. 2002 *Rivista-di-suinicoltura* **43**: 71-80.

Bamburg, J. R. 1976 *Chem. Ser.* **149**: 144-162.

Bamburg, J.R. and F.M. Strong 1971 In: Microbial toxins, A comprehensive treatise, Algal and fungal toxins (eds. S. Kadis, A. Ciegler and S. J. Ajl), Academic Press. New York. pp. 207-292.

Bammler, T. K., D. H. Slone and D.L. Eaton 2000 *Toxicological Science* **54:**30-41.

Badei, D. P., E. J. Neumann, W.F. Hall, K.S. Harin and R.L. Slife 1992 *Mycopathologia* **117:** 121-124.

Barbieri, G., C. Bergamini, E. Ori and P. Resca 1994 *J. Food Sci.* **59**: 1313-1331.

Barber, J., A. C. Chapman and T.D. Howard 1987 *J. Antibiot.* **40:** 245-248

Barna-vetro, T., A. Gyongyosi and L. Solti 1994 *Appl. Envir. Microb.* **60:** 729-731.

Basappa, S. C. and T. Shantha 1996 *J. Food Sci. and Technol. India* **33:** 95-107.

Basappa, S. C., V. Jayaraman, V. Sreenivasamurthy and H. A. B. Parpia 1967 *Indian J. Exp. Biol.* **5 :** 262-263.

Basilico, M. Z. and J. C. Basilico 1999 *Lett. Appl. Microbiol.* **29**: 238-241.

Bassa, S., C. Mestres, K.Hell, P. Vernia and K. Cardwell 2001 *Plant Dis.* **85**: 1032.

Bata, A., A. Vanyi and R. Lasztity 1983 *J. AOAC.* **66:** 577-581.

Bata, A., J. W. Bollwahn, M. Garier, B. Gedek and K. Heinritzi 1985 *Appl. Envir. Microb.* **49:** 678.

Battaglia, R., T. Hatzold, R. Kroes 1996 Conclusion from the workshop on ochratoxin in food, organized by ILSI Europe in Aix-en-Provence pp. 1-3.

Banduin, H., C. Stock, D. Vincent and J. F. Grenier 1975 *J. Cell Biol.* **66 :** 165-181.

Bauer, J., K. Heinritzi, M. Garies and B. Gedek 1987 *Tierdrztiiche Praxis* **15:** 33-35.

Bauer, J., M. Gareis and B. Gedek 1984 *Berliner* and *Miincherer, Tierarzthichi* Wocherschrift. **97**:437-441.

Bauer, J., M. Garies and B. Gedek 1986 Proceedings of the 2nd world congress on Food-borne infection and Intoxications, Berlin 26-30 May 1986 pp. 907.

Baxter, E.D., R. Battaglia, T. Hatzold and R. Kores 1996 *Food Addi. & Contam.* **13:** 23-24.

Beasely, J.N., T. S. Nelson, S. J. Hollingswo and S. C. Goeke 1986 *Arkansas Form Research* **35:** 2-6.

Beasley, V. R., B. L. Bunner and R. H. Dopperga 1989 In Trichothecenes Mycotoxicosis, Pathophysiological effects Vol. II (ed. V. R. Bearley), CRC Press, Boca Raton Florida USA.pp. 27-35.

Becker, B., H. Bresch, V. Schillinger and P. G. Thiel 1997 *World J. Microb. and Biotech.* **13:** 539-543.

Beebe, R. M. 1978 *J. AOAC.* **61**: 1347-1352.

Beekrun, S., R. Govindan, T. Padayachoe and B. Odhav 2003 *Food Addi. and Contam.* **20:** 490-493.

Begum, F. and S. Samajpathi 2000 *Naturwissenschaften* **87**: 275-277.

Beier, R. C. and L. H. Stanker 1996 *American Chemical Society* ACS Symposium, Washington, DC, pp. 519-528.

Bejaoui, H., F.P. Mathieu, A. Thaillandier and A. Lebrili 2005 *J. Appl. Microbiol.* **97**:1038-1044

Bekes, D. 1988 *Verterinarski Arhiv.* **58:** 133-137.

Belcheva L., I. Petrova and I. Nedyalkov 1985 *Rarteniev Dni Nauki* **22:** 10-12.

Bennett, G. A., A. A. Lagoda, O. L. Shotwell and C. W. Hesseltine 1981 *J Amer. Oil Chem. Soc.* **58**: 974-976.

Bennett G. A. and J. L. Richard 1994 *J. AOAC.* **77:** 501-506.

Bennett, J.W. and M. Klich 2003 *Microbiology* Reviews **16:** 497-516.

Bennett, G. A., T. C. Nelson and B. M. Milley 1994 *J. AOAC.* **77**: 1500-1508.

Beno, M. A., R. H. Cox, J.M. Wells, R.J. Cole, J.W.Kirksey and G. G. Christopher 1977 *J. Am. Chem. Soc.* **99 :** 4123-4130.

Bergel, F., A. L. Morrison, A. R. Moss and H. Rinderknecht 1944 *J. Chem. Soc.* 415.

Betina, V. 1964 *Chem. Zvest.* **18**: 128-139.

Betina, V. 1984 In Mycotoxins: Production, Isolation, Separation and Purification. (ed V. Betina), Elsevier, Amsterdam, pp. 217-236.

Betina, V. 1985 *J. Chromatagraphy* **334:** 211.

Betina, V. 1989 In: Mycotoxins, Chemical, Biological and Environmental Aspects (ed. V. Betina), Elsevier Publication, pp. 172.

Betina, V., D. Janstova and J. Splsiakova 1975 *Folia Microbiol.* **29:** 340.

Betina, V., D. Micekova and P. Nemec 1972 *Microbiol.* **71:** 343-349.

Betina, V. and E. Rockova 1972 *Biologia (Bratislova)* **27**: 463-468.

Betina, V. and H. Barthova 1968 *J. Antibiot.* **21:** 628-629.

Betina, V., S. Balint, V. Hanjicka and A. Nadova 1973 *Folia Microbiol.* **18:** 40-48

Betina, V. and Z. Baruth 1964 *J. Antibiot.* (Tokyo) **17**: 127.

Betina, V. and Z. Binovska 1979 *Biologia (Bratislova)* **27:** 463-468.

Bezuidenhout, S.C., W.C.A. Gelderblom, C.P. Gorst-Allman, R.M. Horak, W.F.O. Marasas, G. Spiteller and R. Vleggaar 1988 *J. Chem. Soc. Chem. Commun.* Pp.743-745.

Bhadraiah, B. and P. Rama Rao 1979 *Nat. Acad. Sci. Letters* **2:** 257-259.

Bhandari, N. and R. P. Sharma 2000 *Chem. Biol. Interaction* **139**: 317-331.

Bhandari, P. C. and B. Bhandari 1980 *Indian Pediatr.* **17:** 593-596.

Bharat and R. Mair 1948 *Experientia* **4:** 277-288.

Bhat, R. V. 1983 Mycotoxin health hazards in India. Proceedings of the Japanese International Conference on Indoor Air Quality and Climate, (ed. Y. Ueno), Elsevier *Science* Publishers *B.V,* Amsterdam. 42. Vol 5, 51454.

Bhat, R. V. 1988 *Intern. J. Food Microbiol.* **7:** 219-225.

Bhat, R. V. 1989b *Lancet* **2** (8662): 575-576.

Bhat, R. V., D. N. Roy and P. G. Tulpule 1976 *Toxicol. Appl. Pharmacol.* **36**:11-77.

Bhat, R. V. and P. G. Tulpule 1983 In: Trichothecene hemical, Biological and Toxicological Aspect (ed. Y. Ueno), Elsevier Science Publication. pp. 285.

Bhat, R. V., P. H. Shetty, R. P. Amruth and R. V. Sudershan 1997 *Clinical Toxicology* **35 :** 249-255.

Bhat, R.V., R. B. Sashidhar, Y. Ramakrishna and K. L. Munshi 1989 *Appl. Envir. Microb.* **55:** 35-37.

Bhat R. V., S. Vasanthi, B. S. Rao, R. N. Rao, V. S. Rao, K. V. Nagaraja, R. G. Bai, C. A. Prasad, S. Vanchinathan, R. Rooy, S. Saha, A. Mukherjee, P. K. Ghosh, S. Toteja and B. N. Saxena 1996 *Food Addi.and Contam.* **13:** 325-331.

Bhat, R. V., V. Nagarajan and P.G. Tulpule 1978 Health hazards of mycotoxin in India. *Indian Council of Medical Research* New Delhi. pp.58.

Bhat, R. V., Y. Ramakrishna, S. R. Beedu and K. L. Munshi 1989 *Lancet.* **7:** 35-37.

Bhatnagar, D., E. B. Lillehoj and J. H. Bennett 1994 In, Mycotoxins and Animal foods (Ed. J. E Smith and R. S. Henderson), CRC Press Boston, pp. 816-826.

Bhavanishankar, T. N. and T. Shantha 1987 *J. Sci. and Food and Agric.* **40:** 327-332.

Biancardi, A. 1997 *Indust. Aliment.* **36:** 870-876.

Biehl, M. L., D. B. Prelusky, G. D. Kritz, K. E. Hartin, W. B. Bua and H. L. Trebolm 1993 Toxicol. *Appl. Pharmacol.* **121**: 152-159.

Bijl, J. P., C. H. Van Peteghem and D. A. Dekayser 1987 *J. AOAC.* **70**: 472.

Bieai, I. 1955 *Publ. Acad. Sci. Ukr*, S. S. R., Kiev, pp.32.

Bilgrami, K. S. 1984 Mycotoxins in *Food J. Ind. Bot. Soc.* **63:** 109-120.

Bilgrami,K. S. 1984a Mycotoxins in dry fruits and speices. Technical Report of UGC Project New Delhi. pp. 1-97.

Bilgrami, K. S., A. Singh and K. S. Ranjan 1984b *Nat. Acad. Sci. Letters* **7:** 273-274.

Bilgrami K. S.and K. K. Sinha 1983 *Phytopathol.* **73**: 253-263.

Bilgrami, K. S. and K. K. Sinha. 1984d *Rev. Trop. Pl. Path.* **1:** 335-374.

Bilgrami, K. S. and K. K. Sinha 1984c *Ind. Rev. Life Sci.* **4:** 19-36.

Bilgrami, K. S., K. K. Sinha and P. L. Singh 1981 *Curr. Sci.* **50:** 997-998.

Bilgrami K. S., K. S. Ranjan and A. K. Choudary 1991 *Indian Phytopath.* **44:** 529-531

Bilgrami, K.S., K.S. Ranjan and S. P. Sinha 1986 *Proc. Nat. Acad. Sci. India.* **52:** 280-283.

Bilgrami, K. S., K. K.Sinha and A. K. Sinha 1992 *Indian J. Med. Research* **96**: 171-175.

Bilgrami, K. S., R.S. Misra, K. K. Sinha and Premlatha Singh 1980 *Indian J. Bot. Soc.* **59:** 123-126.

Bilgrami, K.S., R.S. Misra T. Prasad and K.K. Sinha 1982 *Indian Phytopath.* **35:**376-378.

Bilgrami K.S.,T. Prasad, Jamaluddin and A.K. Roy 1979 *Indian Phytopath.* **29:** 374-377.

Bilgrami, K.S., T. Prasad, R.S. Mishra and K.K. Sinha 1981 *Indian Phytopath.* **34:** 67-68.

Bilgrami, K.S., T. Prasad, R. S. Misra and K.K. Sinha 1980 Survey and study of mycotoxin producing fungi associated with grains in the standing maize crop ICAR final report, Bhagalpur University, Bhagalpur, India pp. 98.

Binder, M. and Ch. Tamm 1973 *Helv. Chim. Acta* **56:** 2387-2396.

Binder, M., Ch.Tamm, W.B. Turner and H.Minato 1973 *J. Chem. Soc. Perkin. Trans.* 1146-1147.

Biollaz, M., G. Buchi and G. Milne 1970 *J. Amer. Chem. Soc.* **92**: 1035-1043.

Birkinshaw, J. H., S. E. Michel, A. Bracken and H. Raistrick 1944 *J. Chem. Soc.* 415-425.

Birzele, B., A. Prange and J. Kramer 2000 *Food Addi.and Contam.* **17**: 1027-1035.

Bitonti, A. J. and M.C. Mc. Cann 1982 *Cell Physiology* pp. 243-213.

Blackwell, B. A., O.E. Edward, J.W. Apsimon and A. Fruchier 1994 *Tetrahedron Lett.* **36**: 1973-1976.

Blaney, B. J., M. O. Ramsey and A. L. Tyler 1986 *Aust. J. Agric. Res.* **37** : 235-244.

Blaney, B. J., R. C. Bloomfoeld and C. J. Moore 1984 *Aust. J. Vet.* **61**: 24-27.

Blera, J., J. M.Sirano, J.C.Molto and J. Manes 2004 *J. Chromatogra.* **1054**: 397-401.

Bodan, L. and L. Zoldac 1974 *Acta. Vert. Acad. Sci. Hung*: **24:** 451.

Boenke, A. 1997 *Food Chem.* **60**: 255-262.

Bohm, J., J.Grajewski, H. Asperger, B. Cocon, B. Rabus and E. Razzazi 2000 *Mycotoxin Res.* **16:** 70-74.

Bohni, J. 2000 Ubersichten-Zur-Tierernahrung **28:** 95-132.

Boikov, Y. and E. Khasanova 1986 *Plitesevods to No. 41.*

Bondy, G. S., J. J. Pestka 2000 *J. Toxicol. and Envir. Health* Part B; Critical Reviews. **3**: 109-143.

Bongiovanni, A. M. 1983 *J. Pediatr.* **103**: 245-246.

Booth, C. 1971 In Methods in Microbiology Vol. 4 Academic Press, London 75:1-115.

Booth, C. 1971a In The genus *Fusarium.* Commonwealth Mycological Inst., Kew.pp. 237

Borner, H. 1963 *Phytopathol. Z.* **48**: 370-396.

Bourgeade, M. F. and C. Chaney 1976 *Proc. Soc. Exp. Biol. Med.* **153:** 501-504.

Boutibones, P. 1979 *Canad. J. Microbiol.* **25**: 421-423.

Bower, A . and M. E. Hadley 1975 *Endocrinology* **96:** 431-439.

Boyd, P. A . and J. L. Wittlift 1978 *J. Toxicol. Environ. Health* **4**:1-8.

Bracken, A . and H. Raistrick 1947 *J. Biochem.* **41**: 569-575.

Bradley, M., Ch.Tamm, W.B. Turner and H. Mato 1973 *J. Chem. Soc. Perkin. Trans.* **2:** 1146-1147.

Brakhage, A. A., M. G. Barger, E. E. Creeppy, G. Mrncimer and R. J. Roschenthaler 1988 *Arch. Toxicology* **12**: 341-346.

Branham, B. E. and R. D. Plattner 1993 *J. Natural Prod.* **56:** 1630-1683.

Branton, S. L., J. W. Deaton, W. M. Hagler, W. R. Mashin and J. M. Hardin 1989 *Avians Dis.* **33**: 804-808.

Breitholtz, A., M.Olsen, A.Dahback and K.Hult 1991 *Food Addi.and Contam.***8**:183-192.

Breitholtz-Emanuelssan, A., I. H. Palminger, O.P. Wohhlu, A.Oskarssen, K. Hult and M. Olsen 1993 *Nat. Toxins* **1:** 347-352.

Brekke, O. L., R. O. Sinnhunder, A. J. Peplinski, J. H. Wales, G. B. Putnam, D. J. Lee and A. Ciegler 1977 *Microbiol.* **34**: 34-37.

Bresch, H., M. Urbank and M. Nusser 2000 *Nahrung* **44:** 276-278.

Brester, G., G.Vaamonde, C.Degrossi and P.V.Frnandez 1998 *J. Food Micorbiol.* **42:**57-61.

Brewer, D., J. M. Dunean, W. A. Jerram, C. K. Leach, S.Safe, A.Taylor, L. C. Vinning, R. Archilbald, R.G. Stevenson, C. J. Mirocha and C. M. Christensen 1972 *Canad. J. Microbiol.* **18:** 1129-1136.

Brewer, D., W. A. Jerram and A. Taylor 1978 *Can. J. Microb.* **14 :** 861-866.

Brian, P. W., A. W. Dawkins, J.F. Grove, H.G. Hemming, D.Lowe and G.L. Norris 1961 *J. Exp. Bot.* **12 :** 1-2.

Brian, P. W., G. W. Elson and D. Lowe 1956 *Nature* **178:** 263-264.

Brian, P. W. and H. G. Hemming 1945 *Ann. Appl. Biol.* **32:** 214-222.

Brigys, I. R., N. R. Towers and P. C. Molan 1994 *J. Agri. Food Chem.* **42 :** 2769-2777.

Broce, D., R. M. Grodner, R. L. Killerbrew and L . Bonner 1970 *J. AOAC.* **52**: 616.

Broggi, L. E., S. L. Resnik, A. M. Pacon, H. H. L. Gonzales, G. Cano and D. Taglieri 2002 *Food Addi. and Contam.* **19:** 465-469.

Broster, W. H. 1973 *Veterinary Record* **93:** 417-420.

Brown, J. R., N. J. Cartwright, A. Robertson and W.B. Whalley 1948 *Nature (London)* **162**: 76-77.

Brown, Q. L., P. G. Cotty and T. E. Cleveland 1991 *J. Food Prot.* **54**: 623-626.

Brown, R. L., T. E.Cleveland, G. A.Payne, C.P.Woloshuk, K.W. Campbell and D. G. White 1995 *Phytopath.* **85:** 983-989.

Brown, T. 1996 In T-2 toxin fed to broilers at 3 mg/kg level resulted in lower bone ash content.Published by W. B. Sauders Company Ltd.pp 290-308.

Brown, T. P., R. O. Manning and O. J. Fletcher 1986 *Avian Dis.* **30**: 191-193.

Bruno Doko, M., S.Rapior, Angelo Visconti and Johanne E. Schjoth 1995 *J. Agri. Food Chem.* **43:** 429-434.

Bryden, W. L. 1982 *Food Technol. Augt.* **34:** 216-223.

Bu'Lock, J. D., J. P. Mooney and C. E. Weight 1986 *Biotechnol. Lett.* **8**:323.

Bubbers, J. E. and C. S. Henney 1975 *J. Immunol.* **115 :** 145-149.

Bubien, J. K., G. Lundeen, C.Templeton and W.T.Woods 1989 In Trichothecenes mycotoxicosis : Pathophysiological effects, Vol. I (ed. V.R. Beasley), CRL, Publication, Boca Raton, Florida, USA, Pp.91-111.

Buchanan, J. R., N. F. Somer, R. J. Fortlage, E.C. Maxic, F. G. Mitchell and S. P. H. Hsich 1974 *J. Am. Soc. Hort. Sci.* **99:** 262-265.

Buchanan, R. L. and J. G. Ayres 1979 *J. Food. Sci.* 128-132.

Buchi, G., D.W. Foulkes, H. Kurono and G. F. Mitchell 1996 *J. Am. Chem. Soc.* **87 :** 882.

Buchi, G., P.M.Muller, B. D. Roebuck and G. N.Wogan 1974 Res. *Commun. Chem. Pathol. Pharmacol.* **8:** 585-592.

Buchi, G., Y. Kitaura, S. Yuan, H. E. Wright, J. Clardy, A. L. Demain, T. Glinsukon, N. Hunt and G. N. Wogan 1973 *J. Am. Chem. Soc.* **95 :** 5423-5425.

Buck, W.B. and L.M.Cole 1991 In Toxicology of Plant and Fungal Compounds Vol. 6 (ed. R. F. Kocher and Anthoni T. Tu) Marcel Decker Inc, New Yark **11:**135-140.

Buckelew, A. R., A. Chakravarthi Jr., W.R. Burge, V.M. Jr. Thomas and M. Ikowa 1972 *J. Agric. Food Chem.* **20 :** 431.

Buckle, A. E. 1983 *Vet. Res. Comm.* **7:** 171-186.

Beuther, T. M., D.Slone and D. L.Eaton 1992 *Biochemical and Biophysical Research Communications* **188:** 597-603.

Bullerman, L. B., D. Ryu and L. S. Jackson 2001 In Mycotoxins and Food safety (eds J. W. Devries., M. W. Trucksess and L. S. Jackson), Kluwer Academic / Plenum Publishers, New York, pp. 195-204.

Bullerman, L. B., R.G. Labbe and S.Garcia 1997 *Fusarium* Guide to food-borne pathogens In Annual Report pp. 87-98.

Bulock, J. D. 1980 In The biosynthesis of mycotoxins. A study in secondary metabolism (ed. P. S. Steyn), Academic Press, New York. pp. 1-16.

Burdaspal, P. T., M. Legarda and J. Gilbert 2001 *J. AOAC.* **84**: 1445-1452.

Burditt, S. J., W. M. Hagler and P. B. Jr. Hamilton 1983 *Poult. Sci.* **63:** 2167-2191.

Burgers, D. R. 1975 *J. Ernbryol-Exp. Mondiol.* **34**: 723.

Burgers, D. R. and R. D. Grey 1974 *J. Cell Biol.* **62**: 566-579.

Burroughs, R., L.M. Seitz, D.B. Sauer and H.E. Mohr 1978 *Appl. Environ. Microbiol.* **31**: 685-690.

Burka, L. T., M.Ganguli and B.J.Wilson 1983 *J. Chem. Sec. Chem.Commun.* pp. 544-545.

Burmeister, H. R. and C.W. Hesseltine 1970 *Appl. Microbiol.* **20**: 437-440.

Burns, R. B. and P.D. Dwivedi 1986 *Poult. Sci.* **42**: 48-55.

Busby, W. F. Jr. and G.N.Wogan 1981 In Mycotoxins and N-Nitroso Compounds, Vol. II pp. *Environmental Risks.Boca Raton, Florida:CRC Press*, pp. 29-42.

Butcher, F. R. and R.H. Goldman 1974 *J. Cell Biol.* **60**: 519-523.

Byers, J. H. and H. P. Broquist 1960 *J. Dairy Sci.* **43**: 873.

Cahagnier, B., D.Melcion and D.Richardmolard 1995 *Lett. Appl. Microbiol.* **20**: 247-251.

Caldwell, R.D. 1983 *Nature* **301:** 651.

Camaron, S. and E. W. Colvin 1989 *J. Chem. Sol. ParkerTransactia* **2:** 365-370.

Campbell, H., T.M. Choa, B.Bigier and L.Underhill 2003 *Candian J. Bot.* **50:**526-532.

Cameron, A. F., A. A.Freer, B.Hesp and C. J.Strawson 1974 *J. Chem. Soc. Perkin. Trans* **2** **:**1741-1744.

Candlish, A. A. G., M. S.Wibowo and J. E.Smith 1997 *Biotechnol. Tech.* **15:** 21-24.

Cardeithae, P. T., K. D. C. Nair and W. M. Colwell 1972 *J. AOAC.* **55:** 1120-1121.

Carlton, W. W., G. M. Szezech and J. Tuite 2001 *Toxicol. Appl.Pharmacol.* **25:** 457-459.

Carlton, W. W. and J. Tuite 1986 In Diagnosis of Mycotoxicoses (ed. J. L. Richard and J. R. Thurstom) Dordneche, Netherlands, pp. 195-204.

Carlton, W. W. and P. Krogh 1979 *Food Drug Admin. Bureau. Vet. Med.* National Technical Information Service, U. S.FDA/BVM-79/139 pp. 165-278.

Carson, M. S. and T. M. Smith 1983 *J. Amer. Sci.* **56:** 1469-1478.

Carter, A. 1981 Ph.D. thesis, University of Toronto.

Carter, S. B. 1997 *Nature* (London) **213:** 261-264.

Carvzal, M., F. Muljoland and R. C. Garner 1990 *J. Chromatogr.* **511**: 379-383.

Casadio, S. and I. Niola 1993 *Industrie dello Bevavde* **22:** 10-11.

CAST 1989 Mycotoxins Economical and Health Risk. Task force report No. **116**: 1989.

CAST 2003 Mycotoxins: Risks in Plant, Animal. and Human Systems (eds. J. L. Richard and G. A. Payne), Council for Agricultural Science and Technology (CAST), Task Force Report No. 139, Ames, IA, USA. pp. 48-58.

Castegnaro, M., I. N. Chemozemsky, E.Hietanen and H.Bartsch 1990 *Arch. Geschwulst Forsch.* **60**: 295-303.

Castegnaro, M., J. Basek, J.M.Fremy, M.Lafontuine, M.Mirglia, E.B. Sansone and G.M.Telling 1991 IARC Scientific Publication 113, Lyon, France, pp. 17-23.

Castella, G., M.R. Bragulat and F. J.Cabanes 1999 *J. Agri and Food Chem.* **47**: 4707-4710.

Castella, G., M. R. Bragulat and F. J. Cabanes 1999 *J. Food Prot.* **62:** 811-813.

Castellari, M., S. Fabbri, A. Fabiani, A.Amati and S.Galassi 2000 *J. Chromatogr.* **888:** 129-136.

Castelo, M. M., L. S.Jackson, M.A.Hanna, B.H.Reynolds and L.B. Bullerman 2001 *J. Food Sci.* **66 :** 416-421.

Castelo, M. M., S. S. Summer and L. B. Bullaman 1998 *J. Food Prot.* **61:** 1030-1033

Castoria, R. 2002 *J. Shtienberg. Bulletin SROP.* **25**: 41-44.

Cavangh, J. B., J.C. Holton, C.C. Nolan, D.E. Ray, J.T.Naik and P.G. Mantle 1998 *Veternary Pathology* **35:** 53-63.

Cawood, M. E., W.C.A. Gelderblom, J.F. Alberts and S.D.Snyman 1994 *Food Chem. Toxicol.* **32 :** 627-632.

Cawood, M. E., W.C. A. Gelderblom, R. Vleggaar, Y. Behrend, P. G. Thiel and W. F. O. Marasas 1991 *J. Agri. and Food Chem.* **39 :** 1958-1962.

Cazzaniga, D., J.C. Basilico, R.J. Gonzalez, R. L.Torrens and D. M. de Greef 2001 *Letters in Applied Microbiology* **32:** 144-147.

Celik, I., H. Oguz, O. Dunet, M. Boy Dak, H.H. Donmez, E.Surand and F. Nizamlioglu 2000 *British Poultry Sceince* **41:** 430-439.

Ceynowa, J. 1986 Mykologische Untersuchungen and luftdient gelagertem Getriede. Dissertation, Christian-Albechts Universitat, Kiel. Germany.pp.21.

Chain, E., H. W. Florey and M. A. Jenning 1942 *Brit. J. Exptl. Pathol.* **23:** 202-205.

Chakrabarti, A. B. 1987 *J. Food Prot.* **44:** 591-592.

Chakrabarti, D. K. and S. Ghosal 1986 *Mycopathologia* **97 :** 69-76.

Chalam, R. V. and H. M. Stahr 1979 *J. AOAC.* **62:** 570-572.

Chandrashekar, S., T.S.Suryanarayana and V.A.Murthy 1990 *Curr. Sci.* **59:**108-109.

Chang, H. L. and J. W. Devires 1984 *J. AOAC.* **67***:* 781-794.

Chang, C. F., J. A. Doerr and P. B. Hamilton 1981 *Poult. Sci.* **60**: 144-149.

Chang, C. F. and P. B. Hamilton 1980 *Appl. Environ. Microbiol.* **39**: 572-575.

Chang-yen, I. and K. Bidasse 1990 *J. AOAC.* **73**: 257-259.

Chanh, T.C., R.C.Kennedy and J.F.Hewetson 1992 *Int. J. Clin. Lab. Res.* **2:** 28-35.

Charlubruns, H. A. 2003 *J. Toxicol. Toxin Rev.* **22:** 153-173.

Chary, M. P., S.Girisham and S. M. Reddy 1986 *Food Sci. and Technol.* **23:** 160-162.

Chary, M. P. and S. M. Reddy 1987 *Nat. Acad. Sci. Letters* **10:** 129-132.

Chary, M. P. and S. M. Reddy 1987a *Indian Bot. Reptr.* **6:** 101-103.

Chatterjee, D. and S. K . Mukherjee 1994 *Lett. Appl. Microbiol.* **18:** 251-253.

Chaturvedi, K. and S. N. Chaturvedi 1992 *J. Ind. Bot. Soc.* **71:** 209-212.

Chao, R., W. Khan and Y. A. Hannun 1992 *J. Biol. Chem.* **267:** 23459-23462.

Chao, T. C., S. M. Maxwell and S. Y. Wong 1991 *J. Pathol.* **164**: 225-233.

Chelkowski, J. 1998 In: Mycotoxins in Agriculture and Food Safety. (eds .K.S. Sinha and D.Bhatnagar). Marcel Dekker, New York, N.Y. 45-64.

Chen, J., C. J.Mirocha, W. Xie, L.Hogge and D.Olson 1992 *Appl. Envir.Microb.* **58:** 3928-3931.

Chenulu, V. V. and Dharamvir 1979 *Bulletin of Grain Technol.* **17**: 148-157.

Chertow, B. S., G. A. Williams, R. Kiani, K. L. Stewart, G.K.Hargis and R. L. Flayter 1974 *Proc. Soc. Exp. Biol. Med.* **147:** 16-19.

Chi, M. S., C. J. Mirocha, H.J.Kurtz, G.Weaver, F.Bates, F.Shimoda and H. R. Burmeister 1977 *Poult. Sci.* **56 :** 103-116.

Chiavaro, E., A. Lepiani, F. Colla, P. Bettoni, E. Pari and E. Spotti 2002 *Food Addi. and Contam.* **19**: 575-581.

Chiba, J., N. Nakano, N. Mororka, S. Nakazawa and Y. Watanaba 1972 *J. Meed. Sci. Biol.* **25:** 291-296.

Chohan, J. S. 1983 In mycootxins in food and feed (eds. K.S.Bilgrami, T.Prasad and K.K.Sinha), *Allied Press, Bhagalpur*, pp. 147-154.

Choudhary, A. K .1992 *Letters in Applied Microbiol.* **14:** 143-147.

Choudhary, A.K. and K. K. Sinha 1993 *J. Stored Products Res.* **29**: 75-80.

Choudhary, H., C. W. Carlson and G. Semenwick 1971 *Poult. Sci.* **50:** 2639-2640.

Chourasia , H. K. 2001 *Indian J. Exper. Biol.* **39:** 493-495.

Chourasia, H. K. and A. K. Roy 1993 *J. Indian Bot. Soc.* **72:** 131-134.

Chourasia, H. K. and R. Shelby 1996 *J. Expt. Biol.* **34**: 103-106.

Christensen, C. M. 1973 *Seed Sci. & Tech.* **1:** 547-562.

Christensen, C.M. 1981 In The fungal community (eds. D. T. Wicklow and G. C. Caroll), Marcel Dekker New York, pp.201-232.

Christenson, C. M. 1981 *Mycologia* **73:** 1056-1084.

Christensen, C.M., C. J. Mirocha and R. A. Meronuck 1988 *Minn. Ext. Serv. Bull. AG-FO* 3538-14.

Christensen, C. M., G. H. Nelson, C. J. Mirocha, F. Bates and C.E. Dorworth 1966 *Appl. Microbiol.* **14 :** 774-777.

Christensen, C. M. and H. H. Kaufmann 1969 Grain storage, The role of fungi in quality loss, University of Minnesota Press, Minneapolis **7 :** 153.

Chu, F. S. 1984 *J. Food Protec.* **47**: 562

Chu, F. S. 1991 In *Hand Book of Applied Mycology* (ed. D. Bhatnagar, F. B. Lillehoj and D. K. Arora) Marcel Dekker, Pub. New York pp. 87-136.

Chu, F. S. 1991 In Immunoassays for Trace Chemical Analysis (eds. M. Vanderlaan, L.H.Stanker, B.E.Watkins and D.W.Roberts), ACS Symposium series 451. American Chemical Society, Washington DC, pp. 140-157.

Chu, F. S. 1995 In Analyzing Food for Nutrition labelling and Hazardous contaminants (eds. I. J. Jeon and W. G. Ikins), Marcel Dekker, New York, pp. 283-332.

Chu, F. S. 1996 In Fumonisims in Foods (eds. L.Jackson, J.W.Deveries and L.B. Bullerman), Plenum Press, New York, pp. 123-134.

Chu, F. S. 1997 *Hum. Biol.* **8**: 511-522.

Chu F. S 2000 In Food borne disease Handbook, Volume 3: Plant toxicant (Eds Y. H. Hui, R. A Smith and D. G. Spoerke), 2nd Edition, Marcel Dekker, New York. pp. 683-713.

Chu, F. S. 2003 Mycotoxins/Toxicology. *Encyclopedia of Food Science and Nutrition,* 2nd edn Elsevier Science Ltd, London. pp. 4096-4108.

Chu, F. S. and G.Y. Li 1994 *Appl. Envir. Microb.* **60**: 847-852.

Chu, F. S. and S. C. Bennett 1981 *J. AOAC.* **64**: 950-954.

Chu, F. S., S.Grossmann, R.D. Wei and C.J. Mirocha 1979 *Appl. Envir. Microb.* **45:**104.

Chu, F. S., X. Huang and C. M. Maragos 1995 *J. AOAC.* **43**: 261-267.

Chulze, S. N., M. L. Ramirez, M. C. Farnochi, M. Pascale, A. Visconti and G. March 1996 *J. Agric. Food Chem.* **44:** 2797-2807.

Chung, C. W., M. W. Truckesess, A.L. Giles Jr. and L.Friedman 1974 *J. AOAC.* **54:** 1121-1127.

Ciacci, Z., J. R. Jones 1999 *Food Chem. Toxicol.* **37**: 703-712.

Ciegler, A. 1969 *Appl. Microbiol.* **18:** 128-129.

Ciegler, A. 1977 In Mycotoxins in Human and Animal Health (eds. J. V Rodricks C. W. Hesseltine and M. A. Mehlman), Pathotox, Park-forest South, III, pp. 609-624.

Ciegler, A. and J. I. Pitt 1969 *Mycopath. Myco. Appl.* **42:** 119-124.

Ciegler, A. and P. E. Peterson 1967 *Bacteriol Proc.* **8:** 15-17.

Ciegler, A. and P. E. Peterson 1968 *Appl. Micrbiol.* **16**: 665-668.

Ciegler, A., R.F.Vesonder and L.K. Jackson 1977 *Appl. Envir. Microb.* **33:** 1004-1006.

Cirillo, T., A. Ritieni, F. Galvano and R. A. Cocchieri 2003 *Food Addi. and Contam.* **20:** 566-571.

Cirillo, T., A .Ritieni, M.Visone and R. A.Cocchieri 2003 *J. Agric. Food Chem.* **51:** 8128-8131.

Cladas, E. D., K. Sadikova, B. L. Ward, A. D. Jones, C.K. Winter and D.G. Gilchrist 1998 *J. Agric. and Food Chem.* **46 :** 4734-4743.

Clarke, J. R., R.R.Marquardt, A. A. Frohlich and R. J.Pitura 1994 *J. Food Protec.* **57:** 991-995.

Clavero, M. R. S., Y.C.Hung, L.R.Beuchat and T. Makayania 1993 *J. Food Protec.* **56:** 130-133.

Cockrum, P. A., C.C.J. Culvinor, J.A. Edgar and A.L.Payne 1979 *J. Nat. Prod.* **42:** 534.

Codifier, L.P., G. E. Mann and F. G. Dollear 1976 *J. Am. Oil Chem. Soc.* **53:** 204-206.

Cohen, H., W. R. Bidlack, Y. Dragan, T. Goldsubrthy, G. Hard, P. C. Howard, R. Riley and K.Voss 2001 *Food Addi. and Contam.* **18:** 187-201.

Cohen, S.M. and M. Lapointe 1984 *J. AOAC.* **67**: 1105-1107.

Cohn, R. H., S. D. Banerjee, E. R. Shelton and M. R. Bernfield 1972 *Proc. Nat. Acad. Sci. USA* **69:** 2865-2869.

Cole, M. and G. N. Rolinson 1972 *Appl. Microbiol.* **24:** 660-662.

Cole, R. J. 1986 *Mycotoxin Res.* **2**: 3-7.

Cole. R. J., D. J. Taylor, E. A. Cole, S. Sukspath, G. G. Mc Dowell and W. L. Bryder 1988 *Proc. Nut. Soc. Australia* **13**: 134. 32.

Cole, R.J. and J.W.Dorner 1985 In: Mycotoxins and Phycotoxins (eds.P.S. Styen and R-Vleggar), Elsevier Science Publishers B. V. Amsterdam pp. 397-411.

Cole, R. J., J. W. Dorner, J. A. Landsen, R. H. Cox, C.Pape, B. Cunfer, S. S. Nicholson and D. M. Bedwell 1977 *J. Agric. Food Chem.* **25:** 1197-1201.

Cole, R. J., J. W. Dorner and P. D. Blackenship 1994 US Patent **5**: 292-662.

Cole, R. J., J. W. Dorner, R. H. Cox and L. W. Raymond 1983 *J. Agric. Food Chem.* **31:** 655-657.

Cole, R. J., J. W. Kriksey and G. Morgan-Joneswells 1974 *Can. J. Microbiol.* **20:** 1159-1162.

Cole, R. J., J. W. Kriksey and G. Morgan-Joneswell 1975 *Toxicol. Appl. Pharmacol.* **31:** 465-468.

Cole, R. J., J. W. Kirksey, H. G. Cutler, B. C. Doupnik and J. C. Peckhan 1973 *Science* **179:** 1324-1326.

Cole, R. J., J. W. Kirksey, H. G. Cutler and E. E. Davis 1974 *J. Agric. Food Chem.* **22:** 517-520.

Cole, R. J., J. W. Kriskey, J. H. Moore, B. R. Balakenship, U. L. Diener and N. B. Davis 1972 *Appl. Microbiol.* **24:** 248- 256.

Cole, R. J. and R. H. Cox 1981 In Handbook of Toxic fungal metabolites, Academic press, New York. pp. 67-93.

Cole, R. J., T. H. Sanders, J. W. Dorner and P. D. Blackenship 1989 In Alfatoxin Contami-nation of Groundnut Process,International Workshop (ed. S. D. Hall) 1987 ICRISAT, Pattancheru, India pp. 277-287.

Colvin, B. M., A. J. Cooley and R. W. Beaver 1993 *J. Vet. Diagn. Invest.* **5:** 232-241.

Colvin, B. M. and L. R. Harrison 1992 *Mycopathologia* **117:** 79-82.

Constable, P. D., G. W. Smith, J. H. Foreman, A. L. Waggoner, R. M. Eppley, M. E. Tumbleson, G. J. Benzon and W. M. Haschek 2000 Proceedings of fumonisins risk assessment workshop, January, 10-12, College Park, Maryland USA.

Cooper, R.H., M.Mc Pherson and J.G. Schofield 1972 *Biochem. J.***127:**143-154.

Cooney, J. M., D.R. Lauren and M.E. di Menna 2001 *J. Agric. Food Chem.* **49**: 522-526.

Coppoeck, R. W., M.S.Mostrom, C.G.Sparling, B. Jacobsen and S.C.Ross 1990 *Vet-Hum Toxicol.* **32**: 246-248.

Corneli, S. and C. M. Maragos 1998 *J. Agric. Food Chem.* **46:** 3162-3165.

Corry, J. E. L. 1978 In Food and Beverage Mycology (ed. L. R. Buchat), Avi. Publishing Co., *Inc. Connecticut,* pp. 45-82.

Cottey, P. J. 1992 US Patent **5:** 171-686.

Cottey, P. J. 1994 *Phytopathology* **84:** 1270-1277.

Cottey, P. J. and D. Bhatnagar 1994 *Appl. and Envir. Microb.* **60:** 2248-2251.

Coulombe, R. A. 1994 In Nonhepatic Disposition and Effects of Aflatoxin B1 (eds D. L. Eaton and J. D. Groopman), Published by Academic, Inc, San Diego. pp. 89-101.

Coulambe, R.A. and A. Rother 1993 *J. Dairy Sci.* **76:** 880-891.

Coulter, J. B. S., R. G. Hendrickse, S. M. Lamplugh, S. B. J. Macfarlane, J. B. Moody, M. I. A. Omer, G. I. Sulliman and T. E. Williams 1986 *Transactons of the Royal Society of Tropical and Medical Hygiene* **80**: 945-951.

Cox, R. H., O.Hernandez, J.W.Dorner, R.J.Cole and D.I.Fennel 1979 *J. Agric. Food Chem.* **27:** 999-1000.

Crawley, W. E., P. H. Mortimer and J. D. Smith 1961 *Newzealand J. Agr. Res.* **4:** P. 532.

Creppy, E. E. 2002 *Toxicology letters* **127:** 19-28.

Creppy, E. E., G. Lorkowski, G. Beec, K . Roschenthaler and G.Dirheimer 1980 *Toxicol. Lett.* **5:** 375-380.

Creppy, E.E., K.Chakor, M.J.Fisher and G.Dirheimer 1990 *Archives of Toxicology* **64**: 279-284.

Croft, W. A., B. B. Jarvis and C. S. Yatwara 1986 *Atmosphere Environment* **20:** 549-552.

Crooks, R. F. and W. O. Mc Clure 1972 *Brain Res.* **45**: 643-646.

Cruickshank, R. H. and J. I. Pitt 1987 *Microbiol. Sci.* **4:** 14-17.

Cruickshank, R. H. and J. I. Pitt 1990 In Modern concepts in *Penicillium* and *Aspergillus* Classification (eds. R. A. Samson and J. I. Pitt), Plenum Press, New York, pp. 259-265.

Crump, M. H., E.B.Smalley, J. N. Henning and R.E.Nichles 1963 *J. Am.Vet. Med Assoc.* **143**: 996.

Crump, M. H., E. B. Smalley, R.E.Nicholse and D.P.Rainey 1967 *Am. J. Vet. Res.* **28**: 865-874.

Culler, J. M., M. Wilson, W.M.Hagles, J.F. Orti and R. J. Cole 1988 *Amer. J. Vet. Res.* **49**: 728-731.

Culler, J. M. and P. M. Newborne 1994 In The Toxicology of Aflatoxins: Human Health, Veterinary and Agricultural Significance (eds D.L. Eaton and J.D. Groopman,), Academic. Press Newyork- pp. 3–26.

Cuero, R. G., E.B. Lillehoj, T.E. Cleveland and A.W. Reine 1988 *Proc. Jpn. Assoc. Mycotoxicol.* (Suppl.) **1:** 194-198

Cutler, H. G., F. G. Grumley, R. H. Coz, J. W. Dorner, J. P. Springer, F. M. Latterell, J. M. E. Theanon and A. E. Rossi 1980 *J. Agri. Food Chem.* **28:** 138-142.

Curtni, V. G., M. Garies, E. Usleber and E. Martlbauer 2001 *Food Addi. and Contam.* **18:** 730-738.

Cuccullu, A. F., K. S. Lee, W. A. Pans and J. B. Standley 1976 *Agric. Food Chem.* **24**: 408-410.

De Vries, J. W., M. W. Trucksess and L. S. Jackson 2002 Mycotoxins and food safety, kluwer academic plenum publisher, New York. pp.277-285.

D'Mello, J. P. F., C. M. Placinta and A.M.C. Macdonald 1999 *Animal Feed Science and Technology* **80**: 183-205.

Dailey, R.E., A. M. Blaschka and E. A. Brouwer 1977 *J. Toxicol. Environ. Health* **3:** 479-490.

Dailey, R. E., R. E. Reese and A. Broumer 1980 *J. Agric. Food Chem.* **28**: 286-291.

Dalcero, A., C. Magndi, C. Hallak, S. M. Chiacchiera, G. Pabcio and C. A. Rosa 2002 *Food Addi.and Contam.* **19**: 1065-1072.

Dalton, J. E. 1952 *Arch. Dermatol Syphilol* **65:** 53-55.

Dalvi, R. R. and C. Mc Gowan 1984 *Poultry Sci.* **63:** 485-491.

Damodaran, C., C. S. Ramdoss and E. R. M. Shanmugasundaram 1973 *Anal. Biochem.* **52** : 482-488.

Danicke, S. 2002 *Worlds Poult. Sci.* **58**: 451-467.

Dange, S. R. S. and V. J. Patel 1984 *Bull. Grain Tech.* **22:** 225-231.

Danks, C., J. Ostoja Starzewska, J. Flint and J.N.Banks 2003 *Aspects of Appl. Bio.* **68**: 21-28.

Dashek, W. V., J.F.Mayfield, G.C. Liewellyn and C. E. O. Realanl 1985 In Biodeterioraton Research (ed. A. Bata), *Bio. Essays.***4**:27-30.

Davan, A. S. and M. R. Choudary 1995 *J AOAC.* **78:** 693-698.

David M. J. 2008 *Food Addi and Contam.* **25**: 219-230.

Davis, N. D., D.K. Dalby, U.L.Diener and G.A. Sansing 1975 *Appl. Microbial.* **29:** 118-120.

Davis, N. D. and U. L. Diener 1987 *Van. Mostraval. Reonhold, New York,* pp.517.

Davis N. D., U. L. Diener and G. Morgan-Jones 1977 *Appl. Envir. Microb.* **34**: 155-157.

Davis, P., A. C. Allison and A. D. Haswell 1973 *Biochem. J.* **134:** 33.

De Jesus, A. E., P. S. Steyn, E.R.Van Heerden, R.Vleggaar and P. L. Wessels 1980 *J. Chem. Soc.* **6:** 289-291.

De Jesus, A.E., P. S. Steyn E. R. Van Heerden and R. Vleggaar 1984 *J. Chem. Soc. Perkin. Trans.* 1003.

De Jesus, A.E., P. S. Steyn, E. R. Van Heerden, R. Vleggaar, P. L. Wessels and W. E Hull 1983a *J. Chem. Soc. Perkin Trans.* 1847-1856.

De Jesus, A.E., P. S. Steyn, E. R. Van Heerden, R. Vleggaar, P. L. Wessels and W. E. Hull 1983b *J. Chem. Soc. Perkin. Trans* 1857-1861.

De Jesus, A. E., P. S. Steyn, E. R. VanHeerdan, R. Vleggaar P. L. Wessels and W. E. Hull 1983c *J. Chem. Soc. Perkin Trans.* 1- 69.

De Jesus, A. E., P. S. Steyn, E. R. VanHeerdan, R. Vleggaar P. L. Wessels 1980 *J.Chemical Soc.* **6:** 289-291

De Nijs, M., H.P.Van Egmond, M.Nauta, F.M.Rombouts and S.H.W. Notermans 1998 *J. Food Protection* **61**: 879-884.

De Rosnay, G.D.C., C.Martin-Dupont and R.Jensen 1952 *J. Med. Board.* **129:** 189-199.

De Saeger, G.D.C. and C.Van Peteghem 1999 *Appl. Envir. Microb.* **62:** 1880-1884.

De-Saeger,G.D.C. and C. Van Peteghem 1999 *J. Food Protection* **62:** 65-69.

De Vries, J. W., M. W. Trucksess and L. S. Jackson 2002 Mycotoxins and Food safety: advances in experimental medicine and Biology in vol 504. (ed.C.Whipple), Kluwer academic / Planum Publishers, New York pp.95-105.

Delongchamp, R.R. and J.F. Young 2001 *Food Addi. and Contam.* **18:**187-201.

Demain, A. L., N. A. Hunt, V. Malik, B. Kobbe, H. Hawkins, K. Matsuo and G. N. Wogar 1976 *Appl. Envir. Microb.* **31:** 138-140.

Deng, W. L., X. M. Tang, L. Chen and M. H. Li 1984 *J. Vet. Sci. and Tech.* **9:** 22-25.

Desai, M. R. and S. K. Ghosh 2003 *Cellular and Molecular Biology* **49:** 529-535.

Desaiah, D., A.W. Hayes and I. K. Ho 1977 *Toxicol. Appl. Pharmacol.* **39:**71-79.

Desjardins, A. E., H. K. Manandhar, R. D. Plattner, G. G. Manandhar, S. M. Polling and C. M. Maragos 2000 *Appl. Envir. Microb.* **66:** 1020-1025.

Desjardins, A.E., R. D. Plattner and F. V. Worth 1986 *Appl. Envir. Microb.* **51:** 493-497.

Desylos, C. M. and D. B.Rodrigez Amaya 1999 *Food Addi .and.Contam.* **16**: 71-74.

Detroy, R. W. and C. W. Hesseltine 1969 *Can. J. Microb.* **15:** 495.

Detroy, R. W. and P. E. Still 1979 *Microbiol.* **92:** 167 -174.

Dett, M.P.K., S.J. Haswell, O.G. Roch,R. D. Coker, V. F. P. Medlock and K. Tomlines 1990 *Analyst* **115**: 1435.

Devaraj, H. and N. Devaraj 1987 *Indian J. Expt. Biol.* **25:** 637-638.

Devaraj, R. H., K. R. Shanmugasundarm and E. B. Shanmugasundaram 1983 *J. Ind. Bot. Soc.* **64**: 65-41.

Devaraj, H., R. E. Suseela and N. Devaraj 1986 *Curr. Sci.* **55:** 998-999.

Dhuley, J.N. 1997 *J. Ethnopharmacol.* **58:**15-20.

Dickens, F. and H. E. H. Jones 1961 *Br. J. Cancer Research* **15:** 85-100.

Dickens, J. W. and R. E. Welty 1968 Council for Agric. Sci. and Tech. Rep. No. 80.

Dietrich,R., E. Schneider, E.Usleber and E.Martlbauer 1995 *Natural Toxins* **3**: 288-293.

Diekman, M. A. and M. L. Green 1992 *J. Anim. Sci.* **70**: 1615-1627.

Diener, U. L. and N. D. Davis 1967 *J. Am. Oil Chem. Sec.* **44**: 259-263.

Diener, U. L. and N. D. Davis 1972 *Highlights Agri. Res.* **19**: 241

Dimic, G. and M. Skrinjar 1995 *Technoligija Mesa.* **36:** 302-305.

Dimroth, P., H. Walter and F. Hynen 1970 *European J. Biochem.* **13:** 98-110.

Deng, W.L., X.M.Tang, L.Chen and M.H.Li 1984 *J.Vet Sci. and Tech.* **9:**22-25.

Dingley, J. M. 1962 *Newzealand J. Agric. Res.* **5:** 49-61.

Dirheimer, G. 1996 *Food Addi.and Contamin.* **13:** 45-48.

Dirherimer, G. 1998 *Revenc. Med. Ver.* **149**: 605-616.

Dix, R. D. and R. J. Courtrey 1976 *Virology* **70**: 127-135.

Doherty, W. P. and T. C. Campbell 1973 *Chem. Biol. Interact.* **7:** 63-77.

Doko, M. B., S.Rapior, A.Visconti and J.E.Schjoth 1995 *J.Agri. and Food Chem.* **43**: 429-434.

Dollear, F. G., G. E., Mann, L. P. Codifer, H. K.Gurdner, S.P.Koltun and H. L. E. Vix. 1968 *J. Amer. Oil Chem. Soc.* **45:** 862-865.

Dombrink-Kurtzman, M. A., T.J.Dvorak, M.E. Barron and L.W. Rooney 2000 *J. Agric. Food Chem.* **48 :** 5781-5786

Dominguez, L. J., L.Blanco, E.Goemez-Lucia, E.F.Rodrignez and G.Saurez 1987 *J. AOAC.* **70**: 470-472.

Dorner, J. W. 2008 *Food Addi. and Contam.* **25**: 203-208.

Dorner, J. W. and R. J. Cole 1988 *J. AOAC.* **71**: 43-47.

Dorner, J. W. and R. J. Cole 1993 *J. AOAC.* **76**: 983-987.

Dorner, J. W., R. J.Cole, D. J. Erlington, S. Suksupath, G. H. Mc Dowell and W. I. Bryden 1994 *J. Agric. Food Chem.* **42:** 1516-1518.

Dorner, J.W., R. J. Cole and D. T. Wicklow 1996 *J. Food Protect.* **67**: 650-656.

Dorner, J.W., R. J.Cole, L.G. Lomax, H.S.Gorres and U.L.Diner 1983 *Appl. Environ. Microb.* **46**: 693-703.

Dorner, J.W., R. J. Cole and R. F. Vesonder 1988 *J. Antibiotics* **41**: 1868-1872.

Dovido, K. L., M.W.Trucksess, J.W. Devries and G. Bean 2007 *Food Addi. and Contam.* **24**: 735-743.

Dowd, P. F., F.C.Peng, J.W. Chem and K. H. Ling 1992 *Entom. Experm. Appl.* **65**: 57-64.

Dowd, P. F., R. J. Cole and R. F. Vesonder 1988 *J. Antibiotics* **41**: 1868-1872.

Doyle, M. P. and E. H. Marth 1979 Eur. *J. Appl. Microb. Biotech.* **7**: 281-307.

Doyle, M. P., R. S. Applebaum, R. E. Brackett and E. H. Marth 1982 *J. Food Prot.* **45**: 964-971.

Dragacci, S., E. Gleizes, J. M. Fremy and A. A. G. Candlish 1995 *Food Addi.and Contam.* **12**: 59-65.

Dube,S. 1989 *Canadian J. Botany* **67**: 2085-2087.

Dube, S.1991 *Indian Phytopath.* **44**: 101-105.

Dugyala, R. P., R. P. Sharma, M. Tsundo and R. T. Riley 1998 *J. Pharmacol. Exp. Ther.* **285**:317-324.

Duncan, K., S. K. Ruger, N. Zabe, B. Kolin and R. Prioli 1998 *J. Chromatogr.* **815**: 41-47.

Durackova, Z., V.Betina, B.Hornikova and P.Nemec 1977 *Bacterial, Parasitunked, Infektionskr Hyg.* **132**: 294-299.

Durackova, Z., V. Betina and P. Nemec 1976 *J. Chromatogr.* **116**: 141-154.

Duran, R. M., J. W. Cary and A. M. Calvo 2007 *Appl. Microb. and Biotech.* **73**: 1158-1168.

Dustin, P. Jr. 1963 *Pharmacol. Rev.* **15** : 449-480.

Dutton, M. F. 1996 *Pharmacal. Ther.* **70**: 137-161.

Dutton, M. F. and J. G. Heathcote 1968 *Chem. Ind. (London)* p. 418-421.

Duvick, J. 2001 *Environ. Health Perspect.* **109:** (suppl. 2) 337-342.

Dwarakanath, C. T., E. T. Rayner, G. E. Mann and F. G. Dollear 1968 *J. Am. Oil chem. Soc.* **45**: 93-95.

Dwivedi, P. and R. B. Burns 1984 *Res. Vet. Sci.* **36:** 92-103.

Eaton, D. L. and J.D. Groopman 1994 The Toxicology of Aflatoxins, *Academic Press, San Diego.* Pp.233-258.

Edrington, T. S., L.F.Kubena, R. B. Harvey and G. E. Rotting laus 1997 *Paultry Sci.* **79**: 1205-1211.

Efimento, O. M. and P. A. Yakimov 1960 *Chem. Abstr.* **55**:2147.

El-Banna, A. A., J. I. Pitt and L. Leistner 1987 *Appl. Microbiol.* **10:** 42-46.

El-Banna, A. A. and L. Leistner 1988 *Int. J. Food Microbiol.* **7:** 9-17.

Elghor, A. A. 1994 *Proc. Zool. Soc.* A. R. Egypt. **25**: 183-193.

Elimsukon, T., R.C.Shounk, G.N. Wognand and P.M. Bewborne 1975 *Toxicol. Appl Pharmacol.* **32:** 135.

Elling, F. and B. Hald 1975 *Acta Path. Microbiol. Scan Scet. A.* **83**: 739.

Ellis, W. O., J. P. Smith, B. K. Simpson and J. H. Oldham 1991 *Crit. Rev. Food Sci. Nutr.* **30:** 403-439.

Ellison, R. A. and F. N. Kotsonis 1974 *Appl. Micrbiol.* **27:** 423.

El-Mezami H. A.Chrevatidis, S.Auriola, S.Salminen and H. Mykkänen 2002 *Food Addi. and Contam.* **19**: 680-686.

El-Mezami, H., H. Mykkanen, P. Kankanpaa and S. Salminen, and J. Ahokas 1999 *J. Food Protec.* **63**: 549-552.

Endo, A. and M. Kuroda 1976 *J. Antibiot.* **29:** 841-843.

Endo, H. 1983 *Proc. Jpn. Assoc. Mycotoxicol.* **18**: 21-23.

Engel, G., E.H. Reimerdes, A. Lembke, H. Rlostermayer and J. Behnert 1975 *Milchariss-enchaft* **30:** 129.

Engel, G. and M. Tenber 1984 In Mycotoxin (ed. V. Betina) *Elsevier Publications Amsterdam* pp-291.

Entwisle, A. C., A. C. Williams, P. J.Mann, J. Russell, P. T. Slack and J. Gilbert 2001 *J. AOAC.* **84**: 444-450.

Entwisle, A. C., A. C. Williams, P. J. Mann, P. T. Slack and J. Gilbert 2000 *J. AOAC.* **83**: 1377-1383.

Eppley, R. M. 1974 *J. AOAC.* **57:** 618-620.

Epstein, S. M., B. Bartus and E. Faber 1969 *Cancer Research* **29:** 1045-1050.

Eriksen, L.G.S. and J. Alexander 1998 Nordic Council of Ministery, Jema Nord. **502:** 7-27.

Escoula, L. 1975 *Ann. Rech. Vet.* **6:** 303-310.

Estensen, R. D. 1971 *Proc. Soc. Exp. Biol. Med.* **136:** 1256.

Etcheverry M., C. Magnolic, A. Dalcero, S. Chulze and S. Lecumberry 1998 *Mycopathalogia* **142**: 37-42.

Etienne, M. and J. Y. Dourmad 1994 *A Rreview Livest. Prod. Sci.* **40:** 99-113.

Ewaidah, E. H. 1988 *Ara. Gulf Journal of Scientific Research* **6:** 1-7.

Ewart, A. J. 1933 *Ae. J. Bot.* **47**: 913-915.

Ewaskiewiok, J.I., T.M.Devlin and J.J. Chi 1991 *Biochemical and Biophysical Research Communications* **179:** 1095-1100.

FAO 1997 World Wide Regulations for Mycotoxins. A Compendium, FAO Food and Nutrition paper **64**, Rome, FAO.

FAO 1998 Animal Feeding and Food safety, Report of an FAO Expert consultation, Rome 10-14 March 1997, FAO. Food & Nutriton Paper 69, FAO. Rome.

Farber, F. E. and R. Eberle 1976 *Exp. Cell Res.* **103:** 15-22.

Fasiha Rehana and S. C. Basappa 1990 *J. Food Sci. and Technol. India* **27:** 327.

Fayos, J., D. Lokensgard, J. Clardy, R. J. Cole and J. Kriskey 1974 *J. Amer. Chem. Soc.* **96**: 6785-6787.

Fazekas, A., K.B. Tar and M. Zomborszky-Kovacs 2002 *Acta Vet. Hung.* **50:** 177-188.

Fazekas, B., M.Kis and E.T.Hajdu 1996 *Acta Vet. Hung.* **44**:25-37.

FDA, Food and Drug Administration 1979 Conference on Mycotoxins in Animal Feeds and Grains Related to Animal Health, June, 8. 1979 Rockville, Maryland.

Fehr, T. H. and W. Acklin 1966 *Helv. Chem. Acta* **49**: 1907-1910.

Fenell, A. J. 1966 *Trop. Sci.* **8:** 61-70.

Fenerstein,G., R.H.Zorenzana and V. R. Barley 1989 In Trichotehcene mycotoxicosis. Pathophysiological effect, Vol-I (ed. V.R.Barley), CRC Publication,BocoRaton.Florida., USA, pp. 2-26.

Ferreira, N. P. 1968 *J. Microbiol. Sero.* **34**: 433-440.

Fex, T. 1981 *Tetrahedron Lett.* **22:** 2703-2706.

Field, D. J., J. M. Bowen and R. J. Cole 1978 *Toxicol. Appl. Pharmacol.* **46:** 529-541.

Figueira, A. C., K. D. A. Taylor, P. J. Barlow 1990 *Food and Agric. Immun.* **2:** 125-131.

Figueira, A.C., K.D.A.Taylor, P.J.Barlow and M.R. A. Morgan 1991 *Food Agric. Immun.* **3**: 13.

Filali, A., A. M. Louammi, J. Betbeder, R. Baudrimant, A. Soulaymani Benayada and E. E. Creeppy 2001 *Food Adit.and Contam.* **18:** 565-568.

Fink, Gremmels, J. 1990 *J. Archives of Toxicology* **64:** 72-76.

Fink-Gremmels, J. 1999 *Vet. Quart.* **21**: 115-120.

Fink Gremels, J.2008 *Food Addi. and Contam.* **25**: 172-180.

Fitzpatrick, D. W., C. A. Picken, L. C. Nurphy and M. M. Buhr 1989 *Comp. Biohem. Physiol. Comp. Phenomeol. Toxicol.* **94:** 691-694.

Flannigan, B., J.E. Smith and R.S. Henderson 1991 *Mycotoxins and Animal Foods* **147:** 57-80.

Fletchers, L.R. and I. C. Harhey 1981 *New Zealand Vet. J.* **29:** 185-186.

Fliege, R. and M. Mezler 1999 *Chem. Biol. Interact.* **123:** 85-103.

Flones, H.E., C.H.Proticio and M.W.Singh in Recent advances in phytochemistry nitrogen metabolism (eds. J. E.Pollution, J. T. Romeo and E. E. Conn), Plenum Press, New York **23:** 329-393.

Florey, H. W., M. A. Jennings and F. J. Philpot 1944 *Nature* **153:** 139.

Foreman, J. H., P.D.Constable, A.L.Waggnor, R.M.Eppley, G.W.Smoth, M. E. Tumbleson and W. M. Haschek 2001 *Food Addit. and Contam.* **18:** 182-210.

Forgacs, T. and W. T. Carll 1962 *Adv. Vet. Sci.* **7:** 273-293.

Forrester, P. I. and G. M. Gaucher 1972 *Biochem.* **11:** 1102-1107.

Fortuskny, V.A., A.M.Govrou, I. Z. Tebyhenyko, A. S. Biocheukno and E. T. Kalitunko 1959 *Veterninaria* **36:** 67.

Foster, P. M., P.F. Shater and D. S. Patterson 1975 *Biochem. Soc. Trans.* **3 :** 875.

Franceschi, S., E. Bidoli, A. E. Buron and C. La Vecchina 1990 *J. Nat. Cancer. Inst.* **82**: 1407-1411.

Francis, E., R. Rahman, S. Safe and A. Taylor 1972 *J. Chem. Soc. Perkin Trans.* **1** : 470-472.

Francis, O. J., G. P. Kirschenhanler, G. M. Ware, A. S. Carman and S. S. Kaun 1988 *J. AOAC.* **77**: 637.

Frank, M. 1999 Third Joint FAO/WHO/UNBP *Intrne. Confe. On Mycotoxins* Tunis. pp. 1-11.

Frank, H. K., R. Orth and A. Z. Figg 1976 *Lebenson, Unters, Forsch* **163:** 111.

Frank, H. K., R. Orth, S. Ivankavic, M. Kuhlman and D. Schamath 1977 *Experientia* **33:** 515-516.

Frayssient, C. 1980 *Gazette Medicale de France* **95:** 82-83.

Freeman, G.V. and R. I. Morrison 1949 *Bio. Chem. J.* **44**: 1.

Freerksen, E. and R. Bonicke 1951 *Z. Hyg.* **132**: 274

Frisvad, J. C. 1983 *J. Appl. Bacteriol.* **54:** 409-416.

Frisvad, J. C. 1985 In Advances in *Penicillium* and *Aspergillus* Systematics (eds. R. A. Samson and J. I. Pitt), Plenum Pres, New York, pp. 437.

Frisvad, J. C. 1986 In Modern Methods in the Analysis and Structural Elucidation of Mycotoxins, (ed. R. J. Cole), Academic Press, Orlando, FL. pp. 415-457.

Frisvad, J. C. and O. Filtenborg 1995 In Introduction to food-borne fungi (eds. R. A. Samson, Van Reenen-Hackstra, J. C. Frisvad and O. Filtenborg), 4th Edit. CBS, Baarn, E. S pp. 251.

Frisvad, J. C. and R. A. Samson 1990 In Modern concepts in *Penicillium* and *Aspergillus* Classification (eds. R. A. Samson, and J. I. Pitt),Plenum Press, New York, pp. 201.

Frisvad, J. C. and R. A. Samson 1991 In *Cereal Grain Mycotoxins, Fungi* and *Quality in Drying* and storage (ed. J. Chelkowski),Elsevier, Amsterdam, Pp. 441.

Frisvad, J. C. and U. Thran 1987 *J. Chromatogra.* **404:** 195-214.

Frohilich, A. A., R. R. Marquardt and A. Bernatsky 1988 *J. AOAC.* **71**: 949.

Fung, F. and R. F. Clark 2004 *J. Toxicology Clinical toxicology* **42**: 217-234.

Furlong, E. B. and L. M. V.Soare 1995 *J. AOAC. Int.* **78**: 386.

Gabal, M. A., Y. L. Awad, M. B. Marcos, A. M. Barkat and G. Malik 1986 *Vet. Hum. Toxicom.* **28** : 207-712.

Gafoor, I. and F.Trail 2006 *Appl. Envir. Microb.* **72**: 1793-1799.

Gajendra Prasad, R.Niranjan Kumar and K. K. Sinha 1993 *J. Ind. Bot. Soc.* **72**: 305-306.

Gale, E.F., E.Cundliffe, P.E. Reynolds, M. Richmond, M.J. Waring *(eds)* The molecular basis of antibiotic action. *John* Wiley *and Sons, London,* New York, *Sydney, Toronto,* pp 402-547

Gallagher, R. T. and B. J. Wilson 1978 *Mycopathologia* **66** : 183-185.

Gallagher, E. P., L. C. Wienkers, P.L.Stapleton, K. L. Kunze and D. L. Eaton 1994 *Cancer Res.* **54**:101-108.

Gallagher, R. T., G. C. M. Latch and R. K. Krogh 1980 *Appl. Envi. Microb.* **39**: 272-273.

Gallagher, R. T., J. Finer, J. Clardy, A. Leutwiler, F. Weibel, W. Acklin and D. Arigoni 1980a *Tetrahedron Lett.* **21**: 235-245.

Gallagher, R.T., J. L. Richard, H. M. Stahs and R. J. Cole 1978 *Mycopathoogia* **66**: 31-36.

Galtin, R. P. and R. F. Widerman 1987 *Poultry Sci.* **66**: 1316-1328.

Galvano, F., A. Campisi, A. Russo, G. Galvano, M. Palumbo, M. Renis, M. L. Barcellona, J. R. Perezpolo and A. Vanella 2002 *Neuro Chemical Research* **27**: 345-351.

Galvano, F., A. Piva, A. Ritieni and G. Galvano 2001 *J. Food Prot.* **64** : 120-131.

Galvano, F., A. Russo, V. Cardile, G. Galvano, A. Vanella and M. Renis 2002 *Food and Chem. Toxicology* **40** : 25-31.

Galvano F., V.Galfaro, A. Ritieni and M.Bognanno, A.De Angelis and G.Galvano 2001 *Food Addi.and Contam.* **18**: 644-646.

Gambeh, S. V., R. L. Brown, C. Grimm and T. E. Cleveland 2001 *J. Agri and Food Chemistry* **49**: 4635-4641.

Gambrone, J. J., N. D. Davis and L. Diener 1978 *Poult. Sci.* **57**: 1554-1558.

Gambrone, J. J., U. L. Diener, N.D. Davis, V.S. Panagala and F.J. Hoerr 1985 *Poultry Sci.* **64**: 1678-1684.

Gams, W. and E. Muller Veth 1980 *J. Plant Path.* **86**: 45-53.

Gams W., M. Christensen, A. H. S. Onions, J. I. Pitt and R.A. Samson 1985 In Advances in *Penicillium* and *Aspergillus* Systematics", (eds. R.A. Samson and J.l. Pitt) Plenum Press, New York pp. 55-62.

Gangopadhyay, S. and N. K. Chakrabarti 1981 *Curr. Sci.* **50**: 272-275.

Garies, M. 1993 In Slough, Central Science Laboratory (ed. K.A Schdamene) London pp.7-15.

Gareis, M., R. Bataglia, T.Hatzold and R.Kores 1996 *Food Addi.and Contam.***13**:35-37.

Garner, R. C. 1989 In Mycotoxins and Phycotoxins (eds. K. Hashimato, Y. Ueno and S. Natoxi) Elsevier, Amsterdam. pp. 29-35.

Garner, R. C., M. M. Whattmam, P. J. L. Taylor and M. W. Stow 1993 *J. Chromatogr.* **648**: 485-490.

Garthwaite, I., J. Sprosen, L.Briggs, R.Collin and N.R. Jowers 1994 *Food Agric. Immun.* **6**: 123-129.

Garza, S., R.Caneia, I.Vinas and V.Sanchia 1993 *Zentralbatt-Fuer-Microbology* **148**: 345-350.

Gaumy, J. L., J.D.Bailly, V.Burgatet and P.Gume 2001 *Review Med. Vet.* **152**: 219-234.

Gaur, V. K. and B. S. Siradhana 1989 *Indian Phytopath.* **42**:589-590

Gause, E. M., M.A. Buck, Check Gayse and M. G. Donglan 1981 *J. Bio. Chem.* **256**: 557-559.

Gayatri Murjani 2000 In shaping the future, Global dialogue Expo. 2002, Aug. 29th to 31st, 2002 Supported by Volkswagen foundation, Hannover, Germany.

Gedek, B. and J.Bauer 1983 In Trichothecenes Chemical, Biological and Toxicological Aspects (ed. Y. Ueno), Elsevier Publication pp. 301-356.

Geiger, W. B., J. E. Conn and S. A. Waksman 1944 *J. Bacteriol.* **48**: 531-536.

Gelderblom, W. C. A., C.M. Smuts, S.D.Abel, L. Shyman, L. Vander West-Thuzen, W.W. Huber and S.Swanevelder 1997 *Food Chem Toxicol.* **35**: 361-369.

Gelderblom, W.C.A., C.M.Smuts, S.D.Abel, S.D.Shyman, M.E.Cawood, L. Vander Westhuizen and S.Swanevelder 1996 *Food Chem. Toxicol.* **34** : 361-369.

Gelderblom, W. C.A., K. Jaskiewicz, W.F.O. Marasas, P.G.Thiel, R. M. Horak, R. Vleggaar and N. P. J. Kriek 1998 *Appl. Envir. Microb.* **54**: 1806-1811.

Gelderblom, W. C. A., M.E.Cawood, D.Shyman, R. Vleggaar and W.F.O. Marasas 1992 *Food Chem. Toxico.* **31** : 407-414.

Gelderblom,W.C.A.,M.E.Cawood, D.Shyman and W.F.O.Marasas 1994 *Carcinogenesis* **15** : 209-214.

Gelderblom, W. C. A., P.G.Thiel, K. Jaskiewicz and W.F.O.Marasas 1986 *Carcinogenesis* **7**: 1899-1901.

Gelderblom, W.C.A., P. G. Thiel, W. F. O. Marasass and K.J.Vander Merwe 1984 *J. Agric. Food Chem.* **32** : 1064-1067.

Gelderblom, W. C. A., W. F. O. Marasas, P. S. Steyn, P. G. Thiel, K. J.Vandermerwe, P. H. Vanrooyam, R. Vleggaar and P. L. Wessels 1988 *J. Chem. Soc. Chem. Comm.* Pp.122-124.

Gelderblom, W.C.A., W.F.O.Marasas, R. Vleggaar, P.G.Thiel and M.E. Cawood 1991 *Mycophathology* **117**: 11-16.

Gendloff, E. H., J. J.Pestka, D. E. Dixon and L. D. Hart 1987 *Phytopathology* **77** : 57-59.

Gerlach, W. 1970 In: A Modern System of *Fusarium* Taxonomy (ed.E.A. Jamalainen), *Welsinki Ann. Acad. Sci. Jenn. . A, IV Biologica* **168**: 37-49.

Gerlach, W. and D. Ershad 1970 *Nova Hedwigia* pp. 725-784.

Gerlach, W. and H. Nirenberg 1982 *Mitt. Biol. Bundesanst*, Land Forster, *Berlin-Dahlem* **209**: 1-406.

Ghambir, S. P. and D. N. Kharinar 1986 *Geiobios New Rep.* **5**: 190-192.

Ghedria-Chekir, L., K. Maaroufi, A. Zakhama, F. Ellouz, S. Dhouib, E. E. Creppy and H. Bacha 1998 *Chem. Biol. Interact.* **113**: 15-25.

Ghewande, M. P. and G. Nagaraj 1987 *Mycotoxins Res.* **3**: 19-24.

Ghosal, S., D. K. Chakrabarti, A. K. Srivastava and R. S. Srivasatva 1982 *J. Agric. Food Chem.* **30**: 106-109.

Ghosal, S., D.K.Chakrabarti and K.C.Chaudary 1977 *Experientia* **33**:574-575.

Ghosh, J. and P.Haggblom 1985 *J. Food Microbiol.* **2**: 323-330.

Ghosh, R. C., H.V.S. Chauhan and G.J.Jha 1991 *Veternary Immunology and Immunopathology* **28**: 165-172.

Giber, W., K.Wenger and G. P. Wildner 1971 *Arch. Geschwulstoforsch* **38**: 1-6.

Gifford, L. A., C. Wright and J. Gilbert 1990 *Food Addi.and Contam.* **7**: 829-836.

Glinsukon, T., R. C. Shank and G. N. Wogan 1975 *Toxicol. Appl. Pharmacol.* **32**: 158-167.

Gillepsic, A. M. and G. M. Schenk 1977 *Analytical Letters* **10**: 161-172.

Gimeno, A. 1979 *J. AOAC.* **62**: 579-585.

Gimeno, A. 1983 *J. AOAC.* **66**: 565-569.

Gimeno, A. and J. A. Quintanilla 1981 International Symposium and Workshop on Mycotoxins, Cairo, Dokki, Egypt (Abstracts) pp.50.

Gimeno, A. and J. A Quintailla 1983 *Proc. Int. Symp. Mycotoxins.* pp 387-392.

Giovani, C. 1983 Trichothecene Chemical, Biological and Toxicological Aspects (ed. Y. Ueno), Elsevier Publication pp. 254.

Giridhar, P., G. Laxma Reddy, S.M.Reddy and M.Ather 2000 *Indian J. Polultry Sci.* **35**:45-49.

Giridhar, P. and S.M. Reddy 1996 *J. Indian Bot. Soc.* **75**: 153-154.

Giridhar, P. and S.M. Reddy 1996a. In: Frontiers in Plant Science (ed. Irfan and A. Khan), Ukajaz Publishers, Hyderabad. **101**: 607-612.

Giridhar, P. and S.M. Reddy 1997 *Indian Phytopath.* **50**: 1-5.

Giridhar, P. and S.M. Reddy 1997a *J. Indian Bot. Soc.* **76**: 161-164.

Giridhar, P. and S.M. Reddy 1997b *Nat. Acad. Sci. Letters* **20**: 62-63.

Giridhar, P. and S. M. Reddy 1997c *Seed Res.* **25**: 92-93.

Giridhar, P. and S. M. Reddy 1997d *Indian Phytopath.* **50**: 33-36.

Giridhar, P. and S. M. Reddy 1997e *Ad. Plant Sci.* **10**: 125-128.

Giridhar, P. and S. M. Reddy1997f *Nat. Acad. Sci. Letters* **20**: 62-63.

Giridhar, P. and S. M Reddy 1997g *Indian Phytopath.* **50**:1-5.

Giridhar,P., S.M.Reddy and M.V.Rajam 1997 *Indian J.Plant Sci.***32**:165-172.

Giridhar, P. and S. M. Reddy 1998 *Nat. Acd. Sci. Letters* **21**: 5-8.

Giridhar, P. and S. M. Reddy 1999 *Adv.Plant Sciences* **12**: 85-88.

Giridhar,P.and S.M.Reddy 1999a *J.Mycology and Plant Pathol.* **28**:335-337.

Giridhar, P. and S. M. Reddy 1999b *Adv. Plant Sciences* **12**: 249-254.

Giridhar, P. and S. M. Reddy 1999c J. *Mycol. and Plant Pathol.* **28**: 347-349.

Giridhar, P. and S. M. Reddy 2001 *Adv. Plant Sciences* **14**: 291-294.

Giridhar, P. and S.M. Reddy 2001a *J. Food Sci & Technol. India* **38**: 397-399.

Giridihar, P. and S.M.Reddy 2001b *J. Food Sci & Technol. India* **38**: 419-421.

Giridhar, P. and S.M. Reddy 2002 *Adv. Plant Sciences* **15**: 331-334.

Girish, G. K. and R. K. Goyal 1986 *Bull. Grain Tech.* 24: 157-77.

Girisham, S. and S. M. Reddy 1986 *Geobios New Reptr.* **5**: 121-125.

Girisham, S. and S. M. Reddy 1987 *J. Indian Bot. Soc.***66**. 66.

Girisham, S. and S.M. Reddy 1988 *J. Indian Bot. Soc.* **67**: 59-60.

Girisham, S. and S. M. Reddy 1989 *J. Indian Bot. Soc.* **68**: 42-43.

Girisham, S. and S. M. Reddy 1989a *Geobios* **16**: 235-236.

Girisham, S., V. Krishna Reddy and S.M.Reddy 1989 *J. Indian. Bot. Soc.* **68**: 35-37.

Glahn, R. P., R.F.Widerman, J. R.J.W. Evanelist and W. E. Huff 1988 *Poultry Sci.* **67**: 1034-1042.

Glatin, R. P. and R. F. Widerman 1987 *Poultry Sci.* **66**: 1316-1328.

Glavits, R. and A. Vany 1988 *Acta Vet. Hung.* **36** : 37-41.

Glinsukon, T., R.C. Shank, G.N.Wogan and P.M.Newberne 1976 *Toxicol.Appl. Pharmacol.* **32**: 135-146.

Glinsukon, T., S.S.Yuan, R.Wightman, Y.Kitaura, G.Buchi, R.C.Shank, G.N.Wogan and C. M. Christenson 1974 *Plant Food and Man* **1:** 113-119.

Godtfredsen, W.O., J.F.Grovo and Ch.Tamm 1967 *Helv.Chem. Acta* **50**: 1666-1668.

Goldblatt, L.A. 1971 *J. Am. Oil Chem. Soc.* **48:** 605-610.

Goldblatt L. A. and F.G. Dollear 1977 In Mycotoxins in Human and Animal Health (eds. J. Rodricks, C.W.Hesseltine and M. A. Mehlamam), Park Forest South. IL, Pathiotox Publishers Inc. pp. 139-150.

Golinski, P., K.. Hult, J. Grabarkiewicz-Szczesna, J. Chelkowski and K. Szebiotko 1985 *Appl. and Envir. Microb.* **49**: 1014-1015.

Gonzalez,A.M.,M.Presa, M.G.Latorre and M.C.Lura 2007 *Revista Iberoamericana de Micologia* **24**: 59-61.

Gopal, T. 1985 *Poultry Adviser* **18**: 41-44.

Gordon, S. and Z. Werb 1976 *Proc. Nat. Acad. Sci.* USA **73**: 872.

Gordon, W. L. 1960 *Can. J. Bot.* **38**: 643-658.

Gorelick, W. J. 1990 *Risk Anal.* **10**: 539-559.

Gorst-Allman, C. P. and P. S. Steyn 1979 *J. Chromatogra.* **175**: 325-331.

Goto, T., D. T. Wicklow and Y. Ito 1996 *Appl. Envi. Microb.* **62:** 4032-4038.

Gould-Somero, M., L. Holland and M. Paul 1977 *Dev. Biol.* **58**: 11-22.

Gourama, H. and L. B. Bullerman 1995 *J. Food Prot.* **58**: 1395-1404.

Goyal, R. K. 1991 In: Mycotoxin Prevention and control in Food Grains (eds. R. L. Sample A. S Frio, P.A.Hicks and J. V.Lozare), Marcel Dekker Inc. NewYork. pp. 275-281.

Granholm, N.H. and G. M. Brenner 1976 *Exp. Cell Res.* **101**: 143-153.

Greene, W.C., C.M.Parker and C.W.Parker 1976 *Exp.Cell Res.* **103**: 109-117.

Greenhalgh, R., B.A. Fielder, L. A. Morrison, J. P. Charland, B. A. Blackwell, J. D. Miller, M. E. Savaral and J. W. Apsimon 1989 In Mycotoxins and Phycotoxins (eds. S. Natori, K. Hashimoto and Y. Ueno), Elsevier Publication, Amsterdan Vol. 10. pp.223-232.

Greenway, J. A. and R. Puls 1976 *Can. J. Comp. Med.* **40**: 12-15.

Greenway, J. C., T.H.Shephard and J.Kuc 1977 *Nature New Biol.* **236**: 246-247.

Greenway, J. C., T. H. Shephard and J.Kuc 1977a *Proc. Suc. Exp. Biol. Med.* **155**: 239-242.

Griffiths, J. B. and S. H. Done 1991 *Veterinary Res.* **129**: 113-117.

Grigo-stockli, D. M., C. W. Peyoe, R. F. Satumbaga and J. R. Pedersen 1996 *Cereal Chem.* **73**: 388-391.

Groot Wassink, J. W. D. and G. M. Gaucher 1980 *Bacteriol.* **141**: 443-455.

Groopman, J. D., J. A. Haster, L. J. Trudel, A. Pikul, P. R. Donahue and G. N. Wogun 1992 *Cancer Res.* **52** : 267-274.

Grosse, Y., L. Chekir-Ghedria, A. Hue, S. Dbrecht-pflumino, G. Dirhemer, H. Bacha and A. Pthl-Leszkowics 1997 *Cancer Letter* **114**: 225-229.

Grove, J. F. and P. H. Mortimer 1969 *Biochem. Pharmacol.* **18**: 1473-1478.

Groves, F. D., L. Zhang, Y. S. Chang, P. Frank Ross, C. Howard, W. P. Norred, W. C. You and J. E. Franmeni 1999 *J. AOAC.* **82**: 657-661.

Gruber-Schley, S. and A. Thalmann 1988 *Landwirt schaftliche Forschung Sonder heft* **41**: 11-29.

Gu, T., H. A. Koshinsky and G. G. Khunchatony 1991 *Biotech. and Biochem.* **13**: 212-216.

Guengerich, F. P. and H. P. Broquist 1973 *Biochemistry* **12**: 4270-4274.

Guengerich, F. P., W. W. Johnson, Y.F.Ueng and T.Shimade 1996 *Envi. Health Perspect.* **104**: 557-562.

Gumprecht, L. A., G.W.Smith, P. C.Constable and W.M. Haschek 2001 *Toxicology* **160**:71-79.

Gupta, M., A. K. Dolui, S. N. Dey, S. Mukherjee, U.K. Majumdar and S. K. Batabyal 1986 *Toxicon* **24**: 519-523.

Gupta, P. S. and L. Siminovitch 1978 *Somatic Cell Genet.* **4**: 353

Gupta, R. P. and U. B. Pandey 1986 *Indian Hort.* **31**: 13.

Hackler, W. M., J. E. Hutchinson and L. M. Hutchinson and P. B. Hamilton 1982 *J. Food Prot.* **15**: 1281-1291.

Hagler, W. M. 1990 In perspective on aflatoxins in field crops and animal food products in USA. A symposium USDA, Springfields, ARS-83, NTIS, pp. 35-42.

Hagler, W. M., C. J. Mirocha, S.V. Pathre and J. C. Behrens 1979 *Appl. Envir. Microb.* **37**: 849-853.

Hajjaj, H., A. Klaebe, G. Goma, P. J. Blanc, E. Barbier and J. Francois 2000 *Appl. Envi. Microb.* **66**: 1120-1125.

Hald, B. and D. Krogh 1973 *J. AOAC.* **56**: 1440-1443.

Hald, B., D. H. Christensen and P. Krogh 1983 *Appl. Envir. Microb.* **46**: 1311-1317.

Hall, A. J. and C. P. Wild 1994 In: The toxicology of Aflatoxin (eds. CD L. Eaton and J. D. Groopeman) Academic Press, New York, pp. 233-258.

Hamilton,P.B., W.E.Huff, J.R.Harris and R.D.Wyatt 1982 *Poult. Sci.* **61**:1832-1841.

Hanika, C., W.W.Calton, E.J.Hinsman and J.Tuite 1986 *Veterinary Pathology* **23**: 245-253.

Hansen, D., H. P. Weber and H. P. Weiss 1945 *Proc. Nat. Acad. Sci.USA* **55**: 1161.

Hansen, T. J. 1990 *J. Food Protect.* **53**: 75-77.

Hao, Y. Y. and R. E. Brackett 1988 *J. Food Sci.* **53**:1384–1386.

Haraguchi, H., K. Hartimoto, K. Shibata, M. Tamuguchi and O. Oi 1987 *Agri. Bio. Chem.* **51**: 1373-1378.

Hareeshvardhan Rao, P. 1988 Ph.D. Thesis, Kakatiya Universit, Warangal.

Hareeshvardhan Rao, P. and S. M. Reddy 1990 International Symposium on Seed and Seed Technology, University of Jodhpur.

Hareeshvardhan Rao, P., P.Giridhar and S. M. Reddy 1996 *Indian Phytopath.* **49**: 37-40.

Hareeshvardhan Rao, P., P.Giridhar and S. M. Reddy 1996a *J. Ind. Bot. Soc.* **75**:313-314.

Hareeshvardhan Rao P., P.Giridhar and S. M. Reddy 2001 *Proc. Nat. Acad. Sci.* India **71**:10-12.

Hareeshvardhan Rao, P., P.Giridhar and S.M. Reddy 2001a *Indian Phytopath.* **53**:102-104.

Harrach, B., A.Bata, E. Balmocy and E.Bont 1983 *Appl. Envi. Microb.* **45**:1419-1422.

Harris, E. D. Jr., J.J. Reynolds and Z.Werb 1975 *Nature* (London) **257**: 243-244.

Harrison, L. R., B. M. Colvin, J. T. Green, L. E.Newmar and J. R. Cole 1990 *J. Vet. Diagn. Invest.* **2**: 217-221.

Harvan, D. J. and R. W. Pero 1976 *Advan. Chem. Ser. No*: **149:** 344-355.

Harwing, J. and C. Munro 1975 *J .Can. Vet.* **16**: 125-141.

Harwing, J. and P. M. Scott. 1971 *Appl. Microbiol.* **21**: 1011-1016.

Harwing, J., P. M.Scott, D.R. Stoltz and B. J. Blanchfield 1979 *Appl. Envi. Microb.* **38**: 267-274.

Hasan, H. A. B. 1993 *Folia Microbiol.* **38** : 295-290.

Haschek, W. M., A. L. Waggoner, M. M. McAllister, S. H. Hsiao, G. W. Smith, J. H. Foreman, M. E. Tumbleson, R. M. Eppley and P. D. Constable 1992 *Food Addi.* And *Contam.* **18**: 203.

Haslam, R. J., M. M. L. Davidson and M. D. Mc Clenaghan 1975 *Nature* (London) **253**: 455-457.

Hassanane, M.S., A.El-Sayed, M.A.Abdalla, E.Fiky, S.Amer and A.Handy 2000 *Mycotoxin Res.***16**: 53-63.

Haubeck, H. A., G.Lorkowski, E. Kolsch and R.Roschenthaler 1981 *Appl. Envi. Microb.* **41**: 1040-1042.

Hauser, D., H. P.Weber and H. P. Sigg 1970 *Helv. Chim. Acta* **53**: 1061-1073.

Hawkins, D. 1973 *J. Immunol.* **110**: 294-296.

Hayakawa, S., T. Malisushima and M.Kimura 1973 *Annu. Rep. Shinogi Res.* Lab. No. **23**: 2-3.

Hayes, A.N. 1994 Principles and Methods of toxicology, 3rd ed. Raven Press, New York. pp. 343.

Hayes, A. W. and R. D. Hood 1978 *Toxicology* **16**: 92-96.

Hayes, A. W., R. D. Phillips and L. C. Wallau 1977 *Toxicol.* **15**: 293-300.

Hayes, M. A. and H. B. Schiefer 1979 Ed. *Cosmet Toxicol.* **17** : 611-621.

He, P., L.G. Young and C. Foresberg 1992 *Appl. Envi. Microb.* **58**:3857-3863.

Heathcote, J.G. 1984 In: Mycotoxins Production, Isolation, Separation and Purification (ed. V. Betina), Elsevier Science Publishers, B.V. Amsterdan. pp. 89-130.

Heathcote, J.G. and J.R. Hibbert 1978 Aflatoxins: Chemical and Biological Aspects, Elsevier Publication, New York. pp.55.

Herberle, W., W. Loffler and W.A. Konig 1974 *Arch. Microbiol.* **100**: 73-95

Heisler, E.G., J. Siciliano, E.E. Stinson, S.F.Osman and D.D.Bills 1980 *J. Chromatogr.* **194**: 89-94.

Hetherington, A.C. and H.Raistrick 1931 *Thom. Philos. Trans. R. Soc.* London Ser. B. **220**: 269-297.

Hendrickse, R.G. 1991 *Pediatrics* **88**: 376-379.

Henney, C.S. 1978 in Cytochalasins: Biochemical and Cell Biological Aspects, (ed.W Tanenbaum), North-Holland, New York. pp.1564.

Hennigen, M., R. S. Sanchez, N. H. Di Benedetto, A. Lorghi, J. E. Torroba and L. M. V. Soares 2000 *Food Addi. and Contamin.* **17**: 55-58.

Henry, S. H. and F. Bosch 2001 In Food-borne Disease Handbook (ed. Hui, Yu, R. A. Smith and D. G. Spoerke) Marcel Decker, Inc. New York, pp. 593-626.

Henry, S.H., T. Whiltaker, I. Rabbani, J. Bowers, D. Park, W. Price, F. X. Basch, J. Pennington, P. Verger, T. Yoshizawa, H.Van Egmond, M.A. Jonker and R.Cooker 2001 WHO Additives Series **47**: 1-103.

Hermansen, K., J.C. Frisvad, C. Emborg and J.Hansen 1984 *J. FEMS Microbiol Letters* **21**: 2101-2119.

Hesseltine, C. W. 1969 *Mycopathol Mycol. Appl.* **39**: 371-383.

Hesseltine, C. W., J. M.Visnick and A. B. Smith 1998 *J. Organic Chem.* **3**: 660-662.

Hilakivi, C.L., I.Onojate, M.Raygada, E.Eho, T. Skaas, I. Russo and R. Clarke 1999 *Br. J. Cancer* **80**: 1682-1685.

Hintika J.G.and J.R.Hibbert 1983 Aflatoxms, Chemical and Biological Aspects. Elsevier Public. New York.

Hirooka, B. Y., M.M. Xamagushi, S. Aoymma, Y. Sugiura and Y. Ueno 1996 *Food Addi. and Contam.* **13** : 173-183.

Hisata,K.,T.Terada, K.Yamam, H.Tsubouchi and Y.Sakabe 1984 *J.AOAC.* **7**: 601.

Hilokoto, H., S.Morozumi, T. Wanke, S. Sakai and Y.Ueno 1978 *Mycopathologia* **66**: 161-165.

Hantow, L. L. and R. F. Jr. Widermen 1985 *Poultry Sci.* **64**: 1553-1556.

Ho, J. A. and R. M. Durst 2000 *Anal. Chem. Acta* **414**: 61-69

Hoehler, D. and R. R. Marquardt 1996 *Poultry Sci.* **75**: 1508-1515.

Hocking, A. D., K. Holds and N. F. Tobin 1988 *Aus. Vet. J.* **65**: 82-85.

Hodges, R. and J. S. Shannon 1966 *Australian J. Chem.* **19**: 1059-1066.

Holocomb, M., H. C. Thompson Jr., G. Lipe and L. J. Hankins 1993 *J. Liq. Chromatogr.* **17**: 4121-4129.

Holzapfel, C. W. 1968 *Tetrahedron* **24**: 2101-2119.

Holzapfel, C.W. and D. C.Wilkins 1971 *Tetrahedron* **10**: 351-358.

Holzapfel, C. W., R.D.Hutchinson and D.C.Wilkins 1970 *Tetrahedron* **26**: 5239-5246.

Horn, B. W. and J. W. Dorner 2002 *Mycologia* **94**: 741-751.

Hosae, J., K. Nozawa, S.Udagawa, S.Makasima and K. Kawai 1990 *Chemical and Pharmaceutical Bull.* **38**: 3473-3475.

Hook, W. A. and R. P. Siraganian 1977 *J. Immunol.* **118**: 679-684.

Horne, C. W., L.L.Boleman and C.G.Coffman 1991 Mycotoxions in feed and food producing crops. Texas Agriculture extension service. The Texas A. Y. M. University System. College, Texas. pp.17-24.

Hou, C. T., A. Ciegler and C. W. Hesseltine 1971 *Anal. Chem.* **37**: 422-428.

Howard, P. C., R. M. Eppley, M. E. Stack, A. Warbritton, K. A. Voss, R. J. Lorentzen, R.M. Kovach and T.J.Bucci 2001 *Environ.Health Perspect.* **109**: 277-282.

Hsich, D. P. H. 1973 *J. Agric. Food Chem.* **21**: 468-470.

Hsu, I. U., E. B. Smalley, F.M.Strong and W.E.Ribelin 1972 *Appl. Microbiol.* **24**: 284-290

Hsu, K. H. and F. S. Chu 1992 *Food Agric. Immunol.* **4**: 83-91.

Hua, A.U., S.S.T. Baker, J.L. Flores and M.P.Y. Espiritu 1999 *Appl.Envir.Microbiol.* **65**:2738-2740.

Huang, X.and F. S. Chu 1993 *J. Agric. Food Chem.* **41**: 329-333.

Hutchins, J. E., Y. J.Lee., K.Tyczkowska and W. M. Hagler Jr. 1991 *Arch. Envir. Contam. Toxicol.* **18**: 319-326.

Huff, W. E. 1980 *Cereal Chemistry* **59**: 236-238.

Huff, W. E. and P.B. Hamilton 1979 *J. Food Protec.* **42**: 815-820

Huff, W. E. and W.M. Hagler 1982 *Cereal Chem.* **59**: 152-154.

Hukovic, S. and N. Mulabegovic 1986 *Acad. Sci.and Ams of Bosniov and Hercejovina*.pp. 11-15.

Hult, K. and S. Gatenbeck 1976 *J. AOAC.* **59**: 128-129.

Hunt, D.C., L.A. Philip and N. T. Crosby 1979 *Analyst* **104**: 1171-1175.

Hussein, H.S. and J.M. Brasel 2001 *Toxicol* **167**: 101-134.

Hwang C.A., F.A. Dranghon 1994 *J.Food Protect.* **57**: 410-414.

IARC (Internation Agency for Reseach on Cancer)1986 Monographs on the Evolution of the Carcinogenic Risk Chemicals to Human, IARC, Lyon, France.Vol **40**: 83-98.

IARC (Internation Agency for Reseach on Cancer)1993 Monographs on the Evaluation of Carcinogenic Risks to Humans, IARC, Lyon, France. Vol 56. pp. 245-395.

Ichnoe, M., H. Kurata, Y. Sugiura and Y. Ueno 1983 *Appl. Envir. Microb*.**46**: 1364-1369.

IPCS 1990 Selected Mycotoxins: Ochratoxins, Trichothecenes, Ergot Environmental Health Criteria 105, Published WHO, 1211 Geneva 27, Switzerland.

Ishikawa, M., T.Oki and F.Sawada 1973 *Rept. Tottori.Mycol. Inst.*.**10**:637-645.

Ishii, K.M. Y.Sawano, Y.Ueno and T.Tsunoda1974 *Appl.Microbiol*.**27**:625-628.

Ito, Y. and K. Obtsubo 1994 *J. Vet. Med. Sci.* **56**: 1155-1159.

Ito, Y., K.Obtsubo, K.M.Ishii and Y. Ueno 1986 *Mycotoxin Res.* **2**: 71- 77.

Ito, Y., Y. Nishiyama, K.Shimokata, Y.Kimura, I.Nagata and A. Kunii 1976 *J. Gen. Virol.* **33**: 1-5.

Itoh, Y. and Aibara 1992 *J. Agri. Chem. Soc. Jap.* **66**: 1335-1345.

IUPAC 1974 *Inform. Bull Tech. Dep.* **8**:1.

Ivanics, E., R.Glavits, A. Vany, L.Etter and E.Ezabo 1990 *Magyar Allatosvosok Lapia* **45**: 205-209.

Jabbar, A. and A. Rahim 1962 *J. Pharma. Sci.* **51:** 595-596.

Jackson, L. S., J. J. Hlywka, R. Kannaki, B. Sen Thil, L. loyd, L.B. Bullerman and S. M. Musser 1996a *J. Agric. Food Chem.* **44** : 906-912.

Jackson, L. S., J. W. De Vries and L. B. Bullerman 1996 Fumonisins in *Food. Plenum Press* New York, Pp.345-353

Jackson, R.A., S.W. Fenton, C. J. Mirocha and G. Davis 2004 *J. Agric. Food Chem.* **22**: 1015-1019.

Jacobsen, B. J., K. L. Bowen, R.A. Shelby, U.L Diener, B.W.Kemppainen and J.Floyd 1993 Mycotoxins and mycotoxicosis circular ANR-767. Auburn Univ. Auburn, Alabama.

Jamieson, W. D., R. Rahman and A. Taylor 1969 *J. Chem. Soc.* p. 1564-1567.

Jansen, H., R. Jansen, U. A. Th.Brinkman and R. W. Frei 1987 *J. Chromatogra.* **24**: 555.

Janardhana, G. R., K. A. Raveesha and H. A. Shetty 1999 *Food Chem. Toxicol.* **37**: 863-868.

Jarvis, B.J., G.Pavanasasivam, C.E.Holmlund, T. Desilvia, G. P. Stahly and E. P. Mazzola 1981 *J. Amer. Chem.* **103**: 472.

Jarvis, B. J., J. Salemmcand and A. Morais 1995 *Nat. Toxins* **3**: 10-16.

Jarvis, B. J., R. M. Eppley and E. P. Mazzola 1983 Trichothecene, Chemical, Biological and Toxicological Aspects (ed.Y.Ueno), Elsevier Publication Amsterdam. pp. 20.

Jaskiewiez, K., W. F. O. Marasas and J. J. F. Taljaard 1987 *J. Comp. Pathol.* **97**: 281-291.

Javed, T., G. A. Bennett, J. L. Richard, K. M. A. Dombrink, L. M. Cote and W. B. Buck 1993 *Mycopathologia* **123**: 171-184.

Javed, T., R.M.Bunte, G.A.Bennett, J.L.Richard, M. A. Dombrink-Kurtzman, L. M. Cote and W. B. Buck 1992 In Abstracts of 106th AOAC International Annual Meeting 1992 Cincinatti.

Jayaram, P. and I. Kalyanasundaram 1990 *Mycopathologia* **110**: 81-85.

Jayashree, T. and C.Subramanyam 1999 *Letter in Applied Microbial.* **28**: 179-183.

JECFA 1998 Safety evaluation of certain food additives. WHO food additive serries 40. International Program on Chemical Safety, World Health Organization, Geneva.

JECFA 2000 Joint FAO / WHO Expert Committee in food additives, 53rd report, Safety Evaluation of Certain Food Additives, WHO Food Additives Series. 44.

Jelineck, C. F., A. E. Pohland and G. E. Wood 1989 *J. AOAC.* **72**: 223-230

Jemmali, M. 1987 Trade and economic implications of mycotoxins need for greater uniformity Mye 87/6. Joint FAD/WHO UNEP Second International Conference on Mycotoxins Bangkok, Thailand, 26 Sept.-3 October.

Jemmali, M. and C. Mazerand 1980 *Anna.ide Microbiology* (Paris) **31**: 319-321.

Jemmali, M., Y. M. Ueno, K. Ishii, C. Frayssinet and M. Etienne 1978 *Experientia* **34** : 1333-1334.

Jesus, A. E., P. S.Steyn, F. R.Van Herden, R.Vleggar, P. L.Wessels and W. E. Hull 1981 *J. Chem. Soc. Chem. Commun* **6:** 289-291.

Jimenez, A. M., A. Lopez de Cerain, E. Gonzalez – Penas and J. Bello 2001 *Food Addi. and Contam.* **18**: 559-563.

Jimenez,M., T.Huerta and R.Mateo 1997 *Appl.and Envir.Microbiol.***63**:364-369.

Jimenez, M.,V.Sanchis, R.Mateo and E.Harnandez 1988 *Mycotoxin Res.***119:**59-66.

Jindal, N.,S.K.Mahipal and G.E.Rotting Haus 1999 *Mycopathologia***148**:37-40.

Jacquet, J., P. Boutibonnes and J. P. Cicile 1963 *Bull Acad. Vet. Fr.* **36** :199.

Jodlbauer, J., N. M. Maier and W.Lindner 2002 *J. Chromatogra. A.* **945**: 45-63.

Joffe, A. Z. 1960 *Bull.Res.Coun.* **Isr. 8**: 81-95.

Joffe,A.Z. 1978 In Mycotoxic Fungi, Mycotoxins, Mycotoxicoses. An Encyclopedia Handbook, Vol. 3, (eds. T. D. Wyllic and L. G. Morehouse), Marcel Dekker, New York, pp. 21-86.

Joffe, A. Z. and J. Palti 1974 *Mycopathol. Mycol. Appl.* **52**: 209-218.

Joffe A. Z. and J. Palti 1975 *Appl. Microb.* **29**: 575-579.

Joffe, A. Z. and J. Palti 1977 *Phytoparasitica* **5**: 119-121.

Joffe, A. Z., J.Palti and R.Arbel-Sherman 1973 *Mycopathol.Mycol.Appl.***50**: 85-107.

Johnson, J.R. and J. B. Buchanan 1953 *J. Am. Chem. Soc.* **75**: 2103-2109.

Johnson, W.,Y.F.Ueng,M.Widersten, B.Mannervik, J.D.Hayes, P. J. Sherratt, B. Ketlerer and F. P. Guengrich 1997 *Biochemistry* **36**: 3056-3060.

Joint FAO/WHO Expert Committee on Food Additives (JECFA) 2001 Safety evaluation of certain mycotoxins in Food and Agriculture Organization of the United Nations (FAO) food and Nutrition 74. World Health Organisation (WHO) *Food Additive Series.* **47**: 281-415.

Jones, C., J.R.Ciacci and Y.Z.Hang 2001 *Envir.Health Perspect* **109:** 315-320.

Jones, D., B.R.Sexton and G.Rebell 1969 *Trans.Opthalmol.Soc.*U.K.**89**:781-797.

Jones, F., M. Genter, W. Hagler, J. Hanren, B. Mowrey, M. Poore and L. Whitlow 1994 Understanding and coping with effects of mycotoxins in livestock feed and forage. Electronic Publication No. dRo-29, NCCES, North Carolina State Univ., Raleigh.

Jones, F. T. and P. B. Hamilton 1985 *Poult. Sci.* **65**: 1522-1525.

Jonsyn, F.E., S.M.Maxwell and R.G.Hendrickse 1995 *Mycopathlogia* **131**:121-126.

Jorgensen, K. and A.Petersen 2002 *Food Addi. and Contam.* **19**: 562-567.

Jortner, B. S., M.Ehrich, A.E.Katherman, W.R.Huckie and M.B.Carten 1986 *Drug and Clinical Toxicology* **9**: 101-106.

Josephs, R.D., R. Schumcher and R. Krishna 2001 *Food Addi. and Contam.* **18**: 417-430.

Joshi, B.G., S.K. Srivastava, P.K. Dwivedi and S.K. Prajapathi 1987 *Bull.Grain Tech.* **25**: 85-87.

Joshi, B.G., S.K. Srivastava and S.K.Prajapati 1986 *Bull. Grain Tech.* **24**: 206-210.

Joshi, B. G., S. K. Srivastava and S.K. Prajapati 1987 *Bull Grain Tech.* **25**: 188-190

Juglal, S., R .Govindan and B.Odhav 2002 *J. Food Protect.* **65**: 683-687.

Jyothi Saxena, S. K., N. Pande, P. H. Pandey and B. S. Mehrotra 1988 *Proc. Nat. Acad. Sci. India* 58 B: 427-430.

Kaczka, E. A., C. O. Gittermann, E. L. Dulaney, M.C. Smitha, P.Hedlin, H. B. Woodruff and K. Folkers 1964 *Biochem.Biophys. Res. Commun.* **14**: 54-57.

Kadakal, C., M.Sebahatton and E. S. Poyrazogalu 2002 *J.Food Quality* **25**:357-368.

Kady, E. L., I. A. and M. C. Heath 1982 *Expt. Mycol.* **6**: 25-30.

Kale, S., T. Thirumalesh and G.V. N. Reddy 2006 *Indian J. Animal Res.* **40**: 167-169.

Kalra, D.S. and K.C.Bhatia 1990 *J. Environ. Pathol. Toxicol. Oncol.* **10**:132-135.

Kamimura, H., M. Nishizima, K. Saiato, K. Yamada, A. Lbe, T. Nagayama H. Ushiyama and Y. Naoi 1979 *Jpn. J. Food Hyg.* **20**: 352-357.

Kamimura,H., M.Nishijima, K.Yasuda, H.Ushlyama, S.Tabata, S. Matsumoto and T. Nishima 1985 *J. AOAC.* **68**: 458-461.

Kamimura, H., M.Nishijima, K. Yasuda, K. Saito, A. Lbe, T. Nagayama, H. Oshiyama and Y. Naoi 1981 *J. AOAC.* **64**: 1067-1073.

Kamimura, O., S.M.Saiato, S.Nagura, I.Kishimoto, S.Ueno, T.Sato, I. O. Y. Uda and Y. Ueno 1990 *Food Agric. Imm.* **2**: 135-143.

Kaneko, I., H. Hayatsu, H. Satoh and T. Ukita 1975 *Biochem. Biophys. Acta* **411**: 334-348.

Kang, H. J., J. C. Kim, J. A. Seo, Y. W. Lee and D. H. Son 1994 *Agric. Chem. Biotechnol.* **37**: 385-391.

Karfunkel, P. 1972 *J. Exp. Zool.* **181**: 289-302.

Karlovsky, P. 1999 *Nat. Toxins* **7**: 1-23.

Karppanen, E., A.Rizzo, S.Berg, E.Lindfors and R.Aho 1985 *J. Agric. Sci. Final* **57**:195-206.

Karrer, P. and M. Visconti 1947 *Helv. Chem. Acta* **30**: 268-271.

Karuna, R. and R. B. Sashidhar 1999 Talanta **50**: 381-389.

Karunaratne, A., E. Wezenberg and L. B. Bullerman 1990 *J. Food Prot.* **53**: 230-236.

Katagiri, K., K. Saro, S. Hayakawa, T. Matsushima and H. Minato 1970 *J. Antibiotics* (Tokyo) **23A**: 420.

Katoch, R. E., K.B.D.Bhowmik and B.S.Katoch 1975 *Indian Vet. J.***52**:759-762.

Katzmann, P. A. 1944 *J. Biol. Chem.* **154**: 475-486.

Kavita Waghray and M. Uma Reddy 1996 *Indian Food Industry* **15**: 37-39.

Kawabata, Y., F. Tashiro and Y. Ueno 1982 *J. Biochem.* **91**: 801-808.

Kawai, K., T. N. Akamaru, Y. Maebayashi, Y. Nazawa and M.Yamazaki 1983 *Appl.Envir. Microbl.* **46**: 793-796.

Kawamura, O., S. Sato, M. Nagura, S.Kishimoto, I.Ueno, S.Sato, T.Uda, Y.Ito and Y. Ueno 1990 *Food and Agri. Imm.* **2**: 135-143.

Kawashiro, I., H. Tanabe, H.Hakwchi and C.Nishimura 1955 *Eisei Shikenjo Hokoku* **73**:191-196

Kawaski, I.,T. Oki, M. Umeda and M. Saito 1972 *Jap.J.Expt..Med.***42**:327- 340.

Kedara, C. J., R.D. Plattner and A.E. Desjardins 1999 *App.. Envi. Microb* **65**:41-44.

Kellerman, T. S. and W. F. O. Marasass 1981 *J. Vet. Res.* **48**: 129-131.

Kellerman, T. S., W. F. O. Marasass, P. G. Thiel, W. C. A. Gelderblom, M. Cawood, J.A.W.Coetzer and J.Odersteport 1990 *J.Vet. Res.***57** : 269-275.

Keller-Schierlein, W. and E. Kupfer 1979 *Helv. Chim. Acta* **62**: 1501-1524.

Kemper, F.H. and N. P.Lucpke 1986 *Food Chem. Toxicol.* **24**: 647-648.

Kent, J. and N. G. Heatley 1945 *Nature* **156**: 295-296.

Khan, M. S. and M. R. Siddiqui 1980 *Seed Res.* **8**: 15-19.

Kiessling, K. H., H. Petterson, K. Sandholm and M. Olsen 1984 *Appl. Envi Microb.* **47**: 1070-1073.

Kim, J.C., H.J. Kang, D. H. Lee, Y. N. Lee and T. Yoshizawa 1993 *Appl. Envir. Microb.* **59**: 3798-3802.

Kim, K. C., I. S. Chung and D. H. Do 1992 *J. Agric. Sci.* **10**: 135-124.

Kim, M. S., D. Y. Lee, T. Wang and J. J. Sctroeder 2001 *Toxicol. Appl. Pharmacol.* **176**: 118-126.

Kim, Y. K. and J. K. Roh 1986 *J. Anim. Sci.* **28** : 99-104.

King, S. B. and G. E. Scott 1981 *Phytopath.* **71**: 1245-1247.

Kingsland, G. C. 1986 *J. Stored Prod. Res.* **22**: 29-32.

Kinoshita, T., Y. Renbutsu, J. D. Khan, K. Kohmoto and S. Nishimura 1972 *Ann. Phytophathol. Soc. Jap.* **38**: 397-404.

Kirby, L. K., T. S. Nelson, J. T. Halley and J. N. Beastey 1987 *Poultry Sci.* **66**:966-968.

Kirksy, J. W. and R.J.Cole 1973 *J. Appl. Microbiol.* **26**: 827-828.

Kirksy, J. W. and R. J. Cole 1974 *Appl. Mycol.* **54** : 291-296.

Kitchen, D. N., W. W. Carlton and J. Tuite 1977 *Vet. Pathol.* **14**: 154-172.

Kitchen, D. N., W. W. Carlton and J. Tuite 1977a *Vet. Pathol.* **14**: 261-272.

Kiang, D. T., B. J. Kennedy, S. V. Pathre and C. J. Mirocha 1978 *Cancer Res.* **38**: 3611-3615.

Klein, P. J., R.Buckner, J. Kelly and R.A.Coulombe Jr. 2000 *Toxicology and Appl. Pharma.* **165**: 45-52.

Klich, M.A. and J.I. Pitt 1988 A Laboratory Guide to common *Aspergillus* species and their teleomorphs. CISIRO Division of food processing, North Ryde, New South Wales, Australia.

Klittich, C.J.R., J.F.Leslie, P.E.Nelson and W.F.O.Marasas 1997 *Mycologia* **89**: 643-652.

Knight, A. P. 1981 *Compound Contin. Educ. Pract. Vet.* **3:** 112-118.

Kobayashi, A., Y.Nakae, T.Kawasaki and K.Kawazu 1989 *Agri. and Biochem.* **53** : 583-590.

Kobel, H., E. Schreier and J. Rutschmann 1964 *Helv. Chem. Acta* **47**: 162.

Kodell, R.L., J.F. Young, R.R.Delongchamp, A.Turturro, J.J.Chen, D.W.Gaylor P.C. Howard and Q. Zheng 2001 *Food Addi.and Contam.* **18**: 237-253.

Kochler, P. E., L. R. Beuchat and M. S.Chhinnan 1985 *J. Food Prot.* **48**:1040-1043.

Kollarczik, B., M. Garies and M. Hamett 1994 *Nat. Toxins* **2**: 105-110.

Koppl, B. and H. J. Rehm 1979 *Eur. J. Appl. Microbl. and Bio.***6**: 379-401.

Koof Macher, W.A., M. P. Chiarelli, Jr. J.O.Lay, J.Bloom, M. Holcomb and K. J. McManus 1991 *Rapid Commun. Mass Spectrum.***5** : 463-468.

Kordic, B., S. Zeravkovic and M. Alekssic 1990 *Veterinarshi Glasnik* **44**: 145-159.

Kozakiewicz, Z. 1986 In Spoilage and Mycotoxins of Cereals and other Stored Products (ed.B.Flannigan), International Biodeterioration 22, CAB International, Walling ford, UK, pp. 115.

Kozakiewicz, Z.1989 *Mycol. Pap.* **161**: 1-188.

Kralova, J., J.Hajslova, J.Poustka, M.Hochman, M.Bielkova and L.Odstrcilova 2006 Czech. *J. Food Sci.* **24**: 288-296.

Kramer, P. M., Q. Y. Li and B. D. Hammock 1994 *J. AOAC.* **77**: 1275-1287.

Kirek, N. P. J., T. S. Kellerman and W. F. O. Marasas 1981 *J. Vet. Res.* **48:** 129-131.

Krishna Reddy, V., D.Rajakumari and S. M. Reddy 1997 *Nat. Acad. Sci. Letters* **20**: 97-99.

Krishna Reddy, V. and S. M. Reddy 1989 *Nat. Acad. Sci. Letters* **12:** 67-68.

Krishna Reddy, V. and S. M. Reddy 1989a *Indian J. Microbiol.* **29**: 57-60.

Krishna Reddy, V. and S. M. Reddy 1989b *Nat. Acad. Sci. Letters* **12**:107-109.

Krishna Reddy, V. and S.M.Reddy 1990 *Nat. Acad. Sci. Letters* **13**:11-13.

Krishna Reddy, V. and S.M.Reddy 1992 *World J.Microbiol. and Biotech.*8:208-209.

Krishna Reddy, V., S. Girisham and S. M. Reddy 1987 In Perspectives in Mycological Research, Prof. G. P. Agarwal Feschrit Vol 1 (eds S. K. Hasija, R.C. Rajak, S. M. Singh) Today & Tommorrow Printers Publishers, New Delhi, pp. 177-185.

Krishna Kumari, C., M.Nusrath and B.N.Reddy 1984 *Indian Phytopath.***37**: 284-287.

Krishnamachari, K.A., V.Nagarajan and V.B. Ramesh, T.B.G.Tilak 1975 *Indian. J. Med. Res.* **63**: 1036-1048.

Krishnamurthy T., M.B.Wasserman and E.W. Sarver 1986 *Biomed. Envir.. Mass Spectrum* **13** : 503-518.

Kritzinger, Q., T.A. S. Aveling and C. E. Vander Merwe 2006 *Phytoparasitica* **34:** 178-186.

Krogh, P. 1972 *Symp. Inter. Union of Pure and Appl. Chem.* pp.210-222.

Krogh, P. 1976 *Nord Vet. Med.* **28**: 452-458.

Krogh, P. 1978 *Acta Pathol. Microbial Scand. Sect. A. Suppl. No.* **269**: 1-28.

Krogh, P., B. Hald and E. J. Pedersen 1973 *Acta Pathol. Microb. Immunol. Scand.* **81**: 689-695.

Krogh, P., B. Hald, R. Plestina and S.Ceovic 1977 *Acta Pathol. Microbiol. Immunol. Scand.* **85**: 238-240.

Krogh, P., B. Hald, P. Englund, L. Rutqvist and O. Swahn 1974 *Acta Pathol. Microbiol. Immunol. Scand.* **82:** 301-302.

Krogh, P., E.Hasselager and P.Frills 1970 *Acta Pathol. Microbiol. Immunol Scand***78**: 401-413.

Krogh, P., F. Ellong, B. Hald, B. Jyllong, V. E. Petersen, E. Skudmage and C.K. Svendsen 1976 *Acta Pathol. Microbiol. Immunol. Scand.* **84**: 215-221.

Krska, R. 1998 *J. Chromatogra.* **15:** 49-57.

Krska, R., Patricia Schubert-Ullrich, Alexandra Molinelli, Michael sulyok, Susan Macdonald and Colin Craws 2008 *Food Addi. and Contamin.* **25**: 152-163.

Krska, R., R.Schuhmacher, J.Weingartner and M.GrasserBacer 1997 *Cereal Res. Commun.* **25:** 327-329.

Kruger, S.C.,B.Kohn,C.S.Ramrey and R.Prioli 1999 *J.AOAC.* **82:**1364-1368.

Kubachi, S. J. 1985 In Mycotoxins and Phycotoxins, (eds. P. S. Steynl and R. Vleggar), Elseveir Science Publishers, Amsterdam pp. 1293.

Kubena, L. F., R. B. Harvey, W. E. Huff, M. H. Elissalde, A. G. Yersin, T. D. Phillips and G. E. Rottinghans 1993 *Poultry Sci.***72:** 51-59.

Kubena, L. F., R. B. Harvey, D. J. Fletcher, T.D. Phillips, H. H. Mollenhauer, D. A. Witzel and W. D. Heidelhauch 1985 *Poultry Sci.* **64**: 620-628.

Kubena, L. F., R. B. Harvey, W. E. Huff, D. E. Corrier, T. D. Phillips and G. E Rottingh 1993 *Poultry Sci.***69:** 1078-1086.

Kubena, L. F., W. F. Huff, D. B. Harver, D.E. Corrier, T. D. Philips and C. D. Creger 1988 *Poult. Sci.* **67:** 253-260.

Kuiper – Goodman, T. 1991 In: mycotoxins, Endemic Nephropathy and Urinary tract tumors, IARC Scientific Publications, No. 115. International Agency for Research on Cancer, World Health Organizations, Castegharo, (eds.M. Picstina, R. Dirheimes, G. Chernozemsky, I.N Bartsch), Oxford University Press, New York, pp. 307-320.

Kuiper-Goodman, T. 1996 *Food Addi.and Contam.* **13:** 53-57.

Kuiper-Goodman, T. 1998: in Mycotoxins and Phycotoxins developments in chemistry, Toxicology and Food safety (eds. M. Miraghi, H. Van Egmond, C.Bera and J. Gilbert), Allaken, Fort Collin, pp. 25-48.

Kuiper-Goodman, T., J. G. Lemmen, B. Carlesson, J. C. Carton, S. H. Safe, P. T. Vander Saag, B. Vander Burg and J. A. Gustafssen 1998 *Endrocrinol.* **1139**: 4252-4263.

Kuiper-Goodman, T., K. Ominiski, R. R. Marquardt, S. Malcolm, E.M.Mullen G. A. Lambact and T.Morten 1993 In Human Ochratoxicosis (eds. E.E. Creeppy and G. Dirheimer), Motrouse, John Libby, Eurotext pp.167-174.

Kuiper Goodman, T., P.Scott and H.Watanabe 1987 *Regul. Toxicol. Pharmacol.* **7:** 253-306.

Kulisek, E. S. and J.P. Hazebroek 2000 *J. Agri. Food Chem.* **48:** 65-69.

Kulkarni, J. R., D. N. Padule, P. A. Shinde and D. K. Salunkhe 1986 *J. Maharastra Agric.* Unit. **11:** 199-201.

Kumar, R. and C. Balachandran 2005 *J. Indian Veternary* **82:**1138-1140.

Kumar, S. and A. K. Roy 1993 *Letters in Applied Microbiol.* **17**: 112-114.

Kumar S. and G. Prasad 1992 *Letters in Applied Microbiol.* **15**:131-132.

Kumari, C. K and M.Nusrath 1987 *Nat. Acad. Sci. Letters India* **10:** 303-305.

Kupchan, S.M., B.B.Jarvis, R.G.Jr.Bailey, W.Briht, W.R.F.Byran and Y.Shizuni 1976 *Toxicol.* **78:** 70-92.

Kuraishi, H., M.Ito, N.Tsuzaki, Y. Katayama, T. Yokoyama and J. Sugiyama 1990 In Modern concepts in *Penicillium* and *Aspergillus* classification (eds. R. A. Samson and J. I. Pitt), *Plenum Press*, New York, pp. 407.

Kurbatskaya, Z. A. and A. A. Trostanetskii 1987 *Microbiologicheskii Zhurnal* **49:** 62-65.

Kurtzman, C. P., B. W. Horn and C. W. Hesseltine 1987 *Antonie Van Leeuwenhoek* **53:** 147-158.

Kurtzman,C. P., N.J. Smiley, C.J.Robnett and D. T.Wicklow 1986 *Mycologia* **78:** 955-959.

Kussak, A., B. Andersson and K. Andersson 1994 *J. Chromatogra. B. Biomed Appl.* **656:** 329-334.

Kuvaena, B., E. V. Botyanaskaya, K. Eller, V. V. Pimenova and A. Kroyakova 1990 *SSSR Seriya Biologichekaya,* pp. 519-529.

Lacey, J. 1975 *Trans. Br. Mycol. Soc.* **65**: 171-184.

Laetita, P. G., F. Richard Forget, P. Frasse, B.Christian, B.Cahagnier, D.Richard-Molard and B. Bakan 2007 *Mycopathologia* **165**:51-59.

Lafarge, C. and C. Frayssinet 1970 *Int. J. Cancer* **6**: 74-83

Laidou, I. A., C. C. Thanassoupoulos and A.B. Liakopoulos, M. Kyriakides 2001 *J. Phyto-Pathology* **149**: 457-461.

Lamprecht, S. C., W. F. O. Marasas, J. F. Alberts, M. E. Cawood, W. C. A. Gelderblom, G. S. Shephard, P. G. Thiel and F. J. Calitz 1994 *Phytopathology* **84** : 384-391.

Lancova, K., J. Hajslova, J. Poustka, A. Krplova, M. Zachariasova, P. Dostalek and L. Sachambula 2008 *Food Addi. and Contam..*: **25**: 732-744.

Land, C. J. and K. Hult 1987 *Appl. Microbiol.* **4**: 41-44.

Lane, W., E.Hollsten, R.Donath and G.Zapff 1988 *Manatshefte für Veternarmedizin* **43**: 829.

Lang, G. G., J. Turek, M. A. Diekman and A. B. Scheidt 1992 *Veternary Pathoglogy* **29**: 60-67.

Langseth, W., Y.Ellingren, U.Nymoen and E. M. Okland 1989 *J. Chromatogr.* **478**: 269-274.

Lansden, J. A . 1984 *J. AOAC.* **67**: 738-731.

Lansden, J. A. 1986 *J. AOAC.* **69**: 964-966.

Lansden, J. A. and J. I. Davidson 1983 *Appl. Envir. Microb.* **45**: 766-769.

Lash, J. W. R., A. Cloney and R. R. Minor 1973 *Biol. Bull* **145**: 360-372.

Lattanzio, V. M. T., M.Solfrizzo and A. Visconti 2008 *Food Addi. and Contam.* **25**: 320-330

Lau, B. P., P.M Scott, D.A . Lewis and S.R.Kanhere 2000 *J. Mass Spectrum* **35**: 23-32.

Lauren, D. S. and R. T.Gallagher 1982 *J Chromatogra.* **248**: 150-154.

Laws, I. and P. G. Mantle 1989 *J. General Microbiology* **135**: 2679-2692.

Laws, I., P. G. Mantle and B. L. Smith 1987 *Res. Vete. Sci.* **43**: 199-202.

Laxma Reddy, G . and S. M. Reddy 1992 *Nat. Acad. Sci. Letters* **15**: 175-177.

Laxma Reddy, G. and S. M. Reddy 1992a *Indian J. Microb.* **32**: 401-403.

Laxma Reddy, G. and S. M. Reddy 1992b *Nat. Acad. Sci. Letters* **15**: 237-240.

Lazarides, E. 1976 *J. Cell Biol.* **68**:202-209.

Le Bars, J. 1979 *Appl. Envir. Microb.* **38**: 1052-1055.

Lebepe-Mazur, S., H. Bal, E. Hopmans, P. Murphy and S. Henrich 1995 *Vet. Hum. Toxicol.* **37** : 126-130.

Lee, K. S. and R. Roschentholer 1987 *J. Antibiotics* **40**: 692-696.

Lee, S. E., B. S. Park, P. Bayman, J. L. Baker, W. S. Choi and B. C. Campbell 2007 *Food Addit. and Contam.* **24**: 391-397.

Lee, S. J., M.A.Hanna and L. B. Bullerman 1986 *Cereal Chem.* **63**: 1076-1079.

Lee, T., D. W. Oh, H.S. Kim, J. Lee, Y. W. Kim, S. H. Yun and Y. W. Lee 2001 *Appl. Envir. Microb.* **67**: 2966-2972.

Leggott, N. L. and G. S. Shephard 2001 *Food Control* **12**: 73-76.

Leggott, N. L. G. S. Shephard S. Stockenstrom, E. Stav and K. D. J. Van Schatkwy 2001 *Food Addi.and Contam.* **18**: 825-829.

Lehmann, D. and B.Wald 1990 *Dentsche Labensmittel Rundchan* **86**: 216-217.

Leistner, L. and J.I. Pitt 1977 In: Mycotoxins in Human and Animal Health (eds. J.V. Rodricks, C. W. Hessltine and M. Mehlman) Pathotox Publishers, Park Forest South. III. pp. 699-711.

Lamke,S., S. E. Ottinger, C. L. Ake, K. Mayura and T.D.Phillips 2001 *Chem. Res. Toxicol.* **14**: 11-15.

Lemmer, E. R., W. C. Gelderblom, E. G. Shephard, S. Abel, B. L. Seymour, J. P. Cruse, R. E. Kirsch, W. F. O. Marasas and P. M. Hall 1999 *Cancer Letters.* **146**: 207-215.

Leomi, L. A., L.M.Soares and P.L.Oliveira 2000 *Food Addi and.Contam.***17**:867-870.

Lepom, P. 1986 *J. Chromatogra.* **355**: 335-339.

Lesson, S., G. J. Diaz and J. D. Summers 1995 Poultry metabolic disorders and mycotoxins, University Books, Guelph, Ontario, Canada. pp. 326-332.

Levy, J., A. Herchuelz, A. Sener, F. Malaisse-Lagae and W. J. Malaisse 1976 *Endocrinology* **98** : 429-437.

Lewis, C. W., J. G. Anderson and J. E. Smith 1994 In Biotechnology Hand Book (ed. J. E. Smith), Plenum Press, N. Y, London, pp. 219-261.

Li, F.Q., S.Luo and T.Yoshizawa 1991 *Nat. Toxins* **7 :** 93-97.

Li, M., S. Lu, C.Ji, Y.Wang, M. Wang, S. Chen and G. Tian 1980 In Genetic and Environment factor in experimental and human cancer (ed.H.V. Gelboi), Japan Sciences Society Press, Tokyo, pp. 139-148.

Li., S. Z., P.Y. Chen, R.R. Marquardt, Z. K. Han and J. R. Clarke 1996 *J. Agric.Food Chem.* **44**: 372-375.

Liao, L., A. P. Grollman and S. B. Horwitz 1976 *Biochem. Biophys. Acta* **454**: 273-284.

Lillehoj, E. B.1973 *J. Am. Vet. Med. Assoc.* **163**: 1281-1283.

Lillehoj, E. B., A. Ciegler and H. H. Hall 1967 *Experientia* **23**: 187-188.

Lillehoj, E.B., A. Ciegler and R. W. Detroy 1970 In Essays in toxicology, (ed.F.R. Blood) Academic Press, New York.

Lillehoj, E. B., W. J. Garcia and M. Lambrow 1974 *Appl. Microb.* **28**:763-767.

Lillehoj, E. B., W. F. Kwolek, D. I. Fennelli and M. S. Milburn 1975 *Cereal Chem.* **52**: 403-412.

Lilienfed, A.M. and D.E. Lilienfed 1980 Foundations of epidemiology Oxford University Press, New York. pp. 191-225.

Lin, S., D. C. Lin and M. D. Flanagan 1998 *Proc.Nat.Acad.Sci.USA.* **75**: 329.

Lin, Y. C., J. C. Ayers and P. E. Koehler 1980 *J. Food Sci.* **46**: 974-977.

Line, J. E. and R. E. Brackett 1995 *J. Food Protect.* **58**: 1042-1044.

Line, J. E., R.E.Brackett and R. E. Wilkinson 1994 *J. Food Protect.* **57**: 788-791.

Ling, K. H. 1995 *J. Toxicol Rev.* **13**: 243-252.

Ling, K. H., C. K. Yang and F. T. Peng 1979 *Appl. Envir. Microb.* **37**: 355-357.

Ling, K. H., H. H. Liou, C. M. Yang and C. K. Yang 1984 *Appl. Envir. Microb.* **47**: 98-100.

Liu, X., X. Luo and W. Hu 1992 *Biomed. Environ. Sci.* **5**: 161-177.

Lobeau, M., S. Sarger, De L . Sibanda, I.Barna-Vetro and C.V. Peteghem 2007 *Food Addi. and Contam.* **24:** 398-405.

Logrieco, A, B. Doko,A. Moretti,S. Frisullo and A. Visconti 1998 *J. Agric.* Food *Chem.* **46:** 5201-5204.

Lopez, L. C. and C. M. Christensen 1967 *Phytopathology* **57**: 588-590.

Lopez-Garcia, R., D.C.Augusto, C.A. Mallmann and M. Pineiro 2008 *Food Addi and Contam.* **25**: 231-240.

Lopez-Garcia, R., D.L.Park and T.D. Phillips 1999 *Food Nutrition and Agric* **23**: 38-47.

Louria, D.B., J. K. Smith and G. C. Finkel 1970 *Ann. Acad. Sci.* **174**:583-591.

Lovelace, C. E. A. and C. B. Nyathi 1977 *J. Sci. Food Agric.* **28**: 282-292.

Low, W. C., A. Adler and W. Edinger 1991 *Mycotoxin Res.* **7**: 71-76.

Lowe, G., A. Taylor and L. C. Vining 1966 *J. Chem. Soc.*Pp. 17-99.

Lu, P., R. R. Marquardt and D. Kierekjas Zuk 1995 *Lett. Appl. Microb.* **20**:41-45.

Lu, Z. W. R., E. C. Dantzer, V. Hopmans, J. E. Prisk, J.E. Cunnick, P.A.Murphy and S. Hendrich 1997 *J. Agric. Food Chem.* **45**: 803-809.

Luchlel, D., J.G.Bluemink and S. W. de Latt 1976 *J. Ultrastruct. Res.* **54**: 406.

Luo, K.C., B.Kobbe and J.M. Townsend 1977 *Appl. Envir. Microb.* **33**: 211-212.

Luo, U., T.Yoshizawa and T.Katayama 1990 *Appl. Envir. Microb*.**56**: 3723-3726.

Luo, X. Y. 1988 *Proc. Jpn Assoc. Mycotoxicol. Special issue No.* **1**: 97-98.

Luo, Y., T. Yoshizawa and T. Katayama 1990 *Envir. Microb.* **56**: 3723-3726.

Maaroufi, K., A .Achour, M. Hammami, M. May, M. Betbedi, F. Ellouz, E. E. Creepy and H. Bacha 1995 *Hum. Exp. Toxicol.* **14**: 609-614.

Maaroufi, K., A. Achour, A.M. Betheder, M. Hammami, F. Ellouz, E. E. Creppy and H.Bacha 1995a *Arch. Toxicol.* **69**: 552-558.

Mac Kentzie, S.E., M.E. Savard, B.A. Blackwell, J.D. Miller and J.W. Apsimah 1998 *J. Nat. Prod.* **61**: 367-369.

Mac Millan, J.G., J. P.Springer, J.Clardey, R. J.Cole and J.W.Kriskey 1976 *J. Amer. Chem. Soc.* **98**: 236.

Maclead, R.M., J.E. Lehmayer and C.Bruni 1973 *Proc. Soc. Exp. Biol. Med* **144**: 259-267.

Machiwrski, J. Jr and L. M. V. Soares 2000 *Food Addi.and Contam.* **17**: 875-879.

Madhyastha, M. and R. V. Bhat 1984 *Appl. Envir. Microb.* **48**: 376-379.

Mass, C.W., P.S.Steyn and F.R.Van Heerden 1981 *J.Chromatogr.***234**:489-493.

Mag, T., A.E.Melchinger, D.Klein and M.Bohn 2002 *Plant Breeding* **121**:146-154.

Magan, N.,G.R.Cayley and J.Lacey 1984 *Appl. Envir.Microb.***47** : 1113-1117.

Magnoli, C. E., M.A. Saenz, S.M.Chiacchiera and A.M.Dalcero 1999 *Mycopathologia* **145**: 35-41.

Mahadevan, A. and P. T. Kalacheran 1988 *Curr. Sci.* **57**: 193-194.

Majerus, P. and Z. Zakaria 1992 *Z. Lebensm. Untexs Forsch.* **195**: 316-319.

Makela, S., M. Poutanen, J. Lehtimaki, M. L. Kostian, R. Santti and R. Vihko 1995 *Proc Soc Exp Biol Med..* **208**: 51-59.

Malaveille, C., I. G. Brun and H. Bartsch 1994 *Mutat. Res.* **307**: 141-147.

Malawista, S. E. 1971 *Nature* (London) **234**:354-355.

Malaiyandi, M., R.F.Vesonder and A.Ciegler 1976 *J.Envir.Sci.Health Bull.* **11**:139-184.

Malik. R. K., G.Engle and M.Teuber 1986 *Appl. Microb.Biotechnol.* **24**: 71-74.

Mall, G.P., H.M. Pateria and S.K.Chauhan 1986 *Indian Phytopath.* **39**: 409-413.

Mall, O. P., P. Mehta, S. C. Agarwal and V.C. Vora 1979 *Indian J. Microbiol.* **19**: 118-122.

Mallmann, C. A., J. M. Santurio and P. Dilkin 1999 *Revsta-de-Microbiolgia* **30**:249-252.

Manasek, F. J., B. Burnside and J. Stroman 1972 *Proc. Natl. Acad. Sci. USA* **69**: 308-312.

Manickam, A.,A.Palaniswami and S. Neelakantan 1985 *Indian Ver. J.* **62**:711-714.

Mann, G.E., H. K.Jr. Gardner, A.N.Booth and M.R.Gumbmann 1971 *J. Agric. Food Chem.* **19**: 1155-1158.

Mann, G. E., L.P. Codifer, H.K. Gardner, S.P. Koltun and P.G. Dollear 1970 *J.Am. Oil Chem. Toxicology* **35**: 807-810.

Manorama and R.S. Singh 1995 *J. Dairy Foods and Home Sci.* **14**: 101-107.

Mantle, P.G. and J.Penn 1989 *J. Chem. Soc.* **1:** 1539-1540.

Mantle, P. G., J.B.Day, C.R.Harish and R.H.C. Penny 1978 *Vet. Res.* **103**: 403-408.

Mantle, P. G., P. H.Martimer and E. P. White 1977 *Res. Vet. Sci.* **24**: 49-56.

Maragos, C. M. 1997 *Food Agric. Immunol.* **9**: 147-157.

Maragos, C. M. and J. L. Richard 1994 *J. AOAC.* **77**: 1162-1167.

Maragos, C. M., M.E Jolley, R.D. Plattner and M.S. Nasir 2001 *J. Agric. and Food Chem.* **49**: 596-602.

Maragos, C.M. and R.D.Plattner 2002 *J. Agri. and Food Chem.* **50**: 1827-1832.

Marasass, W.F.O. 1974 In Mycotoxins (ed. I.F.A. Purchase), Elsevier Publication, Amsterdam, pp. 111-127.

Marasass, W.F.O. 1991 In Mycotoxins and animal foods (ed J.E. Smith and R.S. Hederson), CRC Press, BocaRaton, FL, pp. 119-139.

Marasass, W. F. O. 1995 *Nat. Toxins.* **3**: 193-198.

Marasass, W.F.O. 2001 *Envir. Health Prospect* **109**: 239-243.

Marasass, W. F. O., E. B. Smalley, J.R.Bamburg and F.M.Strong 1971 *Phytopathalogy* **61**: 1488-1491.

Marasass, W.F.O., F.C. Wehner, S.J. Van Rensburg and D.J. Van Schalkwyte 1981 *Phytopathology* **71**: 791-796.

Marasass, W.F.O. and N.P.J. Krick 1983 In Trichothecenes chemical, Biological and Toxicolgoical Aspects, (ed. Y. Ueno), Elservier Publication. pp. 273.

Marasass, W.F.O., N.P.J. Kriek, V. Wiggins, P.S. Steyn, D.K. Towers and T.J. Hastic 1979 *Phytopathology* **69** : 1181-1185.

Marasass, W.F.O. and P.E. Nelson 1987 Mycotoxicology: Introduction to the mycology plant pathology, chemistry, toxicology and pathology of natural occurring mycotoxicosis in animals and man, Pennsylvania State, University Press, University Park. pp.41-43.

Marasass, W.F.O., P. E.Nelson and T.A.Tousoum 1988 *Mycologia* 80: 407-410.

Marasass, W. F. O., P. E Nelson and T.A. Toussoum 1984 Pennsylvania State Univ., University Park, Pennsylvania, pp. 328.

Marasass, W.F.O., S.J. Van Rensburg and C.J.Mirocha 1979 *J. Agri. and Food Chem.* **27**: 1108-1112.

Marasass, W.F.O., T.S. Kellerman, J.G. Piennar and T.W. Naude 1976 *J. Vet. Res.* **43**: 113-122.

Marasass, W. F. O., T. S. Kellerman, W. C. A. Gelderblom, J.A.W. Coetzer, P.G. Thiel and E. R. Lugt 1976 *J. Vet. Res.* **55**: 197-203.

Moreau, M. and C. Moreau 1960 *Eances. Acad. Sci.* **251**: 1556-1557.

Marin, S., N.Magan, J. Serra, A.J. Ramos and S.V. Sanchi 1999 *J. Food Protec.* **64**: 921-924.

Marin,S., N.Magan, N.Bell, A.J.Ramos, R.Cancl and V.Sanclis 1999b *J.Food Microb.* **51**: 59-67.

Marin, S., V. Sanchis and N. Magna 1996 *Can. J. Microb.* **41**: 1063-1070.

Marquardt, R. R. 1996 *Animal Feed Science and Technology* **58**: 77-89.

Marshall, E. 1982 *Science* **217**: 31-34.

Martin, R. L., M. D. Willson and D. B. Evans 1978 *J. AOAC.* **58**: 1353-1358.

Marth, E. H. 1973 *Special Report* **3**: 8-13.

Martien, C. Spanjer, Peter M. Rensen and M.Jos.Scholten 2008 *Food Addi. and Contam.* **25**: 472-489.

Martins, M. L., H. M. Martins and F. Bernardo 2001 *J. Food Protec.* **64**: 1268-1270.

Mashaly, El-Deeb, A. Ismail and A. Yousef 1981 Proc. Int. symp. Mycotoxins, Sept 6-8th Cairo, Egypt, the General organization for Govt.**10**: 515-522.

Mass, C.W., P.S. Styn and F.R. Van Heerden 1981 *J.Chromatogra.* **234**: 489-493.

Masimango,N., J.Remacle and H.Ramaut 1979 *Ann.Nutr.Aliment.***33**:149-162.

Matsuka, Y. and K. Kubota 1981 *Toxicol. Appl. Pharmacol.* **50**: 87-94.

Matsuka, Y. and K. Kubota 1982 *J. Pharm. Dym.* **5**: 193.

Matsuka, Y., K. Kubota and Y.Ueno 1979 *Toxicol. Appl. Pharm.* **50**: 87.

Matsuka, Y. and T. Yoshizawa 1981 *J. Food Hygie. Soc. Jap.* **22**: 292-298

Machatschka, S., C. Kamrowsks, B.M.Moerschbacher and J. J. Reizenev 1990 *Physio. Mol. Plant Pathol.* **36**: 451-459.

Mathur, S., P.D. Constable, R.M. Eppley, M.E. Tumbleson, G.W. Smith, W.S., Tranquilli, D.E. Morin and W.M. Haschek 2001 *Toxicological Sci.* **60** : 379-384.

Matini, R., T.A.Venkitasubramaniam and K.G. Mukerji 1983 Proc. Symp. Mycotoxin in food and feed (eds. K.S. Bilgrami, T. Prasad and K.K.Sinha), Allied Press, Bhagalpur, pp.37-44.

Maya, S.P. and G.E. Silva 1997 *J. AOAC.* **80**: 825-828.

Mayura, K., M.A. Abdel, K.S. Mc Kenzie, A.B. Sarr, J.F. Edwards, K. Naguib and T.D. Phillips 1998 *Toxicol. Sci.* **41**: 175-182.

Mayura, K., O. Berges, O.Luzon, V.Barrientes and P.Quijoda 1999 *Rev.de Lafac. De. Agro. Uni. del-zulia* **17** : 185-195.

Mc Corkindale, N. J. 1976 In: The filamentous fungi vol. 2. Biosynthesis and Metabolism (eds. J. E. Smith and D. R. Berry), Edward Arnold, London. pp. 362-422

Mc Daniel, M., C.Roth, J.Fink, G.Fyfe and P.Lacy 1975 *Biochem.Biophys. Res. Commun.* **66**: 1089-1096.

Mc Elory, L. J. and C. M. Weiss 1993 *Can. J.Microb.* **39**: 861-863.

Mc Grath, R. M., P. S. Styen, N.P. Ferreira and P. C. Neething 1976 *Biorg. Chem.* **4**: 11-15.

Mc Lean, M. and M.F. Dutton 1995 An update *Pharmacol. Ther.* **65**: 163-192.

Mc Millan, J.A., C.C. Chang, M.K.Grensley, I.C.Paul, S.A.Patwardhan, S.Dev, M.A.Beno and G.G.Christopher 1977 *Tetrahedron Lett.* 2771

Megalla, S. E. and A. H. Hafez 1982 *Mycopathologia* **77**: 89-91.

Megharaj,M.,I.Garthwaite and J.H.Thiele 1997 *Lett.Appl.Microb*.**24**:329- 333.

Mehan, V. K . and J. S. Chohan 1974 *Indian Phytopath.* **27**: 168-170.

Meisner, H. and S.Chan 1974 *Biochemistry* **13**: 2795.

Meky, F. A., L.J.Hardie, S. W.Evans and C. P. Wild 2001 *Food and Chemical Toxicol.* **39**: 827-836.

Melonk, H. A. and C. N. Akem 1987 *Mycopathology* **100**: 91-96.

Merdith, F. I. 2000 *Methods Enzymol.* **311**: 361-373.

Meredith, F. I., O.R. Torres, S. S. De Tejoda, R. T. Riley and A. H. Merrill Jr 1999 *J. Food Protect.* **62**: 1218-1222.

Meronuck. R.A., J.A. Stule, C.J. Mirocha and C.H. Christensen 1972 *Appl .Microbiology* **23**: 613-617.

Meronuck, R.A., K.H. Garren, C.M. Christensen, G.H. Nelson and Fern Bates 1970 *Am. J. Vet. Res.* **31**: 551-555.

Messler, P. and C. Auclair 1974 *Dev. Biol.* **36**: 218-224.

Metcalf, B. W., P. Bey, C. Denzin, M. J. Jung, P. Casara and J. P. Vevent 1978 *J. Amer. Chem. Soc.* **100**: 2551-2553.

Metmarski, M. T. and K. A. Scudamore 1991 *J. Chromatogr.* **588**: 47-52.

Mikami, Y., H. Nishijima, A. Iimura, A. Suzuki and S. Tamura 1971 *Agri. Biol. Chem.* **35**: 611-618.

Miller, D. C., J. L. Richard, F. B. Hembrough, G. D. Osweiller and D. F. Cox 1990 *Amer. J. Vet. Res.* **51**: 836-838.

Miller, E.N.C., S. M. Alcock, H. A. Lee and M. R. A. Morgan 1990 *Food and Agric. Immun.* **2**: 109-118.

Miller, J. D. 1995 *J. Stored Proudcts Res.* **31**: 1-16.

Miller, J. D., A. Taylor and R. Greenhaigh 1983 *Can. J. Microbiol.* **29**: 1171-1178.

Miller, J.D. and H.L.Trenholm 1994 Myotoxins in Grain: Compounds other than Aflatoxin, Eagan Press, St. Paul, Minn., pp. 19-36.

Miller, J. D and R. Greenhalgh 1985 *Mycologia* **71**: 130-136.

Miller, N., H. E. Pretorius and D. W. Trinde 1985 *J. AOAC.* **68**: 136-137.

Miller, P. A., K. P. Milstrey and P. W. Trown 1968 *Science* **159**: 431-432.

Mills, J. T. 1990 *Can. J. Physiol. and Pharmacol.* **68**: 982-986.

Mills, J. T., D.Abramson, A.A. Fronlich and M.K.A.R. Quardt 1989 *Can. J. Plant Pathol.* **11**: 357-360.

Minato, H. and M. Matsumoto 1970 *J. Chem. Soc.* 38-39.

Minato, H., M. Matsumoto and T. Katayama 1973 *Annu. Rep.Shinogi Res.* La No 23: 4-19.

Minato, H and T. Katayama 1970 *J. Chem. Soc Acta* **53**:.44-45.

Minervini, F., F. Fornelli and K. M. Flynn 2004 *Toxicology in vitro* **18:** 21-28.

Mintzlaff, H. J. and W. Christ 1971 *Fleischwirtschaft.* **51**: 1802-1805.

Miraglia, M. and C. Brera 2002 Reports of exports participation in Scoop Task, 327 Rome; Directonate-General Health and Consumer Protection.

Miraglia, M., C.Brera, S.Corneli and R.De Dominics 1993 in Human ochratoxicosis and its pathologies (eds. E. E. Creppy, M. Castegnaro and G. Dirheimer), John Bibbey Eurotext, Montrouge, France, pp. 129-140.

Miraglia, M., R. de dominicis, C. Brera, S. Corneli, E. Cava, E. Menghetti and E. Miraglia 1995 *Nat. Toxins* **3**: 436-444.

Miranda, A. F. and G.C. Godman 1973 *Tissue Cell* **5**: 1-22.

Mirocha, C. J., A. A. Pawlosky, K. chatterjee, S. Watson and W. Hayes 1983 *J. AOAC*. **66**: 1485-1499.

Mirocha, C.J. and C.M.Christensen 1974 In Mycotoxins (ed I. F. H. Purchase), Elsevier, Publication, Amsterdam, pp. 129-148.

Mirocha, C. J., C.M.Christensen and G.H.Nelson 1971 Microbial Toxins, Academic Press, New York. Vol 7, Pp.107-138.

Mirocha, C. J. and D. Dickson 1982 *Nature* **296**: 281.

Mirocha, C. J. and S. V. Pathre 1979 *J. Food Prot.* **42**: 821-824.

Mirocha, C. J., S.V. Pathre and C. M. Christensen 1977 In Mycotoxic Fungi, Mycotoxins, Mycotoxicoses .(eds.T.D.Wyllie and L.G.Morehouse).Marcel Dekker Inc, New York, Vol 1:365.-420

Mirocha, C.J., T. S. Robison, R. J. Pawlosky and N. K. Allan 1982 *Toxicol. Appl. Pharmacol.* **66**: 77-87.

Mirocha, C.J., S.V.Pathre, B.Schauerhamer and C.M.Christensen 1976 *Appl. Envir. Microb*. **32**: 533-556.

Mirocha, C. J., S.V.Pathre and C.M.Christensen 1977 In Mycotoxins in Human and Animal Health (ed. J. V. Rodricks, C. W. Hesseltine and M. A. Mehlmah), Pathodox, Park forest South, III. pp. 345.

Misra, N. 1987 *J. Food Sci. and Technol. India* **24**: 324-325.

Misra, N. and S. Batra 1987 *J. Food Safety* **8:** 157-160.

Misra, N.K. and S.K.Daradhiyar 1991 *Appl. and Envir. Microbiol.*57: 1223-1226.

Mislivec, P. B., C. T. Dieter and V. R. Bruce 1975 *Appl. Microbiol.* **29**: 522-526

Mislivec, P. B., V.R. Bruce, M.E. Stack and R.Bandler 1987 *J. Food Prot.* **50:**38-41.

Misra, R. S., K. K. Sinha and Premalatha 1981 *Nat. Acad. Sci. Letters* **4**: 123.

Mitra, J., E. R. Rati and R. Joseph 2000 *J. Microbiology* **40**: 119-123.

Mitterbauer,R.,H.Weindrofer,N.Safair,R.Krska,M.Lemmens,P. Ruckenbauer, K. Kuchler and G. Adam 2003 *Appl. and Envir. Microb.* **69** : 805-811.

Miura, K., L. Aminova and Y. Murayama 2002 *Toxicology* **172**: 103-112.

Miyake, I., H. Naito and H. Sumeda 1940 Report of the Research institute for Rice Improvement **1**: 1-24.

Moerc, K. E., M.C. Eltnesh, P.Wohlman and B.W. Hilton 1980 *J. Food Protec.* **43**: 571.

Moller, T. E. and E. Josefsson 1980 *J. AOAC.* **63**: 1055-1056.

Moller, T. E. and H. F. Gustavsson 1992 *J. AOAC.* **75**: 1049-1052.

Molnar, O., G. Scahtzmayr, E. Fuchs and H. Prillinger 2004 *Syst. Appl. Microb*.**27**: 661-671.

Montani, M., G. Vaamonde, S.L.Resmik and P.Buera 1988 *J. Food Microbiol.*7:115-122.

Moodley, J., D. Moudley, L. Reddy, M. F. Dutton and A. A. Chuturgoon 2001 *J.Obst. and Gyne.* **21** : 599-600.

Moody, S. F. and B. M. Tyley 1990 *Appl. Envi. Microb.* **56**: 2453-2461.

Moon-Woosung., Kim-Jootteon, Kang-Myoung Jae, Lee-Dohg Geun, Moon-WS,J. H. Kim, M. J. Kang and D. G. Lee 2000 *Yonsei Medical Journal* **41** : 195-204.

Moran, E. T. Jr., P. R. Ferket and A. K. Lun 1987 *Poultry Sci.* **66**:977-982.

Morcho Romo, M. A. and G. S. Fernandez 1986 *Poultry Sci.* **65**: 284-287.

Moreau, C. 1990 *Mycologia* **11**: 89-110.

Moreau, C. and M. Moss 1979 In Moulds, Toxins and Food.(eds C.Moreaux and M. Moss) John Wiley and Sons, Chichester, New York, Brisbane, Toronto. pp. 286319.

Morirry, R. E., R. J. Cole and W. J. Dorner1984 *J. Toxicol. Envi. Health* **14**: 585-594.

Mortimer, D. N., M. J. Shepherd, J. Gilbert and C.Clark 1988 *Food Addi. Contam.* **5**: 601-608.

Moss, M.O. 1984 In : The applied Mycology of *Fusarium*(ed. M.O. Moss and J.E. Smith), British Mycological Society, Symposial Series 7,London, pp.195

Moss, M.O. 1996 In Microbial Food poisoning. Second Edition. (ed A. R. Eley), Published by Chapman & Hall, New York, NY, *pp.* 75–93.

Moss, M.O. 1996 In Microbial Food poisoning. Second Edition. (ed A. R. Eley), Published by Chapman & Hall. pp.

Moss, M. O. 1996a *Mycological Research* **100**: 513-523.

Moss, M.O. 1998 *Appl. Microbial. Symp. Suppl.* **84**: 625-765.

Moss, M. O. and J. E. Smith 1985 Mycotoxins. Formation, Analysis and Significance, John Wiley and Sons, Chichester United Kingdom pp. 214.

Moos, M.O., M.A.Woody and F.S. Chu 2000 *J. Food Saf.* **3**: 39-47.

Moule, Y. and F. Hatey 1977 *FEBS Lett.* **74**: 121-125.

Moule, Y. 1984 In Mycotoxins (ed. V. Betina) Elsevier Publications, Amsterdam.p. 37.

Moule, Y., M. Jemmali and N. Darracq 1978 *FEBS Lett.* **88**: 341-344.

Mull, R. P., R. W. Townley and C. R. Scholtz 1945 *J. Am. Chem. Soc.* **67:** 1626-1627

Muller, H. G. 1989 *Food Policy and Nutrition* 2: 273-291.

Mukherjee, K . and A. V. Lakshminarasimham 1995 *Zentrolbl Bakeriol*. 282: 237-243.

Mukit, A. and M. S. Kwatra 1989 *Indian. J. Poult. Sci.* **24**: 292-294.

Mukt, A., S. Goswami and M.Bhattacharya 1994 *Indian J.Polult. Sci.***29**: 72-74.

Muldevs, E.D. and H.A.M.Van Impelen – Peek 1986 *Z. Lebensm, Unters, Forsch.* **183**: 406.

Mulkherkar, L., P.Navulurkar and M. V. Josh 1978 *Indian J. Exp. Biol.* **16**: 430.

Mullabacher, A. and R.D.Fichner 1984 *Proc.Nat.Acad.Sci.USA* **81:** 3835-3837.

Mullaney, E. J. and M. A. Klich 1990 In Modern concepts in *Penicillium* and *Aspergillus* Classification. (eds. R. A. Samson and J. I. Pitt), Plenum New York. pp. 321-334.

Muller, A. M., J.E. Reimann, U. Schunacher and K. Schwadorf 1998 *Food Addi.and Contam.* **15**: 801-806.

Muller, H. M. 1984 *Animal Research and development* **19**: 7-37.

Muller, H. M. and K. Schwadorf 1993 *Bull Envir. Contam. and Toxicol* (USA). **51**: 532-537.

Muller, H. M. and M. Thaller 1982 *Arch. Ticrenahr.* **31**: 789-792.

Mullett, W., E. P. C. Lai and J. M. Yeung 1998 *Anal. Biochem.* **25**:161-167.

Munimbazie, J. W., T. Saxena, L. B. Tsai and L.B. Bullerman 1997 *J. Food Protec.* **60**:849-852.

Munkvold, G. P. 2003 *European J. Plant Pathology* **109**: 705-713.

Murphy, G., G.Vogel, G. Krippahl and F. Lynen 1974 *Eur. J. Bio.Chem.* **49**: 443-455.

Murphy,G.,G.Vogel,G.Krippahl and F.Lynen 1975 *Eur.J.Biochem.***58**:467-475.

Musser, S. M. and R. D. Plattner 1997 *J. Agric. Food Chem.* **45**:1169-1173.

Myrchink, T. G. 1967 *Antibiotiki* **12**: 762-766.

Nagaraj, G. and K.Kumar 1986 *J.Oil Technologists Assoc.* India **18**:89-91.

Nagarajan, R., L. L. Huckstep, D. H. Lively, D. C. Delong, M. M. Marsh and N. Neuss 1968 *J. Am. Chem. Soc.* **90**: 29-80

Nagare, M., M. M. Alam, A. Tsushima 2001 *Biosci. Biotechnol. Biochem.* **65**: 1741-1747.

Nagashima, H. 1996 *Mycotoxins* **42**: 57-61.

Nahdi, S. and M. Nusrath 1985 *Indian J. Bot.* **8:** 16-24.

Nakagawara, A., K. Kakinuma, H.Shin, S.Miyazaki and S.Minakami 1976 *Clin. Chim. Acta* **70 :** 133-137.

Nakajima, M., H. Tsubonchi, M.Miyabe and Y.Ueno 1997 *Food Agric. Immunol.* **9**: 77-83.

Nakajima, M., H. Terada, K.Hisada, H.Tsubouchi, K. Yamamoto, T. Uda, O. Kawamura and Y. Ueno 1990 *Food Agric. Immunol.* **2**: 189-195.

Nakajima, M., H.Tsubonchi and M.Miyabe 1999 *Food Addi.and Contam.* **13**:5-9.

Nakamura, Y., M. Ohta and Y. Ueno 1977 *Chem. Pharma. Bull.* **25 :** 3410-3414.

Nakazato, Y. and W. Douglous 1973 *Proc. Nat. Acad. Sci.* USA **170**: 1730-1733.

Narsimhan, K.S. and G.Rangaswami 1968 *Indian Phytopath. Soc. Bull* **4:** 57-61.

Nasir, M. S. and M.E. Jolly 2002 *J. Agric. Food Chem.* **50**: 3116-3121.

Natori, S.1977 In Mycotoxins in Human and Animal Health (ed. J. V. Rodricks, C. W. Hesseltine and M. A. Mehlman) pathodox, Park Forest south, III. p. 559.

Niderkorn, V., D. P. Morgavi, E. Pujos, A. Tissandier and H. Boudra 2007 *Food Addi.and Contam.* **24:** 406-415.

Neelakanthan, S., G. Indira Jasmine, R. Balasaraswathi and T. Balasubramanian 1984 *Madras Agri. J.* **71:** 43-45.

Neelakanthan, S., R. Balasaraswathi, T.Balasubramaniayan and Indira Jasmin 1982 *Madras Agric. J.* **69:** 326-328.

Neelakantan,S., R.Swaminathan, T.Balasubramanian, R. Balasaraswathi and G. Indira Jasmine 1978 *Indian Poult.Gaz.* **62:** 40-44.

Neelakanthan, S., T.Balasubramnian, R.Balasaraswathi, G.Indra Jasmine and R. Swaminathan 1978a *J. Food Sci. Tech.* **15:** 125-134.

Neish, G. A., E. R. Farnwarth, R. Greenlaugh and J. C. Young 1983 *Can. J. Plant Pathol.* **5:** 11-16.

Neish, G. A. and H. Cohen 1981 *Can. J. Plant Sci.* **61 :** 811-815.

Neldon-Ortiz, D. L. and M. A. Qureshi 1992 *Veteri. Immun. and Immunopatho.* **31**: 61-76.

Nelson, D. R. and L. F. Jaffe 1973 *Dev. Biol.* **30:** 206-208.

Nelson K. F., K. Huttunen and A. Hyvarinin 2002 *Mycopathologia* **154:** 201-205.

Nelson, P. E., R. D. Plattner, D. D Shackelord and A. E. Desjardins 1992 *Appl. Envir. Microb.* 984-989.

Nelson, P. E., T. A. Toussoun and W. F.O. Marasas 1985 *Fusarium* species, an illustrated Manual for Identification. Penn. State. Univ. Press, Univ. Park.

Nencere, J. N. and T.E. Cleveland 1997 *J. Agric. Food Chem.* **45**: 299-301.

Nenss, N. 1968 *J. Am. Chem. Soc.* **90:** 2980.

Nesheim, S. 1969 *J. AOAC.* **52:** 975-979.

Nesheim, S. 1971 *J. AOAC. Washington* DC. October 11-14.pp. 23.

Nesheim, S. 1973 *J. AOAC.* **56**: 817-821.

Nesheim, S. 1973 *J. AOAC.* **56**: 822-826.

Nesheim, S. 1976 *Adv. Chem. Ser. No.* **1796**: 273-278.

Neve, P., P. Ketelbant-Balasse, C. Willems and J. E. Dumont 1972 *Exp. Cell Res.* **74** : 227-244.

Newborne, P. M. 1987 Mycotoxins in Food, Academic Press, New York. pp. 177-216.

Newnay, J. and G. M. Gaucher 1982 *Can. J. Microbiol.* **97:** 206-208.

Nicholson, P., D. R. Simpson, A. H. Wilson, E. Chandler and M. Thomsett 2004 *European J. Plant Pathology*. **110**:503-514.

Nicklas, W. J. and S. Berl 1974 *Nature (London)* **247:** 471-473.

Nicolas-Bolnet, C., M. A. Qureshi, J. A. Cieszynski and R. L. Taylor 1995 *Poultry Sci.* **74**: 1970-1976.

Niemierko, A. and A. Komar 1976 *J. Reprod. Fertil.* **48:** 279-284.

Niedwetzki, G., G. Lach and K. Geschwill 1994 *J. Chromatogr.* **661**: 175-180.

NIN 1992 National Institute of Nutrition, Hyderabad, Annual Report 1996-1998. pp. 35-36.

Nirenberg, H. 1976 Mitt. Biol. Bundsanst. *Hand. Forstw. Berlin. Dahlem* **169:**1-117.

Nishic, K. R., J. Cole and J. W. Dorner 1987 In 71st Annual meeting of the Federation of American Societies for Experimental Biology, Washington D. C. *Fed Proc.* **46:** 559.

Nisonoff, A. 1991 *J. Immunol.* **147:** 2429-2438.

Nagawa, K.and A. Ishizaki 1979 *Envir. Res.* **18**:410-420.

Nordin, N. and R. H. Luchese 1998 *Boletin SBCTA* **32** : 35-39.

Norred, W.P., C.W. Bacon, R.T. Riley, K. A. Voss and F. I. Meredith 1999 *Mycopathologia* **146** : 91-98.

Norred, W. P., K. A. Voss, C. W. Bacon and R. T. Riley 1991 *Food Chem. Toxicol.* **29:** 815-819.

Norred, W. P., R. D. Plattner, R. F. Vesonder, R. F. Hayes, C. W. Bacon and K. A. Voss 1990 *Toxicologist* **10** : 165-171.

Norred, W. P., R. J. Cole, D. J. Erlington, S. Suksupath, G. M. Mc Dowell and W. L. Bryden 1987 *J. AOAC.* **70**: 121-123.

Norred, W. P., T. K. Porter, J. W. Dorner and R. J. Cole 1988 *J. Agric. and Food Chem.* **36:** 113-116.

Norris, P. J., C.C.T.Smith, J. De Bellorche, H. F. Bradford, P.G. Mantle, A. J. Thomas and R. H.C. Penny 1980 *J. Neurochem.* **34**: 33-42.

Normann, J. 1968 *Archivfiir Mikrobiologie* **61:** 128-142.

Norstadt, F. A. and T. M. Mc Calla 1969 *Appl. Enivir. Microb.* **17:** 193-196.

Norstadt, F. A. and T. M. McCalla 1971 *Plant & Soil* **34:** 97-108.

Northolt, M.D. and H.P.Van Egmond 1981 *Proceedings of the 4th meeting on mycotoxins and animal disease*. Weybridge, England, pp. 106-108.

Northolt, M.D., H. P. Van Egmond and W. E. Paulsch 1979 *J. Food Prot.* **42**: 485-490.

Northolt, M.D. and L. B. Bullerman 1982 *J. Food Prot.* **45:** 519-526.

Nowar, M. E., 1975 Moldy Corn Naturally Contaminated with Mycotoxins in Feeding Animals and Poultry. Ph. D. Thesis, Navisad University, Yuogoslavia.

Nummi, M., M. L. Niku Paavala and T. M. Enari 1975 *Extrailt De Bios*. 266.

Nusrath, M. and S. Nahdi 1983 *Indian Phytopath*. **36** : 106-109.

Nusarath, M. and V. Ravi 1984 *Indian J. Bot.* **7:** 94-97.

NO'Neill, F. J . 1972 *Chromosoma* **47:** 453-462.

Obzazhal, A. F. 1987 *Vetunariva Moscow* USSR. Pp. 58-61.

Odhav, O., K. Permaul, S. Rama Sunder, T. Padayachee, V. Mohanlal and L. Reddy 2001 Biological control of fungal and bacterial plant pathogens (ed. Y. Blad and S. Freeman), pp. 113-136.

Odhav, O. and V. Naickar 2002 *Food Addi.and Contam*. **19:** 55-61.

Ohonomo, S., M. Sugita and M. Abe 1973 *J. Agric. Chem. Soc.* **47:** 57-63.

Ohta, M., Ishii and Y. Ueno 1977 *J. Biochem.* (Tokyo) **82:** 1591-1598.

Ohtsubo, K. and M. Saito 1970 *Jpn. J. Med. Sci. Bio*. **23** : 217-225.

Ohtsubo, K., M.Saito, H.Kimmura and O.Tsuruta 1978 *Food Cosmet Toxicol.* **16**:143-150.

Ohtsubo, K., M. Saito, S. Sekita, K. Yoshihira and S. Natori 1978 *Jpn. J. Exp. Med.* **48** :105-110.

Ohtsubo, K., M. Yamada and M. Saito 1968 *J. Med. Sci. Biol.* **21**: 185-194.

Okoye, Z. S . 1987 *Food Addi.and Contam*. **15**: 801-806.

Oldenberg, E. 1993 *Mycotoxin Res.* **9:** 72-78.

Oldenberg, E. 1995 *Mycotoxin Res.* **9**: 72-78.

Oliveira, C. A. F., E. Kobashigawa, T. A. Reis, L. Mestieri, R. Albuquerque and B. Correa 2000 *Food Addi. and Contam.* **17**: 459-462.

Olsen, M., C. J. Mirocha, H. K. Abbas and B. Johansson 1986 *Poultry Sci.* **65:** 1905-1910.

Onji, Y., Y. Aoki, N. Jani, K. Umebayashi, Y. Kitada and Y. Dohi 1998 *J. Chromato*gr. **A 815:** 59-65.

Ono, E. U. S., E. Y. Sasaki, E. H. Hashimoto, L. N. Hara, B. Correa, E. N. Itano, T. Sugiura, Y. Ueno and E. Y. Hirroka 2002 *Food Addi. and Contam.* **19:** 1081-1090.

Ono, H., A. Kataoka and M. Koakutsn 1995 *Mycotoxins* **41**: 47-51.

Ono, M., S. Sakuda, A. Suzuki and A. Isoga 1997 *J. Antibiotics* **50**: 111-118.

O'Neill, F. J. 1972 *Natl. Cancer Inst.* **49**: 1733-1737.

Osterhass, H. K. 1946 *Antibiotica Vit Schimmels. Med. Lab. Physiol. Chem.* Univ. Amsterdam, p. 116.

Orange, R. P. 1975 *J. Immunol.* **114** : 182-186.

Orci, L., K. H. Gabbay and W. J. Malaisse 1972 *Science* **175:** 1128-1130.

Orr, T. S. C., D. E. Hall and A. C. Allison 1972 *Nature (London)* **236:** 350-351.

Orth, R. 1977 *Ann. Nutr. Aliment.* **31:** 617-624.

Orzaez, D., A. J. de Jong and E. J. Woltering 2001 *Plant Molecul. Biol.* **46**:459-468.

Osborne, D. J., W. E. Huff, P. B. Hamilton and H. R. Burmeister 1982 *Poultry Sci.* ***61***: 1646-1652.

Ostry, V. and M. Polster 1989 *Verterinarie Medicina* **34**: 421-430.

Osweiler, G. D. and D. W. Trampel 1985 *J. Am. Vet. Med. Assoc.* **189:** 636-637.

Osweiler, G. D., P. F. Ross, T. M. Wilson, P. E. Nelson, S. Witte, T. L. Carson, L.G Rice and H. A. Nelson 1992 *J. Vet. Invest.* **4:** 53-59.

Oyelami, O. A., S. M. Maxwell and E. Adeoba 1996 *Annals of Tropical Pardiatrics* **16**: 137-140.

Ozcelik, S., N. Ozcelik and L. R. Beuchat 1990 *Ing. J. Food Microbiol.* **11**: 187-194.

Padye, S. B., C. R. Joshi and G. S. Pendse 1995 *J. Chromatogr.* **108:** 405-408.

Pallavi, R. M. V., D. Ramana, R. B. Sashidhar 1997 *World J. Microb. and Biotech.* **13:** 713-714.

Palle, K. and H.Bendicte 1983 In: Trichothecenes Chemical, Biological and Toxicological, Aspects (ed. Y. Ueno), Elsevier Publication pp.151.

Palti, J.1978 *Acta Phytomed.* **6**: 1-110.

Panneruselvan, P., K. Velmonrougane, D. R. Shanmukhappa, T. N. Gopinandan and R. Naidu 2000 *J. Coffee Res.* **28**: 40-48.

Pande, M., J. Saxena and H. Pandey 1990 *Mycoses* 33: 126-128.

Panem, S.1977 *Virology* **76:** 146.

Park, D. L. and L. S. Lee 1990 *Agricultural Res. Ser.* pp. 127-737.

Park, D.L., S. Nesheim, M. W. Trucksess, M. E. Stack and R. F. Newell 1989 *J. AOAC.* **73:** 260-266.

Park, J. and F. S. Chu 1996 *J. AOAC.* **79:** 365-371.

Park, J. W., E. K. Kim, D.H. Shan and Y. B. Kim 2002 *Food Addi. and Contam.* **19:** 158-162.

Park, K. J., A. R. Park and X. W. Lee 1992 *Food Addi. and Contam.* **9:** 639-645.

Parker, C. L., R. A. Finch and W. C. Hooper 1978 *In vitro* **14:** 606-615.

Parkhouse, R. M. E. and A. C. Allison 1972 *Nature (London) New Biol.* **235:** 220-222.

Pascale, M., M. B. Doko and A. Visconti 1995 In; *Atti 2nd Congresso Nazional di chimica degli Alimenti, Giardini-naxos, Italy,* pp. 1067-1071.

Passone, M. A., S. Rensik and M. G. Etchevery 2007 J. *Food Microb.* **118**: 8-14.

Paster, N., D. Huppert and R. Barkai-Golan 1995 *Food Addi.and Contam.* **12:** 51-58

Patel, U. D., P. Govindarajan and P. J. Dave 1989 *Appl. Envi. Microb.* **55**: 465-467.

Patey, A. L., M. Sharman and J. Gilbert 1991 *J. AOAC.* **74**: 76-81.

Patkar, K. I., C.M.Usha, H.S.Shetty, N.Paster and J.Lacey 1994 *Crop Protect.* **13:** 519-524.

Patterson, D. S. P.1983 *Vert. Res. Comm.* **7:** 135-140.

Patterson, D. S. P. and B. A. Roberts 1980 *Vet. Rec.* **10:** 249- 252.

Patterson, M. and A. P. Damoglon 1986 *Lett. Appl. Microbiol.* **3 :** 123-125.

Patterson, M. and A. P. Damoglon 1983 *Appl. Microb. Biotech.* **26:** 574-578.

Patwardhan, S. A., R. C. Pandey, S. Dev and G. S. Pendre 1974 *Phytochemistry* **13**:1985-1987.

Paulsch, W. E., H. P. Egmond Van and P. L. Schuller 1982 In: 5th International IUPAC Symposium on Mycotoxins and Phytotoxins, Vienna, pp. 40.

Payne,L. A and J. M. Walls 1984 *Envir. Entomol.* 1609-1612.

Pero, R. W., D. Harvan, R. G. Owens and J. P. Snow 1972 *J. Chromatogra.* **65:** 510-511.

Pena, G. D., D. Mathieu and J.J. Pena-Cabriales 1985 *Rev. Latinoam. Microbiol.* **27:** 249-252.

Peng, F. C., K. H. Ling, Y. Wang and G. H. Lee 1985 *Appl. Environ, Microbiol.* **49**: 721-723.

Pepeljnjak, S. and Z. Cventic 1984 *Mycopathologia* **86:** 83-87.

Percival, J. C. and R. H. Thornton 1958 *Nature* **182**: 1095-1096.

Pero, R. W., D. Harvan, R. G. Ownes and J. P. Snow 1972 *J. Chromatogr.* **65:** 510.

Pero, R. W., H. Porner, M. Blois, D. Harvan and J.W. Spalding 1973 *Envir.Health Perspectives* pp. 87-94.

Pero, R. W., R. G. Ownes and D. Harvan 1971 *Anal. Biochem.* **43**: 80-86.

Perez, R. A., J. Tuite and K. Baker 1982 *Cereal Chem.* **59:** 205-209.

Pestka, J. J. 1994 *Food Agric. Immunol.* **6:** 219-234.

Peteri, Z., J. Teren, C. Vagvolgi and J. Varga 2004 *Food Microbiology* **24:** 265-210.

Peterson, A. and I. Thorup 2001 *Food Addi.and Contam .* **18:** 221-226.

Peterson, D.W., R.H.C.Penny, J.B.Day and P.G.Mantle 1997 *Res.Vet.Sci.* **33**.183-187.

Petersen, E. J., P. Larsen and P. M. Bell 1980 *Tetrahedron Lett.* 5079-5082.

Peterson, R. E. and A. Ciegler 1978 *Appl. Environ. Microbiol.* **36**: 613-614.

Peterson, R. E., G. M. Shanon and O. L. Shotwel 1989 *J. AOAC.* **72**: 332-355.

Phillips, T.D. and A. W. Hayes 1977 *J. Pharmacol. Exp. Ther.* **205:**606-616.

Phillips,T.D., A.W. Hayes and A. W. Heys 1979 *Toxicology* **12:** 285-298.

Phillips, T.D., A.W. Hayes and W. O. Barndt 1980 *J. Chromatogra.* **190:**419-427.

Phillips, T.D., E. Afriyie-gyawu, J. Williams, H. Huebner, N. A. Ankrah, D.Ofori-Adjei, P. Jolly, N. Johnson, J. Taylor, A. Marroquin-Cardona, L. Xu, L. Tang and J.S. Wang. 2008 *Food Addi. and Contam.* **25**: 134-145.

Phillips, T.D., L.F.Kubena, R. B. Harvey, D. R. Taylor and N. D. Heidelbaugh 1988 *Poultry Sci.* **67:** 243-247.

Pohland, A. E. and R. Allen 1970 *J. AOAC.* **53:** 686-697.

Piemontese, L., M. Solfrizzo and A. Visconti 2005 *Food Addi. and Contam.* **22:** 437-442.

Pier, A.C., 1991 In *Mycotoxins and Animal Foods* (eds. J.E Smith and R.S. Henderson), Rc Press, Boca Raton, Florida, pp. 489-497.

Pincire, M. S. and G. E. Silva 1997 *Res. Commun.* **25:** 805-806.

Pitt, J. I.1979 *Mycologia* **71:** 1166-1171.

Pitt, J. I. 1979 "The genus *Penicillium* and its teleomorphic states *Talaromyces* and *Eupenicilium*", Academic Press, London. pp. 634

Pitt, J. I. 1988 *A Laboratory Guide to Common Penicillium species*, 2nd ed. CSIRO Division of food processing, North Ryde, N. S. W., Australia.

Pitt, J. I. 1989 *Appl. Envir. Microb.* **53:** 266-269.

Pitt, J. I. 1996 *Mycotoxon News Letters* **7**:1-3.

Pitt, J. I. 1997 In: Food Microbiology Fundamental and Frontiers. (eds. M. P. Doyle, L. R. Burchat and T. J. Mont ville), Washington, DC: ASM Press. pp. 406-418.

Pitt, J. I. 2000 *Med. Mycol.* **38:** 17-22

Pitt, J. I. and A. D. Hocking 1985 In: Fungi and Food spoilage ESS CSIRO Division of food Res. Academic Press, Sydney. pp. 1-413.

Pitt, J. I. and A. D. Hocking 1997 *Fungi* and *Food Spoilage,* 2nd ed., Chapman & Hall, London.

Pitt, J. I. and J. H. B. Christian 1970 *Appl. Microbiol.* **20**: 682-686.

Pittet, A. 1998 *Rev. Med. Vet.* **149:** 479-492.

Pittet, A., V.Parlsod and M.ShellenBergh 1992 *J. Agri. Food Chem.* **40**: 1352-1354.

Plaa, G. L. and H. Witsch 1976 *Ann. Rev. Pharmacol. Toxicol.* **16 :** 125-141.

Placenta, C. M., J. P. D'Melo and A. M. Mc Donald 1999 *Animal Feed Science and Technology* **78 :** 21-37.

Plagemann, P. G. W. and J. Erbe 1973 *Cancer Res.* **33**: 482-492.

Plattner, R. D., W. P. Norred, C. W. Bacon, K. A. Voss, R. Peterson, D. D. Shackelford and D. Weisleder 1990 *Mycologia* **82 :** 698-702.

Pohland, A. E., K. Senders and C. W. Thorpe 1970 *J. AOAC.* **53:** 692-695.

Polacco, J. C. and D. C. Stands 1977 *Plant Sci. Lett.* **9:** 121.

Poling, M. and R. D.Plantner 1996 *J. Agric. Food Chem.* **44:** 2792-2796.

Pollard, T. D. 1976 *J. Supranol. Struct.* **5**: 317-334.

Pong, S. and L. Devine 1976 *Prostaglandins* **12:** 803-812.

Popken, A. M. and K. Dose 1983 *J. Anal. Chem.* **316**: 47-50.

Porter, J. K., W. P. Norred, R. J. Cole and J. W. Dorner 1988 *Proc. Soc. Experimental Biology & Medicine* **187**: 335-340.

Prasad, B. K. and N. Narayana 1994 *Geobios* **8 :** 47-48.

Prakash, K., B. Umakantha, G. Devagowda, C. V. Gowda and G. Krishnappa 2000 *Indian J. Animal Sci.* **70:** 197-199.

Prasad, B., B. K. Sinha and A. K. Sinha 1987 *Indian J. Comp. Microbiol. Immunol. and Inf. Dis.* **8**: 135-136.

Prathap Kumar, S.H., V. S. Rao, R.J. Param Kishan and R.V. Bhat 1997 *British Poultry Science.* **38:** 475-479.

Pravindrachary, M. and S. M. Reddy 1986 *Acta Botanica. India* **14:** 79-82.

Pravindrachary, M. and S. M. Reddy 1986a *Biol. Bull. Indica* **8:** 81-83.

Pravindrachary, M.and S. M. Reddy 1986b *Nat. Acad. Sci. Letters* **10:** 129-132.

Prawat, C.K. and Tanboon 1991 (eds.R.L.Semple, A.S.Frio and P.A.Hiecks,J.Loware) Bangkok, pp. 166-172.

Preis, R . A. and E. A. Vargas 2000 *Food Addi.and Contam.* **17:** 463-468

Prelusky, D.B., B.A.Rotter and R.G.Rotter 1994 In: mycotoxins in grain: compounds other than aflatoxins (eds. J.D.Muller and H.L.Trenhorm), Egan Press, S. E. Paul M.N., pp. 359-403.

Prelusky, D. B., H. L. Trenholm, G. A. Lawrence and P. H. Scott 1984 *J. Envir. Sci. Health B* **19**:593-609.

Premalatha Singh 1983 *Proc. Symp. Mycotoxin in Food and Feed* (eds. K.S. Bilgrami, T. Prasad and K.K. Sinha), Allied Pub.pp. 207-325.

Premalatha Singh, Sita Bhagat and S.K. Ahmad 1988 *J. Indian Bot. Soc.* **67:** 201-202.

Premalatha Singh, Sita Bhagat and S.K. Ahmad 1990 *J. Food Sci and Technol.* **27:** 60-62.

Premalatha Singh and S. K. Ahmed 1989 *Anals of Agricultural Science* **34**: 805-812.

Priest, J. W. and R. J. Light 1990 *J. Chromatogra.* **513:** 237-246.

Prieta, J. M. A. Moreno, J. L. Blanco, G. Suarez and L.Dominguez 1992 *J. Food Protect.* **55**: 1001-1002.

Prieta, J., M. A. Moreno, J. Bao, S. Diaz, G. Suarez, L. Dominguez, R. Canela and V. Sanchis 1993 *Analyst* **118**: 171.

Prieta, R. and C. P. WoloShuk 1997 *Appl. Envir. Microb.* **63:** 1661-1666.

Prireia, A., J. Tomko and L. Dolejs 1975 *Phytochemistry* **14**: 285.

Prior, M. G., J. B. O. Noili and C. S. Sisodia 1980 *Poultry. Sci.* **59**: 1254.

Probst, A. and Ch.Tamm 1981 *Chem. Helv. Acta* **64**: 2056-2064.

Probst, A. and Ch.Tamm 1982 *Chem. Helv. Acta* **65:** 1543-1546.

Proctor, R. H., A. E. Desjardins, S. P. Mc Cormick, R. D. Plattner, N. J. Alexander and D. W. Brown 2000 *European J. Plant Pathology* **108:** 691-698.

Pruthi, J. S. 1978 *Bull Grain Tech.* **16:** 5168.

Pumplin, D. W. and Mc Clure 1974 *Eur. J. Pharmacol.* **28:** 316-325.

Purchase, I. F. H. 1971 *Toxicol. Appl. Pharmacol.* **18**: 114-123.

Purchase, I. F. H. and J. J. Theron 1968 *Food Cosmet. Toxicol.* **6**: 479-483.

Purchase I. F. H. and W. Nel 1967 In Biochemistry of some Food-borne Microbial Toxins (eds. R. I. Matels and G. N. Wogan), B M I T and Cambridge, Mass. pp.153-156.

Qian, G. S. and G. C. Yang 1984 *J. Agric. Food Chem.* **32**: 1071-1074.

Qiu, M. and X. Liu 2001 *Food Addi.and Contam.* **18 :** 263-269.

Quiroga, N. M., S. L. Sola and E. Varsavsky 1994 *J. AOAC.* **77** : 939-941.

Radostits, O. M., D. C. Blood and C. C. Gay 1994 Veterinary Medicine, 8th ed., W. B. Saunders Co. Ltd., London. pp. 1604-1605.

Raff, R. A. 1972 *Exp. Cell Res.* **71:** 455459.

Rahman, R., S. Safe and A. Taylor 1969 *J. Chem. Soc.* **18**:1665.

Raillo, A. I. 1959 *Fusarium. Publ. Agric. Lit.* Moscow. pp. 415.

Raina, G. L. and J. S. Chohan 1976 *Indian Phytopath.* **29 :** 105-110.

Raistrick, A. and G. Smith 1941 *Chem. Ind.* (London) **6:** 828-830.

Raisuddin, S. and J. K. Misra 1991 *Food Addi. and Contam.* **8**: 707-712.

Raisuddin, S., K.P.Singh, S.I. Zaidi, A.K. Saxena and P.K. Ray 1990 *J. Appl. Toxicol.* **10**:245-250.

Raisuddin, S., K. P. Singh, S. I. Zaidi and P.K. Ray 1994 *Inter. J. Immunopharma.* **16**: 977-984.

Rajakumari Kaushal, Ashwini Kumar and Surendra Singh 1994 *J. Indian Bot. Soc.* **73 :** 131-134.

Rajam, M. V. 1993 *Curr. Sci.* **65**: 461-469.

Rajam, M. V. and A. W. Galston 1985 *Plant Cell Physiol.* **26:** 638-692.

Rajam, M. V., L. H. Wenstein and A. W. Galston 1985 *Proc. Natl. Acad. Sci.* USA. **82:** 6874-6878.

Rajender Singh, D. S. R., B. Bhadraiah and P. Rama Rao 1991 *Indian J. Microbial. Ecol.* **2:** 133-135.

Rajendran, G., Anthoni Raj and N.G.C. Lenger 1987 *Bulletin of Grain Technol.* **25**: 90-91.

Rajendran, M. P., M. J. Hussain and K. Ramani 1975 *Indian Vet. J.* **52:** 234-235.

Rajesh Chahota, R. C., S. P. Katoch, U.P.Singh, S. Verma, Arvind Mahajan and K. B. Nagpal 2000 *Indian J. Animal Sci.* **70:** 22-24.

Raju, M. V. L. N. and G. Devegowda 2000 *Br. Poult. Sci.* **41:** 640-650.

Ramakrishna, Y. and R. V. Bhat 1987 *Curr. Sci.* **56**: 524-526.

Ramakrishna, Y., R. V. Bhat and S. Vasanthi 1990 *J. Agric. Food Chem.* **338:** 1857-1859.

Ramakrishna, Y., R. V. Bhat and V. Ravindranath 1989 *Appl. Environ. Microbiol.* **55 :** 2619-2620.

Ramdoss, C. S. and A. Mukherjee 1972 *J. Antibiot.* **30:** 172-174.

Ramdoss, C. S. and E. R. B Shanmugsundaram 1973 *Indian J. Biochem. Biophys.***10**:296-297.

Ranfft, K., R. Gerstl and G. Mayer 1990 *Zeitschrift fur Lebensmittet Unterscuhing and Forschung* **19:** 449-553.

Ranjan, K. S., S. S. Sahay and A. K. Sinha 1992 *J. Food Sci. and Technol.* India **28**: 221-224.

Rao, A. G., P. K. Dehuri, S. K. Chand, S. C Mishra, R. K. Mishra and B. C. Das 1985 *Indian J. Poult. Sci.* **20:** 240-244.

Rao, B. L. and A. Hussain 1985 *Mycopathologia* **89:** 177-180.

Rao, B. L. and A. Hussain 1987 *Mycopathologia* **97**: 89-92.

Rao, G. V., P. S. Rao, S. Girisham and S. M. Reddy 1985 *Curr. Sci.* **54:** 507-509.

Raper, K. B. and D. I. Fennel 1965 The genus *Aspergillus*. Rober E. Kriegger Pub. Co. Inc. pp. 686.

Rathinavelu, A. and E. R. B. Shanmugasundaram 1984 *J. AOAC.* **67:** 38-40.

Rathket, P. C., E. Scieb, K. Weber, M. Osborn and W. W. Franke 1977 *Exp. Cell Res.* **105:** 253.

Rati, E. R., S. C. Basappa and V. S. Murthy 1979 *J. Food Sci. and Technol. India* **16:** 113-114.

Rati, E. R., T. Shantha and H. P. Ramesh 1991 *J. Exp. Biol.* **29**: 813-817.

Rati, E. R. and T. Shantha 1994 *J. Food Sci. and Technol. India* **31:** 327-329.

Rati, E. R., V. Prema and T. Shantha 1987 *J. Food Sci.* and *Technol.* India **24**:90-91.

Reddy, A. S. and S. M. Reddy 1982 *Proc. Symp. Mycotoxin in Food and Feed* (eds.K. S. Bilgrami, T. Prasad and K. K. Sinha), Allied Press, Bhagalpur, pp. 293-296.

Reddy, A. S. and S. M. Reddy 1983 *Curr. Sci.* **52:** 613-614.

Reddy, B. N. and M. Nusrath 1985 *Curr. Sci.* **54**:31-32.

Reddy, B. N. and M. Nusrath 1986 *Science and Culture* **52**: 31-32.

Reddy, B. N., M. Nusrath, C. K. Kumari and S. Nahdi 1983 *Indian Phytopath.* **36:** 683-686.

Reddy, M. R. S. and J. Subbayya 1985 *Seed Tech.* News **15:** 5.

Reddy, P. S., C. V. Reddy, V. R. Reddy and P. V. Rao 1984 *Indian J. Anim. Sci.* **54**: 235-238.

Reddy, S. M. and M. Surekha 1994 In Proceedings of ISCA Symposium 1991 (ed G.P. Agarwal and S.K. Hasija), Narendra Publishing House, Delhi, pp. 133-144.

Reddy, S. M., D. Rajakumari and V. Krishna Reddy 1997 *J. Poult. Sci.* **32**:169-172.

Reddy, S. V., D. K. Mayi and M. U. Reddy 2001 *Food Addi.and Contam.* **18**: 553-558.

Reddy, V. K. and S. M. Reddy 1987 *Hindustan. Antibiot. Bull.* **29:** 10-12.

Reddy, V. K. and S. M. Reddy 1987a *Nat. Acad. Sci. Letters* **10**: 413-415.

Reddy, V. K. and S. M. Reddy 1988 *Nat. Acad. Sci. Letters* **11:** 133-134.

Reddy, V. K. and S. M. Reddy 1988a *Acta Botanica India* **16:** 122-124.

Reddy, V. K. and Reddy S. M. 1990 *J. Food Sci. and Technol*. India **27:**180-181.

Reddy, V. K., M. Surekha, S. M. Reddy and D. Rajakumari 1989 *Proc. Conf. on Cytol and Genetics* **2:** 271-275.

Redman, C. M., D. Banerjee, K.Howell and O. E. Palade 1975 *J. Cell Biol.* **66 :** 42-59.

Reen, R. K., F. J. Wiebel and J. Singh 1997 *J. Ethnopharma.* **58**: 165-173.

Reif, K. and W. Metzger 1995 *J. Chromatogra. A.* **692**: 131-136.

Reiss, J. 1971 *Z. Pflanzenphysiol.* **64**: 260.

Reiss, J. 1972 *Naturwissenschaften.* **59:** 37.

Reiss, J. 1973 *Chem. Microbiol. Technol. Lebensum* **2:** 171-173.

Reiss, J. 1973a *Z. Lebens, Unterrs – forech,* **151 :** 256.

Reiss, J. 1975 *J. AOAC.* **53**: 624-625.

Reiss, J. 1977 *Toxicology* **8:** 121-123.

Reiss. J. 1976 *Experientia* **31**: 168-169.

Reiss, J. 1979 *Experientia* **32**: 168-169.

Reiss, J. 1981 Z. Deut. Lebenem-Rundsch. **77**: 27.

Reshik, S., A. Neare, A. Pacin, E. Martinez, N. Apro and S. Latreite 1996 *Food Addi. and Contam.* **13:** 115-120.

Reynolds, G. and J. J. Pestka 1991 *J. Food Protect.* **54:** 105-108.

Reynoso, M. M., A. M. Torres, M. L. Ramirea, M. I. Rodriguez, S. N. Chuize and N. Magan 2002 *Mycological Res.* **106:** 1093-1099.

Rheeder, J. P., W. F. O. Marasas, P. G. Thiel, E. W. Sydenham, G. S. Shephard and D. J. Van Schlakwyk 1992 *Phytopathology* **82:** 353-357.

Rice, L. G. and P. F. Ross 1999 *J. Food Protect.* **57:** 536-540.

Rice, S. L., L. R. Beuchat and R. E. Worthington 1997 *Appl. Environ. Microbiol.* **34:** 791-796.

Richard, J. L. and L. H. Arp 1979 *Mycopathologia.* **67**: 107-109.

Richard, J. L., P. Bocchelli and L. H. Arp 1981 *Mycopathologia* **76:** 107-109.

Richard, J. L., W. M. Peden and J. R. Thurston 1986 in proceedings *symposium on die diagnosis of mycotoxisis of importance in USA and Japan held at* Ames, October 11-13, (eds. J. L.Richard and J.R. Thurston), Dordrecht Netherlands, Martinum Nijin off Publ.pp. 51-59.

Richard, M.D., L. Learge and B. Cahagneir 1985 Applied Sciences, No. 90, Properties of Water in foods in relation to quality and stability meeting (eds. D. Simatos and J. L. Multon), pp. 273-292.

Richardson, K. E., W. M. Hagler Jr. and P. B. Hamilton 1984 *Appl. Envir. Microb.* **47**: 643-646.

Riley, R. T., C. W. Bacon, F. I. Meredith, O. Torres, S. Tyada, E. Wang and A. H. Murrill Jr. 2001 *Food Addi. and Contam.* **13:** 197-198.

Riley, R.T. and D.E.Gorger 1992 In Hand Book of Applied Mycology Vol 5. Mycootxins in Ecological System (eds. D. Bhatnagar, E. B. Lillehoj and D. K. Arora), Marcel Dekkar, Inc., New York. pp. 385-402.

Riley, R. T., D. M. Hintous, J. L. Showker, W. Rigbry and W. P. Norred 1990 *Toxicology and Appl. Pharmacol.* **102:** 128-141.

Riley, R. T. and J. L. Showker 1991 *Toxi. Appl. Pharm.* **109:** 108-126.

Riley, R. T., W. P Norred, E. Wang and A.H. Merrill 1999 Toxicology and Mycotoxin in Research Unit, USDA –ARS PO Box 5677, Ga. USA. pp. 5604-5677.

Rizzi, L., L.Lambertini, L.Merchesin and L.Gramenzi 1995 *Vet. Grafiche Scuderi, Messina, Italy.* pp. 591-592.

Robens, J., R.T. Riley and J.Robens 2002 *Mycopathologia* **155**: 1-127

Roberts, A. C. and D. J. Me Weeny 1972 *J. Food Tech.* **7:** 221.

Roberts, B. A. and D. S. P. Patterson 1975 *J. AOAC.* **58**: 1178-1181.

Roch, O. G., G. Blunden, R. D. Coker and S. Nawaz 1992 *J. Chromatogra.* **33**: 208-212.

Roinestad, K. S., T. J. Montville, J.D. Rosen 1994 *J. Agric. and Food Chem.* **42**: 2025-2028.

Roland, J. O. and L. R. Beuchat 1984 *Appl. Envir. Microb.* **47 :** 205-207.

Romer, T. 1990 Advances on Mycotoxin Analysis: Making *Fusarium* Feed Management. **41**: 30-32.

Romer, T. 1994 Labs In Mycotoxins Control Consultants. Quantitative methods for DON, aflatoxin B_1 and zearalenone. **14 :** 333-339.

Ronaldson, J. W. and A. A. Fyvie 1974 *Lab. Pract.* **22**: 734-735.

Ronaldosn, J. W., A. Taylor, E. P. White and R. J. Abraham 1963 *J. Chem. Soc.* **3:** 1372-1380.

Rood, H. D. Jr., S. D. Swanson and W. B. Buck 1986 *J. Chromatogr.* **378:** 375.

Rosen, J. D. 1984 Archs. Beleces. Suppl. 173.

Rosenberg, E., R. Krska, R. Wissiack, V. Kmetov, R. Josephs, E. Razzazi and M. Grasserbauer 1998 *J. Chromatogr.* **A819:** 277-288.

Rosiler, M. R., J. Bautista, V. O. Fuetes and F. Ross 1998 *Zentralbl. Veterinarmed.* **A 45:** 299-302.

Ross, P. F., L. G. Rice, G. D. Osweiler, P. E. Nelson, J.L. Richard and T. M. A. Wilson 1990 *Mycopathologia* **117:** 109-114.

Rothweiler, W. and Ch. Tamm 1966 *Experientia* **22:** 750-752.

Rothweiler, W. and Ch. Tamm 1970 *Helv. Chem. Acta* **53**: 696-724.

Rotter, B. A., D. B. Prechisky and J. J. Pestka 1996 *J. Toxicol. Envir. Health* **48**:1-34.

Rottinaham, G. E., C. E. Coatney and H. C. Minor 1992 *J. Vet. Diagn. Invest.* **4:** 326.

Roy, A. K. and H. K. Chourasia 1989 *Appl.Envir. Microb.* **55:** 531-532.

Roy, A. K. and S. Kumar 1993 *Vistas in Seed biology* **I:** 381-393.

Rubin, L. L., G. G. Agorio Selman, J. Jacob and A. Mauro Perry 1976 *J. Embryol. Exp. Brain Res.* **36:** 321-341.

Ruhr, L. P. 1987 *Arkansas Agric. Exp. Stn. Spec. Rep.* **123:** 37-41.

Rukmini, C. and R. V. Bhat 1978 *J. Agric. Food Chem.* **26 :** 647-649.

Russell, G. R. 1962 *Nature* **186:** 788-789.

Rustom, I. Y. S. 1997 *Food Chem.* **59**: 57-67.

Rutqvist, L. N., L. Bjorklund, K. Hult, E. Hokby and B. Carlison 1978 *Appl. Environ. Microbiol.* **36**: 920.

Rytych, B. J. and B. J. Zyska 1977 *Indian Boil. Bull.* **13:** 89-93.

Ryu, J. C., J. S. Yang, Y. S. Song, O.S Kwon, J. Park and J.M.Chang 1996 *Food Addi. and Contam.* **13:** 333-341.

Sabbioni, G.and O. Sepai 1998 In Mycotoxins in Agriculture and Food Safety (eds. K. K. Sinha and D. Bhatnagar) Marcel Decker, New York, pp. 183-226.

Sabbioni, G., S. Aphbs, G. N. Wogan and J. D. Groopman 1990 *Carcinogenesis* **11**: 2063-2066.

Sabbioni, M. and D. B. Rodriguez-Amaya 1993 *Cionciae Culture* **45:** 359-371.

Saeniz de Rodrigues, C. A. 1984 *The New England Journal of Medicine* **310:** 1741-1742.

Saha, D., D. Acharya and T. K. Dhar 2006 *Analytical and Bio analytical Chemistry* **385:** 847-854.

Sahay, S. S. 1988 *J. Indian Bot. Soc.* **67:** 188-190.

Sahay, S. S. and T. Prasad 1990 *Food Addi.and Contam.* **7:** 509-513.

Sahoo, P. K. and S. C. Mukherjee 2001 *Comparative Immunology, Microbiology and Infectious Diseases* **24**: 143-149.

Saito, M. and K.Ohtsubo 1973 In: Mycotoxins (ed.I.F.H.Purchase), Acad. Press, London. pp. 263.

Sakai, R. K., D. H. Gelf, S. Stoffel, S. J. Scharf, R. Higuchi, G. T. Horn, K. B. Mullis and H. A. Erbich 1988 *Science* **239:** 487-491.

Sakthisekaran D. and B. R. B. Shanmugasundaran 1990 *Biochemistry International* **21**:117-134.

Sakuda, S., M. Ono and K. Furith 1996 *J. Am. Chem. Soc.* **118**: 7855-7856.

Sakuda, S., M. One, H. Ikeda, T. Nakamura, Y. Inagaki, R. Kawachi, J. Nakayama, A. Suzuki, A. Isogai and H. Nagaswa 2000 *The Journal of Antibiotics* **53**:1265-1271.

Samapundo, S., B. Demeulenaer, B. Atukwase, A. Debevere and J.Derlieghere 2007 *J. Food Microbiology* **118**:139-150.

Samapundo, S., B. Demeulenaer, N De Muer, J. M. Debere and F. Derlieghere 2002 *J. Chromatogr. A.* **1109:** 312-316.

Samson, R. A., A. C. Stolk and R. Hadlok 1982 *Stud. Mycol. (Baren)* **11:** 1047-1049.

Samson, R. A., E. S. Hockstra, J. C. Frisvod and O. Filtenborg 1995 In *Introduction to food-Borne fungi.* Centralbureau voor schimmel cultures, Baarn, The Netherlands pp.239-249.

Samson, R. A., P. V. Nielsen and J.C. Frisvad 1990 In: *Modern concepts* in *Penicillium* and *Aspergillus* classification (eds. R. A. Samson and J. I. Pitt), Plenum Press, New York, pp 455.

Samuels, G. J. and I. C. Hallett 1983 *Trans. Br. Mycol. Soc.* **81**:473-483.

Sanchis, V., M. Abadias, L. Oncins, N. Sala, I. Vinas and R. Canela 1994 *Appl. Envir. Microb.* **60:** 2147-2148.

Sang, N. K. and D. I. Yoo 2001 *J. Antibiotics* **54:** 831-835.

Santos, E. A. and E. A. Vargas 2002 *Food Addi.and Contam.* **19:** 447-458.

Sargeant, K., A. Sheridan, J. O. Kelley and R. B. A. Carmagam 1961 *Nature* (London) **192:** 1096-1097.

Sarr, A. B., K. Mayura, L. F. Kubena, R. B. Harvey and T. D. Phillips 1995 *Toxicol.Lett.* **75**: 145-151.

Sashidhar, R. B. 1993 *Food Chem.* **48**: 349-352.

Sashidhar, R. B., M. V. V. Murthy and R. V. Bhat 1992 *Food Chem.* **43:** 403-405.

Sashidhar, R. B., R. V. Bhat and S. Vasanthi 1989 *Curr. Sci.* **58:** 882-885.

Sashidhar, R. B., V. Sudershan Rao, Y. Ramakrishna and R. V. Bhat 1989 *Food Chem.* **31:** 51-56.

Sashidhar, R. B., Y. Ramakrishna and R. V. Bhat 1988 *Analytical letters* **21:** 507-518.

Sashidhar, R. B., Y. Ramakrishna and R. V. Bhat 1992 *J. Stored Products Res.* **28:** 257-260.

Sato, C., K. Kojima and K. Nishizawa 1975 *Biochem. Biphys. Res. Commun.* **67**: 22-27.

Sauer, D. R., L. M. Seitz, R. Burroughs, H.E. Mohr, J. L. West, R. J. Milleret and H. D. Anthony 1978 *J. Agric. Food Chem.* **26**: 1380-1383.

Saxena, J. and B. S. Mehrotra 1990 *Acta Alimentaria* **19:** 221-224.

Saxena, J. and C. S. Mathela 1996 *Appl.Envir. Microb*. **62:** 702-704.

Schroeder, H. W. and R. J. Cole 1977 *J. Agr. Food Chem.* **25**: 204-206.

Schroeder, H. W., R. J. Cole, J. Hein and J. W. Kirksey 1975 *Appl. Microbiol.* **29:** 857-858.

Schroeder, P. G. 1973 *Experientia* **29**: 617-618.

Schabort, J. C., D. C. Wilkins, C. W. Holzapfel, D. J. J. Porgieter and A. W. Neitx 1971 *Biochem. Biophys. Acta* **250**: 311-328.

Schaeffer, J. L., J. K. Tyczkowski, J. E. Riviere and P. B. Hamilton 1988 *Poultry Sci.* **67**: 619-625.

Schanerr, A. W., D. C. Hooker, T. S. Baute and I. L. Tamburic 2002 *Plant Disease* **86**: 1123-1126.

Schappert, K. T. and C. G. Khachatourians 1986 *Curr. Genet.* **10 :** 671-676.

Schatzki, T. F. and J. L. Pan 1996 *J. Agric. Food Chem.* **44**: 1076-1084.

Schepertz, A. I., R. A. Fleischman and J. H. Cisle 1972 *J. Chromatogr.* **69:** 411-443.

Schindler, D., P. Grant and J. Davies 1974 *Nature* **248 :** 535-536.

Schisler, D. A., N. I. Khan, M. J. Bochm and P. J. Sliniger 2002 *Plant Dis.* **86:** 1350-1356.

Schlechter, M., W. F. O. Marasas, E. W. Sydenham, S. Stockenstrom, H. F. Vismer and J. P. Rheeder 1998 *S. Afr. J. Science* **94:** 185-187.

Schmitt, K., E. E. Martlbauer, R. Usleber, J. Gessler, J.Lepschy and D. Abranson 1996 In Immunoassays for residue analysis: Food safety (eds. R. C. Beier and L. H. Stanker), ACS symposium series book no. 621. American chemical Society: Washington, DC. pp. 314-321.

Schneider, E., E. Usleber and E. Martlbauer 1995 *J. Agri. Food Chem.* **43 :** 2548-2552.

Schnieder, E., R. Dietrich, E. Martlbauer, E. Usleber and G. Terphan 1991 *Food. Agric. Immunol.* **3:** 185-193.

Schoenhard, G.L., D.J. Lee, S.E. Howell, N.E. Pawlowski, L.M. Libeey and R.O. Sinhubber 1976 *Cancer Res.* **36**: 2040-2045.

Schoental, R. 1983 *Vet. Res.* **112**: 594.

Schoental, R. and A. F. Wite 1965 *Nature* **205**: 57-58.

Schoental, R., A.Z. Joffe and B. Yageen 1979 *Cancer Res.* **39:** 2197-2220.

Schroeder, H.W. and J. B. Storey 1976 *Horticultural Science* **11:** 53-54.

Schroeder, T.E. 1978 In *Cytochalasins: Biochemical and Cell Biological Aspects* (ed. S. W. Tanenbaum), North-Holland, New York, pp.99-112.

Schuster, E., N. Dunn-columan, J.C. Frisvad and P.W.M. Van Dijck 2002 *Appl. Microbiol. Biotechnol.* **59**: 426-435.

Schumacher, R., R. Kriska, J. Weingaertner and M. Grasserbauer 1997 *Fresenius J. Anal. Chem.* **359 :** 510-515.

Schumacher, R., M. Kriska, W. Grasserbauer, R.Edinger and H. Lew 1998*J. Anal. Chem.* **360:** 241-246.

Schumacher, D.M.I., I. Lehmann and M.I. Metzler 2003 *Naunyn-Schmiedebergs Archivs of Pharmacology* **367:** 166.

Schwadorft, K. and H. M. Muller 1992 *J. Chromatgr.* **595:** 259-267.

Scott, De B. 1965 *Mycopathol. Mycol. Appl.* **25**: 213-222.

Scott, G. E. and S. B. King 1984 *Plant Dis.* **68:** 804-806.

Scott, P. M.1998 *Revue Med. Vet.* **149:** 543-548.

Scott, P. M.1974 In *Mycotoxin* (ed. Elscrter), Pub.Co. pp 383-403.

Scott, P. M.1977 *Forschr. Chem. Org. Naturst.* **31:** 119-151

Scott, P. M.1983 In Trichothecenes, Chemical Biological and Toxicological Aspects. (ed. Y. Uneo) Published by Koddnsha Ltd. Japan and Elsevier Amsterdam Chey VI. pp. 218-221.

Scott, P. M.1984 In: Toxigenic fungi, their Toxins and Health Hazard (ed. H. Kurata and Y. Ueno), Elsevier, Amsterdam, pp. 182-189.

Scott,P.M.1985 In Mycotoxins (eds. P. M. Scott, H.E. Iranholmand and M. D. Sultan) A Canadian Prerspective, Publication, NRCC/CNRC: Ottawa, pp. 29.

Scott, P.M.1991 In: Fungi and Quality in Drying and Storage, Amsterdam. Elsevier pp. 529-572.

Scott, P. M.1995 in 16th ed. International Arngton, VA Sec. 986.

Scott, P. M.1996 *J. AOAC.* **79:** 875-882.

Scott, P.M.1999 Natural toxins In Official methods of analysis of AOAC International (ed. P. Cuniff), Gaithersburg, pp. 45-46.

Scott, P. M. and B. P. C. Kennedy 1973 *J. AOAC.* **56:** 813-816.

Scott, P. M., B. Kennedy and W. Vanwalbeck 1971 *J. AOAC.* **54**: 1445.

Scott, P. M. and G. A. Lawrence 1988 *J. AOAC.* **71**: 1176-1179.

Scott, P. M. and G. A. Lawrence 1992 *J. AOAC.* **75 :** 829-834.

Scott, P.M. and G. A. Lawrence 1994 *J. AOAC.* **77:** 541-545.

Scott, P. M. and G. A. Lawrence 1997 *J. AOAC.* **80**: 1229-1234.

Scott, P. M., G. A. Lawrence and T. J. Mutala 1986 In Mycotoxons and Phycotoxins. (eds. P. S. Steyn and R. Vleggwar), Elsevier, Amsterdam, pp. 305-315.

Scott, P.M., J. Harwig, Y.Chen and B.P.C. Kennedy 1975 *J. Gen. Microbiol.* **87**: 177-180.

Scott, P. M., J. W. Lawrence and W. Van Walbeek 1970 *Appl. Microbiol.* **20:** 839-842.

Scott, P. M. and M. W. Trucksess 1997 *J. AOAC.* **80:** 941-949.

Scott, P. M. and S. R. Kanhere 1980 *J. AOAC.* **63:** 612-621.

Scott, P. M., S. R. Kanhere and D. Weber 1993 *Food Addi. Contam.* **10:** 381-389.

Scott, P. M., S.R. Kanehere, R.Canela, G.A.Lombeart and S.Bacler 1990 *Prehrambeno Technol. Biotechnol. Rev.* **29:** 61-64.

Scott, P. M., T. Fuleki and J. Haswig 1978 *J. Agric. Food Chem.* **25:** 434-437.

Scott, P. M., W. F. Miles, P. Toft and J. G. Dube 1972 *J. Agr. Food Chem.* **20**: 450-451.

Scott, P. M., W. Vanwalbeck, B. Kennedy and D. Aryeti 1972 *J. Agr. Food Chem.* **20**: 1103-1109.

Scott, T. W. and L. B. Bullerman 1975 *J. AOAC.* **58**: 297-299.

Scott, T. W. and L. B. Bullerman 1976 *J. Food Sci.* **41:** 201-203.

Scudamore, K. A. and S. Patel 2000 *Food Addit. and Contam.* **17:** 407-416.

Scudamore, K. and P. Atkins 1985 In proc. 8th meeting on mycotoxins in Animal and Human health (eds. M. O. Moss and M. Frank), Edinburgh, U. K. pp. 30-36.

Scudamore, K. A., R. Battaglua, T. Hatzold and R. Krores 1998 *Food Addi. and Contam.* **13:** 39-42.

Seefelder, W., H. U. Humpf, G.Schwerdt, R. Freudinger and M. Gekle 2003 *Toxicol. Appl.Pharm.* **192:** 146-153.

Seefelder, W., M. Hartl and H. U. Humpf 2001 *J. Agric. Food Chem.* **49**: 2146-2151.

Seegers, J. C., M. L. Lottering and P. J. Garlinski 1994 *Med. Sci. Res.* **22**: 417-419.

Seemuller, E. 1968 *Mitt. Biol. Bundesanst, Berlin-Dhalem* **127:** 93.

Segal, R., I. Milo-Goldzwig, A.Z. Joffe and B.Yagen 1983 *Toxicol. Appl. Pharmacol.* **172:**71-132

Segvic, M. and S. Pepeljnjak 2003 *Periodica-Biologerum*, pp. 275-279.

Seitz, L. M. and H. E. Mohr 1926 *Anal. Biochem.* **70:** 224-230.

Seitz, L.M., D.B. Sauer, H.E. Mohr and R. Burroughs 1975 *Phytopathology* **65**: 1259-1263.

Seitz, L. M., D. B. Sauer, H. E. Mohr, R. Burroughs and J. V. Paukstelis 1975 *J. Agr. Food Chem.* **23**: 1-4.

Seitz, L. M. and H. E. Mohr 1976 *Cereal Chem.* **54:** 179-183.

Sekiguchi, J. 1973 *Hakkakajaki* **61**: 129.

Sekiguchi, J. and G. M. Gaucher 1979 *Can. J. Microbiol.* **25:** 881-887.

Sekita, S., K. Yoshihira, S. Natori and H. Kuwano 1973 *Tetrahedron Lett.* 1351-1354.

Sekita, S., K. Yoshihira, S. Natori and H. Kuwano 1976 *Tetrahedron Lett.* 2109-2112.

Sekita, S., K. Yoshihira, S. Natori and H. Kuwano 1977 *Tetrahedron Lett.* 2771-2774.

Selala, M.T., F.Daelemans and P. J. C. Schepens 1989 *Drug and Chemical Toxicology* **12:** 2372-257.

Selala, M. T., G.M.Lackeman, B. Loenders, A.Musuker, A.G. Herman and P. Schepens 1991 *J. Natural Products* **54:** 207-212.

Solman, G. C., J. Jacob and M. M. Perry 1976 *J. Embryol. Exp. Morphol.* **36:** 321.

Sen, A. C., C. I. Wei, S. Y. Fernando, J. Toth, E.M. Ahmed and G.E. Dunaif 1988 *Food Chem. Toxicol.* **26:** 745-752.

Seo, J. A., R. H. Proctor and R. D. Plattner 2001 *Fungal Genetics and Biology* **34:** 155-165.

Seo, J. A. and Y. W. Lee 1999 *Appl. Envir. Microb.* **65:** 1331-1334.

Sergeen, L. N., N. M. Piliya, V. A. Tulelyan and V. B. Spinchav 1990 *Voprosy Meditsinskoi Khinmi.* **36:** 345-48.

Serralheiro, M. L. and M. L. Quinta 1985 *J. AOAC.* **68**: 952-954.

Shank, R. C., N. Bhamaropravati, J. E. Gordan and G. N. Wogan 1972 *Food Cosmet. Toxicol.* **10:** 171-181.

Shankaran, R., H. C. Raj and T. A. Venkatasubramanian 1970 *Enzymologia* **39:** 371-378.

Shantha, T. 1991 *Indian J. Microb.* **31:** 113-130.

Shantha, T. 1999 *Natural toxins* **7:** 175-178.

Shantha, T. and E. R. Rati 1990 *Curr. Sci.* **59**: 326-327.

Shantha, T., E. R. Rati and R. Joseph 1996 *Letters in Applied Microbiology* **23:** 437-438

Shantha, T., M. E. R. Sreenivasa, E.R.Rati and V. Prema 1986 *J. Food Saf.* **7**: 225-232.

Shantha, T. and V. Sreenivasa Murthy 1981 *J. AOAC.* **164:** 291-293.

Sharma, A. and S. R. Padwal Desai 1989 *J. Food Sci. and Techno. India* **26**: 366-367.

Sharma, A., S. R. Padwal Desai and G. B. Nadkarni 1988 *J. Agri. and Food Chem.* **36**: 546-548.

Sharma, R. P. 1993 *J. Dairy Sci.* **76:** 892-897.

Sharma, R. P., Neetesh Bhandari, M. Tsunoda, R. T. Riley, K. A. Voss, F. I. Meredith and N. Bhandari 2000 *Envir. Toxic. and Pharm.* **8:** 173-182.

Sharma, V. K. and N. Singh 1971 *Indian Vet. J.* **48:** 235-238.

Sharma, Y. P. and G. Sumbali 1999 *Mycopathologia* **148:** 103-107.

Sharman, M., A. L. Patey and J. Gilbert 1989 *J. Chromatogra.* **474**: 457-461.

Sharman, M. and J. Gilbert 1991 *J. Chromatogra.* **543:** 220-225.

Sharman, M., S. Mc Donald and J. Gilbert 1992 *J. Chromatogra.* **603:** 285-289.

Shashi, P. L. and J. N. Kapoor 1980 *Indian Phytopath.* **33 :** 336-338.

Shelby, R. A. 1996 In Immunaassay for Residue Analysis: Food Safety (eds. R. C. Beier and L. H. Stonker), ACS Symposium Series Book no. 621. American Chemical Society: Washington DC. pp. 231-242.

Shelby, R. A., D. G. White and E. M. Bauske 1994 *Plant Dis.* **78 :** 582-584.

Sheldon, J. L. 1904 Agric. Exp. Stn. Nebraska, *Annu. Rept.* **17:** 23-43.

Shen, H. M., C.Y. Shi, Y.Shen and C.N. Ong 1996 *Free Rado. Biol. Med.* **21**: 139-146.

Shephard, G. S. 2008 *Food Addi. and Contam.* **25**: 146-151

Shephard, G. S., A. Fabiani, S. Stockenstrom and V. Sewram 2003 *J. Agric. Food Chem.* **51**: 1102-1106.

Shephard, G. S., E. W. Sydenham, P.G. Thiel and W. C. A. Gelderblom 1990 *J. Liquid Chromatogr.* **13:** 2077-2087.

Shephard, G. S., P. G.Thiel, S. Stockenstrom and E. W. Sydenham 1996 *J. AOAC.* **79 :** 671-687.

Shephard, G.S., W. F. O. Marasas, N. L. Leggott, H.Yasdanpancin, H. Rahimian and N. Safavi 2000 *J. Agri. and Food Chem.* **48**: 1860-1864.

Shephard, G. S., W.F.O. Marasas, N.L.Leggott, H. Yazdanpanab, N. Safari, A. Zarghi, A. Shataati and H. R. Resekh 2002 *Food Addi.and Contam .* **19:** 676-679.

Shephard, M. J. and T. Gilbert 1984 *Food Addi.and Contam.* **1**: 325-335.

Shephard, T. H. and J. C. Greenway, S. A. Patwardhan, R.C.Pandey and S.Dev 1977 *Teratology* **16:** 131-136.

Shepro, D. F.A., B. Eamarich, L.Robblee and F.C.Chao 1970 *J.Cell Biol.* **47:** 544-547.

Shetty, P. F. and R.V. Bhat 1997 *J. Agric. Food Chem.* **45:** 2170-2173.

Shetty, P. F. and R. V. Bhat 1998 *Food Chem.* **66:** 371-374.

Shetty, P.K., V.S.Rao, R. J.Paramkishan and R.V. Bhat 1997 *Br. Poultry Sci* **38**:475-479

Shivefer,H.B.and R.Beasley 1989 In Trichothecene mycotoxicosis, pathophysiological effects Vol. II (ed. V. R. Bearley),CRC press,Boco Raton, Florida,USA. pp. 61-89.

Shier, W. T. 1998 *Rev. Med. Vet.* **149:** 599-604.

Shier, W. T. 2000 *Journal of Toxicology* **19:** 161-187.

Shim-Won, B.O., C.P.Woloshuk and W.B. Shim 1999 *FEMS-Microbiology Letters* **177:** 109-116.

Shinozuka, J., L. Guanmin and K. Vetsuka 1997 *Exp. Anim. Tokyo* **46**: 117-126.

Schoental, R. 1983 *Vet. Res.* **112:** 594.

Shotwell, O. L., M. L. Goulden and G. A. Benett 1976 *J. AOAC.* **59**: 666-670.

Shotwell, O. L., C. W. Hesseltine and M. L. Goulden 1969 *Appl. Microbiol.* **17**: 765-766.

Snyder, W. C. and H. N. Hansen 1945 *Mycologia* **38:** 455-462.

Sibanda, L., K.S. Desaeger, I. Barna – Vetro and C. Van Peteghem 2002 *J. Agric. Food Chem.* **50:** 6964-6967.

Silva, C. M. G. and E. A. Vargas 2001 *Food Addi.and Contam.* **18 :** 39-45.

Silva, L. J. G., C. M. Lino, A. Pena and J. C. Molto 2007 *Food Addi. and Contam.***24:** 381-390.

Silverton, J. V., T. Akiyama, C. Kabuto, S. Sekita, K. Yoshihira and S. Natori 1976 *Tetrahedron Lett.* 1349-1350.

Simonart, P. and R. de Dehatowar 1956-1957 *Zentrabl. Bakteriol. Parasitenkd. Abt.* II **110**: 107.

Simpson, T. J., A. E. Dejesus, P. S. Steyn and R. Vleggaar 1983 *J. Chem. Soc. Chem. Commun.* 338-340.

Singh, A., K. C. Satija and S. K. Mahipal 1988 *Indian J. Poult. Sci.* **27:** 153-156.

Singh, A., M. S. J. Oberoi and S. K. Singh 1994 *Rev. Sci. Tech.* **15:**1053-1060.

Singh, B. K. and N.K. Saha 1994 *Indian Phytopath.* **47:** 266.

Singh, P. P., P. S. Bedi and D. Singh 1987 *J. Stored Prod. Res.* **22:** 29-32.

Singh, S. N. and L. Chand 1993 *Crop Research* **6:** 149-154.

Singh, T., R. P. S. Tyagi and B. K. Varma 1982 *J. Food Sci. and Tech.* **19:** 35-36.

Singh, V. and V. Singh 1998 *J. Phytopath. Res.* **11:** 7-10.

Singh, Y. P. and P. M. K. Gupta 1983 *J. Vet. Physiol. Alli. Sci.* **2:** 17-25.

Singhania, S., T. Satyanarayana and M. V. Rajam 1991 *Mycol. Res.* **95:** 915-917.

Sinha, A.K., K.S. Bilgrami, K.S. Ranjan and T. Prasad 1988 *Indian Phytopath.* **41**:434-437.

Sinha, A.K. and K. Ranjan 1989 *Indian Phytopath.* 42: 514-518.

Sinha, B. K., K. Ranjan and J. N.Pandey 1999 *Ind. J. Veterinary Research* **8**: 31-38.

Sinha, B. K., K. S.Ranjan and T. N. Pandey 2002 *J. Food Sci. and Technol.* **39**: 38-41.

Sinha, K. K. 1987 *Appl. Envir. Micro.* **53:** 1391-1391.

Sinha, K. K. 1988 *Food Lab. News lett.* **14:** 25-26.

Sinha, K. K. 1990 *Food Addi. and Contam.* **7:** 55-61.

Sinha, K. K. and A. Singh 1987 *J. Food Sci. and Technol.* India **24**: 45

Sinha, K. K. and A. K. Sinha 1992 *Journal of Stored Prod. Res.* **28:** 211-219.

Sinha, K. K., A. K. Sinha and Gajendra Prasad 1993 *Letters in Applied Microbiol.* **16**: 114-117.

Sinha, K. K. and A. Singh 1987 *J. Food Sci.* and *Technol. India* **24:** 4-45.

Sinha, K. K. and P. Kumari 1990 *Mycopthologia* **110:** 274-279.

Sinha, R. C. and M. E. Savard 1997 *Can. J. Plant Pathol.* **19:** 8-12.

Sinha, R. C., M. E. Savard and R. Lau 1995 *J. Agric. Food Chem.* **4**

Siraj, M. Y., T. D. Philips and A. W. Hayes 1981 *J. Toxicol. Envir. Health* **8 :** 131-140.

Sivaswamy, S. N., B. Balachandran and S. Balanchru 1991 *Bull. Envir. Contamin. Toxicol.* **47:** 251-260.

Skringar, M., J. L. Rasic and V. Stojicic 1996 *Folia Microbiol.* **41**: 26-28.

Sle, W., C. J. Mirocha and J. Chen 1997 *J. Agri. Food Chem.* **45** : 1251-1255.

Sleiman,R.F.,L.Scigle-Morandi,L.Sage and S.Kribobok1989 *Mycopathologia* **105:**129-133.

Smalley, E. B. and F. M. Strong 1974 In Mycotoxins (ed. I. F. H. Purchase), Elsevier Scientific Publishing Co., New York pp. 199-228.

Smela, M.E., S.S. Currier, E.A. Bailey and J. Messigmann 2001 *Carcinogene.* **22**: 535-545.

Smiley, R. D. and F. A. Dranghan 2000 *J. Food Protection* **63:** 415-418.

Smith, A. B., J. M. Hesseltine and M. Visnick 1989 *Tetrahedron* **45:** 2431-2449.

Smith, J., C. Wesselink, J. Parr, J. M. Sprosen, E.A. Fowke, N. R. Towers and D. Laboyric 1995 In : *Toxicology and Food Safety"* John Wiley and Sons, Inc.pp. 41-43.

Smith, J. D., W. E. Crawley and F. T. Lees 1961 *Newzeland J. Agr. Res.* **4**: 538-551.

Smith, J. D., W. E. Crawley and F. T. Lees 1961a *Newzeland J. Agr. Res.* **4:** 725.

Smith, J.E. and G.L. Solomons 1994 Mycotoxins in Human Nutrition and Health. Agro Industrial Research Divison, European Commission Directorate-General XII. pp.1-300.

Smith, J.P., H.S.Ramaswamy and B.K.Simpson 1990 *Trends in Food Sci.Technol.* **1**:111-118

Smith, J. E. and K. Ross 1991 In Mycotoxins and Animal Foods (eds. J. E. Smith and R. S. Henderson), CRC Press,Boco Raton, FL pp. 553-565.

Smith, J. F., M. E. Di. Menna, L. T. Mc Gowan and M. E. Di Menna 1990 *J. Reprod. Fert.* **89 :** 99-106.

Smith, J. E. and M. O. Moss 1985 Mycotoxins, Formation, analysis and significance, John Wiley and Sons, Ltd., pp. 31-50.

Smith, J. E. and R. S. Handerson 1991 Mycotoxins and Animal Foods (eds. J.E. Smith, R. Henderson), CRC Press, Boca Raton, Florida, pp.363-397.

Smith, J.W., C.W. Lewis, J.G. Anderson and G.L.Solomons 1994 Mycotoxins in Human and Animal Health. Technical Report EUR16048 EN of the Agro-Industrial Research Division of Directorate XII: Science, Research and Development, European Commission, Brussels, Belgium. pp.487-540.

Snijders, C. H.A. and J.Perkowski 1990 *Phytopathology* **80 :** 566-570.

Snyder, W. C. and H. N. Hansen 1965 *Am. J. Bot.* **32:** 657-666.

Snyder, W. C. and T. A. Tousson 1965 *Phytopathology* **55:** 833-837.

Soares, L. M. V. and D. B. Rodrigues – Amaya 1989 *J. AOAC.* **72**: 22-26.

Sobotka, T. J., R. E. Broadic and S. L. Spaid 1978 *Pharmacology* **16**: 284-294.

Soenike, A. 1997 *Food Chem.* **60**: 255-262.

Sohn, H. B., I. H. Seo and Y. W. Lee 1999 *J. Food Addi.and Contam.* **16**: 153-158.

Soleas, G. J., J. Yan and D. M. Goldberg 2001 *J. Agric. Food Chem.* **49**:2733-2740.

Solovey, M. M., C. Somoza, G. Cano, A. Pacin and S.Resnik 1999 *Food Addi.and Contam.* **16:** 325-329.

Somsyn, F. E., S. M. Maxwell and R. G. Hendrickse 1995 *Mycioatgikigua* **131:**121-126.

Sorenson, W. G., D. G. Frager, B. B. Jarvis, J.Simpson and V.A.Robinson 1987 *Appl. Envi. Microb*. **53**: 1370-1375.

Sorenson, W. G., J. Simpson and V. Castranova 1985 *Environmental Res.* **38:** 407-416.

Spanjer, P. M., C. Martien and M. Rensen and Jos M. Scholten 2008 *Food Addi. and Contam.* **25:** 472-489.

Spector, W. S. 1957 Handbook of Toxicology, Vol. II. Antibiotics, W.B.Saunders Co., Philadelphia, London, pp. 56.

Speijers, G. J. A. 2004 In Mycotoxins in food Detection and Control (eds. N. Mayam and M. Olsen), Wood head publishing limited, Cambridge, England, pp. 339-352.

Speijers, G.J. A. and H. P. Van Egmond 1993 In: Human Ochratoxicosis and its pathologies (eds.E. Creppy, M. Castegnaro and G. Dirheimer),Paris, John Libbey Eurotext. pp. 85-100.

Speijers, G. J. A., M. A. M. Franken and F. X. R. Van 1988 *Food and Chem. Toxicol*. **26:** 23-30.

Sphan, J. A., P. A. Dreifuss and H. R. Schulten 1977 *J. AOAC*. **60 :** 73-82.

Spooner, B. S. and N. K. Wessells 1972 *Dev. Biol.* **27 :** 38-54.

Springer, J. P., J. Clardy, J. M. Wells, R. J. Cole, J. W. Krikrey, R. D. Macfarlane and D. F. Torgerson 1976 *Tetrahedron Lett.* 1355-1358.

Sprosen, J.M. and N.R.Towers 1995 In *Toxicology and Food Safety* Hamilton, New Zeland pp. 45-46.

Sreenivasa Murthy, N., H.A.B. Parpia, S.Srikanta and A.Shankar Murthy 1967 *J. AOAC.* **50**: 350-354.

Srivastava, J.L. 1987 Paper presented at the Joint FAO/WHO/UNDP Second International Conference on Mycotoxins held at Bangkok, Thailand from September 28- October 3, 1987.

Srivastava, L. S. and R. N. Verma 1987 *Indian J. Mycol. and Pl. Pathol.* **17:** 46.

Srivastava, P. M. 1986 *Bull Grain Tech.* **24:** 64-69.

Stack, M. E. and M. J. Prival 1986 *Appl. Envi. Microb.* **52**: 718-722.

Stack, M. E., P. B. Mislivec, J. A. G. Roach and A. E. Pohland 1985 *J. AOAC.* **68**: 640-642.

Stack, M.E. and R. M. Epply 1992 *J. AOAC* **75 :** 834-837.

Stahl, C., L. V. Vanderloef, N. Siegel and J. P. Helgeson 1973 *Plant Physiol.* **52** : 663-666.

Stanley, N. F. 1964 *Australian J. Exptl. Biol. Med. Soc.* **24:** 133.

Stark, A. A. 1980 *Ann. Rev. Microbial.* **34:** 235-262.

Stark, A. A., B. Kobbe, K. Matsuo, G. Buchi, G. N. Wogan and A. L. Demain 1978 *Appl. Envir. Microb.* **36**: 412-420.

Stark, A. A., J. M. Townsend, G. N. Wogon, A. L. Demain, A. Manmade and A. C. Ghosh 1978 *J. Envir. Pathol. Toxicol.* **2:** 313-324.

Steele, J. A., J.L. Lieberman and C.G. Mirocha 1974 *Can. J Microbilogy 20*: 531-534.

Steiner, W. E., K. Brunschweiler, E. Leimbacher and R. Schneider 1992 *J. Agric. Food Chem.* **40**: 2453-2457.

Steiner, W. E., R. H. Reiker and R. Battaglia 1988 *J. Agric. Food Chem.* **36**: 88-91.

Stern, P. 1971 *Yugoslov Physiol. Pharmocol. Acta* **7:** 187-196.

Stewart, R. G., J. K. Skeeles, R. D. Wyatt, J. Brown, R. K. Page, I. D. Russel and P. D. Lukert 1985 *Poultry Sci.* **64**: 616-619.

Steyn, P. S. and C. J. Rabie 1976 *Phytochemistry* **15:** 1977-1979.

Steyn, P. S. and R. Vleggaar 1985 In Progress in the Chemistry of Organic Natural Products Vol. 3 k (eds. W. Her, H.Grisebach, H.Kirby and Ch. Tamm), Spring-Verlag, New York.

Steyn, P. S., P. G. Theil and D. W. Trinder 1991 In Detection and Quantitation of Mycotoxins by Chemical Analysis, (eds.J. E. Smith and R. S. Henderso),*Boca Raton, F.L.*, pp 165-221.

Steyn, P. S., R. Vleggaar, P. L. Wessels and D. B. Scott 1979 *J. Chem. Soc. Perkin Trans;* **1**: 460-463.

Steyn, P. S., R. Vleggaar, P. L. Wessds and M. Woudenberg 1982 *J. Chem. Soc. Chem. Perkin Trans.* **1**: 2175-2178.

Still, P. B., C. Eckardt and L. Leistner 1978 *Fleischwirschaft* **58:** 876-877.

Stinson, E. E., D.D.Bills, S.F.Osman, J.Siciliano, M.J. Ceponis and E.G. Heisler 1980 *J. Agri. Food Chem.* **28**: 960-963.

Stinson, E. E., S. F. Osman and E. G. Heisler 1981 *J. Agri. Food Chem.* **29**: 790-792.

Stockmann-Jlivala, H., H. Alenins and K. Savolainen 2008 *Food and Chem. Toxicol.* **46:** 1444-1451.

Stob, J., R. S. Bardwin, J. Tuite, F. N. Andrews and K. G. Gillette 1962 *Nature* **196:** 1318.

Stoloff, L., S. Henry and O. J. Jr. Francis 1976 *J. AOAC.* **59:** 118-121.

Stoloff, T. W. and L. B. Bullerman 1976 *J. Food Sci.* **41:** 201.

Strachen, N. J. C., P. G. John and I. G. Millar 1997 *Food Agric. Immunol.* **3:**177-183.

Stray, H. 1978 *J. AOAC.* **61:** 1359-1362.

Subramanian, C. V. 1952 *Proc. Nat. Inst. Sci. India* **18:** 557-584.

Subrahmanyam, P. and A. S. Rao 1974 *Curr. Sci.* **43**: 707-710.

Subramanianyam, P. and A. S. Rao 1974a *Indian J. Microb.* **64**: 31-34.

Subramanian, S. and S. Govindaswamy 1985 *Curr. Sci.* **54**: 860-862.

Subramaniam, S., S.Kathaswamy, S.William and S. Govindaswamy 1986 *Indian J. Poult. Sci.* **21** : 354-356.

Subramanian, T. 1982 *J. AOAC.* **65:** 507.

Subramanian, T., K. M. Namasivayan and E. R. B. Shanumugasundram 1978 *J. AOAC.* **61:** 581-583.

Sudarshan Singh 1987 *Poultry Adviser* **20** : 25-29.

Sudershan, R. V., G. S. Prasad, T. P. Krishan and R. V. Bhat 1992 *J. Food Protection* **55:** 392-394.

Sudhakar, B. V. 1992 *Indian Veter .J.* **69**: 1061-1062.

Sudhakar Rao, G. V. S. and T. V. George 2000 *Indian J. Animal Sci.* **70**: 759-760.

Sunija, S. K., D. S. Wagle and G. C. Ran 1987 *Archives of Toxicology* **60** :382-387.

Suman, M. and Kishan Singh 2000 *Indian J. Animal Sci.* **70**: 766-768.

Suman, Rai and S. N. Chaturvedi 1993 *J. Indian Bot. Soc.* **72**: 189-190.

Sunija, S. K., D. S. Wagle, D. P. Monga and G. C. Ram 1988 *Indian J. Microbiol.* **28:** 1-18.

Surekha, M. and S. M. Reddy 1990 *Animal Feed Sci & Technol.* **31**: 167-171.

Surekha, M. and S. M. Reddy 1991 Proc. Symp. Mycotoxin Incidence and Human Health (eds K.S.Bilgrami and K.K.Sinha), CBS Publications, New Delhi pp.181-185.

Surekha, M. and S.M. Reddy 1992a *Acta Botanica Indica* **20:** 317-318.

Surekha, M. and S. M. Reddy 1992 *Nat. Acad. Sci. Letters* **15**: 67-71.

Surekha, M. and S. M. Reddy 1991 *Nat. Acad. Sci. Letters* **14**: 337-378.

Surekha, M. and S. M. Redyy 1991a *Nat. Acad. Sci Letters* **15**: 463-465.

Susheela, R., E.R.B. Shanmugasundaram and K. R. Shanmugasundaram 1986 *Curr. Sci.* **55:** 99-100.

Sutton, P., J. Beaver and P. Waring 1995 *Biochem. Pharma.* **50**: 2009-2014.

Suzuki, S., T. Kumura, F. Saito and K. Ando 1971 *Agric. Biol. Chem.* **35** : 287.

Swanson, S. P. and R. A. Corley 1989 In Trichothecene mycotoxicosis: Pathophysiological effects, Vol. I (ed. V. R. Beasley),CRC press,Boca Raton, Florida, USA pp. 37-61.

Swanson, S. P., J. Nicoletti, H. D. Rood, W. B. Buck, L. M. Cote and T. Yoshijawa 1987 *J. Chromatogr.* **414:** 335.

Swanson, S. P., W.M. Hagler and H.D. Rood 1984 *Ann Meet. Am. Soc. Microbiology* **84**:192.

Swasoni, S. P., R. A. Corley, D. G. White and W.B. Buck 1984 *J. AOAC.* **67:** 580-582.

Sydenham, E. W. 1989 The Chromatographic Determination of *Fusarium* toxins in Maize Associated with Human/Oesophagial cancer M. Sc. thesis, University of Cape Town, South Africa.

Sydenham, E. W., G. Pieter, P.G. Thiel, W. F. O. Marasas, G. S. Shephard, D. J. Van Schalkwyk and Klaus R. Koch 1990 *J. Agric. Food Chem.* **38** : 1900-1903.

Sydenham, E. W. and G. S. Shephard 1996 In: Progress in Food Contaminant Analysis (ed. J. Gilbert), Blackie Academic and Professional, London, pp. 65-146.

Sydenham, E. W., G. S. Shephard, P. G. Thiel, W. F. O. Marasa, J. P. Rheeder, L. E. P. Peralta Sanhueze, H. H. L. Gonzalez and S. L. Resnik 1993 *J. Agri. Food Chem.* **41:** 891-895.

Sydenham, E. W., G. S. Shephard, P. G. Thiel, S. Stockenstrom, P. W. Snijman and D. J. Van Schalkhyk 1996 *J. AOAC.* **79:** 688-696.

Sydenham, E. W., G. S. Shephard, S. Stockenstrom, J. P. Rheeder, W. F. O. Marasas and M. J. Vander Merwe 1997 *J. Agric. Food Chem.* **45:** 4004-4010.

Sydenham, E. W., L. Vander Westhuizen, S. Stockenstrom, G. S. Shephard and P.G. Thiel 1996 *Food Addi.and Contam.* **11:** 25-32.

Sydenham, E.W., M.Walter, W.F.O. Marasas, S. Gorden, G. S. Shephard, G. Pieter, P. G. Thiel and E. X. Hirooka. 1992 *J. Agri. Food Chem.* **40:** 994-997.

Sydenham, E. W. and P. G. Theil 1987 *Food Addi. and Contam.* **4:** 277-284

Sydenham, E. W., W. C. A. Gelderblom, P. G. Thiel and W. F. O. Marasas 1990 *J. Agric. Food Chem.* **38 :** 285-290.

Sydenham, E. W., W. F.O. Morassas, P. G. Thiel, G. S. Shephard and J. J. Nieuwenhusis 1991 *Food Addi.and Contam.* **8:** 31-41.

Sylvia, V. L., T. D. Phillips, B. A. Clement and J. L. Green 1986 *J. Chromatogr.* **362:**79-85.

Synge, R. L. M. and E. P. White 1959 *Chem. and Ind.* (London) pp.1546

Szezech, G. M., W. W. Carlton, J. Tuite and R. Caldwell 1972 *Vet. Pathol.* **10**: 347-364.

Szigeti, G. J. 1976 Ph. D. Thesis, Miskok, Hungary.

Szurdoki, F., E., Trousdale, B. Ward, S. J.Gee, B.D.Hammock and D.G. Gilchrist 1996 *J. Agric. Food Chem.* **44:** 1796-1803.

Szurdoki, F., E.Trousdale, S. J.Gee, B.Ward, B.D.Hammock and D.Gilchrist 1996 In Immunoassays for Residue analysis: food safety. (eds. R. C. Beier and L. H. Stanker) ACS symposium series book no. 621. American Chemical society: Washington, DC pp. 330-340.

Tabata, S., H. Kamimura, Y.Tamura, K. Yasuda, H. Ushiyama, H. Hashimoto, M. Nighijima and T. Nishima 1987 *Food Hyg .Soc. Japan* **28 :** 395-401.

Tabbanche, S. 1981 *Bull. Ec. Natl. Super. Agron. Ind. Aliment* **23:** 63-70.

Taira, T. and S. Yamatodani 1947 *Penishirin Sono Ta Koseibusshitsui* **1**:275-276.

Takahashi, D. M. 1977 *J. Chromatogr.* **131**: 147-156.

Takahashi, T. Y., On One and M. Mori 1982 *J. Food Hyg. Soc. Japan* **23:** 73-80.

Takahashi, T. Y., N. Ando, S. Ohsato, S. Takehito, H. Hiroshi, I. Yamaguichi and M. Kimura 2004 *Appl. Envir. Microb*. **70:** 3239-3245.

Takahashi, U., M. W. Trucksess, W. R. Beaver, D. M. Wilson, J. W. Dorner and F. E. Dowell 1992 *J. AOAC.* **75**: 838-841.

Tamalaiven,E. A.1943 Uber die Fusarien Finlands. II. Valt. Maatalusk. Jult. Helsinki **122:** 1026.

Tamm, Ch. 1977 In: Mycotoxins in Human and Animal Helath (eds. J. V. Rodricks, C. W. Hesseltine and M. A. Mehlman), Pathodox, Park Forest South III. pp.209-228.

Tamm, Ch. and M.Tori 1984 Mycotoxins Production, Isolation, Separation and Purification, Elsevier, Amsterdam, Oxford, New York, Tokyo, pp. 131

Tamm, Ch. and S.W.Tannenbaum 1978 *Cytochalasins: Biochemical and cell Biological Aspects, (*ed. S.W.Tannenbaum), North-Halland, New York. pp.15-51.

Tamm, Ch. and W. Breitensetin 1980 In Biosynthesis of Mycotoxin: A study in secondary metabolism (ed. Picter Steyn), Academic Press, pp. 69.

Tanabe, A . 1977 Abstract 8c 206. 26th IUPAC Symposium, Tokyo.

Tanaka, T., A. Hasegawa and Y. Matsuki 1985 *J. Chromatogra.* **328:** 271-278.

Tanaka, T., A. Hasegawa, Y. Matsuki, U.S. Lee and Y.Ueno 1973 *Food Addi and. Contam.* **3:** 247-252.

Tanaka, T., A. Hasegawa, S.Yamamoto, U.S. Lee, Y. Sugiura and Y.Ueno 1988 *J. Agric. Food Chem.* **36** :979-983.

Tanaka, T., A.Hasegawa, Y.Matsuki and Y.Ueno 1985 *Food Addi. & Contam.* **2:**259-265.

Tanaka, T., A. Yoneda, S. Inove, Y. Sugiura, Y. Ueno 2000 *J. Chromatogra.* **A882:**23-28.

Tanaka, T., J.C. Kim, J.A. Seo, Y. W. Lee and D.H. Son 1994 *Agric. Chem. Biotechnol.* **37:** 385-391.

Tanaka, T., R. Teshima, H. Ikebuchi, J. Sawada, T. Tetrao and M. Ichinoe 1993 *J. AOAC.* **76**: 1006-1009.

Tanaka, T. and Y. Ueno 1988 In : Mycotoxins and Phycotoxins (eds. S. Natori, K. Hashmoto and Y. Neho), Elsevier Science Publishers, Amsterdam, pp. 88.

Tanaka, T., R. Teshima, H. Ikebachi, J. Sawada and M. Ichinoe 1995 *J. Agric. Food Chem.* **43:** 946-950.

Tapai, K., J. Teren and A. Mesterhazy 1997 *Cereal Res. Commun.* **3:** 307-308.

Tarkowski, A. K., A.Witkowska and J.Opas 1977 *J. Embyol. Exp. Morphol.* **41:** 47-64.

Tartar, E. J., J.P. Harchay and P. M. Scott 1984 *J. AOAC.* **67**: 597-600.

Tatu, C.A., W. H.Orem, R.B.Finkelman and G. L. Feeder 1998 *Envir. Health Perspect.* **106**: 689-700.

Taylor, A. 1971 Personal Communication.

Taylor, A. 1967 In Biochemistry of some food-borne *Microbial* toxins (eds. R. T. Mateles and G. N. Wogan), Press Cambridge Masschusect, pp. 69-107

Teplitz, R. L., J. Mazie, I. Gerson and K. J. Barr 1975 *Exp. Cell Res.* **90 :** 393.

Theron, J. J., K. J. Vander Merwe, N. Liebenberg, J. J. B. Joubert and W. Nel 1966 *J. Pathol. Bacteriol.* **91**: 521-529.

Thiel, P. G. 1978 *Biochem. Pharmacol.* **27**: 483-486.

Thiel, P. G., C.J. Meyer and W.F.O. Marasas 1982 *J. Agri. and Food Chem.* **30**: 308-312.

Thiel, P. G., E.W.Sydenham and G. S. Shephard 1996 In Fumonisins in Food, (eds. L. Jackson, J. W. De Varies and L.B. Bullerman), Plenum Press, New York, pp. 145-151.

Thiel, P. G., E.W. Sydenham, G. S. Shephard and D. J. Van Schlkwyk 1993 *J. AOAC.* **76:** 361-366.

Thiel, P. G., S.Stockenstrom and P.S.Gather Cole 1986 *J. Liq. Chromatogr.* **9**:103-112.

Thiel, P. G., W. F. O. Marases, E.M. Sydenham, G. S. Shephard and W.C.A. Gelderblom 1992 *Mycopathologia* **117:** 3-9.

Thiel, P. G., W.F.O. Marasas, E. W. Sydenham, G. S. Shephard, W. C. A. Gelderblom and J. J. Nieulwenhuis 1999 *Appl. Envir. Microb.* **57:** 1089-1093.

Thirumala Devi, K., M. A. Mayo, G.Reddy, S. V. Reddy, P. Delforse and D. V. Reddy 2000 *J. Agric. Food Chem.* **48:** 5079-5082.

Thirumala-Devi, K., M. A. Mayo, G. Reddy, K. E. Emmannal, X. Larondella and D. V. Reddy 2001 *Food Addi.and Contam.* **18:** 830-835.

Thoa, N. B., G. F. Wooten, J. Axelrod and I. J.Kopin 1972 *Proc. Nat. Acad. Sci.* USA **69:** 520-522.

Thomas, D. and S. Des 1978 *In cytochalasin: Biochemical and cell Biological aspects*. (ed. Tanenbaum), North-Halland, New York, Chap. II.

Thomas, D., S.Des, M. Lutzac and E. Manavathu 1974 *Nature (London)* **249:** 140-142.

Thompson, D. P., L. Metevon and T. Vessel 1993 *J. Food Prot.* **56:** 134-138.

Thompson, V. S. and C. M. Marogas 1996 *J. Agric. Food Chem.* **44:** 1041-1046.

Thompson, W.F. and R.W. Wannemachus 1990 *Toxicology and Applied Pharmacology* **105** : 483-491.

Thorton, S.T., 1958 *Chem. Absts*. **50:** 72296.

Thuvander, A., A. E. Breitholtz, D. Brabencova and I. Gadhasson 1996 *Food Chem. Toxical.* **34:** 547-554.

Taira, T. and S. Yamatodani 1947 *Penishrin Sono takosci busshitsu.* **1:** 275-276.

Tiemann, V. and S. Danicke 2007 *Food Addi. and Contam.* **24**: 306-314.

Tomaszewski, J., R.Mniturski, A. Semczuk, J. Kotarski and J. Takowicki 1998 *Ginekol. Pol.* **69:** 363-366.

Torres, A. M., M. M. Reynoss, F. G. Rojo, M. L. Ramirez and S. N. Chulze 2001 *Food Addi. and Contam.* **18** : 836-843.

Torres, A., S.Chulze, E.Varsavaky and M.Rodriguez 1993 *Mycopathologic* **121:**17-20.

Tou, J. and R. L. Stjernholm 1974 *Biochem. Biophys. Acta* **393:** 1-11.

Toussoun, T. A. and P.E. Nelson 1968 A pictorial Guide to the identification of *Fusarium* species According to the taxonomic system of Synder and Hansen"pennsylvania state University Press, University park. Pennsylvania.

Towers, N. R. and J.M. Sprosen 1992 *Fusarium* mycotoxins in pastoral.In Recent Advances in Toxicology Research Vol. 3 (eds. P.Gopalkrishnakone and C.K.Tan) Venom & Toxin *Research* Group, National University of Singapore, pp. 272-284.

Towers, N. R., J. M. Sprosen and W. Webber 1995 In: Toxicology and Food safety" Toxicology and Food Safety Research Groups, Ruakura Research Centre, Hamilton, New Zealand.

Townsend, C. A 1997 *Pure Appl. Chem.* **58:** 227-238.

Townsend, C. A., S. M. Mc Guire, S. W. Brobst, T. L. Graybill, K. Pal and C. E. Barry 1992 in secondary metabolites. Biosynthesis and Metabolism (eds. R. J. Petroski and S. P. Mc Cormick), Plenum press, New York, pp. 141-153.

Toyoda, M., K. Saisho, G. Aoki, K. Kobayashi, Y. Saito and M. H. Martinez 1994 *Int. Dairy J.* **4**: 369-375.

Trenholm, H. L., L. L.Charmely, D. B. Prelusky and R. H. Warner 1991 *J. Agric. Food Chem.* **39** : 356-361.

Trenholm, H. L., L. L.Charmely, D.B.Prulusky and R.H.Warner 1992 J. *Agric. Food Chem.* **40:** 2147-2151.

Tseng, T. 1983 In Trichothecenes, Chemical, Biological and Toxicological Aspects (ed. Y. Ueno), Elsevier Publication, New York, pp. 290.

Trenholm, H. L., L.L.Charmley, K.L.Underhill and D.B.Prelusky 1996 In: Proceedings of the 4th International Feed Production Conference. Catholic University of Sacred Heart, Piacenza, Italy, pp. 69-86.

Trigo-Stokli, D. M., C. W. Deyoe, R. F. Salimbaga and J. R. Pedersen 1996 *Cereal Chemistry* **73 :** 388-391.

Tripathi, R. K. 1973 *Indian J. Expt. Biol.* **11**: 361-362.

Tripathy, S. B., N. C. Nayak and K. M. Mohanty 1967 *Indian Vet. J.* **44:** 813-814.

Trucksess, M. W. 1995 *J. AOAC.* **77:** 135-142.

Trucksess, M. W. and A.E. Pohland 2001 *J. AOAC.* 84:941-949.

Trucksess, M. W., P. B. Mislivec, K. Yang, V.R. Bruk and S. W. Page 1987 *J. AOAC.* **70**: 123-126.

Trucksess M. W., J.Giller, K. Young, K.D.White and S.W. Page 1997 *J. AOAC.* 82-89.

Trucksess, M.W., J. Giler, K. Young, K.D. White and S.W. Page 1999 *J. AOAC.* **82:** 85-89.

Tseng, T. 1983 In Trichothecenes, Chemical, Biological and Toxicological Aspects (ed. Y. Ueno) Elesevier Publication. New York. pp. 135-146.

Tuason, M. M. S. 1993 Philippines Univ. Los Bands Colleen, Laguna. pp.52.

Tubajika, K. M. and K. E. Damann 2001 *J. Agri. Food Chem.* **49:** 2652-2656.

Tubaki, K. 1981 In Hyphomycetes. Their Imperfect-perfect connections *J. Cramer, Vaduz.* pp. 181

Tuinstra, L. G. M., P. G. M. Kienhuis and P. Dols 1990 *J. AOAC.* **73**: 969-973.

Tulloch, M. 1972 The genus *Myrothecium* Tode ex Fr.Mycol. Papers No.130.1-42.CMI, Kew, Surrey, England.

Turner, W. B. 1971 *Fungal Metabolites,* Academic Press, London.

Turner, W. B. and D. C. Aldride 1983 *Fungal metabolites vol II*, Academic Press, London.

Towner, P.C., S. E. Moore, A.J.Hall, A.M. Prentice and C. P. Wild 2000 *Environmental Health Perspectives* **111:** 217-220.

Tutour, B. Le., T. Elaraki and A. Aboursalim 1984 *J. AOAC.* **67**: 611-612.

Tyagi, R. P. S., T. Singh and S. K. Srivastava 1986 *Bull. Grain Tech.* **24:** 36-58.

Udagawa, S., T. Muroi, H. Kurata, S. Sekita, K.Yoshihira, S.Natori and M. Umeda 1979 *Can. J. Microbiol.* **25:**170-177.

Ueno, Y. 1977 Trichothecenes, Overview . In Mycotoxins in Human and Animal Health (eds. J.V.Rodricks, C.W.Hesseltine and M. A. Mechanman), Pathotox Public.

Ueno, Y. 1983 *Developments in Food & Feed-IV.* **23**: 110-127.

Ueno, Y., F.Tashiro and H.Nakaki 1985 *Gann Monograph Cancer Res.***30**: 111-124.

Ueno, Y. and H.Matsumoto 1975 *Chem. Pharm. Bull.* **23 :** 2439-2442.

Ueno, Y. and I. Ueno 1972 *J. Exper. Med.* **42:** 91-105.

Ueno, Y. and K. Kuboto 1976 *Cancer Res.* **36**: 445.

Ueno, Y., K. Iijima, D.S. Wang, Y. Sugiura, M. Tanaka, C. Chen and S. Z. Yu 1997 *Food Chem. Toxicol.* **35**: 1143-1150

Ueno,*Y.*,K.Umemori,E-C.Niimi,S-.Tanuma,S.Nagata,M.Sugamata,T.Ihara,M.Sekijima, K-I. Kawai and I .Ueno 1995 *Natural* Toxins 3:129-137,

Ueno, Y., H. Matusmoto, K. Ishii and K. Kotia 1976 *Biochem. Pharmacol.* **25:** 209-2095.

Ueno, Y., I. Ueno, Y. Iitoi, H.Issumodo., M.Ehamoto and K.Ohtsubo 1971 *J. Exp. Med. Japan.* **41:** 521-532.

Ueno, Y., K.Ishii, K.Sakai, S.Kanaeda, H.Tsunoda, T.Tanaka and M. Enomoto 1972 *Jap. J. Exp. Med.* **42:** 187-208.

Ueno, Y., M. Sato, K.Ishii, K.Sakai and M.Enomoto 1972 *Jap. J. Exp. Med.* **42:** 461-472.

Ueno, Y., M.Sawano and K.Ishii 1975 *Appl. Microbiol.* **30**: 699-704.

Ueno, Y., S. Aoyama, Y. Sugiura, D. S.Wang, U.S.Lee, E. Y. Hirroka, S. Hara, T. Karki, G. Chen and S. Z. Yu 1993 *Mycotoxin Res.* **9**: 27-34.

Ukau, T., Y.Yamamoto and T.Yamamoto 1954 *J.Phar Soc.J.Kpn.***74:** 450-454.

Umeda, M., K. Matsumoto, M. Siato, S. Sekita, K. Yoshilira, S. Natori, S. S. Adgaura, F. Sakabe and H. Kurata 1975 *Experientia* **31:** 435-438.

Umeda, M., T. Tsutavi and M. Saito 1977 *Gann.* **68:** 619-626.

Umeda, M., T. Yamamoto and M. Saito 1972 *Jpn. J. Exp. Med.* **42:** 527-535.

Umetsu, N., J. Kaji and K. Tamari 1972 *Agri. Biol. Chem.* **36**: 859-866.

Umetsu, N., J. Kaji, K. Aoyama and K. Tamari 1974 *Agri. Biol. Chem.* **38:** 1867-1874.

Umezawa, H., Y. Mizhuhara. K. Uekane and M. Haagihara 1947 *J. Penicillin* (Japan Chem. Abstr) **41:** 6918F, 1-6.

Urano, T., M.W.Truckesess and S.W.Page 1993 *J.Agric.Food Chem.***41**: 1982-1985.

Urano, T., M. W. Trucksess, R.W. Beaver, D. M. Wilson, J. W. Dorner and F. E. Dowell 1992 *J. AOAC.* **75**: 838-841.

Urry, W.H., H. L. Wehrmeister, E. B. Bodge and P. H. Hidy 1966 *Tetrahedron Lett.* **27 :** 3109-3114.

Usha, C. M., K. L. Patkar, H. S. Shetty, R. Kenedy and J. Lacey 1994 *Tropical Science* **34**: 353-360

Usleber, E., E.Martlbauer and D.Abramson 1996 In *Immunoassays for Residue analysis: Food Safety* (eds. R. C.Beier and L. H.Stanker), ACS symposium series book no. 621. American chemical society: Washington, DC pp. 322-329.

Usleber, E., E.Schnider, E.Martlbauer and G. Terplan 1993 *J. Agric. Food Chem.* **41:** 2019-2023.

Usleber, E., M.Straka and G.Terplan 1994 *J Agric. Food Chem.* **42:** 1392-1396.

Uzal, F.A., C.A.Robles, M. A. Scuteri, M. Sala-de-miguel and L. H. Lischinsky 1992 *Rivista-de-Medicna-veterinari-Buenos-aics* **73**: 177-182.

Vail, R. B. and M. J. Hormann 1990 *J. Chromatogra.* **535**: 317-323

Val Walbeek, W., P.Scott and F.S.Thatcher 1969 *Can.J.Microbiol.***14:**131-137.

Valdivia, A.G., A. Martinez, F.J.Damian, T.Quezada, R. Ortiz, I. C. Martinez, J. Llamas, M. L. Rodriguez, L. Yamatgo, F. Jaramillo, M. G. Loarca-Pina and J. L. Reyes 2001 *Poultry Sci.* **80**: 727-734.

Valenta, H. 1998 *J. Chromatogra.* **815:** 75-92.

Valvano, Fabio, Audrea Piva, Alberto Ritieni and Glacono Gaivano 2001 *Journal of Food Protect.* **64:** 120-131.

Vanderstegan, G. H. D. 2003 *Food Control* **14:** 245-249.

Van Rensburg, S. J. 1984 *Food Chem. Toxicol.* **22**: 993-998.

Van Wert, S. L. 1984 *Plant Dis.* **68 :** 1036-1038.

Vandegraft, E. E., C. W. Hesseltine and O. L. Shotwell 1975 *Cereal Chem.* **52:**79-84.

Vander Merve, K. J., P. S. Steyn and L.F. Fourie 1965 *J. Chemical Society* 7083-7088.

Vander Merve, K., J., P. S. Steyn, L. F. Fourie, D. W. Scott and J. J. Therom 1965 *Nature* **205 :** 1112-1113.

Vander Stegen, G.H., P. J.Essens and H. Van de Voorde 2001 *J. Agric. Food Chem.* **49**: 4313-4315.

Vanwalbeek, W., P. M.Scott, J.Harwing and J.W.Lawrence 1969 *Can. J. Microbiol.* **15:** 1281-1285.

Varga, J., B. Toth, E. Kevei, A.Palagyi and Z. Kozakiewiez 2000 *Antonie van Leeuwenhoek* **77:** 83-89.

Varga, J., B.Toth, K.Rigo, J.Teren, R.F.Hoekstra and Z. Kozakiewics 2002 *Fungal Genetic and Biology* **30:** 71-80.

Vasan, B. S. 1980 *Bull. Grain Tech.* **18:** 223-232.

Vasanthi, S. and R. V. Bhat 1997 *J. Sci. Food and Agric.* **73:** 226-230.

Vedpal Singh. and V. Singh 1998 *J. Phyto. Res.* **11:** 7-10.

Velaziquez, C. and C. Monserrak 1999 *Food Addi.and Contam.* **10 :** 125-128.

Veldman, A. J., A. C. Meijs, G. J. Borggreve and J. J. Heeres-Vander Tol 1992 *Anim. Prod.* **55:** 163-168.

Verma, R. J. and A. Nair 1999 *Medical Science Research* **27**: 223-226.

Verma, R. J. and A. S. Kolhe 1998 *Medical Science Research* **26**: 427-428.

Verma, R.J., A.S. Kolhe and S. B. Chaudhari 1998 *Medical Science Research* **26**: 339-341.

Verma, R.J. and P. J. Raval 1991 *Bulletin of Environmental Contamination and Toxicology* **47**: 428-433.

Verma, R. J. and P. J. Raval 1997 *Medical Science Research* **25**: 655-657.

Verma, S. H., S.Pandey and R. A. B. Verma 1988 *National Symposium on Environmental Importance on Animals and Aquaculture*, University of Kalyani, W.B. India.

Vesela, D., D. Vesely and R. Jelinck 1983 *Appl. Environ. Microbiol.* **45:** 91-93.

Vesonder, R.F. 1983 In:Trichothecenes, Chemical, Biological and Toxicological Aspects (ed Y. Ueno), Elsevier Publication, p p 210-217.

Vesonder, R. F. and B. W. Horn 1985 *Appl. Envi. Microb.* **49**: 234-235.

Vesonder, R. F. and C. W. Hesseltine 1981 *Process Biochemistry Journal* **16** :12-14.

Vesonder, R.F., J.J. Ellis and W. K. Rohwedder 1981 *Appl. Envir. Microb.* **42:** 1132-1134.

Vesonder, R. F., L. Tjarakh, W. Rohweddar and D. O. Kieswetter 1980 *Experientia*; **36**.562-564.

Vesonder, R.F., A. Ciegler, R. E. Rogers, K. A. Burbridge, R. J. Bothest and A. H. Jensen 1978 *Appl. Envir. Microb.* **36:** 885-888.

Vesonder, R.F., P.Golinski, R. Plattner and D.L. Zietkiewiez 1991 *Mycopathologia* **113:**11-14.

Vijayalakshmi, M. and A.S.Rao 1985 In: Trichothecenes and other mycotoxins (ed. J. Lacey), John Wiley and Sons Ltd. pp. 33-45.

Vikram Reddy, M. 1990 Proceedings of Summer Institute on molds, *Mycotoxins in Poultry Diseases* A. P. A. U., pp. 154.

Viani, R., R. Battaglia, T. Hatzold and R. Kores 1996 *Food Addi.and Contam.* **13:** 29-33.

Vincelli, P. and G.Paker 1999 Mycotoxins in corn produced by *Fusarium* fungi ID-121 Electronic file created in 1999 Corn Growers Association, Kentucky, Frankfort.

Visconti, A. and M. Pascale 1998 *J. Chromatogr.* **815**: 133-140.

Visconti, A., A. D.E. Girolamo 2005 *Food Addi. and Contam..* **22**: 37-44.

Visconti, A., A. Logrieco and A.Bottalico 1986 *Food Addi.and Contam.* **3**: 323-330.

Visconti, A., A. Sibilia and F. Palmisano 1991 *J. Chromatogra.* **540:** 376-382.

Visconti, A., A. Boenke, M. Saltrizzo, M. Pascale and M. B. Doko 1996 *Food Addi.and Contam.* **13:** 909-927.

Visconti, N. and M.B. Doko 1994 *J AOAC.* **77 :** 546-550

Visconti, A., M.Solfrizzo, M.B.Doko, A.Boenke and M.Pascle 2008 *Food Addi.and Contam.* **13:** 929-938.

Voelker, H.H., S. S. Ray and R. M. Luther 1986 *J. Dairy Sci.* **69:** 139.

Vora, V.C. 1978 Central Drug Research Institute, Lucknow. pp. 13.

Vorster, L.J. 1985 E'tudes sur la de' detoxification des arachides contamines par I' aflatoxine et destinecs a l" huilerie, *Rev. Frane Corps. Res.* **13**: 7-12.

Voss, K.A., J.K. Portor and C.W.Bacan 1999 *Food Chem. Toxicol.* **37**: 853-861.

Voss, K.A., S.M. Poling, F.I. Meredith, C.W. Bacon and D.S. Saunders 2001 *J. Agri. Food Chemistry* **49:** 3120-3126.

Voss, K.A., W.J.Chamberlian, C.W.Bacon, R.A.Herbert, D.B.Walters and W.P. Norred 1994 *Fund. Appl. Toxicol.* **24**:102-110.

Voss, K. A., W. P. Norred, D. M. Hinton, P. J. Cole and J. W. Dorner 1990 *Mycopathologia* **110:** 11-18.

Vrabcheva, T. M. 2000 Mycotoxins in species .*Voprosy Pitaniya* **69:** 40-43.

Vrabcheva, T.M, E. Usleber, R. Dietrich, E. Martlbauer 2000 *J. Agri. and Food Chem.* **48:** 2483-2488.

Wagener, R. E., N. D. Davis and V. L. Diner 1980 *Appl. Envir. Microb.* **39**: 882-887.

Waghray, S., C. S. Reddy and A. P. K. Reddy 1988 *Indian Phytopath.* **41:** 492-494.

Wang, C. R. and F. S. Chu 1991 *Appl. Envir. Microb.* **57:** 1026-1030.

Wang, E., F.Ross, T.M.Wilson, R. T. Riley and A. H. Morril 1992 *J. Nutrition* **122:** 1706-1716.

Wang, E., W. P. Norred, C. W. Bacon, R. T. Riley and A. H. Morril 1991 *J. Biol. Chem.* **266 :** 14486-14490.

Wang, J.S., T.W.Kensler and J.D.Groppman 1998 *Current Toxicology Nutrition and Chemical Toxicity* New York; Wiley pp. 29-57.

Wang, H., C. Jones and J. Ciaccii-Zanela 1996 *Proc. Nat. Acad. Sci.* USA **93**: 3461-3465.

Wannematchar, R. W., D. L. Bunner and H.A. Meufeld 1991 in Mycotoxins and Animal foods (eds. J. E. Smith and R. S. Henderson), CRC Press, Boco Raton,Florida, pp.499-552.

Warden, B. A., K.Allam, A.Sentissi, D. J.Cecehini and R. W. Giese 1987 *Anal. Biochem.* **162:** 363-369.

Ware, G. M. 1975 *J. AOAC.* **58 :** 754-756.

Ware, G. M., P.P.Umrigar, A.S.Carman and S.S.Kuan 1994 *Anal.Lett.* **27:**693-715.

Ware, G. M.,Y. Zhao, S. S. Kuan and A. S. Carman 1999 *J. AOAC.* **82:** 90-94.

Warlg, L.Y., M. Hatch, C. J. Chen, B. Levin, S. L. You, S. N. Lu, M. H. Wu, L. W. Wang, Q. Wang, G. Y. Huang, P. M. Yang, H. S. Lee and R. M. Santella 1996 *Int. J. Cancer* **67:** 620-625.

Wassarman, P. M., W. J. Josefowicz and G. E. Letourneau 1976 *J. Cell Sci.* **22 :** 531-545..

Watanabe, C. M. H., D.Wilson, J.E. Linz and C.A.Townsend 1996 *Chem. Biol.* **3:** 463-469.

Wawrzkeiwicz, K., A.Ghich, B.Rubaj and M.Wrobel 1989 *Medycyna Weterynary* **45**:27-30.

Webster, W. and J. Langman 1978 *Am. J. Anat.* **152** : 209-222.

Wegst, W. and F. Lingens 1983 *FEMS Microbiol. Lett.* **17:** 341-344.

Weihing, R. R. 1976 *J. Cell Biol.* **71**: 303-307.

Weiblug, T. S., D. R. Ledoux, A. J. Bermudez, J. R. Turk, G. E. Pottinghaus, E. Wang and A. H. Merrill Jr. 1993 *Poult. Sci.* **72:** 456-466.

Weidenborner, M. 2001 Encyclopedia of Food Mycotoxins, Springer Publishers, Berlin, New York and London. pp.

Weiser, M., J.Fink-Gremmels and L. Leistnes 1988 *Proce. of the Japanese Association of Mycotoxicology* **1**: 125-126.

Wells, J. M., H. G. Cutler and R. J. Cole 1976 *Can. J. Microb.* **22:** 1137-1143.

Westlake, K., R.I.Mackic and M.F. Dutton 1987 *Appl. Envir. Microb.* **53:** 613-614.

Westlake, K. and M. F. Dutton 1985 *S. Afr. Tydskr. Veekd.* **15** : 175-177.

White, A. G. and B. Truclove 1972 *Can. J. Bot.* **50:** 2659-2664.

White, E. P. 1958 *Newzealand J. Agr. Res.* **1:** 433-446.

White, E. P. 1959 *Newzealand J. Agr. Res.* **1**: 859-865.

Whitehair, C.K., H.C. Yang, M. E. Gibson and G. E. Short 1951 *Agric. Exp. Station Pub.* No. **22**: 57-59.

W.H.O.1993 Some Naturally Ocurring Substances: Food items and constitutents heterocyclic aromatic amines and mycotoxins IARC *manager. Eval. Carcinogen.* **56:** 397-444.

W.H.O. 2000 Safety Evaluation of Certain Food Additives and Comtaminants, WHO *Food Addiive Series* 44.

Wicklow, D.T., C.W.Hesseltine, D.L.Shotwell and G.L.Adams 1980 *Phytopathol.* **70:**761-764.

Wicklow, D. T. and R. J. Cole 1984 *Mycologia* **76:** 959-961.

Widiastuti, R., R. Maryam, B.J. Blaney, S. Stoltz and D.R. Stoltz 1988 *Mycopath.* **102** : 45-49.

Widstrom, N. W. 1996 *Adv. Agron.* **56**: 219-250.

Wild, C. P. and A.J. Hall 2000 *Reviews in Mutation Research* ***462*:** 381-393.

Wild, C. P., S. M. Shrestha, W. A. Anwar and R. Montesano 1992 *Toxicol. Lett.* **64/65**:455-461.

Wiley, B. J. and E. G. Simmons 1973 *Mycologia* **65:** 934-938.

Wilkes, J. G., O. L.Lay Jr. 2001 In Mycotoxin Protocols (eds.M. W Trucksess, A. E Pohland), Humana Press, Totowa, N J. pp 37-48.

Wilkins, A. L., C. O. Miles and R.M. Ede 1992 *J. Agr. Food Chem.* **40**:1307-1309.

Williams, B. A., K. T.Mills, C. D. Burroughs and H. A. Bern 1989 *Cancer Lett.* **46** : 225-230.

Williams, B. C. 1985 In Mycotoxins (eds. P. M. Scott, H. L. Trenholm and M. D. Suttan),Canadian Perspective, Publication, NRCC /CNRC, pp. 49.

Williams, J. A. and J. Wolff 1971 *Biochem. Biophys. Res. Commun.* **44:** 422-425.

Williams, S. S. 1984 *J. AOAC.* **60**:1125-1130.

Wilson, B. J. 1971 In Microbial Toxins vol VI (eds. A. Ceigler, S. Kadis and S. J. Aji) Academic Press, New York, pp. 207-295.

Wilson, B. J. and C. H. Wilson 1964 *Science* **144**: 177-178.

Wilson, B.J., C.H.Wilson and A.W. Hayes 1968 *Nature (London)* **220:** 77-78.

Wilson, B. J. and R.R. Maronpot 1971 In Mycotoxins in Human Health (ed. I. F. H. Purchase), Mc Millan, London, pp. 223-229.

Wilson, D. M., N. W. Widstrom, L.R. Marti and B. D. Evans 1981 *Cereal Chem.* **58**: 40-42.

Wilson, D. M., W.H. Tabor and M. T. Trucksess 1976 *J. AOAC.* **59:** 125-127.

Wilson, T. J. and T. R. Romer 1991 *J. AOAC.* **74**: 951-956.

Wilson, T. M., P.E. Nelson and C. R. Knepp 1985 *Carcinogenesis* **6:** 1155-1160.

Wilson, T. M., P. F. Rose, D. L. Owens, L.G. Rice, S. A. Green, S. J. Jenkins and H. A. Nelson 1988 *Mycopathologia* **117:** 115-120.

Windels, C. E., C.J.Mirocha, H. K.Abbas and W. Xie 1989 *Mycologia* **81 :** 272-277.

Windham, G. L. and W. P. Willaims 2002 *Plant Dis.* **86:** 232-234.

Wiseman, A., I. E. Selman, C.O. Dawson, R. G. Breeze and H. M. Pirle 1973 *Vet. Res.* **93 :** 410-417.

Wollenweber, H. W. 1935 *Phytopathology* **3**: 24-50.

Wollenweber, H.W. and O.A. Reinking 1935 Die fusarien, inhre Beschreibung, Schadwirkung and Bekampfung. Paul Parrey, Berlin, pp. 35-53.

Wolf, G. 1994 *Nutr. Rev.* **52 :** 246-247.

Wolf, J. C. and C. J. Mirocha 1973 *Can. J. Microbiol.* **19:** 725-734.

Wood, G. E. and M.W. Trucksess 1998 In: Mycotoxins in Agriculture and Food Safety (eds. K. K. Sinha and D. Bhatnagar) Marcel Dekker, New York, pp. 461.

Woody, M. A and F.S. Chu 1992 In *Alternaira* Biology plant disease and Metabolites (eds. J. Chelkowski and A. Visconti), Elsevier, Amsterdam, pp 409-433.

Wu, M. T., J. C. Ayres and P. E. Kochler 1974 *Appl. Microb.* **27**: 427-428.

Wu, S. G., Yang and T. Sun 1983 *J. Clon. Oncol.* **5** :81-84.

Wyatt, R. D. 1979 *Nat. Acad. Sci. Washington*, U. S. A. pp.87-95.

Wyatt, R. D. 1991 In Poultry (eds. J. E. Smith and R. S. Hunderson) CRC Press, Boca Raton, Florida, pp. 553-578.

Wyatt, R. D., P.B. Hamilton and H. R. Burmeister 1973 *Poultry Sci.* **52:** 1853-1859.

Wyllie, T.D. and L.G.Morehouse 1978 In: Mycotoxigenic fungi, Mycotoxins, Mycotoxicoses, an Encyclopedia Hand Book, Volume 2, Marcel Dekker, Inc, Newyork, pp. 570.

Xie, W., C. J. Mirocha and J. Chen 1997 *J. Agri. Food Chem.* **45 :** 1251-1255.

Yabe, K., P. S. Yan, Y.Song, M. Ichinomiya, H. Nakagawa, Y. Shima, Y. Ando, E. Sakuno and H. Nakajima 2008 *Food Addi. and Contam.* **25:** 1111-1117

Yacoub-George, E., H.Wolf, S.Koch and P.Woias 1996 *Sensors Acturators B. Chem.* **34:** 429-434.

Yadgiri, B. and P. G. Tulpule 1974 *J. Dairy Sci.* **27**: 293-297.

Yagen, B., M. Bialer and A. Sintove 1985 *J. Chromatogr.* **343:** 67-75.

Yamada, K. M., B. S. Spooner and N. K. Werrells 1970 *Proc. Natl. Acad. Sci.* USA. **66 :** 1206-1212.

Yamgish, D., A. Visconti, G. Avantaggiato and S. Ouchi 2001 *In Bulletin of the Institute of Comprehensive Agricultural Sciences,* Kinki University, **9 :** 13-28.

Yamaguchi, T., K. Zozawa and T. Hosoe 1992 *Phytochemistry* **32:** 1177-1181.

Yamakawa, H. and Y. Ueno 1970 *Chem. Phas. Bull* (Tokyo) **18:** 177-181.

Yamamoto, S., M. W. Greaves and V. Plumer 1973 *Immunology* **24:** 1007-1012.

Yamamoto, T. 1954 *J. Pharm. Soc. Jpn.* **74 :** 810-812.

Yamashita, A., T. Yoshizawa, Y. Aiura, P. C. Sachez, E. I. Dizon and R. H. Arim Sardjono 1995 *Biosci. Biotech. Biochem.* **59 :** 1804-1807.

Yamato, T., J.Sugiyama and H. Kuraishi 1990 In Modern concepts in *Penicillium* and *Aspergillus* Classification (eds. R. A. Samson and J. I. Pitt), Plenum Press, New Yrok, pp. 395-405.

Yamazaki, M. 1978 In Toxicology, Biochemistry and pathology of mycotoxins (eds. K. Uraguchi and M. Yamazaki), Halsted Press, pp. 65.

Yamazaki, M., K. Sasago and K. Miyanki 1974 *J. Chem. Soc. Chem. Commun.* 401.

Yamazaki, M., S. Suzuki and K. Kuriki 1979 *J. Pharm-Dyn* **2**: 119-125.

Yamazaki, M., S. Suzuki and N. Ozaki 1983 *J. Pharm. Dyn.* **6**: 748-753.

Yang, C. S. 1980 *Cancer Res.* **40:** 2633-2644.

Yang, C. Y. 1972 *Microbiol.* **24**: 885-890.

Yang, G., Li Shiguang, J.J. Pestka, G.H. Yang and S.G. Li 2000 *National Food Safety* and *Toxicology Center, Michigan state University* East Lansing, MI 48824, USA. pp.177.

You, J. O., S. T. Lee and Y. B. Kim 1988 *J. Korean Agric. Chem. Soc.* **31:** 189.

Yao, R. G. and D. P. H. Hsieh 1967 *Appl. Microbiol.* **28:** 52-57.

Yates, S. G. 1971 In Microbiol Toxins, (eds. S. Kadis, A. Ciegler and S. J. Ali), Vol. 7, Academic Press, New York, London.

Yatim, A. M. and D. S. Sanchan 2001 *Journal of Nutrition* **131**: 1903-1908.

Yeh, F. S., M. C. Yu, C. C. Mo, S. Luo, M. J. Tong and B. E. Henderson 1989 *China. Cancer Res.* **49:** 2506-2509.

Yoneyama, M. and R.P.Sharma 1987 *Archives of Environmental Contamination and Toxicology* **16:** 765-770.

Yoneyama, M., R. P.Sharma and S. J. K. Leinschuster 1986 *Ecotoxicology and Environmental safety* **11:** 110-111.

Yoshinaga, M., B. H. Waksman and S. E. Malawista 1972 *Science* **176:** 1147.

Yshino, N., M. Takizawa and H. Akiba 2000 *Nat. Toxins* **4:** 234-241.

Yoshizawa, T. 1983 In Trichothecenes-Chemical, Biological and Toxicological Aspects (ed. Y. Ueno) Amsterdam, Elsevier. pp. 195-209.

Yousef, A. E. and E. H. Marth 1981 *J. Food Prot.* **44:** 736-741.

Yu, Fing-Yih and F. S. Chu 1998 *J. AOAC.* **81:** 749-756.

Yu, Fung-Yih and F. S. Chu 1999 *Food Agric. Immunol.* **11:** 297-306.

Yu, J., R. D. Wei, C. J. Mirocha and F. S. Chu 1990 *Food Agric. Immunol.* **2:** 55-64.

Yu, W. and F. S. Chu 1999 *J. Agric. Food Chem.* **47**: 4815-4820.

Yu, W., J. W. Dorner and F. S. Chu 1998 *J. AOAC.* **81**: 1169-1175.

Zaki, N. 1992 *J. Agus. Sci.* **1:** 1069-1079.

Zaky, Z. M., D. A. Salem, S. S. El-Ballal and A. M. A. Mekki 1998 *Winer Tierarztliche Monatschrift* **85**: 43-48.

Zamir, L. O. 1980 *In The Biosynthesis of Mycotoxins* (ed. P. S.Steyn) Academic Press, New York pp. 224-268.

Zhang, Y., C. Jones and M. B. Dickman 2001 *Food Chem. Toxicol.* **39**: 45-53.

Zhang, Y., M.B. Dickman and C. Jones 1999 *J. Biol. Chem.* **274**: 12367-12371.

Zheng, Z., J. Hannaken, D. Houchins, R. King, P.Lee and J. Richard 2005 *Mycopathologia* **159:** 265-272.

Zigmond, S. H. and J. G. Hiresh 1972 *Expt. Cell Res.* **73**: 383-393.

Zimmerli, B. and R. Dick 1995 *J. Chromatogr.* **666** : 85-99.

Zinedine, A., J. M. Soriano, C. Juan, B. Mojemmi, J. C. Molto, A. Benklonze, X. Cherrah, L. Indrissi, R. E.L. Aouad and J. Manes 2007 *Food Addi.and Contam.* **24:** 285-291.

Zismek, C. A. and D. Epel 1975 *Science* **189:** 139.

Zollner, P., J. Jodlbauer and W. Lindner 1999 *J. Chromatogra. A* **858:** 167-174.

Zollner, P., J. Jodlbauer, M. Kleinova, H. Kahlbaher, T. Kuha, W. Hoschsteiner and W. Linder 2000 *J. Agric. Food Chem.* **50** : 2494-2501.

Zomo, M. C. and M. Vurro 1999 *Weed Res.* **39:** 15-20.

Zuber, M.S. 1977 *Mycotoxins in Human and Animal Health* (eds. J.I Kodrides, C.M. Hesseltine and M.A. Mehlman) *Pathodox Publ. Inc. forest. Park. South* IV, pp. 173-179.

Zuber, R. B., S.Hoffstein and G. Weissmann 1973 *Proc. Nat. Acad. Sci.* USA **70** :844.